PURCHASING, COST CONTROL, AND MENU MANAGEMENT

THE
INTERNATIONAL
CULINARY
SCHOOLS [SM]
at The Art Institutes

John Wiley & Sons, Inc.

This custom textbook includes content from the following books:

Purchasing: Selection and Procurement for the Hospitality Industry, Sixth Edition by Andrew Hale Feinstein and John M. Stefanelli (ISBN: 978-0-471-46005-3)

The Book of Yields: Accuracy in Food Costing and Purchasing, Seventh Edition by Francis T. Lynch (ISBN: 978-0-471-74590-7)

Management by Menu, Third Edition by Lendal H. Kotschevar and Marcel R. Escoffier (ISBN: 978-0-471-41319-6)

Food and Beverage Cost Control, Third Edition by Jack E. Miller, Lea R. Dopson, and David K. Hayes (ISBN: 978-0-471-27354-7)

Library of Congress Cataloging-in-Publication Data:

ISBN: 978-0-470-17916-1

Printed in the United States of America

10 9 8 7 6 5 4

CONTENTS

PREFACE

We are pleased to present this first edition of The International Culinary Schools at The Art Institutes' *Purchasing, Cost Control, and Menu Management*. This unique combination of materials reflects a philosophy of culinary education carried out at The International Culinary Schools at The Art Institutes' programs across the country. We are committed to collaborative, learner-centered instruction that encourages students to combine effective business leadership skills with excellent culinary fundamentals. This new book brings the management skills that we believe are absolutely essential to a contemporary chef into a single text for our students. These are the skills that distinguish our program and that will allow our graduates to reach their full potential as leading culinarians.

As a group of professional chefs and educators, we undertook this project with a strong sense of purpose and focus. One of the hallmarks of our program is its integration of classical culinary technique with core employability skills such as communication, time management, problem solving, teamwork, diversity appreciation, leadership and social responsibility. We wanted to provide a text which would support the development of these skills in a context specific to the world of the professional chef—and that would be relevant to our students from their first jobs as line cooks through their development to executive positions. We present the project with grateful appreciation to the many chefs, industry leaders, educators and students who contributed to its content. May their work and generosity of spirit be repaid by the contributions of a new generation of culinary leaders.

Michael Nenes, MBA, CEC, CCE
Assistant Vice President, Culinary Arts, The International Culinary Schools at The Art Institutes

Matthew Bennett, M.Ed. CEC, CCE, CWPC, CFBE
The International Culinary Schools at The Art Institutes, Colorado

Gary Eaton, MA
The International Culinary Schools at The Art Institutes, Houston

Sarah Gorham, MS, CEC
The International Culinary Schools at The Art Institutes, Atlanta

Walter Leible, CMC
The International Culinary Schools at The Art Institutes, Phoenix

Robynne Maii, MA
The International Culinary Schools at The Art Institutes, New York

1

PURCHASING AND PRODUCT IDENTIFICATION

1

THE PURCHASE SPECIFICATION: AN OVERALL VIEW*

The Purpose of this Chapter

After reading this chapter, you should be able to:

- List the information included on the purchase specification.

- Identify factors that influence the information included on the purchase specification.

- Explain the potential problems related to purchase specifications.

- Describe how quality is measured, including the use of government grades and packers' brands.

INTRODUCTION

A "product specification," sometimes referred to as "product identification," is a description of all the characteristics in a product required to fill a certain production and/or service need. It typically includes product information that can be verified upon delivery and that can be communicated easily from buyers to suppliers.

Unlike the product specification, which includes only information about the product, the "purchase specification" implies a much broader concept. The purchase specification includes all product information, but, in addition, it includes information regarding the pertinent supplier services buyers require from suppliers who sell them products.

* Authored by Andrew Hale Feinstein and John M. Stefanelli.

Large hospitality companies normally prepare purchase specifications. They usually seek long-term relationships with several primary sources and intermediaries and, before entering into these relationships, want to iron out every detail concerning product characteristics and desired supplier services. Smaller hospitality firms, on the other hand, tend to shop around for products on a day-to-day basis. These companies concentrate their efforts on preparing and using product specifications. If, for example, a particular supplier's supplier services are found lacking, these buyers will seek an alternative supplier who provides at least some of the desired supplier services.

Preparing detailed purchase specifications is not an easy task. It can be time-consuming, and a shortage of time is a major obstacle to getting this work done. If you plan to invest the time, money, and effort needed to develop adequate purchase specifications, you must be prepared to study the product's characteristics. Among the best sources here are the references the U.S. Department of Agriculture (USDA [www.usda.gov]) distributes. Many libraries carry these materials, or you can procure them from USDA offices or state agriculture offices. (These materials are particularly attractive because you can reproduce them without violating a copyright.) Further, many of these materials are now available online at the USDA's Website.

Other references are available as well. The U.S. government publishes purchasing guidelines for use by school food services that participate in the subsidized school lunch program (www.fns.usda.gov/cnd/Lunch). The various product industries, like the apple growers, also publish literature depicting characteristics of their products (www.bestapples.com). Industry associations, such as the Produce Marketing Association (PMA [www.pma.com]), similarly publish and distribute a significant amount of information that you can use to prepare specifications for fresh produce. And you can always find a supplier waiting to help you, especially if you buy from that supplier.

One decision you must usually make for yourself when preparing specifications is to choose the quality and supplier services you want. You cannot always expect to find a neat formula to guide you. This book offers several considerations that you should examine. But eventually you must make your own decisions concerning these other variables. You must also keep in mind that a purchase specification should contain more than just a brief description of a product.

WHY HAVE SPECS?

"Specs," or specifications, have several basic purposes and advantages, the primary ones being that: (1) they serve as quality control standards and as cost control standards (in these respects, specifications are important aspects of a hospitality operation's overall control system); (2) they help to avoid misunderstandings between suppliers, buyers, users, and other company officials; (3) in a buyer's absence, they allow someone else to fill in temporarily; (4) they serve as useful training devices for assistant buyers and manager trainees; and (5) they are essential when a company wants to set down all relevant aspects of something it wants to purchase, to submit a list of these aspects to two or more suppliers, and to ask these suppliers to indicate (bid) the price they will charge for the specific product or service.

In short, a specification is a sounding board for your ideas through which you detail every relevant consideration. By contrast, a purchase order is much less involved. After you know what you want and from whom you want it, completing the purchase order is a formality. But it is a legal formality: a contract between you and a supplier that he or she will deliver goods at a specific time, for a specific price, to a specific place. The specification lays out the parameters of what you must have. The purchase order is a written or sometimes verbal—for example, over the telephone—contract that arranges an actual transaction.

WHO DECIDES WHAT TO INCLUDE ON THE SPECS?

Four potential decision-making entities are involved here: (1) the owner-manager or another top management official, (2) the buyer, (3) the user, or (4) some combination of these three. It is unlikely that the buyer would write the specs alone, without the advice of the supervisor and of the users of the items to be purchased. All companies seem to approach this issue differently, but the buyers and users do most of the legwork, all the while staying within overall company guidelines. That is, a top company official normally sets the tone for the specs, and the buyers or users complete the details. The biggest problem with this participatory approach is agreeing on what is a main guideline and what is a minute detail.

WHAT INFORMATION DOES A SPEC INCLUDE?

A spec can be very short; it might include only a product's brand name—nothing else. Alternately, it might include several pages of detailed information, which is often the case with equipment specifications.

Be aware that specifications are sometimes categorized as either "formal" or "informal." A formal specification is apt to be extremely lengthy, perhaps several pages of information. Government agencies typically prepare formal specifications. The average hospitality enterprise owner-manager may prepare informal specifications, perhaps just a bit of information regarding product yield, quality, and packaging. You should not assume that the person preparing an informal specification is not cognizant of all of the other information normally found on a formal one. It is just that the typical operator does not spend so much time writing.

The buyer is apt to include at least some of the following pieces of information on a spec:

1. **The Performance Requirement, or the Intended Use, of the Product or Service.** This is usually considered the most important piece of information. You must have a clear idea of what is supposed to happen.

2. **The Exact Name of the Product or Service.** You must note the exact name, as well as the exact type of product you want. For example, you cannot simply note that you want olives; you must note that you want black olives, green olives, or anchovy-stuffed olives, or whatever. In some instances, you must be extremely careful to indicate the correct name and/or type of merchandise desired, or you are apt to be disappointed at delivery time.

3. **The Packer's Brand Name, If Appropriate.** Packers' brands are an indication of quality. Some items, such as fresh produce, do not normally carry instantly recognizable brand name identification. Many other items do, however, and a buyer may be interested primarily in only one or two brands and not any others. If you do indicate a brand name on the spec, you may want to add the words "or equivalent" next to it. This ensures that more than one supplier can compete for your business. By noting merely the brand name, you may reduce the opportunity to shop around, since usually only one supplier in your area will carry that product.

   In lieu of the words "or equivalent," some buyers prefer to add the words "equal to or better" to their brand name preferences noted on the specs. This phrase is used in conjunction with a brand name to indicate that the product quality characteristics desired must be similar or "superior" to the brand identified. The drawbacks with these words are that there may be several superior brands and that it may be very difficult for buyers to make a sound purchase decision if they are unfamiliar with some of them.

   At times, it is very important to insist on a certain brand name and avoid all other comparable brands. For instance, if a recipe has been developed that calls for a certain brand of margarine, the buyer should not purchase another brand unless it is compatible with that recipe. In this case, the finished product may be unacceptable if a different brand is used.

4. **U.S. Quality Grade, If Appropriate.** The federal government has developed U.S. grades to allow the buyer the option of using an independent opinion of product quality when preparing specifications. A good place to view U.S. quality grades

online is at www.ams.usda.gov / howtobuy. This site provides information on how to select numerous food products.

Unfortunately, since grading generally is a voluntary procedure, many items in the channel of distribution may not be graded. However, you can, at least, indicate a desired grade, along with the notation "or equivalent." This will enable suppliers who do not have graded merchandise to bid for your business. Also, these suppliers then have a quality standard to guide them. Some states also have grading systems. For example, Wisconsin has a grading procedure to use for some dairy items (www.wisdairy.com).

5. **Size Information.** In most instances, buyers must indicate the size desired for a particular item. For some products, such as portion-cut steaks, buyers can indicate an exact weight. For other items, though, such as large, wholesale cuts of beef or whole chicken, usually buyers can only indicate the desired weight range. In some instances, the size of an item, such as lemons or lobster tails, is indicated by its "count," that is, the number of items per case, per pound, or per 10 pounds.

6. **Acceptable Trim, or Acceptable Waste.** For some products, including many fresh foods, you may need to indicate the maximum amount of waste you will tolerate. Another way to say this is to note the minimum edible yield of a product you will accept. For instance, fresh lettuce may have varying degrees of waste, depending on how the food distributor processes it. Some lettuce is a cleaned and chopped, ready-to-serve product, whereas a typical head of lettuce has an edible yield of much less than 100 percent. Of course, you expect to pay much more for the product that has little or no waste.

7. **Package Size.** In most situations, you will need to indicate the size of the container you desire. For instance, the can size must be noted when purchasing canned vegetables.

8. **Type of Package.** In some cases, the type of packaging materials used is highly standardized. For example, dairy products packaging must meet minimum standards of quality. This is not the case for other items. Frozen products, for instance, should come in packaging sufficient to withstand the extreme cold without breaking. Some suppliers scrimp on this, and, while the quality of the product may meet your specification, the poor packaging will result in a rapid deterioration of this once-acceptable item.

Packaging can add considerable cost to the items you purchase. In some cases, the value of the packaging may exceed the value of the item. The cost of packaging of single-serve packets of salt, for instance, can easily be higher than the cost of this food ingredient.

When specifying the desired type of packaging, some buyers may require suppliers to use recyclable packaging materials. Alternatively, buyers may request reusable packaging, such as the plastic tubs some suppliers use to deliver fresh fish.

9. **Preservation and/or Processing Method.** For some products, you will be able to identify two or more preservation methods. For instance, you could order refrigerated meats or frozen meats, canned green beans or frozen green beans, and refrigerated beer or nonrefrigerated beer.

   You also could specify unique types of preservation methods, such as smoked fish instead of salted fish, irradiated poultry instead of nonirradiated poultry, oil-cured olives instead of brine-cured olives, and genetically altered tomatoes instead of natural tomatoes.

   The type of preservation and/or processing method selected often influences the taste and other culinary characteristics of the finished food product. Consequently, it is important for you to be familiar with recipe requirements before altering this part of the spec.

10. **Point of Origin.** You may want to indicate the exact part of the world that a specific item must come from. This is a rather important consideration for fresh fish. For instance, you may need to specify that your lobster must come from Maine and not from Australia.

    Buyers may want to note the point of origin on some specs for several reasons. One is that the flavor, texture, and so forth of an item can differ dramatically among growing regions. Another reason is that the menu may state that an item comes from a particular producing region in the world; if so, it would be a violation of truth-in-menu regulations to serve an alternative product. Freshness can be another important consideration because buyers may specify nearby points of origin in order to ensure product quality. And, finally, buyers may indicate where a product cannot come from, instead of where it must come from, in order to adhere to various company policies and/or legal restrictions. For instance, for political reasons, some companies may refuse to purchase products that come from certain parts of the world.

11. **Packaging Procedure.** Some products are wrapped individually and conveniently layered in the case. Others are "slab-packed," that is, tossed into the container. The more care taken in the packaging procedure, the higher the as-purchased (AP) price is apt to be. However, carefully packaged products will have a longer shelf life. In addition, they will tend to maintain their appearance and culinary-quality characteristics much longer than those products that are packaged indiscriminately.

Another packaging consideration concerns the number of individual containers that normally come packaged in a case lot. For instance, it is traditional for No. 10 cans of foods to come packed six to a case. However, some buyers cannot afford to purchase six cans, or they cannot use six cans. Will the supplier sell fewer than six cans; that is, will he or she "bust" the case? Buyers who request busted cases run the risk of having few suppliers willing to compete for their business.

12. **Degree of Ripeness.** This is important for fresh produce. The same concept applies to beef items; for example, you may desire a specific amount of "age" on the item. Wines have a similar system that reflects, among other pieces of information, the year of production.

13. **Form.** This is an important consideration for many processed items. For example, do you want your cheese in a brick, or would you rather have it sliced? Do you want your roast beef raw, or do you want it precooked?

14. **Color.** Some items are available in more than one color. For example, buyers can order fresh red, green, or yellow peppers.

15. **Trade Association Standards.** Some trade associations establish minimum performance standards for items. This information can be commonly found in their trade publications (see Figure 1.1). For instance, the National Sanitation Foundation (NSF) International certification (www.nsf.org) seal on a piece of food-production equipment testifies to the equipment's sanitary acceptability.

16. **Approved Substitutes.** Some buyers make it a habit to include on some specs a list of acceptable substitutes that the suppliers can deliver if they are out of the normal item. This can save a great deal of time and effort over the long haul since suppliers would not have to call buyers every time a product shortage occurs. Buyers also may like this convenience. Unfortunately, before determining approved substitutes, buyers must ensure that they are compatible with production and service needs. So, while this notation on each spec saves time and trouble eventually, it can be more difficult in the short run to spend the time needed to test all potential substitute items.

17. **Expiration Date.** Many buyers will not accept products if they are concerned about possible quality deterioration. To avoid this problem, they may indicate on some specs that suppliers must prove that the products delivered are not too old. For instance, some product labels list "sell-by" dates; these are sometimes referred to as "pull dates," "best-if-used-by dates," or "freshness dates." For such items, buyers may want to add to their specs some reference to these dates.

UPDATE

October 2003, Volume 13, Issue 10

The Wheat Foods Council is an industry-wide partnership dedicated to increasing grain foods consumption through nutrition education and promotion programs.

Political cartoon #1 released

WheatFoods
COUNCIL

WFC's political cartoon series began with two scientists discussing what will be the newest "scapegoat" for selling more fad diet books.

In September, the first of four WFC cartoons was distributed to newspapers nationwide. Cartoons will be released quarterly and will depict clever messages to get consumers thinking about the truth and benefits of grain foods.

10841 S. Crossroads Dr., Suite 105 Parker, Colo. 80138

Phone: 303/840-8787

Fax: 303/840-6877

E:Mail: wfc@wheatfoods.org

URL: www.wheatfoods.org www.homebaking.org

Judi Adams President

Lori Sachau Communications Specialist

Vikki Berry Office Manager

Sharon Davis Charlene Patton HBA/WFC Consultants

Council members are encouraged to reprint articles from this publication.

Grain foods to be featured in "Easy Home Cooking"

WFC recipes, photographs, and grain food information will be featured in the February/March issue of "Best Recipes - Easy Home Cooking" magazine. Since the magazine is sold at grocery checkout stands across the country, it provides an excellent opportunity for the Council to reach it's target audience. Magazines will go on sale February 10, 2004 and will include grains and breads in a special multi-page spread in the magazine's "Heart Healthy" section.

Approved by the WFC Executive Board, the project was made possible because of unanticipated funds that became available after the 2003-04 budget process. The project was included as an add-on for the 2003-04 WFC Communication Plan.

The chapter insert will be approximately eight pages and will feature the headline "A Lovin' Spoonful of Grains." The insert will also include text promoting the healthful benefits of eating grain foods and a description of the WFC. Recipes utilizing pasta, cereal, crackers, tortillas, flour, and bread products were submitted to the publisher. Additionally, the Home Baking Association's "Bake for Family Fun Month" (designated for February) logo will be included to encourage families to bake at home.

FIGURE 1.1 Trade associations provide information useful to buyers. Courtesy of the Wheat Foods Council.

18. **Chemical Standards.** Buyers might decide to specify a particular level of acceptable chemical use for some of the items they purchase. For instance, it is possible to purchase organic produce that is grown in chemical-free soil. Meat and poultry products raised without added chemicals in the animals' diets are also available in the marketplace.

19. **The Test or Inspection Procedure.** This is the procedure you intend to use when checking the items delivered to you or the services performed for you. Generally, this is the logical outcome of specifying the intended use. After you note the intended use, you should be prepared to indicate the tests or inspection procedures you will use to see whether your purchases will perform adequately.

20. **Cost and Quantity Limitations.** Buyers might indicate how much of the item or service is to be purchased at any one time. In addition, they might require an item to be removed from production and a substitute item to be sought when the cost limits are approached.

21. **General Instructions.** In addition to specific details, buyers might include such general details as: (a) delivery procedures, if possible; (b) credit terms; (c) the allowable number of returns and stockouts; (d) whether the product purchased must be available to all units in the hospitality company, regardless of a unit's location; and (e) other supplier services desired, like sales help in devising new uses for a product.

22. **Specific Instructions to Bidders, If Applicable.** Suppliers who bid for your business may want to know: (a) your bidding procedures, (b) your criteria for supplier selection, and (c) the qualifications and capabilities you expect from them.

WHAT INFLUENCES THE TYPES OF INFORMATION INCLUDED ON THE SPEC?

Several factors must be assessed before determining what information to include on a specification. Eight of these are:

1. **Company Goals and Policies.** These are probably most important. Overall managerial guidelines must be consulted before buyers write specs.

2. **The Time and Money Available.** Industry members continually argue the costs and benefits of written specifications. Obviously, we consider the time and money preparing specifications well spent.

3. **The Production Systems the Hospitality Operation Uses.** If, for example, a restaurant broils its hamburgers instead of grilling them, the fat content in its ground beef should be a bit higher than usual to compensate for the additional loss of juices that can occur if meat is broiled to the well-done state.

4. **Storage Facilities.** If, for example, freezer space is limited, a buyer may have to purchase larger amounts of fresh vegetables; a specification might carry this reminder.

5. **Employee Skill Levels.** Generally, the lower the skill level, the more buyers must rely on portion-controlled foods, one-step cleaners, and other convenience items. The trade-off is between a higher AP price and a lower wage scale. The balance in these issues is not always clear-cut; this is a good example of the trade-off concept that usually arises in value analysis.

6. **Menu Requirements.** For example, live lobster on the menu forces a buyer to include the words "live lobster" on the specification.

7. **Sales Prices or Budgetary Limitations.** If, for example, a restaurant is located in a very competitive market, its menu prices may be fixed by its competition. This fact may force a buyer to include cost limits for some or all food specifications.

8. **Service Style.** A cafeteria, for example, needs some food items that have a relatively long hot-holding life since the food may remain on a steam table for a while. This type of information might be included on the specs, especially the specs for preprepared food entrées.

WHO WRITES THE SPECS?

Generally, four options are available to the hospitality operation, including:

1. Company personnel can write the specs. This option assumes that the necessary talent to write them exists in the company somewhere.

2. Many specs can be found in industry publications, CDs, online services, and in government documents. Although they may not fit your needs exactly, they are at least a good starting point (see Figures 1.2 and 1.3).

3. You can hire an expert to help you write your specs. This is a reasonable alternative, as you can control the amount of money you care to spend for this service.

   The USDA operates an "Acceptance Service" (www.ams.usda.gov/gac/) that permits hospitality operators to hire USDA inspectors to help prepare specs. The inspectors check the products you buy at the supplier's plant to make

American Bakers Association	http://www.americanbakers.org/
American Beverage Institute	http://www.abionline.org/
American Egg Board	http://www.aeb.org/
American Institute of Baking	http://www.aibonline.org/
American Meat Institute	http://www.meatami.com/
American Poultry Association	http://www.ampltya.com/
American Seafood Distributors Association	http://www.freetradeinseafood.org/
Beer Institute	http://www.beerinstitute.org/
Canned Food Alliance	http://www.mealtime.org/
Florida Fruit and Vegetable Association	http://www.ffva.com/
Food and Drug Administration	http://www.fda.gov/
Foodservice Equipment Distributors Association	http://www.feda.com/
International Beverage Dispensing Equipment Association	http://www.ibdea.org/
International Dairy Food Association	http://www.idfa.org/
International Foodservice Manufacturers Association	http://www.ifmaworld.com/
National Cattlemen's Beef Association	http://www.beef.org/
National Fisheries Institute	http://www.nfi.org/
National Food Processors Association	http://www.nfpa-food.org/
National Frozen and Refrigerated Foods Association	http://www.hffa.org/
National Pasta Association	http://www.ilovepasta.org/
National Poultry and Food Distributors Association	http://www.npfda.org/
National Soft Drink Association	http://www.nsda.org/
National Turkey Federation	http://www.eatturkey.com/
North American Association of Food Equipment Manufacturers	http://www.nafem.org/
North American Meat Processors Association	http://www.namp.com/
NSF International	http://www.nsf.org/
Produce Marketing Association	http://www.pma.com/
Quality Bakers of America Cooperative	http://www.qba.com/
Retail Bakers of America	http://www.rbanet.com/
United Egg Producers	http://www.unitedegg.org/
United Fresh Fruit and Vegetable Association	http://www.uffva.org/

FIGURE 1.2 Some government and private agencies that provide product information useful to buyers.

U.S. Department of Agriculture	http://www.usda.gov
Agricultural Marketing Service	http://www.ams.usda.gov/
Dairy Market Branch	http://www.ams.usda.gov/dairy/index.htm
Fruit and Vegetable Branch	http://www.ams.usda.gov/fv/
Livestock and Grain Branch	http://www.ams.usda.gov/lsg/
Poultry Market News Branch	http://www.ams.usda.gov/poultry/
U.S. Department of Commerce	http://www.commerce.gov/
National Marine Fisheries Service	http://www.nmfs.noaa.gov/
Wine and Spirits Wholesalers of America	http://www.wswa.org/
World Association of the Alcohol Beverage Industries	http://www.waabi.org/

FIGURE 1.2 (Continued)

sure they comply with your specs (see Figure 1.4). They then stamp each item or sealed package to certify product compliance. This is often done for meat products. The acceptance service is provided for a fee, which the supplier usually pays. Although this expense may be included in your AP price, the service could save you money by assuring you that you receive exactly what you want.

4. The buyer and supplier can work together to prepare the specifications. The problem with this arrangement is that the buyer usually neglects to send the specs out to other suppliers for their bids. Also, the cooperating supplier may help slant the specs so that only he or she can provide the exact item wanted. Nevertheless, this is the option most independents find realistic, given their limited time resources and prospective order sizes.

The question of who writes the specs is important to hospitality operators because few part-time buyers have enough time to learn this task thoroughly. If operators want to prepare their own specs, they often consult outside expertise.

The reasonable compromise seems to be to hire someone on a consulting basis to help write the specs or, if this is too expensive, to work with the specs found in various trade and governmental sources. The usual approach, to huddle with a supplier, may actually be least advantageous, but it does allow operators to spend more time in other business activities.

A. M. Pearson and Tedford A. Gillett, *Processed Meats,* 3rd ed. (New York: Aspen Publishers, 1998).

Arabella Boxer, *The Herb Book: A Complete Guide to Culinary Herbs* (Berkeley, CA: Thunder Bay Press, 1996).

ComSource Canned Goods Specifications Manual (Atlanta: ComSource Independent Foodservice Companies, Inc., 1994) [out of print].

ComSource Frozen Food Specifications Manual (Atlanta: ComSource Independent Foodservice Companies, Inc., 1994) [out of print].

Elizabeth Schneider, *Vegetables from Amaranth to Zucchini: The Essential Reference* (New York: William Morrow, 2001).

Ian Dore, *Shrimp: Supply, Products, and Marketing in the Aquaculture Age* (Toms River, NJ: Urner Barry Pub. Co., 1993).

Ian Dore, *The Smoked and Cured Seafood Guide* (Toms River, NJ: Urner Barry Pub. Co., 1994).

Ian Dore, *The New Fresh Seafood Buyer's Guide,* 2nd ed. (New York: Van Nostrand Reinhold, 1991) [out of print].

James A. Peterson, *Fish & Shellfish: The Definitive Cook's Companion* (New York: William Morrow, 1996).

John R. Romans, *The Meat We Eat,* 14th ed. (Upper Saddle River, NJ: Prentice Hall, 2000).

Kenneth T. Farrell, *Spices, Condiments, and Seasonings,* 2nd ed. (New York: Aspen Publishers, 1999).

Lewis Reed, *SPECS: The Comprehensive Foodservice Purchasing and Specification Manual,* 2nd ed. (New York: John Wiley & Sons, 1993).

North American Meat Processors Association, *The Meat and Poultry Buyers Guide on CD-ROM* (McLean, VA: North American Meat Processors Association, 2002).

North American Meat Processors Association, *The Meat Buyers Guide* (McLean, VA: North American Meat Processors Association, 1997).

North American Meat Processors Association, *The Poultry Buyers Guide* (McLean, VA: North American Meat Processors Association, 1999).

Seafood Business, *Seafood Handbook* (Portland, ME: Diversified Business Communications, 1999).

The Produce Marketing Association Fresh Produce Manual (Newark, DE: Produce Marketing Association, 2002).

FIGURE 1.3 Some comprehensive reference materials buyers can use to prepare product specifications.

POTENTIAL PROBLEMS WITH SPECS

As in most business activities, you should consider several costs in addition to benefits in specification writing. There are, for example, a number of clearly identifiable

FIGURE 1.4 An inspector employed by the USDA Acceptance Service will inspect the buyer's order on the supplier's premises. If the order meets the buyer's specifications, the government inspector will apply a stamp, such as the one shown here, to the package. *Source:* United States Department of Agriculture.

costs, and there are some cleverly hidden problems. Some potential problems with specs include:

1. Delivery requirements, quality tolerance limits, cost limits, or quantity limits may appear in the specs. If these are unreasonable requirements, they usually add to the AP price, but it may be questionable whether they add to the overall value.

2. Some inadvertent discrimination may be written into the specs. For example, the spec may read, "Suppliers must be within 15 miles to ensure dependable deliveries." Dealing with a supplier 16 miles away could cause legal trouble because of this.

    Worse, if a spec effectively cuts out all but one supplier, you will have wasted your time, money, and effort if your intention was to use the spec to obtain bids from several sources. You do, of course, still have the benefit of having specified very precisely what you want. This gives you a receiving standard and a basis for returning unacceptable product.

3. The specifications may request a quality difficult for suppliers to obtain. This situation adds to cost, but not always to value. In some situations, the quality you want cannot be tested or inspected adequately without destroying the item. In these cases, however, a sampling approach may ensure the requisite quality. Before you embark on such an expensive process, careful consideration is called for.

4. Some specs rely heavily on government grades. Unfortunately, some may not be specific enough for a foodservice operator's needs. For example, USDA Choice beef covers a lot of possibilities: There are high-, medium-, and low-choice grades. Also, grades do not usually take into account packaging styles, delivery schedules, and so forth. Thus, U.S. grades alone are not adequate for most operations.

5. Food specifications are not static; they usually need periodic revision. For instance, a spec for oranges might include the term "Florida oranges," a perfectly reasonable requirement at certain times of the year. But during some seasons, Arizona oranges might be preferable. It costs time, money, and effort to revise specifications. Moreover, if you cannot determine exactly when to revise, not only might you receive what you do not want, but your customer might also become dissatisfied if you are forced to serve the food because you have no acceptable substitute.

6. The best specs in the world will be of no use to you if the other personnel in the hospitality operation are not trained to understand them and to use them appropriately. For example, a buyer may be adept in the use of specs, but if the receiving agent does not have similar expertise, he or she may accept the delivery of merchandise that is not in accord with the properly prepared specifications.

7. The potential problems and costs multiply quickly if the spec is used in bid buying. Some of these additional problems include the following:

Getting Hit with the "Lowball"

The term "lowball" refers to a bid that is low for some artificial or possibly deliberately dishonest reason. For instance, bidders may meet a buyer's spec head on; that is, they may hit the minimum requirements and might even reduce their normal profit levels in order to win the bid. Once they are in, they may try to trade up the users.

Lowballing is a fairly standard way of doing business for suppliers trying to woo buyers away from their regular suppliers. These suppliers are willing to sacrifice a bit of revenue in the short run for the opportunity to establish a long-term and potentially more profitable arrangement. Suppliers know that once they get their foot in the door, buyers may get comfortable and stick with them through force of habit.

To avoid falling for lowball prices, buyers need to shop around frequently, which means they need to keep their specs current. This tends to keep suppliers competitive and more responsive.

Inequality Among Bidders

If your specs are too loose, that is, if too many suppliers can meet the specs, you run the risk of finding several suppliers of differing reliability bidding for your business. Choosing one of the less reliable suppliers can result in serious operational problems.

This problem is particularly prevalent in the fresh produce trade simply because the available qualities of fresh produce change continually and some buyers do not know exactly when to revise their specs. Several suppliers may bid for your lettuce contract, and several qualities may meet your specifications. Suppose one supplier has a good product, and he bids 60 cents per pound. Suppose, too, that another supplier has lettuce that she could gain good profits on even if she sold it for 55 cents per pound. What she probably will do, though, is enter a bid for 59 cents per pound because she has discovered that the other supplier will bid 60 cents. You gladly accept the 59-cent bid, and—who knows?—the quality may be satisfactory. To avoid this problem, do not use the costly bidding procedure unless you are willing to expend a great deal of effort to keep your specifications current.

A related problem occurs whenever an inexperienced buyer accidentally rigs the procedures by asking a supplier who has a high AP price and high quality to bid for the business. In the preceding example, the 59-cent-per-pound bidder is very happy to include the 60-cent-per-pound bidder in the process. The wise buyer strives to include in the bidder pool only responsible, competent, and competitive suppliers who are able to follow through if they win the buyer's business.

Sometimes, good suppliers may unintentionally differ significantly from others bidding for your business because of unanticipated changing business conditions. For instance, some suppliers who bid for your business may do so only when their regular business is slow. Consequently, you may be forced to continually change suppliers, which could cost you time, trouble, and money in the long run. In addition, although you may indeed receive an AP price break, when their regular business picks up, these suppliers may decide to stop bidding for yours.

Specifications that Are Too Tight

Tight specifications tend to eliminate variables and allow a buyer to concentrate on AP prices. Unfortunately, if only one supplier meets the buyer's specifications, that buyer will end up spending a lot of time, money, and effort to engage in specification writing and bidding, and still find there are only two choices available: take it or leave it.

Large hospitality companies sometimes run into a similar problem when they demand items that only one or two suppliers are able to deliver. For instance, a typical large hospitality firm wants to purchase products that are available nationally; this ensures that all units in the company use the same products, and this, in turn, ensures an acceptable level of quality control and cost control. The number of suppliers who can accommodate national distribution, though, is limited.

Advertising Your Own Mistakes

Bids may be entered on a three-month contract basis. If your specifications are in error, you can look forward to being reminded of your mistake whenever a delivery comes in.

Redundant Favoritism

The buyer who writes several specs, sends them out for bid, and then rejects all of the bids except the one from the supplier he or she usually buys from anyway is a genuine annoyance. This practice is followed by some operations that must use bid buying. The buyer solicits a bid for, for example, corn chips. Three companies bid. But the buyer decides to buy from supplier A because this supplier's product is always preferred. If this is the case, why seek bids?

Too Many Ordering and Delivery Schedules

Another potential problem with bid buying is the possibility that you will have to adjust to several suppliers' ordering and delivering schedules. A large hospitality organization can handle this extra burden. But if a small firm is accustomed to receiving produce at 10:00 A.M., it can be a difficult readjustment for that operator to receive produce at 2:00 P.M. one week and at 9:00 A.M. the next. (We have seen this need to readjust operating procedures cause a great deal of trouble, especially when a delivery must sit on the loading dock for a while because no one is free to store it. When the receiving routine is broken, ordinary problems multiply.)

And Always Remember . . .

The object of bid buying is to obtain the lowest possible AP price. But if the lowest possible AP price does not, somehow, translate into an acceptable edible-portion (EP) cost, you have gained little or nothing.

The costs and benefits of specification writing are never clear, and the subject becomes more confusing when you complicate it with a bid-buying strategy. We believe writing specs is generally necessary because they help you to clarify your ideas on exactly what you want in an item. We are not so confident about the bid procedures, though. For some items, such as equipment, bids may be economically beneficial to the hospitality operator. But on the whole, the buyer who uses this buying plan had better know as much or more about the items as the supplier. Only large operations consistently approach this requirement.

THE OPTIMAL QUALITY TO INCLUDE ON THE SPEC

You frequently hear references to "quality" products. To most people, a "quality" product represents something very valuable. However, when business persons talk about quality, they are referring to some "standard" of excellence. This standard could be high quality, medium quality, or low quality. In other words, suppliers offer products and services that vary in quality. In most cases, they can sell you a "high quality," "highest quality," "substandard quality," or almost any other quality you prefer.

It is important to keep in mind that quality is a standard: something to be decided on by company officials and then maintained throughout the operation.

We do not intend to second-guess the types of quality standards that hospitality operators develop or decide upon. Rather, our objective is to examine the typical process by which the optimal quality is determined.

WHO DETERMINES QUALITY?

Someone, or some group, must decide on a quality standard for every product or service the hospitality operation uses. If somebody decides to use a low choice grade of beef, this decision should reflect the type of customers the operation caters to, the restaurant type, and its location, among other factors.

Most analysts agree that a hospitality operator can hardly decide on quality standards without measuring the types of quality standards his or her customers expect. As the AP prices are translated into menu prices and room rates, customers are affected. The quality of the product purchased affects customers' perceptions of the operation, too. On the other hand, for the most part, supplier services are apparent principally to management. It is clear that value has many facets.

Most hospitality operations conduct some sort of market research to determine the types of value their customers, or potential customers, seek. The owner-manager's greatest responsibility is to interpret the results of the market research and translate them into quality standards. In other words, he or she must examine: (1) the overall value retail customers expect; (2) "supplier" services, that is, the property's surroundings, service style, decor, and so on; and (3) the typical menu or room price ranges attributable to his or her type of operation. Then the owner-manager must formulate a definition of quality standards. So, in the final analysis, the consumer really has the major say in determining the quality standards an operation establishes for most of its items.

Company officials may have a bit more latitude in determining the quality standards for those operating supplies and services retail customers do not directly encounter—items such as washing machine chemicals and pest control service. In these instances, it is interesting to note the number of people who may become involved in these determinations. A large group of company personnel may help work out these quality standards. The owner-manager, the department heads, and the buyer often influence the decision. Hourly employees may also be consulted since they constantly work with many of the products and services and are, hence, most familiar with them.

Quality standards for supplies, services, and equipment normally come from the top of the company. Buyers exercise a great deal of influence in these areas, though, because they get involved in such technical questions as "Is the quality standard available?", "What will it cost?", and "Can it be tested easily?" The ultimate decision, though, usually rests with the owner-manager or, in the case of chain organizations, an executive officer.

MEASURES OF QUALITY

A buyer is expected to be familiar with the available measures of quality, as well as their corresponding AP prices and ultimate values. Several objective measures of quality exist. Here are some of them.

Federal-Government Grades

Under authority of the Agricultural Marketing Act of 1946 and related statutes, the Agricultural Marketing Service (AMS [www.ams.usda.gov]) of the USDA has issued quality grade standards for more than 300 food products. These grade standards for food, along with standards for other agricultural products, have been developed to identify the degrees of quality in the various products, thereby helping establish their usability or value.

Federal-government grades are measurements that normally cannot be used as the sole indication of quality. This is true because federal-government grading is not required by federal law, except for foods a government agency purchases for an approved feeding program, or for commodities that are stored under the agricultural price support and loan programs; as a result, a buyer must use other measures of quality for ungraded items. Where possible, though, U.S. grades are the primary measures of quality that buyers use most frequently, at least at some point in the overall purchasing procedure.

The federal government, by legal statute, inspects most members of the channel of food distribution. Generally, the federal government's role is to check the sanitation of production facilities and the wholesomeness of the food products throughout the distribution channel. In some instances, states have set up additional inspector-powered agencies that either complement the federal agencies or supplant them.

Ordinarily, to be graded, an item must be produced under continuous federal-government inspection. Meat and poultry items and items that require egg breaking during their production process are always made under continuous inspection, but other types of items may not be.

The federal government will provide grading services for food processors, usually those at the beginning of the channel of distribution, who elect to purchase this service. Some of these producers buy this service, and some do not. Some opt for U.S. government grading because their customers include these grade stipulations in their specifications. Alternately, in some cases, the state requires federal grading. For example, several states require fresh eggs to carry a federal quality grade shield.

The grading procedure usually takes a scorecard approach with the products, beginning with a maximum of 100 points distributed among two or more grading factors. To receive the highest grade designation, a product must usually score 85 to 90 points or more. As the product loses points, it falls into a lower grade category. In addition, graders work under "limiting rules," which stipulate that if a product scores very low on one particular factor, it cannot be granted a high grade designation regardless of its total score. The grader usually takes a sample of product and bases his or her decision on that sample.

Grading can be a hurried process that can tax the resourcefulness of even the hardest grader. Although some food producers accuse graders of being capricious, unreasonable, and insensitive to production problems, the grading system actually functions fairly well.

Some buyers in the hospitality industry have been conditioned to purchase many food products primarily on the basis of U.S. government grades. The effect of government grading has ultimately been to create demand among retail consumers for specific quality levels; for example, consumers are conditioned to buy USDA Choice beef or USDA Select beef in the supermarket.

A major problem with grading is the emphasis graders place on appearance. Although appearance is an overriding criterion used in U.S. government grading, this sole criterion is dangerous for the foodservice industry because our customers are not making a purchase based solely on visual inspection but, rather, are purchasing and almost immediately evaluating the product based on taste and other culinary factors.

A number of other problems are associated with U.S. government grades. These additional difficulties include: (1) the wide tolerance between grades—so much so that buyers quickly learn that when they indicate U.S. No. 1, they must also note whether they want a high 1 or a low 1 (this tolerance gap is especially wide for meat items); (2) grader discretion—graders operate under one or more "partial limiting rules," which allow them to invoke a limiting rule or not; (3) the deceiving appearance of products—for example, some products can be dyed (like oranges), some can be waxed (like cucumbers), and some can be ripened artificially and inadequately (like tomatoes); (4) the possible irrelevance of grades to EP cost—for instance, a vine-ripened tomato may have a high grade and a good taste, but it may be difficult and wasteful to slice; (5) the fact that graders could slight such considerations as packaging and delivery schedules, which are important in preserving the grade—for example, a lemon may look good in the field, but if it is not packaged and transported correctly, it could be dry and shriveled when it arrives; (6) a raw food item is not a factory-manufactured product, and, therefore, its quality, as well as its U.S. quality grade, can fluctuate and may not be consistent throughout the year; (7) the lack of uniformity among terms used to indicate the varying grade levels—for instance, some items are labeled with a letter, some with a number, and some with other terminology; and (8) the lack of a specific regional designation. There is, for instance, a big difference between Florida and California oranges, particularly during certain times of the year.

AP Prices

To some degree, quality and AP prices go hand in hand. The relationship, however, is not usually direct. One notch up in AP price does not always imply that the item's quality has gone up one notch too. AP prices, though, are considered good indicators of quality by many hospitality managers, especially novices.

Packers' Brands

Some food producers resort to their own brand names and try to convince buyers to purchase on the basis of these names.

The terms "brand names" and "packers' brand names," although often used interchangeably, do differ to some extent. For example, the word "Sysco" is a brand name, but the terms "Sysco Supreme," "Sysco Imperial," "Sysco Classic," and "Sysco Reliance" are the company's packer's brand names. In this case, the Sysco supplier offers several levels of quality, with Sysco Supreme representing its highest quality and Sysco Reliance its lowest (see Figure 1.5).

THE TOP TEN BROADLINE DISTRIBUTORS

Food Services of America	www.fsafood.com
Golbon	www.golbon.com
Gordon Food Service	www.gfs.com
J&B Wholesale	www.jbwhsle.com
Performance Food Group	www.pfgc.com
Reinhart FoodService	www.reinhartfoodservice.com
Seneca Foods	www.senecafoods.com
Sysco	www.sysco.com
U.S. Foodservice	www.usfoodservice.com
Zanios Foods	www.zaniosfoods.com

FIGURE 1.5 Some organizations and companies with packers' brands.

Both brand names and packers' brand names are indications of quality standards; however, packers' brand names are very specific quality indicators, whereas most brand names are much more general. A packer's brand system is essentially that food processor's personal grading system; that is, the food processor uses his or her personal "grade" in lieu of a federal quality grade. The companies that use such grades usually offer at least three quality levels: good, better, and best. Food processors typically identify their different quality levels by using a particular nomenclature (such as that Sysco uses) or by using different colored package labels (see Figure 1.6).

Even though they are not widely known in many parts of the country, packers' brands exist for many products. In some cases, the food processor uses the brand name in conjunction with U.S. grade terminology. For example, a fresh-produce packer might stencil on a box the designation "No. 1." This would indicate that the item was not produced under continuous government inspection, and that, in the opinion of the packer, who is not a government grader, the product meets all U.S. requirements for U.S. No. 1 graded products.

Packers' brands, too, present problems when they are designed to overlap U.S. grades. So, for example, a beef product that might be marked a high USDA Select instead of a low USDA Choice might be switched to the packer's brand. This will permit it to carry a quality designation that food buyers might generally associate with USDA Choice. In addition, packers' branded merchandise, unless it is a meat or poultry item or includes egg breaking in its production, may not be under continuous government inspection. However, even if a food processor does not purchase the U.S. government

FIGURE 1.6 An example of packer's brand packaging. Courtesy of Sysco Corporation.

grading service, he or she must still undergo an inspection procedure. But inspection is concerned only with safety and wholesomeness; it makes no quality statement. Only U.S. grading makes quality judgments.

Brand names may possibly be a little more reliable than government grades because the brand extends over several other considerations, not just the food product's appearance. For example, the brand can also indicate a certain size of fruit and a certain packaging procedure. In addition, for products that do not come under the grading system, brand names may be the logical alternative.

Some food buyers also think that packers' brands are a bit more consistent from day to day and month to month, though not everyone feels this is true. Some argue that the U.S. government graders are not always so consistent. A brand's supposed consistency should effect a more consistent and predictable EP cost. It is critical to recall that the EP cost is more important; the AP price represents only your starting point.

Keep in mind that in this text, we use the term "packer's brand" a bit more frequently than the term "brand name" merely because it seems as if our industry uses such terminology more often.

Samples

It may be necessary to rely on samples, and one or more relevant tests of these samples, when assessing the quality of new items in the marketplace. Samples and testing are commonly used to measure the quality of capital equipment.

Endorsements

Several associations endorse items that we purchase. For instance, NSF International attests to the sanitary excellence of kitchen equipment. The Foodservice Consultants Society International (FCSI [www.fsci.org]) is an association of food-service consultants whose members must achieve rigorous standards. We find, however, fewer associations endorsing foods and operating supplies.

Trade Associations

Various organizations, such as the National Cattlemen's Beef Association (NCBA [www.beef.org]), and other trade groups, help set quality standards that the buyer can use.

Your Own Specifications

A buyer may use some combination of all of the measures we have been discussing and work them into an extended measure of quality. This lengthy exercise usually finds its way into the specification. In many cases, particularly when a hospitality operation needs a special cut of meat, a unique type of paper napkin, or special cleaning agents, this extended measure is the only appropriate one.

As we imply throughout this discussion, few buyers consider only one of these quality measures. But, in our opinion, too many operators become overreliant on only one measure when it would be more appropriate to consider two or more.

IS THE QUALITY AVAILABLE?

Another aspect of quality a buyer must know is whether the quality desired is available at all. This is quite a practical question. It is useless to determine quality standards if the quality you want is unavailable. Oddly enough, some types of quality are too often unavailable to the hospitality operation. A chef who wants low-quality apples to make homemade applesauce may find that suppliers do not carry such low quality. (Food canners usually purchase them all.)

In addition, a buyer must pay particular attention to the possibility that the quality desired is available from only one supplier. This may or may not be advantageous. In some cases, an owner-manager may take this opportunity to build a long-standing relationship with one supplier. But some company officials are not especially eager to lose flexibility in their supplier selection.

It is easier than you think to restrict yourself unknowingly to one supplier. If this does not happen because of the quality standards you set, it may happen because of the AP price you are willing to accept.

THE BUYER'S MAJOR ROLE

We have noted that the buyer usually provides his or her supervisors with the information they need to determine quality standards. Buyers normally do not set these standards by themselves, but they do generally participate in these decisions. The buyer's major role here is to maintain the quality standards that someone else has determined. Generally, the standards have some flexibility. But whatever the standards are, and whatever the degree of flexibility, a buyer must ensure that all of the items purchased measure up to company expectations.

THE OPTIMAL SUPPLIER SERVICES TO INCLUDE ON THE SPEC

Buyers normally have a major voice in determining supplier services, though they generally have less to say regarding economic values that the company should bargain for.

If you want specific supplier services, chances are you will severely restrict the number of purveyors who can provide what you want. Consequently, if you like the bid-buying activity, you must be prepared to put up with a variety of supplier capabilities. In our experience, it is the supplier services that we become so attached to since, for many items, not that much difference in quality usually exists.

KEY WORDS AND CONCEPTS

Advantages and purposes of specs

Agricultural Marketing Service (AMS)

Approved substitutes

Best-if-used-by dates

Bid from a supplier

Busted case

Buyer's major role in setting quality standards

Chemical standards

Color of a product

Cost and quantity limitations

Count

Difference between "brand name" and "packer's brand name"

Endorsements

Equal to or better

Expiration dates

Foodservice Consultants Society International (FCSI)

Formal versus informal spec

Form of a product

Freshness dates

General and specific instructions to bidders

Industry and government publications

Information included on a spec

Intended use

Limiting rule

Lowball bid

Measures of quality

National Cattlemen's Beef Association (NCBA)

National Sanitation Foundation (NSF)

International

Optimal quality to include on a spec

Optimal supplier services to include on a spec

Package size and type

Packaging procedure

Packer's brand

Packer's "grade"

Partial limiting rule

Performance requirement of a product

Point of origin

Potential problems with specs

Preservation and/or processing method

Problems associated with the use of U.S. grades as a measure of quality

Produce Marketing Association (PMA)

Product identification

Product specification

Product substitutions

Pull dates

Purchase specification

Returns

Ripeness

Samples

Sell-by dates

Size of a product

Slab-packed

Standards of quality

Stockouts

Test procedures for delivered products

Trade association standards

Trim

Truth-in-menu regulations

USDA Acceptance Service

U.S. Department of Agriculture (USDA)

U.S. government quality grade

Waste

Weight range

What influences the information included on a spec?

Who determines quality?

Who should the spec writer be?

Yield

QUESTIONS AND PROBLEMS

1. What is a purchase specification? How does it differ from a product specification?

2. What are some of the reasons hospitality operations develop purchase specifications?

3. What information is included on a typical purchase specification?

4. Assume you are the owner of a small table-service restaurant.

    (a)   How much time, money, and effort would you spend to develop specifications? Why?

(b)   Assume that you do not want to write specifications; you want to rely strictly on packers' brands and government grades to guide your purchasing. What are the advantages and disadvantages of this strategy?

5. Explain how the following factors influence the types of information included on the specification:
    (a)   Company policies
    (b)   Storage facilities
    (c)   Menu requirements
    (d)   Budgetary limitations
    (e)   Employee skills

6. What are the costs and benefits of hiring an outside consultant to help you write specifications?

7. What are the costs and benefits of writing specifications and using them in a bid-buying strategy?

8. Which items do you think a buyer should receive bids on? Why?

9. Explain how company personnel normally determine quality standards for the food products they use.

10. How does a buyer usually get involved in determining quality standards? What is his or her major role once these quality standards are set?

11. Describe five measures of quality. Name some advantages and disadvantages of each.

12. Why do you think endorsements are used so much in measuring the quality of consulting services?

13. Are AP prices good measures of quality? Why or why not?

14. Some industry practitioners feel that hospitality operators can set quality standards for some non-food supplies without considering their customers' views. Do you think this is true? Why or why not?

15. A product specification for fresh meat could include the following information:
    (a)
    (b)
    (c)
    (d)
    (e)
    (f)

16. Why are expiration dates important to include on fresh-food specs?

17. A food-processing plant normally must undergo continuous federal-government inspection for wholesomeness if:
    (a)
    (b)
    (c)

18. List some problems that the buyer will encounter if he or she is overreliant on U.S. grades.

19. What is the primary difference between a brand name and a packer's brand name?

20. Should a small hospitality operation prepare detailed purchase specifications, or should it prepare product specifications? Why?

21. What is the most important piece of information that can be included on a spec?

22. When should a buyer use packers' brands as an indication of desired quality in lieu of U.S. quality grades?

23. When should a buyer include on the specification "point of origin"?

EXPERIENTIAL EXERCISES

1. **What are some potential advantages of limiting yourself to one supplier, the only one who can meet your quality standards?**
    a. Write a one-page answer
    b. Provide your answer to a hotel manager and ask for comments.
    c. Prepare a report that includes your answer and the manager's comments.

2. **Why would package quality be important to a foodservice buyer? Would you be willing to pay a bit more to ensure high-quality packaging? Why or why not?**
    a. Write a one-page answer
    b. Provide your answer to a foodservice manager and ask for comments.
    c. Prepare a report that includes your answer and the manager's comments.

3. **Develop a purchase specification.**
    a. Write a purchase specification using information provided from one of the agencies in Figure 1.2.
    b. Provide your answer to an executive chef and ask for comments.
    c. Prepare a report that includes your specification and the chef's comments.

4. **Identify an agency that provides product information on a Website and that is not included in Figure 1.2. Write a one-page paper explaining how the Website can be used to help purchasing managers to write specifications.**

2

THE OPTIMAL SUPPLIER*

The Purpose of this Chapter

After reading this chapter, you should be able to:

- Determine a buying plan by selecting a single supplier or bid buying.
- Explain additional criteria used when choosing suppliers.
- Describe the relationship between suppliers and buyers.
- Describe the relationship between salespersons and buyers.

INTRODUCTION

Buyers have a good deal more to do with selecting suppliers than fixing quality standards and economic values. The major exception to this rule occurs when another company official insists that a buyer purchase from a certain supplier. This insistence usually means that some sort of reciprocal buying arrangement has been reached or that the owner-manager has prepared an approved-supplier list without consulting the buyer.

---

* Authored by Andrew Hale Feinstein and John M. Stefanelli.

THE INITIAL SURVEY

The first step in determining the optimal supplier is to compile a list of all of the possible suppliers, or at least a reasonable number of potential suppliers. Local suppliers' names can be gathered from the local telephone directories, local trade directories, local trade magazines, other similar publications, and other hospitality operators.

National suppliers' names can be obtained from similar sources. They also can be gathered from national buying guides and directories such as:

> The *Thomas Food and Beverage Market Place* (see Figure 2.1)
>   (www.tfir.com)
> Restaurants & Institutions Marketplace (www.rimarketplace.com)

FIGURE 2.1 The *Thomas Food and Beverage Market Place*. Courtesy of Grey House Publishing, www.greyhouse.com.

The National Restaurant Association's Online Buyer's Guide
   (www.restaurant.org/business/buyersguide)
Foodtrader (www.foodtrader.com)
Nation's Restaurant News Marketplace (www.nrn.com)
Foodservice Product Link (www.fsdmag.com)

Suppliers can also be found and evaluated at live trade shows and conventions, including:

American Culinary Federation National Convention (www.acfchefs.org)
National Restaurant Association Show (www.restaurant.org)
The Foodservice Symposium (www.thefoodservicesymposium.com)
Fresh Summit (www.pma.com)
Multi-Unit Food Service Operators (MUFSO) Conference (www.mufso.com)
North American Association of Food Equipment Manufacturers (NAFEM)
   Show (www.nafem.org)
Hospitality Information Technology Conference (HITEC) (www.hitec.org)
International Foodservice Technology Expo (FS/TEC) (www.fstec.com)
The International Hotel/Motel & Restaurant Show (www.ihmrs.com)
Multi-Unit Restaurant Technology Conference (www.htmagazine.com)

Large corporations take the time to compile lengthy lists of suppliers. The procedure most small operators follow is to seek out a more limited number of suppliers that carry most of the required items. In some cases operators may contact only one supplier for a particular product line. This is true especially for such items as liquor and dairy products because the middlemen dealing in these product lines usually have few competitors.

Whatever initial survey is undertaken, it can present three major problems. First, it may be difficult to determine which suppliers to include on the initial list. Many potential suppliers carry several product lines; consequently, the list can become larger than you would wish.

A second problem stems from the first. In the buyers' haste to shorten the potential supplier list, they may stop adding suppliers when they reach a certain number. The longer the list, the more time is required for interviewing, checking references, touring plants, and completing the other analytical work involved in culling the list. But indiscriminate culling can eliminate a good potential supplier. Furthermore, it tends to limit the pool of potential suppliers in the future when buyers stick with the original list. It can be costly to a hospitality firm to lock out a good supplier in this way.

The third problem is less common. It occurs when buyers need to purchase a unique item. In such situations, the search for a supplier can be extremely time-consuming.

TRIMMING THE INITIAL LIST

Buyers begin to narrow their initial list into an approved-supplier list by looking closely at each supplier's product quality, as-purchased (AP) price, and supplier services. (We assume that, at this stage, a buyer knows what types of products are wanted.) These factors help to separate acceptable suppliers from the initial list.

It is relatively easy to ascertain the quality standards and AP prices of suppliers. The major obstacle occurs when buyers examine supplier services. What is the best way to evaluate these supplier services? Basically, this process becomes a matter of taste. But the important considerations come under the rubric of "performance." When evaluating performance, buyers should be interested in prompt deliveries, the number of rejected deliveries, how adjustments on rejected deliveries are handled, how well suppliers take care of one or two trial orders, the capacity of their plants, and their technological know-how.

It might be easier to narrow a supplier list by trial and error. But a supplier's poor performance can leave buyers without a product, as well as with disgruntled customers demanding that particular product. It might be best, then, to accept the list-narrowing procedure as an essential aspect of purchasing.

THE RELATIONSHIP OF PURCHASING POLICY TO SUPPLIER SELECTION

The actual selection of the optimal supplier is the next logical step. Buyers cannot do this, however, without considering the type of procurement policies best for them. For instance, buyers might want to work with one particular supplier and negotiate long-term contracts for some items. If this is the case, they must keep these requirements in mind when going over the approved supplier list. Some suppliers may wish to be accommodating; others may not.

Large corporations have a bit more latitude in formulating their preferred buying policies and then convincing suppliers to cooperate. Small operators have less discretion; that is, they may have to accept the buying procedures their suppliers prefer. But at least a few procurement policies are available to any size operator. And, just as important, most suppliers are willing to adjust to more than one policy.

BUYING PLANS

Generally speaking, the hospitality industry has two basic buying plans: (1) the buyer selects a supplier first, and they work together to meet the buyer's needs; or (2) the buyer prepares lengthy specifications for the items needed and then uses bid-buying procedures.

The first plan, which involves selecting one or more suppliers to work with, is not common. Usually, a buyer chooses this plan only when: (1) a reciprocal buying policy is in effect; (2) only one supplier provides the type of item needed; (3) the buyer or owner-manager, for some reason, trusts the supplier's ability, integrity, or judgment; or (4) the buyer, for some reason, wants to establish a long-term relationship with a supplier. This plan is, however, used somewhat more often in small operations in which management, already spread thin with other operational problems, decides to limit the number of suppliers, even in some cases to a single supplier, for as many products as possible. This practice is called "one-stop shopping," and we discuss it below.

Bid buying is more common, particularly for items that several suppliers sell.[1] It works fairly well as long as buyers realize that all suppliers are not created equal. Also, buyers must keep in mind that obtaining the lowest bid may not ensure the lowest edible-portion (EP) cost.

Deciding which plan to use is a matter of judgment for buyers. For some items, the first plan may be appropriate. For example, since the quality of fresh produce tends to vary significantly, buyers may opt to select only one or a few suppliers whom they can trust. However, those same buyers might purchase canned goods strictly on a bid-buying basis.

Buyers who use bid buying generally take two approaches: (1) the "fixed bid," and (2) the "daily bid." Typically, buyers use the fixed bid for large quantities of products purchased over a reasonably long period of time. This is usually a very formal process.

The fixed-bid buying plan usually begins with a buyer sending a "Request for Bid" or "Request for Quote" to prospective suppliers, asking them to submit bid prices on specific products or services. The request includes detailed specifications and outlines the process bidders need to follow, as well as the process the buyer will use to award the contract.

Buyers send bid requests only to eligible, responsible bidders. An ineligible bidder is a company that, because of financial instability, unsatisfactory reputation, poor history of performance, or other similar reasons, cannot meet the qualifications needed in order to be placed on the approved-supplier list.

Responsible bidders usually need to send in sealed bids when participating in the fixed-bid process. A sealed bid is almost always required on major purchases to ensure fair competition among bidders. After the buyer opens the sealed bids, he or she awards the business to the responsible bidder with the lowest bid. The buyer awards the contract to this bidder because the unit price is lower, or the value per dollar bid is higher than what the other bidders quoted. Furthermore, the bid winner's reputation, past performance, and business and financial capabilities are judged best for satisfying the needs of the contract.

The daily bid is often used for fresh items, such as fresh produce. (The daily-bid method is sometimes referred to as "daily-quotation buying," "call sheet buying," "open-market buying," or "market quote buying.") Buyers also use this type of bid when purchasing a small amount—just enough to last for a few days or a week. The daily bid usually follows a simple, informal procedure: (1) the suppliers that form a list of those with whom the buyer wants to do business—the approved-supplier list—are given copies of the buyer's specifications; (2) when it is time to order some items, the buyer contacts these suppliers and asks for their bids; (3) the buyer records the bids or analyzes them electronically; and (4) the buyer usually decides on the supplier selection by choosing the supplier with the lowest AP price quote.

Some sort of value analysis could be used here to determine the optimal plan to use, given the types of items being purchased. The optimal procedure, though, is not an easy formula to develop because several good reasons exist for buyers to choose either plan. Convenience, degree of buyer skill, product availability, and so on, come into play when buyers determine the optimal plan. In the final analysis, the plan used will result from examining several factors.

Regardless of the plan or combination of plans a buyer chooses, he or she must ascertain the suppliers' willingness to participate in the plan. Most suppliers will jump at the chance to be a part of the first plan. Buyers, though, normally start with some type of bid-buying procedure, if only to determine which supplier they want to use all the time. Alternately, at the very least, buyers use bid buying in order to select the suppliers they plan to use for the next three, four, or six months.

Not all suppliers like to become involved with bid buying, especially when they feel that the other bidding suppliers are not in their league. These nonparticipating suppliers frequently balk at bid buying because their AP prices look high, due to the amount of supplier services they include. Generally, high AP prices do not win bids. Furthermore, the competing bidders may inflate their AP prices to fall just under those legitimately high AP prices. High-priced, reputable suppliers do not like to be involved in this type of practice.

Many suppliers try to circumvent a buyer's desire to bid buy by offering various discounts, other opportunity buys, introductory offers, and so forth. In addition, suppliers may try to become exclusive distributors for some items: if a buyer wants to purchase them, he or she will have no choice in supplier selection.

OTHER SUPPLIER SELECTION CRITERIA

Local Merchant-Wholesaler or National Source?

A small operator normally deals with local suppliers. But larger operators sometimes bypass these middlemen and go directly to the primary source; this is especially common with equipment purchases. Large hospitality operations normally require national distribution, so that all of the units in the chain organization can use the same type of products. As a result, they usually seek out the large suppliers who can provide this alternative.

Operations must address the question of providing their own economic values, especially transportation and risk, before they make a decision. And, as we have already pointed out, several advantages and disadvantages must be weighed here.

On a dollars-and-cents basis, small operations find it economical to purchase from local suppliers. Chains and other larger operations might, however, save money buying directly from the primary source. But they must consider the possible enmity engendered among the local suppliers. Disgruntled locals are, perhaps, the biggest, though not an immediately apparent, disadvantage associated with centralized buying and direct purchases. A buyer can expect little sympathy from bypassed local suppliers if he or she needs an emergency order or service on a piece of equipment.

A compromise is possible. For instance, a vice president of purchasing might go directly to the source and negotiate a long-term contract, for, perhaps, six months. Then he or she might "hire" local suppliers to take delivery from the sources and distribute the items to the local unit operations. Parceling out these end-user services is usually an acceptable and profitable compromise for all parties involved in the transaction.

Delivery Schedule

All hospitality operations have preferences regarding the time(s) of day and the day(s) of the week when they accept delivery from their suppliers. For instance, if buyers had their druthers, most of them would demand morning delivery.

Realistically, hospitality operations often must make do with what is available to them. However, this does not mean that they cannot swing their purchase dollars toward the supplier(s) who most closely matches their desired delivery schedule. This is a valued supplier service, and, although buyers often must expect to pay a little more for a preferred delivery routine, the overall effect may prove profitable for both the hospitality operation and the supplier.

Ordering Procedure Required by Supplier

As with the delivery schedule, buyers will be partial to those suppliers who most closely meet their needs. Suppliers who offer very convenient ordering procedures will most likely have a valuable competitive edge in the marketplace.

Credit Terms

Buyers are interested in the credit terms that are available from the various suppliers with whom they might consider conducting business. It is important to note such factors as the availability of cash discounts, quantity discounts, volume discounts, cash rebates, and promotional discounts; when payments are due (i.e., the credit period); the billing procedures; the amount of interest charges buyers may have to pay on the outstanding balance; and the overall installment payment procedure available, if any.

A preferred buying plan often will be bent to accommodate superior credit terms. That is, many hospitality operators are quite enamored with credit terms and will do what is necessary, within reason, to deal with suppliers who offer generous credit terms. This criterion conceivably could be the major consideration in supplier selection.

Minimum Order Requirement

Before a supplier will agree to provide buyers with "free delivery," they must, normally, order a certain minimum amount of merchandise. This is true even when they want to pick up the merchandise on a will-call basis, although the minimum order requirement usually is much lower in this situation.

Most buyers have little trouble in meeting minimum order requirements, so it is unlikely that such a criterion would be a concern. But a small operator might be very concerned with these stipulations; in this case, this aspect becomes an important supplier selection standard.

Variety of Merchandise

This concept is related to the one-stop shopping opportunity discussed earlier. In general, a supplier may or may not have the one-stop shopping capability, but he or she can, at least, offer a reasonable range of options. The supplier may offer a variety of quality grades, brand names, and/or packers' brand names for the merchandise he or she carries.

If a supplier specializes, for example, in fresh produce, he or she could offer tremendous variation even within such a relatively narrow product line. A supplier who can offer buyers a variety of qualities of fresh produce may conceivably be more valuable than a one-stop supplier who carries only one quality level of fresh produce along with several other product lines.

Lead Time

The shorter the lead time, the more convenient it is for a buyer, since he or she can wait until the last possible moment before entering an order for delivery at a predetermined time. All other things being equal, buyers would probably want to deal with a purveyor who offers them the ability to call tonight for an order to be delivered tomorrow morning, rather than a supplier who requires two or three days' notice.

Free Samples

Suppliers will often give buyers one or two free samples for their evaluation, particularly if the buyers represent a potentially large amount of business. However, some suppliers may not want to do this. Also, some buyers may not feel comfortable accepting free samples because it could compromise them.

Returns Policy

This is a very sensitive issue, and buyers should evaluate it well before it ever becomes necessary to return merchandise and/or refuse to pay for goods or services. Needless to say, the more liberal the returns policy, the more buyers expect to pay in the long run.

A related issue is the return of prepayment for merchandise that buyers ordered but for some reason must refuse its delivery. For instance, it often is necessary for buyers to put up a significant deposit for equipment purchases. If they then decide that they do not want or need the item, what happens to their deposit? It would be prudent for buyers to iron out any potential problems early.

Size of Firm

If a buyer has a large amount of business, he or she must be assured that suppliers are large enough to accommodate the buyer. On the other hand, large suppliers may be more impersonal. Perhaps the buyer would prefer dealing with small firms, allowing him or her to talk to the owners regularly. If nothing else, dealing directly with owners generally makes a buyer feel that his or her concerns will be met consistently.

A related issue is the amount of time suppliers have been in business. Some buyers will consider suppliers only after they have established acceptable performance track records that indicate they can handle buyers' needs and will most likely be around for a while.

Number of Back Orders

It seems to us that a supplier who has a history of excessive back orders will not be part of a buyer's approved-supplier list. A buyer can probably forgive a back order once in a while. But if this is a recurring problem, he or she cannot do business with such a purveyor. The buyer will want to do business with suppliers who have very high "fill rates." A fill rate is a ratio calculated by dividing the number of items delivered by the number ordered. Ideally, it would always equal 100 percent.

Substitution Capability

On occasions when back orders cannot be avoided, it is nice if the supplier can provide a comparable substitute. Generally, though, only suppliers who offer a one-stop shopping opportunity are capable of doing this.

A related possibility is the supplier who runs out of an item but who would be concerned enough about buyers personally to secure the products necessary to complete their order from another supplier or from one of his or her competitors. This type of purveyor is rare, but one or two of them may be in your area.

Buyout Policy

We can recall years ago when suppliers who wanted a buyer's business would agree to purchase his or her existing stock of competitors' merchandise. For instance, if a soap salesperson was soliciting a buyer's business, he or she might agree to buy out the existing stock so that the buyer could begin immediately to use the new merchandise. This is an uncommon policy today, but it may exist somewhere. If it does exist in your area, it represents one more criterion on which to judge a potential supplier.

A related issue is the willingness of a supplier to buy back outdated or obsolete merchandise. For example, when buyers purchase replacement equipment, a major supplier selection factor would be the trade-in allowance that competing suppliers offer. All other things being equal, the supplier who has the most favorable policy is apt to have an edge over his or her competitors.

Suppliers' Facilities

Buyers should be particularly concerned with a potential supplier's storage and handling facilities, the delivery facilities, and the facilities' sanitation. For instance, if the supplier uses old, dirty, and uncooled vans to deliver fresh produce, you may want to avoid that purveyor regardless of the AP price and other supplier services provided. Inadequate facilities harm product quality, and this is intolerable. As the industry moves more and more toward e-procurement, it will become increasingly more difficult to evaluate this supplier selection criterion.

Outside (Independent) Delivery Service

Shopping on the Internet may yield several suppliers who do not provide delivery service personally, but who outsource it to an independent service, such as UPS or FedEx. Buyers should be leery about e-suppliers who use no-name delivery services that have no verifiable track record. Delivery inconsistencies will wipe out any good deals that buyers obtain by shopping around. Furthermore, independent drivers, even from major delivery services, are unable to rectify mistakes on the spot.

Long-Term Contracts

Some suppliers are unwilling or unable to enter into long-term contracts for AP price and/or for availability of the product during the contract period. If buyers prefer some type of long-term commitment, they may have to settle for a relatively short approved-supplier list.

Case Price

When buyers purchase a case of merchandise, such as a six-can case of tomatoes, they will pay a certain price for it, say, $12.00. If buyers wish to purchase one can of tomatoes and can purchase it for $2.00, they are receiving what is normally referred to as the "case price" for that can.

Few, if any, suppliers will give a buyer a case price when he or she purchases less than a case. If he or she is lucky, suppliers will "bust" a case for a buyer, but they will usually charge a premium to do this. Typically, either the buyer purchases the whole case or does business elsewhere.

When a buyer purchases some items in small batches, it is important to deal with suppliers who understand his or her needs. In some situations, a buyer cannot afford to purchase a whole case of, for example, soup bases if he or she expects the contents to sit around for a period of time losing flavor and otherwise deteriorating. This buyer will need to look for those suppliers who can and will accommodate him or her. In many cases, he or she may have to settle for warehouse-club suppliers.

Bonded Suppliers

A buyer is concerned about the capability of suppliers to cover the cost of any damage they might inflict on his or her property. Usually, before the appropriate government authority will issue a business license to a supplier, that supplier must display adequate insurance coverage; that is, they must be bonded. However, what is adequate for the licensing bureau may not be adequate for the buyer.

A related issue is the fact that buyers may inadvertently be dealing with an unlicensed supplier. This situation should be avoided because, if, say, a customer gets ill from products this supplier provided, the buyer's organization could become entangled in all sorts of litigation.

Consulting Service Provided

To a great extent, salespersons, and suppliers in general, are the primary sources of product, and related, information for the typical hospitality operator. Buyers are interested in data concerning product specifications, preparation and handling procedures, nutrition, merchandising techniques, and other similar types of advice.[2]

Small hospitality operators are especially loyal to suppliers and salespersons who willingly share their expertise. For example, a small caterer who is bidding for an unusually large banquet contract will appreciate the salesperson who takes the time to help prepare the proposal.

Formal consulting, though, is not something that every purveyor is able or willing to provide. For instance, when purchasing equipment, buyers may find that some dealers stock it, sell it, and deliver it—period. Other dealers provide some additional advice, such as providing blueprints or seeking the appropriate building permits. Buyers pay more for this type of service, but they may be willing to do so. When this is the case, buyers must seek suppliers who can provide for their needs.

Deposits Required

For some products, buyers may need to put up a deposit. For example, if they purchase soda pop syrup in reusable containers, they may need to put up a cash deposit for them.

Usually, deposit requirements are not burdensome, but if they are, buyers probably will want to eliminate such a demanding supplier from their approved supplier list.

Willingness to Sell Storage

Some suppliers will sell storage, which can be a tremendous service if, for example, a buyer needs space to house a large amount of merchandise that he or she purchased through a favorable opportunity buy. A supplier's storage space is usually better than one rented from a generic warehouse or storage locker location because it is apt to be appropriately cooled and/or heated for items the typical hospitality buyer purchases.

A supplier who will sell storage probably is a rare find, but if buyers are fortunate enough to have one in their area, they must be certain to inquire not only about the fees for this service, but also about any other sort of requirements. For instance, to qualify to purchase storage, buyers may need to purchase $1000 worth of merchandise per week. This may or may not be attractive to them, however, and they should be alert to these kinds of restrictions, which could place them in an unprofitable position.

Suppliers Who Own Hospitality Operations

Some hospitality operations own commissaries and/or central distribution centers that sell merchandise to other hospitality companies. For instance, many quick-service restaurants purchase Pepsi-Cola products. In the past (prior to the sale of its foodservice division to Tricon Global Restaurants, Inc., and subsequent renaming of the company to YUM! Brands [www.yum.com]), PepsiCo competed directly with restaurants through company-owned and franchised Pizza Huts, Taco Bells, and KFC outlets. In a similar situation, General Mills, whose foodservice brands include Gold Medal Baking Mixes, Betty Crocker Potato Buds, Cheerios, and Yoplait, also owned restaurants such as Red Lobster and Olive Garden. However, General Mills spun off its restaurant division to Darden Restaurants, Inc., in 1995. The main issue, of course, is: Could such a purchasing strategy be beneficial for the buyer?

At first glance, it would seem foolish to buy from a competitor. Buyers would think that the competitor would learn too much about their business. Furthermore, would the competitor favor his or her hospitality units at the buyers' expense?[3]

On the other hand, some people feel that buying from another hospitality operator has its advantages. For example, the competitor understands the business much better than a conventional supplier and is much more conscious of the required supplier services.

Socially Responsible Suppliers

Some buyers prefer to work with suppliers who promote socially responsible agendas. For instance, some buyers will not purchase from suppliers that sell products manufactured by employees in foreign countries who do not receive a basic level of wages and/or benefits.[4] Alternately, some buyers will not purchase from suppliers carrying products whose processing damages the earth's rain forests. Also, some buyers prefer to purchase from suppliers who employ minorities and deal with minority-owned subcontractors.[5]

Buyers can use subscription services to search for socially responsible firms. For example, for an annual fee, buyers can subscribe to referral services, such as Thomas Publishing Company (www.thomaspublishing.com), that list suppliers that are owned by minorities and/or women. Suppliers are typically listed by product category, and, with little more effort than picking up a Touch-Tone™ phone or visiting the referral service's Website, buyers can quickly secure the information they need.[6]

References

Usually, a large part of a buyer's supplier selection work is devoted to obtaining personal references. This is normally an informal process, whereby the buyer talks with friends in the industry who may be able to provide meaningful input about certain suppliers. The buyer might also consider contacting credit-rating firms to uncover a potential supplier's financial strength. Generally, though, if a friend whose opinion the buyer trusts has had a good experience with a particular supplier, the buyer would want to do business with that firm.

It would appear that the buyer is most anxious about a potential supplier's integrity and overall dependability. These are the characteristics the buyer tries to uncover when conversing with friends. These factors can mean many things to many persons, but if a friend is impressed with a supplier's dependability and integrity, the buyer will probably want that supplier on his or her approved-supplier list.

MOST IMPORTANT SUPPLIER SELECTION CRITERIA

No one can dictate the criteria that buyers should consider when selecting their suppliers. This is something that only buyers can judge for themselves. It is interesting, though, to note those criteria that are most important to members of the hospitality industry.

Generally, most buyers are interested primarily in product quality. Suppliers must be able to consistently provide the quality needed, or else buyers cannot deal with them.

Supplier service is, usually, a close second to product quality. Dependability is critical. Suppliers must ensure that buyers receive what they need when they need it.

The AP price seems to trail quality and supplier services in most buyer surveys. While this does not necessarily imply that buyers are unconcerned with product costs, it does emphasize the point that AP prices do not unduly influence purchase decisions in the hospitality industry.

Typical hospitality buyers seem to follow the supplier selection process that Walt Disney World food services adopted. When selecting its suppliers, Disney is concerned with product quality, supplier service, whether the purveyor is large enough to handle the account, and AP price.[7]

Regardless of the number and type of supplier selection criteria the hospitality operation employs, the common thread running through them is one of consistency, dependability, loyalty, and trust.[8] If suppliers can render consistent value, chances are they will be on the approved-supplier list of several hospitality operations. Furthermore, suppliers who consistently provide acceptable value will continue to grow and prosper.

MAKE A CHOICE

As buyers gradually complete their basic buying plan, they simultaneously reduce the potential supplier pool. Eventually, common sense and company policy guide them toward the optimal suppliers. Buyers do not want too many restrictions placed on their basic buying plan. On the other hand, they do not necessarily want to ignore all of the suppliers' needs. Buyers must strike a balance, within reason, so that both they and the seller feel confident that profit will result from the relationship. The best relationship is one in which both the buyer and the seller are satisfied.

SUPPLIER–BUYER RELATIONS

The buyer's principal contact with suppliers is through salespersons. In the initial stages of supplier selection, the buyer may meet an officer of a supply house. But after this meeting, a supply house officer almost never usurps the salesperson's role. The top management of a supply house is never out of the picture, although it may be out of sight. Those officials work hard to improve business; some of their major activities include those discussed next.

Supply House Officers Set the Tone of Their Business

Usually, supply house officers set this tone by establishing the quality standards of the items they carry, by determining the types of economic values and supplier services they provide, and by planning their advertising and promotion campaigns. While considering these aspects of the business, moreover, suppliers seek a balance between what they want to do and what their customers, the hospitality operations' buyers, need.

Supply House Officers Set the Overall Sales Strategies

Two basic sales strategies exist: (1) the "push strategy," in which suppliers urge their salespersons to do whatever is necessary to entice the buyer to purchase the product—the normal push is AP price discounts of one type or another, and (2) the "pull strategy," in which suppliers influence those who use the items the buyer purchases. For example, suppliers may advertise heavily on television, exhorting ultimate customers to demand the suppliers' product in their favorite restaurant. If they do, the restaurant buyer has little choice but to purchase the product. In other words, the ultimate customer "pulls" the product through the channel. Alternately, if backdoor selling can be implemented successfully, a user in the company "pulls" the product through by influencing the buyer's purchasing decisions (see Figure 2.2).

You have undoubtedly seen many types of pull strategies. If, for example, a restaurant customer orders a Coke®, which is a brand name, what choice does the buyer have? This also applies to catsup or hot sauce on the tables. Restaurant patrons typically prefer Heinz® and Tabasco® brands.

Of course, various shades and combinations of these two basic strategies exist, but, generally speaking, suppliers lean toward one, or, at least, they lean toward one for some items and toward the other for their remaining items.

The pull strategy can be risky and extremely costly for suppliers to implement and maintain. But if it works, the rewards are fruitful indeed.

The pull strategy also is a major weapon suppliers use to steal business from one another.

Supply House Officers Sponsor a Great Deal of Product and Market Research

Suppliers also spend considerable time and effort evaluating the bids they make for buyers' business. Furthermore, they continually prepare and revise files that contain information about current and potential customers. These information files are sometimes referred to as "buyer fact sheets" or "buyer profiles." They constitute a selling tool and contain as much or as little information as thought necessary to facilitate the sales effort.

FIGURE 2.2 The pull strategy. Courtesy of J. R. Simplot and Company.

The following pieces of information are usually found in these files:

1. Does the buyer have a favorable impression of the supplier's reputation? Generally speaking, a favorable impression makes it easier for a salesperson to get his or her foot in the door on the first sales visit.

2. What are the major characteristics of the ultimate customers of the buyer's company? If, for example, the ultimate consumers are price conscious, the buyer will probably adopt a similar posture.

3. Is the buyer concerned with AP prices?

4. Is the buyer concerned with fast and dependable deliveries?

5. Will the buyer take a chance on new products? Does he or she have the authority to suggest new products to the respective hospitality departments?

6. Does the buyer have a great deal of confidence in his or her purchasing skill, or will second-guessing prevail?

7. Does the buyer have other duties? For example, is he or she a buyer and a user? Will these other duties minimize the time he or she spends with salespersons?

8. Does the buyer insist on rigid quality control, or will he or she accept certain exceptions or substitutions from time to time?

9. What is the possibility of setting up a reciprocal buying arrangement?

10. What is the payment history of the buyer's company?

11. How does the buyer treat suppliers and salespersons?

12. Do any little things irritate the buyer? For example, does he or she get annoyed if a salesperson is a few minutes late for an appointment?

Suppliers Train Their Sales Staffs

Suppliers expend tremendous efforts in sales training, for both new salespersons and, continually, for salespersons currently on staff. Quite often, the training materials are based on market research, new products, and buyer profiles.

Suppliers Keep Their Salespersons' Promises

Suppliers must, for example, make sure that orders are handled properly and delivered on time.

SALESPERSON–BUYER RELATIONS

Salespersons, who are sometimes referred to as "distributor sales representatives" (DSRs), are buyers' main contact with supplier firms. Buyers must usually meet several DSRs every week. Many of them are familiar faces; others are new. Establishing firm and fair business relations with DSRs, and particularly setting the ground rules regarding sales visits, is essential to efficient procurement.

Buyers need to be aware of the sales tactics salespersons use. Generally, on the first sales call, salespersons might: (1) make some attempt, however slight, at backdoor selling (i.e., they might try to interest users in the supplier's wares), (2) attempt to use free samples and literature in an effort to interest and possibly to obligate a buyer, (3) try to establish a justification for their presence, (4) try to talk buyers away from the current supplier, or (5) try to be invited to return, thereby starting a nominal business relationship.

Sales professionals are usually adept at practicing what is usually referred to as "relationship marketing."[9] Salespersons after a buyer's business will bend over backward to start some type, any type, of business relationship. They will usually take any order, no matter how small, so that future sales visits are justified. Even if the buyer purchases only one item once, the salesperson still feels, as "one of your suppliers," free to drop in periodically. It may seem ludicrous that salespersons would hang around once a buyer makes it clear that he or she probably will not buy from them again. Also, you would think that supply house officers would prohibit salespersons from taking small orders. But buyers may not always respond the same way. The next buyer, or manager, may be more receptive. Today, it may be a small order; tomorrow, who knows? Hence, salespersons continue their efforts.

Small operators enamored of one-stop shopping like to avoid excessive contacts with salespersons and to minimize their ordering procedures. These preferences turn them into house accounts. House accounts are regular, steady customers for whom suppliers are not always motivated to provide generous supplier services. However, they may continue to provide exceptional supplier services in order to keep these customers happy.

This is a touchy issue. Dealing with many salespersons is time-consuming. But never seeing them at all is poor local public relations, shuts off good sources of information, and prevents them from helping buyers check inventories, production techniques, and any equipment they may have loaned for use with their products. Good trade relations might dictate that buyers spread their orders out a bit more. But this, too, can be costly. Each operation must, therefore, balance the potential ill will with this loss of time and make its decision in the light of such factors as order size and management availability, as well as public relations.

A full-time buyer for a large operation, though, is expected to spend a good deal of time with salespersons. The company pays the buyer to minimize the AP prices. In these large organizations, other people are responsible for the steps the product follows from purchase to use. Someone watches for pilferage, shrinkage, and spoilage in storage. Another individual is responsible for using cooking or other production techniques that prevent waste and shrinkage. Someone else is responsible for minimizing overportioning of finished product. How these responsibilities are distributed is not relevant to the present point. Our point is that, in a large operation, the achievement of a good EP cost results not only from a good AP price, but also from the proper working of a complex, skilled organization.

Several volumes have been written on sales tactics, strategies, and procedures. Buyers would be wise to read some of these materials, paying particular attention to such topics as: (1) personal characteristics of good salespersons, (2) types of salespersons, (3) what to avoid when dealing with salespersons, (4) what to do when dealing with salespersons, (5) types of sales tactics, and (6) techniques for evaluating salespersons.

We do not want to suggest that an adversarial relationship necessarily exists between buyers and salespersons. But buyers must expect salespersons to go to whatever ethical lengths they can to make sales. Salespersons come to sell, not to entertain. They want to meet your expectations, but for a price.

Good salespersons will never sell a buyer something he or she does not need. But keep in mind that the main objective is to convert the buyer into a regular customer, not by pressuring the buyer, but by providing satisfaction. Within reason, then, salespersons do what is necessary to turn a buyer into a house account.

An alert buyer should be able to compete in the game of sales strategy and tactics. Objectivity helps a buyer, as does an understanding supervisor. In most cases, the buyer and salesperson work together for each other's benefit. Remember, though, business being business, a buyer should never become too friendly with a sales representative.

EVALUATING SUPPLIERS AND SALESPERSONS

Suppliers and salespersons sometimes become such integral parts of a business that a buyer starts treating them as he or she would an employee. For this reason, a buyer should periodically evaluate the suppliers' and salespersons' performance and consider disciplining or rewarding them as necessary. The ultimate discipline is to switch to another source of supply. The ultimate reward is to become a house account. (This may be no reward for salespersons, though. Some supply houses pay no sales commissions on house account sales; the theory is that little effort has gone into making the sales. These salespersons may, however, receive a bonus when they obtain a house account for their firm.) Obviously, there is a considerable range between these extremes and several discipline-reward combinations.

Most analysts agree that an operator should rate suppliers and their salespersons as part of the discipline-reward cycle. However, few analysts agree on the criteria to use in these evaluations. For instance, some buyers are appreciative of the salespersons who take the time to listen to what the buyers have to say. Other buyers seek only those salespersons who can answer the buyers' questions completely and correctly. Still other buyers are more enamored with effective and impressive sales presentations.

In any case, once again the common thread running throughout is consistency: consistent quality, consistent supplier services, and so forth. If suppliers and salespersons consistently fulfill their part of the bargain, whether it was made yesterday or last year, buyers should have no complaint. Our suggestion, then, is for buyers to enumerate those factors on which they and the suppliers and salespersons agree and, from time to time, to use a consistency yardstick to measure performance. While buyers look for consistency, though, they must remember that a high AP price often accompanies high levels of consistency, especially consistent supplier services.

If buyers expect suppliers and salespersons to be consistent, they themselves must be consistent. That is, they should never change their evaluation criteria unilaterally. Professional buyers generally try to be consistent, but users who also buy, in contrast to professional buyers, tend to be more subjective about the items they purchase, as well as more abusive toward suppliers and salespersons. User-buyers, therefore, should be especially leery of finding these traits in themselves.

In the final analysis, evaluation is probably a combination of art and science. Having evaluated consistency, buyers could examine other subjective factors. But buyers should resist being too hasty in this process. It is true that, unless buyers have a long-term contract, they can drop their supplier quite abruptly. But this may do more harm than good. If a supplier is deficient, buyers should give him or her a chance to improve, just as they would give a poor-performing employee a chance to improve. Buyers should never "fire" the supplier or salesperson without allowing a second chance. If buyers acquire a reputation for rash decisions, other suppliers or salespersons may become gun-shy, especially those who consider themselves fair and reputable performers. Buyers do not want to be left with only the poorest supply choices.

A step short of cutting a supplier off completely, a step that is often used, is to cut him or her off for a week or so, just to make certain that this supplier realizes that he or she can lose the buyer's business. This discipline supposedly helps keep suppliers in line. Buyers want to be certain, though, that the supplier really did something to deserve this treatment and that the problem is not in their operation rather than the supplier's.

GETTING COMFORTABLE

S upplier selection is not something to be done once and then forgotten. But small operators often seem to think it is, even when competing suppliers and salespersons bombard them with sales pitches.

Salespersons will fight to prevent a buyer from settling in with one or two suppliers unless they are among those he or she has selected. They do not want the buyer to enter into the "comfort stage" of the supplier selection procedure. Their sales efforts will, in fact, become increasingly insistent. On the other hand, a buyer's current suppliers and salespersons will see to it that he or she is satisfied in order to discourage the advances of other suppliers and salespersons. Of course, the buyer's current sources may get comfortable themselves and need to be brought up short now and then.

Many buyers and user-buyers become comfortable with a salesperson, but at least they remain aware of the need to examine alternate suppliers and to make a switch if necessary. However, flitting continually from one supplier to another involves a certain amount of emotional strain, broken loyalties, and disrupted business patterns. The switching becomes particularly difficult when a buyer's favorite salesperson goes to work for another supplier and the buyer wants to continue doing business with him or her. In effect, the buyer allows this salesperson to carry the buyer's business to his or her new employer.

Generally, a supplier who takes good care of buyers' needs deserves some type of reward. Suppliers and salespersons are, indeed, just like partners, or employees. Good employees are rewarded with continuous employment, and a salary raise or a bonus. Suppliers and salespersons should be treated with equal consideration. We are not sure whether it is always a good idea to become a house account, but we do believe that, at the very least, good current suppliers and salespersons deserve first crack at a buyer's business, now and in the future. A restaurant manager once expressed this sentiment precisely when he remarked:

> We don't believe in getting locked into one supplier, because it would make us too vulnerable. But, on the other hand, we don't switch suppliers just to gain a few cents. We try to find suppliers who appreciate our dedication to quality and then stay with them. That doesn't mean we don't check the market every Monday and maintain a continuing check on prices. But we believe in commitment—on both sides.[10]

KEY WORDS AND CONCEPTS

Approved supplier

Approved-supplier list

As-purchased price (AP)

Back order

Barter

Bid-buying procedures

Bonded supplier

Bust a case

Buyer fact sheets

Buyer profiles

Buying clubs

Buying plans

Buying services

Buyout policy

Call sheet buying

Case price

Cash and carry

Cash discount

Cash rebate

Consistency

Consulting service

Contract house

Co-op purchasing

Cost-plus purchasing

Credit period

Credit terms

Daily bid

Daily quotation buying

Delivery schedule

Deposits

Distributor sales representative (DSR)

End-user services

Evaluating suppliers and salespersons

Fill rate

Firing a supplier

Fixed bid

Forward buying

Free samples

House account

Ineligible bidder

Lead time

Long-term contract

Market quote buying

Market research

Minimum order requirement

National distribution

One-stop shopping

Open-market buying

Ordering cost

Ordering procedures

Par stock

Potential supplier

Prime-vendor procurement

Product substitution

Profit markup

Promotional discount

Pull sales strategy

Push sales strategy

Quantity discount

Reciprocal buying

Relationship marketing

Request for bid

Responsible bidder

Returns policy

Route salesperson

Salesperson-buyer relations

Sealed bid

Single-source procurement

Size of a supplier firm

Socially responsible supplier

Sole-source procurement

Standing order

Stockless purchasing

Supplier facilities

Supplier references

Supplier selection criteria

Supplier services

Supplier-buyer relations

Suppliers who own hospitality operations

Trade show

Trade-in allowance

Value analysis

Volume discount

Warehouse wholesale club

Wholesale club

Will-call purchasing

REFERENCES

1. Michael Guiffrida, "Saying What You Mean on Bids," *FoodService Director*, May 15, 1990, p. 78. See also: Patt Patterson, "Bid Buying Saves Time and Cuts Food Costs for Red's Seafood," *Nation's Restaurant News*, October 28, 1991, p. 58; Foster Frable, Jr., "Strange Business: How Kitchen Equipment Is Sold," *Nation's Restaurant News*, September 1994, pp. 183–185.

2. Peter Matthews, "What Constitutes Excellence in In-Store Supplier Service?" *Retail World*,

November 25–December 6, 2002, 55(23), pp. 38, 45. See also: Phil Roberts, "Looking for a Supplier 'Partner,' Not a Salesman," *Nation's Restaurant News*, September 9, 1991, p. 30; Phil Roberts, "WholeSELLERS: Putting Wind in Our Sales," *Nation's Restaurant News*, September 21, 1992, p. 122; Patt Patterson, "Commercial Electric Contest Can Spark Bright Ideas," *Nation's Restaurant News*, January 11, 1993, p. 18.

3. Ted Richman, "Suppliers' Demands Are Too High a Price to Pay," *Nation's Restaurant News*, December 19, 1994, p. 21. See also: Martha Brannigan, "Coke Is Victim of Hardball on Soft Drinks," *The Wall Street Journal*, March 15, 1991, p. B1.

4. G. Pascal Zachary, "Starbucks Asks Foreign Suppliers to Improve Working Conditions," *The Wall Street Journal*, October 23, 1995, p. B4.

5. James Morgan, "How Well Are Supplier Diversity Programs Doing?" *Purchasing*, August 15, 2002, 131(13), pp. 29–35. See also: Carolyn Walkup, "MultiCultural Confab Weighs Supplier Diversity as Rising Bottom-Line Issue," *Nation's Restaurant News*, May 10, 1999, 33(19), p. 46; Dale K. DuPont, "Minority Markets Offer Diverse Opportunities," *Hotel and Motel Management*, April 5, 1999, 214(6), pp. 30–31.

6. Leon E. Wynter, "Supplying a Minority or Female Supplier," *The Wall Street Journal*, September 21, 1993, p. B1.

7. Stephen M. Fjellman, *Vinyl Leaves: Walt Disney World and America* (San Francisco: Westview Press, 1992), p. 390.

8. John R. Farquharson, "And These Doctors Make House Calls!" *Nation's Restaurant News*, September 21, 1992, p. 116. See also: Michael L. Facciola, "Supply & Dementia," *Food Arts*, May 1992, p. 113; Rob Johnson, "A Deal of Time and Effort," *Supply Management*, April 10, 2003, 8(8), pp. 30–31; Vijay R. Kannan and Keah Choon Tan, "Supplier Selection and Assessment: Their Impact on Business Performance," *Journal of Supply Chain Management*, Fall 2002, 38(4), pp. 11–21; Michael Tracey and Chong Leng Tan, "Empirical Analysis of Supplier Selection and Involvement, Customer Satisfaction, and Firm Performance," *Supply Chain Management*, 2001, 6(3/4), pp. 174–188.

9. John Bowen, "Don't Imitate, Differentiate," *Nevada Hospitality*, July/August 1996, p. 20. See also: Susanne Frey, Roland Schegg, and Jamie Murphy, "E-Mail Customer Service in the Swiss Hotel Industry," *Tourism and Hospitality Research*, March 2003, 4(3), pp. 197–212; Isabelle Szmigin and Humphrey Bourne, "Consumer Equity in Relationship Marketing," *Journal of Consumer Marketing*, 1998, 15(6), pp. 544–557.

10. Patt Patterson, "Shere Sets High Standards for Coach and Six," *Nation's Restaurant News*, January 13, 1986, p. 4.

QUESTIONS AND PROBLEMS

1. What major problems are associated with the initial-survey stage of supplier selection? What would you suggest to alleviate these difficulties?

2. The buyer usually has a minor role in selecting equipment suppliers. What do you think is the usual role? Why would an owner-manager take the initiative in selecting equipment suppliers?

3. Identify the two basic buying plans. Suggest items that would be purchased under each plan.

4. Why do you think bid buying is so popular in our industry? What are the costs and benefits of this plan, as opposed to those of the other basic buying plan?

5. What are the advantages and disadvantages of one-stop shopping? Suggest the types of hospitality operations you feel would most likely benefit from one-stop shopping.

6. A buyer often takes a daily bid before placing a meat order. What does this procedure involve?

7. What are the major advantages and disadvantages of co-op purchasing?

8. What are some of the advantages of using a computer online service to search for potential suppliers? Do CDs offer the same advantages?

9. What is the difference between a formal bid procedure and an informal bid procedure?

10. What are the major advantages and disadvantages of will-call purchasing?

11. What are some advantages and disadvantages of purchasing only from socially responsible suppliers?

12. What is the difference between the push sales strategy and the pull sales strategy? Which strategy do you think a supplier prefers?

13. When is it appropriate for a buyer to use a fixed-bid buying procedure?

14. Why might a supplier be reluctant to participate in a cost-plus purchasing procedure?

15. How could a buyer save money by using the stockless purchasing procedure?

16. What are some advantages and disadvantages of the standing-order purchasing procedure?

17. Why is it important to purchase merchandise from licensed and bonded suppliers only?

18. Develop a checklist that you would use to evaluate your suppliers. Assign degrees of importance to each item on the list. If possible, ask a hotel manager and/or a food buyer to comment on your list.

19. Define or briefly explain the following:
    (a) Call sheet buying
    (b) Approved supplier list
    (c) Forward buying
    (d) Cash rebate
    (e) Credit period
    (f) Buying club
    (g) Direct purchase
    (h) Credit terms
    (i) Minimum order requirement
    (j) Lead time
    (k) Returns policy
    (l) Buyout policy
    (m) Case price
    (n) Buyer fact sheet
    (o) House account
    (p) National distribution
    (q) Trade-in allowance
    (r) Ineligible bidder

EXPERIENTIAL EXERCISES

1. Assume that you are very happy with your current supplier, who has been supplying most of your needs for over a year. A new supplier comes along with what appears to be a better deal: a promise of lower AP prices, along with the same quality and supplier services.
    a. Write a one page report on what you would do.
    b. Ask a hotel or restaurant manager to comment on your answer.
    c. Prepare a report that includes your answer and the manager's comments.

2. Create a list of national hospitality suppliers that you might use if you owned a hospitality operation. Prepare a report that explains what type of hospitality operation you would own and how/why you identified specific national suppliers. (Hint: start by searching for suppliers using the Websites discussed at the beginning of this chapter.)

3. Create a list of hospitality suppliers in your local area that you might use if you owned a hospitality operation. Prepare a report that explains what type of hospitality operation you would own and how/why you identified specific local suppliers.

TYPICAL RECEIVING
PROCEDURES*

The Purpose of this Chapter

After reading this chapter, you should be able to:

- Explain the objectives of receiving.

- Explain the essentials of effective receiving.

- Describe invoice receiving and other receiving methods.

- Outline additional receiving duties.

- List good receiving practices and methods to reduce receiving costs.

INTRODUCTION

Someone once said, "Receiving is the proof of purchasing." It's at receiving that you determine what it is you actually got—not what you ordered but what you received. And there could be a lot of difference between the order you placed and the delivery you received. That's why receiving is so important to the proper control of purchasing.[1]

Receiving is the act of inspecting and either accepting or rejecting deliveries. It is an activity with many facets. In any particular hospitality operation, receiving can range anywhere from a buyer letting delivery truck drivers place an order in his or her

---

* Authored by Andrew Hale Feinstein and John M. Stefanelli.

storage facilities, to having various receiving clerks waiting at the delivery entrance to check every single item to see that it meets the specifications set forth in the purchase order (PO).

Although many varieties of practice exist, the variety of correct procedures, though subject to debate, is considerably less. For instance, many operations permit the bread route salesperson to come into the kitchen and storage areas, remove bread left over from the previous delivery, restock the operation, and leave the bill with someone. Purists could argue that this is poor practice on two counts: (1) the delivery agent could cheat on both returns and delivery; and (2) the delivery agent could steal other items while he or she was there.

Operators following this practice, however, would probably argue that the cost of a unit of product—a loaf of bread, for instance—is so small that the cost of a receiving clerk or management time to physically check returns and delivery is unwarranted. Moreover, the delivery agent in this case saves time. As for the theft issue, the delivery agent is no more likely to steal than employees because he or she is subject to the same controls they are. No hard-and-fast rules exist for a case like this. Each operator or company must weigh the advantages and disadvantages in order to arrive at a policy.

On the other hand, consider a situation in which a supplier provides a very large portion of an operation's goods and offers, as a supplier service, to check the storeroom for all of the items this supplier provides and to determine what needs to be replenished. This might be done "to save the operator some time." Most people would argue, here, that both the receiving and the purchasing functions have been turned over to a supplier, and allowing a supplier to do this is inappropriate.

THE OBJECTIVES OF RECEIVING

These objectives resemble the objectives of the purchasing function itself. Recall the main objectives: obtaining the correct amount and correct quality at the correct time with the correct supplier services for the correct edible-portion (EP) cost. The main objective of receiving is to check that the delivered order meets these criteria.

Another important objective of receiving is controlling these received products and services. Once the receiver accepts the items, whatever they are, they become the property of the hospitality organization. Thus a cycle of control begins at this point.[2] For the most part, an owner-manager requires a certain amount of documentation during this receiving function in order to ascertain what was delivered and where the delivered items were sent within the organization. Keep in mind that this control activity can be simple or elaborate, depending on management's policy. At the risk of being redundant, though, we repeat that the best buying plan in the world is useless if someone or something goes awry during the cycle of control and causes a reduction in quality or an increase in costs.

In some large firms, additional control is exerted in the receiving function by placing receiving personnel under the direction of the accounting department. Thus, management minimizes the possibility of any fraudulent relationship between the buyer and the receiver. Small firms cannot afford this luxury.

On the other hand, some small firms often use their very smallness—their lack of personnel—as an excuse to avoid receiving control. Although the small operator may not be able to afford the elaborate receiving department of a large hotel, he or she certainly can adapt the techniques it uses to some degree. For instance, many small- to medium-sized operations can designate a specific employee, perhaps a line cook, to act as receiver and the only person authorized to sign for received products. This person might warrant a modest salary increase for the added responsibility assumed. To make the task easier, he or she should be responsible for verifying counts and weights; quality evaluations can be left to some members of management.

ESSENTIALS FOR GOOD RECEIVING

To ensure that the receiving function is performed properly, several factors must be in place:

1. **Competent Personnel.** Such personnel should be placed in charge of all receiving activity. By "competent," we mean persons, full-time or part-time, who are reasonably intelligent, honest, interested in the job, and somewhat knowledgeable about the items to be received. Once this person is designated, it is necessary to train him or her to recognize acceptable and inferior products and services. However, where the line cook-receiver combination is used, management may need to make quality checks and to supervise the receiving activity more closely.

   It is very important to provide the receiving agent with appropriate training. This can be a time-consuming, costly procedure, but it is absolutely essential. The receiver must be able to recognize the various quality levels of merchandise that will be delivered to the hospitality operation. He or she also must be able to handle the necessary paperwork and/or computerized record keeping adequately. Furthermore, the receiver must know what to do when something out of the ordinary arises. While the training costs may be considerable, they will be recovered many times over if the receiver is able to prevent merely one or two receiving mistakes per month.

2. **Proper Receiving Equipment.** This is a must. Since many deliveries must be weighed, accurate scales are, perhaps, the most important pieces of equipment in the receiving area. Temperature probes let receivers check the temperatures of refrigerated and frozen products. Rule measures are useful in checking trim of, for example, fat on portion-cut steaks. Calculators are needed to verify costs. Cutting instruments, such as a produce knife, are handy when product sampling is part of the receiving process. Conveyor belts, hand trucks, and motorized fork-lift trucks can help transport the received items to the storage areas or, in some cases, directly to a production area. And, where applicable, receiving agents must have the technology to read existing bar codes on product packaging in order to process shipments correctly. In short, receivers should have enough of the proper equipment to do the most efficient, thorough job possible.

   In a smaller operation, a reliable scale is a bare minimum. It is surprising how many small operations "save money" by purchasing an inexpensive scale that is often inaccurate. Then, because "it doesn't work anyway," they do not use it. This is the falsest of economies since, even if the operators trust their suppliers, unintentional errors can still occur.

3. **Proper Receiving Facilities.** If an operation wants a receiver to perform adequately, he or she must have the facilities that will make that possible. By "facilities," we are referring to the entire receiving area. By "proper," we mean, for example, that the area should be well lit, big enough to work in comfortably, reasonably secure, and convenient for both delivery people and receivers.

   In some old buildings, or in a hospitality operation built into another kind of building—for example, an office building—management may not see exactly what we have described, but the closer management can move to this ideal, the better the receiver can do his or her job.

4. **Appropiate Receiving Hours.** Deliveries should be scheduled carefully. If possible, deliveries should be staggered so that a receiver is not rushed. Also, all delivery times should be relatively predictable so that a competent receiver will be on hand and receiving will not be left to whoever happens to be handy. Remember, we mentioned that one of the biggest benefits of one-stop shopping is to minimize any difficulties and expense arising from too many deliveries. Perhaps now you can appreciate why many managers are swayed by this potential benefit. It not only reduces the number of hours a receiver must work, but it also allows for a more secure backdoor routine and, because of fewer transactions, minimal theft opportunities.

5. **Available Copies of All Specifications.** The receiver will need these as references. This can help whenever ambiguity arises, as it sometimes will. When a supplier

is out of a particular brand of soap, for example, he or she might deliver what is believed to be a comparable substitute. When this occurs, it becomes necessary for the deliverer and receiver to have a reference handy, unless the buyer insists on handling any substitutions personally, in which case the receiver will ask him or her to inspect the substitute. In many cases, though, drivers are not eager to wait for this decision; it is expensive for them to leave their trucks idle. As a result, decisions must often be made quickly, and a copy of the specifications can, therefore, be quite helpful.

6. **Available Copies of Purchase Orders.** Most hospitality operators feel that a receiver should know what is due to be delivered so that he or she can be prepared. These purchase order copies are necessary to ensure this preparedness.

INVOICE RECEIVING

The most popular receiving technique is sometimes referred to as "invoice receiving." In this scheme of things, an invoice, or bill, accompanies the delivery (see Figure 3.1). An invoice is an itemized statement of quantity, price, and other information that usually resembles a purchase order. An invoice may, in fact, be nothing more than a photocopy of the original purchase order. The receiver uses the invoice to check against the quantity, quality, and prices of the items delivered. He or she may also compare the invoice with a copy of the original purchase order as a further check. The order is either accepted or rejected. If accepted, it is stored or delivered to a production department. If rejected, appropriate credit must be obtained. Sometimes, an invoice does not accompany the delivery. In these situations, the receiver normally fills out some type of form right there on the spot and treats this completed form as the invoice. The following sections discuss the typical invoice receiving sequence in more detail.

Delivery Arrives

When arriving at a large operation, the delivery person must usually announce that he or she is there, sometimes by ringing a doorbell and asking the receiver for access. These procedures are, of course, less formal in a small operation, but they represent good security precautions.

The receiver opens the receiving area and, using the invoice and, perhaps, a copy of the original purchase order, checks immediately for the proper quantities. The receiver's first step, then, is to check each item's weight, count, and/or volume as quickly and efficiently as possible and compare them to the invoice and the original purchase order, or some other purchase record. The comparisons should match.

Sheldon's Meats

PLEASE PAY FROM THIS INVOICE

22058

100 MAIN ST.   LAS VEGAS, NEVADA 89123
555-1212
Federal Inspected Meat Plant Est. 1000

IMPORTANT
MAIL REMITTANCE TO:
P.O. BOX 1000
LAS VEGAS, NV 89123

SOLD TO _____

ADDRESS _____

☐ C.O.D.   ☐ CASH   ☐ CHARGE   P.O. NO. _____   DATE ___ / ___ / ___

QUAN	DESCRIPTION	WEIGHT	PRICE	AMOUNT
	PATTIES			
	PATTIES			

All claims must be made immediately upon receipt of goods. If any discrepancy, please call (702) 555-1212; otherwise late claims will not be allowed.

TOTAL ⟶

Received by _____

Boxes _____   Pkgs _____   Pcs. _____

All accounts are due the 10th of the month following date of purchase and after such date are past due accounts subject to an interest charge of $1\frac{3}{4}$% per month, which is **AN ANNUAL PERCENTAGE RATE OF 21%**. In the event legal proceedings are instituted to collect any sums due, the purchaser agrees to pay reasonable attorney's fees and costs.

I hereby certify that the above described product, which is offered for shipment in commerce, has been U.S. inspected and passed by the U.S. Department of Agriculture, is so marked, and at this date is not adulterated or misbranded.

FIGURE 3.1 A typical invoice.

Next, where applicable, the receiver checks for the proper quality. Unfortunately, except for a check of the packers' brand names, this is the most difficult kind of check to make and, in some cases, is almost impossible to complete. For example, it is difficult to determine the overall quality of lobster tails. A receiver might be able to tell whether they have been refrozen. But he or she can never be sure if any particular tail is bad until it has been cooked; a bad lobster tail crumbles after it is cooked.

Some establishments expect the receiver to check for quantities only and to call someone else to inspect for quality. In these situations, the receiver calls the chef, housekeeper, maître d', or whatever department head is appropriate to come check the quality of the items that will eventually be used in his or her department. The big drawback of this procedure is the potential time lag in waiting for the department head to arrive. Another problem is the possibility of the items, especially frozen ones, deteriorating while waiting for a quality check.

Large operations, particularly commissaries, usually engage quality control, or quality assurance, inspectors who check deliveries, as well as the products prepared in the commissary. These inspectors do not normally work for the purchasing agent. In fact, they act as a check on the purchasing agent by keeping tabs on the quality that the purchasing agent procures.

If a quality discrepancy is present, the buyer should be notified as soon as possible since it is his or her duty to deal with suppliers and salespersons. Moreover, the buyer may want occasionally to be on hand in the receiving area in order to gain a firsthand impression of the types of items that are actually delivered as compared to what was ordered.

After the quality inspection—unless the entire order has been rejected—the receiver in some operations checks all prices and price extensions. [A price extension for a particular item on the invoice is the as-purchased (AP) price per unit of that product times the number of units purchased.] The receiver might also check all sales tax and other use taxes that are noted on the invoice to ensure accuracy. For instance, in most states, a hospitality operator must pay sales tax for merchandise that he or she will not resell, such as cleaning chemicals, but not for products earmarked for resale to consumers, such as meats and produce.

Unfortunately, checking all of these figures can take too long, usually because a receiver must compare the prices on the invoice with those that the supplier quoted prior to ordering the merchandise. It is necessary for the receiver to compare the invoice with the purchase order, or some other written purchase record, both to check the prices and to note whether the merchandise delivered actually was ordered in the first place. To save time, many operations have the accountant or bookkeeper check invoices later, before paying the bill.

It is probably a good idea to handle any AP price discrepancies as soon as possible. Waiting too long can produce confusion and distrust among business partners. Also, a supplier may be honest with the buyer and would want to know immediately if, for instance, a driver has altered the prices on the invoice. The opposite can also be true; that is, the supplier, for example, may have quoted a much lower AP price over the telephone than the one now written on the invoice. If so, the receiver should notify the deliveryperson so that he or she will be a witness to this discrepancy.

Rejection of Delivery

In some cases, a receiver may merely note a discrepancy regarding prices and/or taxes noted on the invoice. Alternately, he or she may have to reject all or part of an order. When this occurs, the receiver might prepare a Request for Credit memorandum, which is a written statement attesting to the fact that the particular item or items did not meet quality, quantity, or price standards (see Figures 3.2 and 3.3 for typical Request for Credit memoranda). The deliveryperson's signature shows that a representative of the supplier has agreed that the hospitality operation's account must be credited. The objective of this memo is to ensure that your account is credited and that all costs are accurate.

A receiver also may need to prepare a Request for Credit memo if he or she should receive credit for product substitutions, such as when a less expensive item than the one ordered is delivered. Also, the invoice might contain arithmetic errors that require adjustment. A back order may have been charged to the hospitality operator, in which case the receiver might want to ensure that the firm does not pay for the items, thereby tying up its money unnecessarily, until it receives them.

When the accounting office pays the invoices, it will reduce the invoice total by the amount indicated on the credit memorandum itself, which comes from the supplier after the receiver sends him or her a copy of the Request for Credit memo. Alternately, the supplier will give credit to the operator on the next delivery. Some operations eliminate credit memos when the supplier either agrees to give his or her deliverypeople authority to "reprice" the invoice on the spot or allows the deliverypeople to prepare a credit memorandum, or "credit slip," right then and there and give it to the receiver. However, if a common carrier (i.e., an independent trucking firm) delivers the shipment, the receiver must complete a Request for Credit memo because the driver will have no authority to alter the delivery.

When credit paperwork is prepared, the original copy is usually sent to the supplier. In addition, the receiver or buyer might call the supplier; this serves as a check on the deliverypersons, who, for example, may have stolen the original item and substituted an inferior product. Another copy usually goes to whoever pays the bills. Some receivers

```
┌──────────────────────────────────────────────────────────┐
│                                                            │
│              ┌─────────────────────────────┐              │
│              │    NAME OF HOSPITALITY      │              │
│              │        OPERATION            │              │
│              └─────────────────────────────┘              │
│                                                            │
│                 Request for Credit Memorandum             │
│                                                            │
│                                                            │
│   Date _____                                      │
│                                                            │
│   To Supplier:  Please credit our account because:        │
│                                                            │
│              1. _____         │
│                                                            │
│              2. _____         │
│                                                            │
│              3. _____         │
│                                                            │
│                                                            │
│                                                            │
│                                                            │
│   Return ordered by _____        │
│                                                            │
│   Invoice no. _____                       │
│                                                            │
│   Date received _____                       │
│                                  _____  │
│                                       Deliverer's signature│
│                                                            │
│                                                            │
└──────────────────────────────────────────────────────────┘
```

FIGURE 3.2 A Request for Credit memorandum.

might want to keep a copy as a reminder to be especially careful of any future deliveries from this particular supplier. In addition, the buyer may want a copy in order to keep up to date on the supplier's performance.

Whenever rejection is contemplated, the owner-manager must not act too hastily. It may not be a good idea to reject a product that deviates only slightly from the hospitality operation's standard, for at least two reasons: (1) suppliers may not like to do business with a customer who focuses on small details, particularly one who sends back a reasonable substitute that the supplier sent because the ordered item was unavailable; and (2) in many cases, a rejection leaves a receiver short. It might be good business occasionally to accept some slight deviation since the potential ill will generated among suppliers and customers by hasty rejections may be detrimental.

Sheldon's Meats

2336
CREDIT REQUEST

100 MAIN ST.   LAS VEGAS, NEVADA 89123
555-1212
Federal Inspected Meat Plant Est. 1000

DATE

P.O. NUMBER

INVOICE NUMBER	DATE	WEIGHT	ITEM	PRICE	CREDIT AMOUNT	REASON NUMBER

_____
AUTHORIZATION

CODE KEY ➡

1 - REFUSED	5 - NOT ON TRUCK
2 - WRONG PRODUCT	6 - PRICE ERROR
3 - SPOILED	7 - INVOICE ERROR
4 - SHORT WEIGHT	8 - OTHER

FIGURE 3.3 A Request for Credit memorandum used by customers of Sheldon's Meats.

Returning Merchandise

The receiver may have something from a previous delivery that the deliveryperson must return to his or her company's warehouse. For example, say the buyer has arranged to send back an unintentional overbuy of canned pears from the preceding week. Usually, in this situation, the supplier has given the deliveryperson a "Pick-Up" memorandum, authorizing him or her to take back the merchandise (see Figure 3.4).

Whenever a belated return must be arranged, the deliveryperson leaves a copy of the Pick-Up memo with the receiver. This copy serves as a receipt for the returned

Sheldon's Meats

100 MAIN ST.   LAS VEGAS, NEVADA 89123
555-1212
Federal Inspected Meat Plant Est. 1000

CUSTOMER NAME

_____

PICK-UP MEMO

_____

_____

DATE		DRIVER	CUST NO		PICK UP MEMO NO	
QTY	GRADE	PROD CODE	DESCRIPTION		WEIGHT	PRICE

THIS IS NOT AN INVOICE OR CREDIT MEMO. IT IS A RECEIPT FOR MERCHANDISE RETURNED TO OUR PLANT FOR INSPECTION. YOU WILL BE ADVISED OF OUR FINDINGS AND DECISION AT THE COMPLETION OF THE INSPECTION.

DISPOSITION	DATE

FIGURE 3.4 A typical Pick-Up memorandum.

goods. The supplier will issue a credit memo later on, once he or she inspects the returned merchandise and is satisfied that the return is justified.

Acceptance of Delivery

When an order has been accepted, the receiver normally initials some paperwork attesting to the fact that everything is correct. The deliveryperson usually produces a delivery sheet or a copy of the invoice for the receiver to sign.

Unless the items are definitely substandard, the receiver accepts most deliveries. Even if only part of a shipment is acceptable, the customary practice is to keep what is good and return the rest along with a Request for Credit memo. For the most part, an owner-manager is reluctant to send back everything because the resultant shortages can, as we said, lead to dissatisfied customers.

Upon acceptance of the deliveries, the receiver usually places the items in the proper storage location or, in some cases, delivers them to a production department.

To ensure that all pertinent checks have been made, the receiver normally applies an ink stamp with a predetermined format to the invoice (see Figure 3.5 for a typical invoice stamp format stamped on incoming invoices).

This format notes all checks that must be performed and provides a space for those responsible to affix their initials. The receiver normally initials the first three entries; the accountant or bookkeeper, number 4; the buyer, number 5; and the owner-manager, number 6.

After processing the invoice, the receiver may record the delivery on a "receiving sheet," or "receiving log." This sheet is nothing more than a running account of deliveries (see Figure 3.6 for a typical receiving sheet).

---

1. Date received _____

2. Received by _____

3. Prices checked by _____

4. Extensions checked by _____

5. Buyer's approval _____

6. Payment approval _____

FIGURE 3.5 Typical invoice stamp information.

DATE	TIME DELIVERED	QUANTITY	INVOICE NO.	PURVEYOR	DESCRIPTION OF ITEM(S)	UNIT PRICE	EXTENSION	*DIRECT			†STORES			OTHER INFORMATION?
								FOOD	BEVERAGE	NONFOOD	FOOD	BEVERAGE	NONFOOD	

* The receiver notes in this column the amount of food, beverage, and nonfood items that go directly to the production department, bypassing the main storage area. In other words, these items go directly into the in-process inventory.

† The receiver notes in this column the amount of food, beverage, and nonfood items that go into main storage.

FIGURE 3.6 A receiving sheet.

To a certain extent, the receiving sheet is a redundant exercise: it contains a good deal of information already on the invoice or affixed to the invoice via one or more invoice stamps. Large hospitality operations traditionally use the sheet, but the whole process may be avoided without any significant loss of control by merely photocopying invoices or scanning them into the computer for the buyer's and receiver's files.

One reason many operators like the receiving sheet is that it forces the receiver to record information, however redundant. Thus, mistakes previously overlooked sometimes come to light. Also, a copy of the receiving sheet usually stays in the receiver's files, which makes it handy for the receiver if he or she needs to evaluate a certain supplier's past performance. Furthermore, the sheet is useful to cost accountants who prepare daily food, beverage, and nonfood cost reports. Since the receiving sheet notes the deliveries on one page, it is convenient. In addition, the "Other Information" column can contain several comments that are not easily recorded on the incoming invoices. Such elements as the deliveryperson's attitude and the cleanliness of the delivery truck may be important to the buyer in future negotiations with that particular supplier.

Another reason for some managements' continuing desire for the receiving sheet is that it can be treated as the receiver's daily report of activities. This type of report is particularly attractive to the accountant in a large hotel who is responsible for the receiver's actions.

Overall, however, it is far more economical to record all such information on the incoming invoices or on invoice copies, or to attach a small Post-It note to these invoices if space is insufficient, and then make a copy of this completed invoice for the receiver's files.

After making the necessary entries to the records, the receiver normally sends the incoming invoices, any credit slips, and a copy of the receiving sheet if one is used, to the accountant or bookkeeper. If a bill of lading, which is a piece of paper that represents title to the goods, comes with the delivered items, he or she sends this along also.

If the storage areas are supervised and controlled by someone other than the receiver, this person may want a copy of the receiving sheet to compare what is on the sheet with what has been put in storage. This is yet another type of control serving as a check on the receiver, although the buyer's and the accountant's copies can serve as more than sufficient control.

At this point, the receiver normally has stored the items or delivered them to production departments, completed the necessary paperwork, and sent the appropriate paperwork to the right office(s), along with any bills of lading that may have arrived at the receiving dock that day. He or she also keeps a copy of the receiving sheet.

Additional Receiving Duties

As a general rule, the receiving procedure is now complete. But the receiver may have other, less routine duties to perform. He or she may have to do one or more of the following:

Date the Delivered Items If it is too costly to do this, the usual compromise is to date only the perishable items. This dating is usually done with colored tags or with an ink stamp. This can facilitate proper "stock rotation," a process by which older products are used first.

Price All of the Delivered Items Like dating delivered items, this pricing may also be too costly, but it can have such benefits as costing of inventories for accounting purposes and providing an easy cost reference. Some operators like to price the items for the psychological effect it supposedly provides. Items that an operator has priced are no longer just merchandise to employees, but articles of value to be treated accordingly.

Many properties use the "Dot System" to date and price inventories. These are color-coded, stick-on dots (usually a different color for each day) that have sufficient space to pencil in dates, times, and prices. Incidentally, this procedure is also used to identify and code preprepared products that the kitchen staff has made. For instance, grated cheese to be used later on can be coded so that all cooks use any older grated cheese first (see Figure 3.7).

Create Bar Codes In some large hospitality operations, the receiving agent may need to create bar codes and apply them to incoming products that do not have them on their

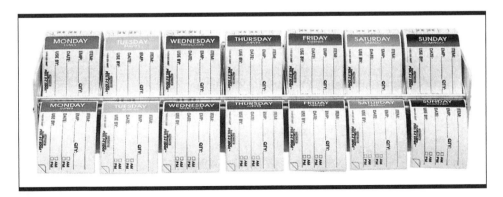

FIGURE 3.7 An example of a Dot System. Courtesy of DayDots.

package labels. This is usually done to enhance the inventory management and control process, in that it makes it very easy to track inventories and their AP prices throughout the operation. While investing in the technology needed to adopt this procedure can be very expensive, in the long run it could prove very cost-effective.

Apply "Meat Tags"   A meat tag contains information similar to that on an invoice stamp. The major difference between the two is that the typical meat tag contains two duplicate parts (see Figure 3.8). During the receiving procedure, one part of the tag is put on an item, and the other part goes to the accountant or bookkeeper for control purposes. When an item moves from the storage area to production, the part on the item is removed and sent to the accountant or bookkeeper, who matches it with the other part and removes it from the inventory file.

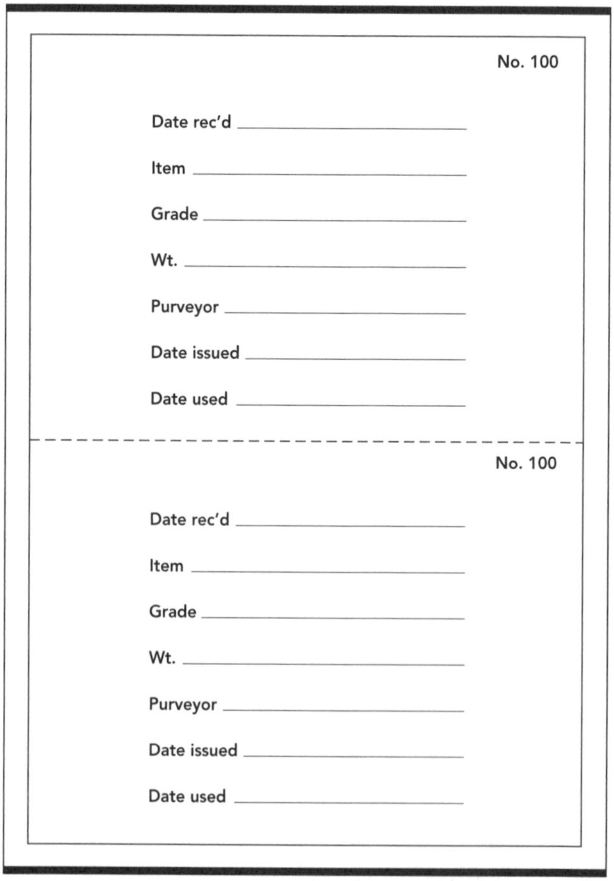

No. 100

Date rec'd _____

Item _____

Grade _____

Wt. _____

Purveyor _____

Date issued _____

Date used _____

- -

No. 100

Date rec'd _____

Item _____

Grade _____

Wt. _____

Purveyor _____

Date issued _____

Date used _____

FIGURE 3.8 A meat tag.

Specifically, meat tags are used as a check on the overall use of an item. For instance, comparisons are made between meat tags and stock requisitions (requests from a production department for items that are held in storage). Also, meat tags are sometimes compared with the service department's record of guest services. For example, with steak items, the meat tags can be compared with the sales of steaks to customers, thereby producing a check on the waitstaff. If everything goes right, all these comparisons will reveal that what was used from storage actually went to the paying customer, with no loss of product along the way. The units recorded on the meat tags should correspond exactly to the amounts used in production and the amounts sold to customers.

The meat tag control is, however, cumbersome, unless it can be computerized and/or bar coded. And, like the receiving sheet, it tends to be redundant as well. If meat tags are used, they tend to be used only for high-cost products. Nevertheless, both meat tags and receiving sheets are used in operations that desire close control over their stock.

Housekeeping Management usually requires the receiver to maintain a clean, efficient workplace. Also, he or she usually ensures that all equipment and facilities are kept in good working order (see Figure 3.9).

Update AP Prices Hospitality operations that use a computerized management information system normally maintain updated AP prices for all merchandise they buy. They also might maintain updated portion factors, portion dividers, and EP costs for all of the ingredients they currently serve, as well as all of the ingredients they might serve in the future. Furthermore, computerized hospitality operations also tend to maintain costed recipes in a recipe file for those menu items that are currently being offered to customers, as well as those that may be offered at a future date.

A computerized system includes the necessary formulas and databases to make the necessary calculations quickly. Since AP prices tend to vary from day to day in our industry, someone must continually "load the computer" with the new, current AP prices. In some operations, this task falls on the receiving agent. He or she, after performing the other required receiving duties, must follow the procedures needed to enter the new AP prices and to remove the old, outdated AP prices. Someone in the accounting department might just as easily do this task, but a receiving agent may be deemed the best person to do this work, especially if the task involves scanning the bar codes of all of the incoming products.

Backhaul Recyclables Some operations save their recyclables, such as corrugated cardboard, glass bottles, and metal cans, and hold them until a common carrier hired by a

FIGURE 3.9 An example of a modern storage area. Courtesy of Summit Foods.

primary source to deliver a product shipment uses the emptied truck to "backhaul" the recyclables on the return trip. In this case, the receiving agent usually needs to help the driver load the truck and see to it that the driver has all of the necessary paperwork and authorizations.

OTHER RECEIVING METHODS

O ccasionally, other receiving procedures are used. For the most part, though, they are variations of invoice receiving. Some of these alternate approaches are described in the following sections.

Standing-Order Receiving

This receiving procedure may not differ at all from invoice receiving. But receivers sometimes tend to "relax" a bit when checking items received on a standing-order basis. Also, delivery tickets rather than priced invoices may accompany the delivery, since the operation may make a regular, periodic payment to the supplier in exchange for the same amounts delivered at regular intervals.

It is really best to use invoice receiving to receive standing orders. Otherwise, deliverypeople, receivers, and bill payers can grow careless. In addition, deliveries may begin to "shrink" in both quantity and quality if strict receiving principles are not maintained.

Blind Receiving

The only difference between blind receiving and invoice receiving is that the invoice accompanying the delivery contains only the names of the items delivered, and no information about quantity and price. A duplicate invoice, which contains all of the necessary information, is usually sent to the accountant or bookkeeper one day before delivery.

Another form of blind receiving involves the need for the receiver to complete a "Goods Received Without Invoice" slip whenever a shipment comes in that does not have an invoice or delivery slip. For instance, a mailed delivery or shipment delivered by a messenger service may not have accompanying paperwork. When this happens, the receiver must check with management and, if the shipment is legitimate, inspect the products and complete the in-house invoice slip.

The whole idea behind blind receiving is to increase the margin of control. The receiver is forced to weigh and count everything and then record this information. Such a procedure effectively prohibits the receiver from stealing part of the delivery and altering the invoice. Also, the procedure precludes any fraudulent relationship between the receiver and the deliveryperson.

A good deal of disagreement exists regarding the benefits of blind receiving. The general feeling in the industry is that the receiving agent should have some idea of what to expect; otherwise he or she might receive the wrong product, too much product, too little product, and so forth. Such unintentional errors can destroy an operation's production planning.

Blind receiving is a time-consuming, costly method of receiving and processing deliveries. Operations can employ technology to speed up the process, but it normally is too expensive to be used for only a short period each day or in a small operation.

Furthermore, drivers do not like to wait for a receiver to record every last detail of information.

Although we can appreciate the control benefits of blind receiving, we consider it an archaic method, similar to receiving sheets and meat tags. We know of few establishments that still use it. It is too expensive; besides, a receiver under suspicion can be checked with the accountant's copy of the original purchase order. The invoice should look just like the original purchase order; if they look different, management should ask for a good explanation—or look for a new receiver.

Odd-Hours Receiving

The major difference with this receiving method is that the regular receiver is not on hand to accept the delivery. In most cases, an assistant manager is then entrusted with this duty. Although the invoice method may be applied during odd hours, an inadequate receiving job may result. The stand-in receiver usually has other pressing duties and, as a result, tends to rush the receiving process. Usually, the owner-manager recognizes this potential danger and tries to arrange for deliveries when the regular receiver is on duty. But some deliveries must be made at odd hours. As a result, it may be a good idea for the owner-manager to print the regular receiving procedure on a poster and hang it in the receiving area to aid the stand-in receiver.

Drop-Shipment Receiving

When a buyer purchases products from a primary source, that source usually hires a common carrier to "drop ship" the merchandise to the hospitality operation. Remember, the common carrier is typically an independent trucker hired to provide only the transportation function.

When a common carrier delivers a shipment, the receiving procedure used is very similar to the standard invoice receiving process. The major difference is that the driver is not involved with any disputes that may arise between the buyer and the primary source, unless he or she is directly responsible for the problem. For instance, if the driver damages the goods along the way, the buyer must deal with the driver or the driver's employer. But, as is more often the case, if the products do not meet the buyer's specifications, the receiver usually must take the shipment from the driver and hold it until the problems are rectified. Ordinarily, the driver is not in a position to take back returned merchandise.

When disagreements arise between the buyer and the supplier in this situation, it is difficult to resolve them. For example, the buyer may have to arrange for another common carrier to return the shipment. Alternately, he or she may have to wait for the supplier's representative to arrive and check the items personally before a settlement can be reached. In addition, if the shipment is insured by an independent insurance company, its representative may need to inspect the claim and monitor the negotiations. When hospitality operators buy directly from primary sources, especially unfamiliar ones on the Web, seemingly little problems can add a great deal of stress to the transaction before they are cured.

Mailed Deliveries

When orders are delivered by mail, or by similar means, such as United Parcel Service (UPS) or FedEx, the invoices that come with them are normally referred to as "packing slips." These slips are treated like any other invoice except when the order does not match the packing slip's description. In this instance, a Request for Credit memo or some similar record must be completed, but usually management, not the receiver, does this. The receiver notes any discrepancy and then turns the shipment over to his or her supervisor.

Cash-on-Delivery (COD) Deliveries

Under this system, the receiver has the added duty of paying the delivery agent or, more commonly, sending the delivery agent to the office for the payment check. It is also possible that the receiver accompanies the delivery agent to the office so that he or she can attest to the adequacy of the delivered items.

GOOD RECEIVING PRACTICES

Receivers should follow a number of sound procedures. Most of them fall mainly under the security category:

1. Receivers should beware of excess ice, watered-down products, wrapping paper, and packaging that can add dead weight to the delivered items. Receivers must subtract the amount of this dead weight, which is sometimes referred to as the "tare weight," from the gross weight in order to compute the net weight of the merchandise.

2. Receivers should always check the quality under the top layer. Make sure that all succeeding layers are equal to the facing layer.

3. Receivers should always examine packages for leakage or other forms of water damage. This could indicate that the package contents are unusable. If the

packages, especially cans, are swollen, the contents are probably spoiled and receivers should reject the shipment.

4. If a package label carries an expiration date, receivers should ensure that it is within acceptable limits. Receivers should also make sure that the dating codes are correct.

5. Receivers should not weigh everything together. For example, they should separate hamburger from steak and weigh each product by itself. If they weigh these items together, they might begin to buy hamburger at a steak price.

6. Receivers should be wary of deliverypersons eager to help them carry the delivered items to their storage areas. Trust is not the issue. The big problems with letting people on the premises are the distraction they cause among employees and the possibility that liability insurance premiums will increase.

7. Receivers should watch for incomplete shipments, as well as for the deliveryperson who asks them to sign for a complete order after telling them that the rest of the order will arrive later. Later may never come.

8. Receivers should spot-check portioned products for portion weights. For example, if receivers buy portioned sausage patties by the pound and sell them by the piece, a 2-ounce sausage patty that is consistently $1/4$ ounce overweight will inflate the food cost. But operators will not reflect this in their sales since their menu price will still reflect a 2-ounce portion. It is equally troubling if the sausage patties are underweight and are purchased by the piece; a short weight of as little as $1/8$ ounce can cost quite a bit of money in the long run.

9. Receivers should be careful of closed shipping containers with preprinted dates, weights, counts, or quality standards. Someone may have repacked these cartons with inferior merchandise. It might be wise to weigh flour sacks, rice sacks, potato sacks, and the like, once in a while. Receivers might even open a box of paper napkins occasionally to count them.

10. Receivers should also be careful that they do not receive merchandise that has been refrozen. In addition, they should be on the lookout for supposedly fresh merchandise that is actually "slacked out" (i.e., has been frozen, thawed, and made to appear as if it is fresh).

11. At times, receivers may confuse brand names and/or packers' brand names. This is easy to do when receivers are in a hurry.

12. When receiving some fresh merchandise, such as meats, fish, and poultry, receivers normally give suppliers a "shrink" allowance. For instance, dehydration might turn 25 pounds of fresh lobster today into 24 pounds of lobster tomorrow. The product

specification normally indicates the minimum weight per case that receivers will accept, but in some circumstances, they may not be able to judge the delivery as closely as they would like. When they weigh the shipment, it might be within the accepted tolerance; however, it is not easy for receivers to determine acceptability quickly when they are busy weighing several packages of various sizes. Delivery drivers know this, and, as such, some may be tempted to test a receiver's skills to see how much shrink he or she is willing to accept.

13. In general, receivers are concerned about any product that they receive that does not live up to their specifications. It is absolutely essential to prepare adequate specifications because this is the only way receivers can ensure that they have the appropriate standard upon which to judge incoming merchandise.

We do not intend to criticize suppliers or delivery agents. A good rule in business is to maintain a cautious optimism when receiving, but remember that it is possible to get "stung" in at least four ways: (1) the unintentional error, (2) the dishonest supplier with an honest delivery agent, (3) the honest supplier with a dishonest delivery agent, and (4) a dishonest supplier with a dishonest delivery agent. Keep in mind that once receivers sign for a delivery, the items are theirs. So receivers must verify that they receive the right quantity, quality, and AP price.

REDUCING RECEIVING COSTS

Receivers can reduce receiving costs a few ways without losing a proportionate amount of control. Some common cost-saving methods are described in the following paragraphs.

1. **Field Inspectors.** Large firms sometimes use field inspectors, which saves some time for receivers in that they do not have to check for quality and quantity. This is because the inspector often seals the packages to be delivered. The overall cost, though, may not decrease; field inspectors, like receivers, must be paid.

2. **Night and Early-Morning Deliveries.** These odd-hour deliveries are often the rule in the downtown sections of many large cities in order to avoid daytime traffic congestion. With fewer distractions, delivery agents can make more deliveries, and part of the lower transportation cost per delivery may be passed on to hospitality operations. A variation of this procedure is the "night drop," in which a delivery agent uses a key to get in, places the items inside the door, locks up, and leaves. Opinion varies among operators regarding the degree of trust required for this practice.

3.   **One-Stop Shopping.** This is, perhaps, the most common method of reducing receiving costs. Although some people are not enthusiastic about this buying method, everyone agrees that it can reduce receiving costs.

In trying to reduce receiving costs, receivers must be careful that they do not simply shift the costs around. For instance, low receiving costs accompany one-stop shopping, but the savings from reducing the number of potential suppliers may be wiped out by higher AP prices from a single supplier.

Certain inescapable costs must be incurred if receivers expect to meet the objectives of the purchasing and receiving functions. It is absolutely essential not to negate the effective job the buyer may have done. In the end, the receiving function affords few cost-cutting possibilities unless receivers are willing to give up a certain amount of control.

KEY WORDS AND CONCEPTS

Accepting a delivery

As-purchased price (AP price)

Backhaul

Bar codes

Bill of lading

Blind receiving

Cash on delivery (COD)

Checking the quantity, quality, AP price, sales tax, and other use tax

Common carrier

Computerized record keeping

Credit memo

Credit slip

Cycle of control

Date and price items that are delivered

Delivery ticket

Dot system

Drop shipment

Early-morning deliveries

Edible-portion cost (EP cost)

Equal to facing layer

Essentials for good receiving

Expiration date

Field inspectors

Good receiving practices that should be followed

Goods Received Without Invoice slip

Gross weight

Handling an invoice discrepancy

Incomplete shipments

Invoice

Invoice receiving

Invoice stamp

Loading the computer with current AP prices

Mailed deliveries

Meat tags

Net weight

Night deliveries

Night drop

Odd-hours receiving

One-stop shopping

Packing slip

Pick-Up memo

Price extensions

Purchase order (PO)

Quality assurance

Quality control

Receiving objectives

Receiving sheet

Rejecting a delivery

Request for Credit memo

Returning merchandise

Route salesperson

Sales tax

Shrink allowance

Slacked out

Specifications

Standing-order receiving

Stock requisition

Stock rotation

Tare weight

Use tax

Water damage

Ways to reduce receiving costs

REFERENCES

1. Patt Patterson, "Checks and Balances Prevent Disputes Over Orders," *Nation's Restaurant News*, November 22, 1993, p. 82.

2. Robert B. Lane, "Food and Beverage Management," in *VNR's Encyclopedia of Hospitality and Tourism*, Mahmood Khan, Ed. (New York: Van Nostrand Reinhold, 1993), p. 39.

QUESTIONS AND PROBLEMS

1. You hear that your competitors are using a control device called "blind receiving." What is blind receiving? Under what conditions would you use this procedure?

2. Explain how one-stop shopping can reduce your overall receiving costs.

3. At 10 A.M. on September 27, the A & H Foods Company delivered the following items:

UNIT	QUANTITY	ITEM DESCRIPTION	UNIT PRICE	EXTENSION
Pound	80	T-bone steaks	$ 6.85	$548.00
Pound	18	Sliced bacon	2.80	50.40
Pound	20	Flank steak	4.25	85.00
Case	2	Floor wax, gallon cans	28.00	56.00
Case	2	Boston lettuce	16.50	33.00
Case	1	Canned green beans, No. 10 cans	17.50	17.50
Total				$789.90

Upon inspection, you determine that the sliced bacon is inferior and that you must return it to the supplier.

(a) Using Figure 3.2 as a guide, prepare a Request for Credit memo for the bacon.

(b) Using Figure 3.5 as a guide, complete the information on the invoice stamp that the receiver usually completes.

(c) Using Figure 3.6 as a guide, transfer the acceptable items to the receiving sheet.
   Note: The flank steak goes directly into the in-process inventory.

4. What type of information would you like to have in the "Other Information" column on the receiving sheet? Why?

5. Many operators feel that the receiving sheet is useful in calculating daily food, beverage, and nonfood costs. How do you think the receiving sheet is helpful in this matter?

6. What should a receiver do when a question arises regarding the quality of merchandise received?

7. What should a receiver do if a delivery is made without an accompanying invoice?

8. A receiver will prepare a Request for Credit memo when:
   (a)
   (b)
   (c)

9. List some objectives of the receiving function.

10. List the primary essentials that are needed for proper receiving.

11. What is the primary difference between invoice receiving and blind receiving?

12. Briefly describe the computation of price extensions.

13. Describe one purpose of using an invoice stamp.

14. What is the primary reason for using meat tags?

15. Briefly describe the concept of stock rotation.

16. What is a bill of lading?

17. What does it mean when we say that a food item has been "slacked out"?

18. Why should you separate meat items before weighing them?

19. What is the primary purpose of the Pick-Up memo?

20. Assume you must pay sales tax for all nonfood items you purchase. If the sales tax rate is 6 percent, recalculate the invoice total for Question 3.

21. What is the significance of the expiration date placed on the package label of some food products?

22. Assume you are checking in a shipment of canned goods. You notice some dried water spots on the bottom of one of the cases. You open the case and notice nothing leaking from the cans. Should you accept the shipment? Why or why not?

23. What are the advantages and disadvantages of the standing-order receiving procedure?

EXPERIENTIAL EXERCISES

1. Visit a local supplier and arrange to ride with a deliveryperson as he or she makes the rounds. Compare and contrast the receiving procedures you see in each hospitality operation. In addition, take the time to examine how the supplier processes his or her copy of the invoice.

2. Arrange to spend one day in the receiving area of a hotel or restaurant. Evaluate the receiving procedures the receiver uses. In addition, try to follow the paperwork, from invoice processing, receiving-sheet completion, and so on, up to the end of the receiver's paperwork duties. If management allows, pick out one invoice and stay with it as it travels from the receiving area to the accounting department. While in the accounting department, see whether you can determine how and when this particular invoice will be paid.

FRESH PRODUCE*

The Purpose of this Chapter

After reading this chapter, you should be able to:

■ Explain the selection factors for fresh produce, including government grades.

■ Explain the process of purchasing, receiving, storing, and issuing fresh produce.

INTRODUCTION

Purchasing fresh produce calls for a great deal of skill and knowledge. Next to fresh-meat procurement, fresh-produce buying is, perhaps, the most difficult purchasing task the hospitality buyer faces.[1] In fact, it can be so difficult that some operators hire professional produce buyers to select and procure these products for them.[2] Fresh-produce buyers must have the wherewithal to purchase products that fluctuate in quality, quantity, and price on a daily basis. The real mark of an amateur in this area is to insist on top quality when none is available anywhere, or to accept poor quality when he or she should know good quality is available.

Fresh-produce buyers, especially those who work for supply houses, are extremely well paid. This fact alone indicates the difficulty and huge responsibility associated with the job. Even assistant fresh-produce buyers for a supply house require about two years of on-the-job training (OJT) before being allowed to make major purchasing decisions.

---

* Authored by Andrew Hale Feinstein and John M. Stefanelli.

When a buyer purchases fresh, natural food products, several quality variations within the same product line can appear daily. Soil and climatic conditions can affect the quality of the product he or she receives. Different geographical areas favor different plant varieties and can have a significant impact on the quality of the crop. Seasonal changes, natural or manmade disasters, or changes in demand have effects on the availability of quality product as well. As a result of variations in quality and quantity, the buyer should expect as-purchased (AP) prices to fluctuate throughout the year for produce, both within a given grade and among grades. For this reason, a single year-round price is unrealistic.

To stay abreast of these changes, the savvy buyer subscribes to trade publications that include information on both price and quality. The *Daily Fruit and Vegetable Report*, available online (http://www.ams.usda.gov:80/fv/mncs/fvdaily.htm) or by written request from the U.S. Department of Agriculture (USDA), Washington, DC, is one such publication. The Agricultural Marketing Service (AMS) also provides market news for fruit and vegetables online at http://www.ams.usda.gov/fv/mncs/fvwires.htm. Other subscription services, such as *The Packer: The Business Newspaper of the Produce Industry* (http://www.thepacker.com), and *The Produce News: National News Weekly of the Produce Industry since 1897* (http://www.producenews.com) are additional sources of current fresh-produce data.

In addition to the natural variations outlined above, another difficulty in buying fresh produce is choosing from the tremendous number of varieties and sources. Several hundred varieties of fresh-produce items are regularly available at any given time from various primary sources and intermediaries. Some supply sources stock more than 500 varieties of fresh produce.[3] Without research, it is difficult to decide which variety to use for a particular purpose.

Another major problem that the buyer may face is a lack of acceptable sources of fresh produce. Different varieties of produce grown in different regions come to market throughout the growing season. At times, it may be impossible to find suppliers that can obtain the quality the buyer wants, the quantity he or she needs, or both. On the other hand, it may be fortuitous if the buyer is located near an orchard or farm where the produce is harvested: he or she may be able to obtain fruits and vegetables at the peak of freshness.

SELECTION FACTORS

The owner-manager usually specifies the quality levels of fresh produce desired. The buyer normally carries out these specifications, as much as possible. Management personnel, often in concert with other individuals in the hospitality operation, usually consider one or more of the following fresh-produce selection factors when determining the quality standards as well as the preferred supplier(s).

Intended Use

As with any product or service a buyer plans to purchase, it is very important to identify exactly its performance requirement or intended use. This could save money in the long run because the buyer will not purchase, say, a superb-quality product to be used for a menu item if a lower-quality, lower-priced product will suffice. For instance, apples that must be on display on a buffet line should be very attractive, and the buyer would probably pay a premium price for this appearance. But if apples were to be used in a fruit cup, where their appearance would be camouflaged to some degree, perhaps lesser-quality apples would be adequate.

Exact Name

With the development of so many new types and varieties of fruit and vegetables and the advent of genetically modified foods,[4] the fresh-produce market is filled with a lot of terminology. Keeping track of the types, varieties, and styles of fresh produce can be a challenge. However, understanding this terminology is an absolute necessity for foodservice operations that prepare many menu items from raw ingredients because each item serves a specific culinary purpose. Therefore, a buyer must stay carefully and closely in tune with the needs of the hospitality operation and must find a supplier who is likewise aware.

It is not sufficient for a buyer to specify only the type of fresh produce required; he or she must also specify its variety. If a particular foodservice operation requires lettuce, then the variety must be known (e.g., romaine, iceberg, or red or green leaf). The same is true for potatoes (e.g., Burbank russet or Norgold russet potatoes), apples (e.g., Jonathan, Winesap, Golden Delicious, or McIntosh), and so forth.

U.S. Government Inspection and Grades (or Equivalent)

Grade standards were developed out of the necessity for common terminology of quality and condition in the produce industry (see http://www.ams.usda.gov/fv/ for more discussion on this topic). The first U.S. grade standard for fresh produce was established for potatoes in 1917. "U.S. No. 1" was the term given to the highest grade. It covered the majority of the crop and meant that the product was of good quality. "U.S.

No. 2" represented the remainder of the crop that was worth packing for sale under normal marketing conditions.

To enforce these standards, the U.S. Inspection Service for fresh produce was established that same year. In 1930, the Perishable Agricultural Commodities Act (PACA) was signed; it prohibited unfair and fraudulent practices in the interstate commerce of fruits and vegetables. By 1946, the Agricultural Marketing Act was signed into law and provided for the integrated administration of marketing programs. This act also gave the Agricultural Marketing Service (AMS) basic authority for major functions, including federal standards, grading and inspection services, market news services, market expansion, and consumer education. Currently, the USDA, through its AMS, Fruit and Vegetables Division, has grading standards for approximately 150 types of fruits, vegetables, and nuts.

Although government grades are used as a quality guideline, the buyer should be aware that each vegetable or fruit might have a different grading schedule. For example, the grades for grapefruit are U.S. Fancy, U.S. No. 1, U.S. No. 2, U.S. Combination, and U.S. No. 3; the grades for carrots are U.S. Extra No. 1, U.S. No. 1, U.S. No. 1 Jumbo, and U.S. No. 2; and the grades for apples are U.S. Extra Fancy, U.S. Fancy, U.S. No. 1, and U.S. Utility (see Figure 4.1).

Since most specifications for fresh produce include some reference to federal grades, buyers need to know where to find this information. In addition to the USDA, they can locate grading data in *The PMA Fresh Produce Manual* (see Figure 4.2). This reference manual is available for purchase at http://www.pma.com.

The U.S. government grader considers several factors when grading fresh produce, but appearance is the most important factor. The critical appearance factors include

FIGURE 4.1 A federal grade stamp used for fresh produce.
Source: United States Department of Agriculture.

Cherries, Sweet

Availability

Some major production areas include:

	January	February	March	April	May	June	July	August	September	October	November	December
California					•	•	•					
Oregon						•	•	•				
Washington						•	•	•				
Chile												•

Variety/Type Descriptions

Bing – Large firm cherry with mahogany skin and flesh; sweet rich flavor.

Chelan – Firm, round, heart-shaped fruit.

Lambert – Dark red, heart-shaped cherry with sweet rich flavor.

Lapins – Mahogany red cherry that exhibits excellent firmness and flavor.

Rainier – Firmly-textured cherry; golden skin with pink-red blush and clear-colored flesh. Sweet delicate flavor.

Sweetheart – Bright red, heart-shaped cherry with mild, sweet flavor and outstanding firmness.

Ordering Specifications

Common packaging:
11- to 20-lb. cartons or lugs
32-lb. crates

Grades:
U.S. No. 1
U.S. Commercial

NOTE: Differences between grades are based primarily on external appearance. Individual growing areas may also set their own grades.

Sizes:
9, 9.5, 10, 10.5, 11, 11.5, and 12 row

Equivalents

80 cherries = 2 cups pitted and sliced
1 pound cherries = 1½ cups juice

Receiving and Inspecting

Look for cherries that are plump with firm, smooth, and brightly colored skins. Avoid cherries with blemishes, rotted or mushy skins. Avoid those that appear either hard and light-colored, or soft, shriveled, and dull. Good quality cherries should have green stems intact.

Storing and Handling

Temperature/humidity recommendations for short-term storage of 7 days or less:
32-36 degrees F/0-2 degrees C
90-98% relative humidity

Retail display tips:
Water sprinkle: No
Top ice: No

Ethylene production/sensitivities:
Produces ethylene: Yes–very low
Sensitive to ethylene exposure: No

Storing tips:
Maintain high humidity while storing cherries. Keep separated from foods with strong odors.

Handling tips:
Cherries bruise easily; handle with care.

Nutrition*

Serving Size 1 cup Cherries (140g)

Amount Per Serving	% Daily Value
Calories 90	
Calories from Fat 5	
Total Fat 0g	0%
Saturated Fat	(Not Available)
Cholesterol 0mg	0%
Sodium 0mg	0%
Total Carbohydrate 23g	8%
Dietary Fiber 3g	12%
Sugars 20g	
Protein 2g	
Vitamin A	2%
Vitamin C	15%
Calcium	2%
Iron	2%

*These values are based on the proposal published by FDA in the Federal Register of March 20, 2002. While PMA believes that use of these data should not result in FDA regulatory action, such a result can never be assured. Consultation with Company counsel is suggested before the data is used in conjunction with the marketing of specific products.

PMA

Produce Marketing Association

Troubleshooting

Pitted skin:
Pitting is the result of damage caused by rough handling. To prevent pitting, keep handling to a minimum and do not dump cherries from shipping containers.

Shriveling; dry, dark stem:
These are indications of moisture loss due to low humidity. For best quality, maintain humidity level of 90-98% during storage.

Loss of flavor; dull color:
These are indications of age. For best quality, inspect cherries carefully upon arrival and use soon after receiving.

Hard, light-colored cherries with dry, acidic flavor:
These are indications of immature fruit; do not use.

FIGURE 4.2 An example of the type of product information noted in *The Produce Marketing Association Fresh Produce Manual*. Courtesy of the Produce Marketing Association. © November 2002.

size, size uniformity, maturity, shape, color, texture, and freedom from disease, decay, cuts, and bruises. Other factors sometimes come into play. For example, if produce is to be shipped long distances, say from California to Chicago, a more stringent examination may be applied to ensure that it represents the grade stated at the delivery point (as opposed to the shipping point). The wise buyer is aware of the potential product change during transit and will insist that the fresh produce meet the specified grade at the time of delivery, not at the time the products were shipped from the supplier's warehouse.

Several grading terms exist in the marketplace for fresh-produce items. The most commonly used terminology for fresh fruit, vegetables, and nuts are as follows:

- Fancy—the top quality produced; represents about 1 percent of all produce
- No. 1—the bulk of the items produced; the grade that most retailers purchase
- Commercial—slightly less quality than U.S. No. 1
- No. 2—much less quality than U.S. No. 1; very superior to U.S. No. 3
- Combination—usually a mixture of U.S. No. 1 and U.S. No. 2 products
- No. 3—low-quality products just barely acceptable for packing under normal packing conditions
- Field run—ungraded products

The grades most commonly used in food service are the top grades since the low grades yield less, require additional labor for trimming, and often have a shorter shelf life. Thus, they are not generally a good buy, even at a low price.

The grade most often ordered is the high end of U.S. No. 1, or the equivalent. Clubs, hotels, and restaurants normally order the high end of U.S. No. 1 or the U.S. Fancy grade; the low end of U.S. No. 1 usually is reserved for supermarkets and grocery stores. The few items that fall into the lower-grade categories may not make it to market in any fresh form. Typically, they find their way to some of the food-processing plants that produce juices, jams, and generic-brand canned fruit and vegetables.

Generally, purveyors are knowledgeable about the grading system. If a buyer expresses interest in U.S. No. 1, they will know what he or she wants. However, grades are sometimes unavailable for some fresh produce because: (1) the buyer may be purchasing an item for which there is no grading standard; (2) the suppliers refuse to have an item graded; or (3) more commonly, the produce may come from a foreign country that may not carry a grade, though it must be inspected before it is allowed to enter the United States.

Since considerable problems exist regarding variation in quality due to several seasonal factors, the use of government grades in buying fresh produce typically is not the sole selection criterion. The grade is just one factor, and buyers usually do not base their purchase strictly on it. However, if they bid buy and U.S. grades are a major criterion in their specifications, they must be sure to use the appropriate grade terms.

Packers' Brands (or Equivalent)

Because branded fresh produce is not as commonplace as branded canned and frozen goods, some buyers commonly use both U.S. government grades and packers' brands when developing their fresh-produce buying procedures. Perhaps the most familiar brands are the Sunkist® brand used for citrus fruits and the Blue Goose® brand used for several high-quality fruits and vegetables.

With fresh produce, a packer's brand may also indicate that a particular packing process has been used or that a particular cleaning and cooling process has been followed in the field (see Figure 4.3). U.S. grades may or may not indicate these attributes, depending on the area of the country the products come from.

Consistency, which should be a hallmark of packers' brands, is particularly important with fresh produce. It can result in a much more predictable edible-portion (EP) cost, which is always difficult to calculate for fresh produce in the best of situations. Consistency helps to minimize the variation both between and within case packs. For example, many products, such as whole lettuce, are sold by the case. Some types of lettuce are firm and weighty; others have a lot of space between the leaves. Citrus fruit varies in juice content from one crop to the next; fruits carrying a packer's brand may be more consistent. Asparagus may be old and woody or young and pleasantly crisp, but a packer's brand usually is consistently sound.

Even though a packer may not purchase the U.S. government grading service, government inspections, which consist of random visits by an official inspector, are mandatory. In addition, if a packer wants to use a brand name, most states require that this name be registered with the state's department of agriculture.

Packers' brands exist for many varieties of fresh produce. Packers' fresh-produce brands that do not include some reference to federal grades make up only a small portion of the fresh-produce business, and they are not widely known in most parts of the country. In most situations, packers use the brand name in conjunction with the U.S. grade designation (e.g., U.S. No. 1). Beware of a stencil on the box with the designation "No. 1." This sign indicates that the product is not under continual government inspection, nor has it been graded; but, in the opinion of that packer, the product meets all U.S. requirements for U.S. No. 1 graded products.

FIGURE 4.3 A packer's brand is a useful selection factor when the buyer wants to purchase precut fresh produce. Courtesy of Fresh Western Marketing Inc.

A major problem found with packers' brands for produce is that packers sometimes put out two categories of the same brand. For example, some packers have been accused of putting the same brand name on demonstrably different qualities of the same product—and trying to imply that the qualities are the same since they both carry the same name. For instance, on a high No. 1 and on a low No. 1, the high No. 1 produce typically goes to hospitality operations, while the low No. 1 produce goes primarily to supermarkets. The possibility of a switch should disturb and alert hospitality buyers.

Product Size

Product size is a very critical selection factor for at least two reasons. First, it would be embarrassing to serve guests items of varying sizes. Second, if the products were sold by the piece rather than by weight, buyers would find it impossible to achieve effective cost control if varying sizes are served.

Many buyers, especially novices, are unaware of the many fresh-produce sizes available. Indicating this selection factor is sometimes overlooked when buyers purchase fruits and vegetables. Buyers should not merely order a box of lemons; they should indicate the size of the lemon wanted by specifying the count per box—the lower the count, the larger the lemon. A great many types and varieties of fresh fruits are sold this way.

Still another common way to indicate produce size is to indicate the desired number of pieces per layer of the "lug," or box. This is particularly true when you buy a lug of tomatoes. For example, a "4 by 5" preference indicates a specific tomato size: the lug has layers of tomatoes with 4 on one side and 5 on the other, or 20 tomatoes per layer.

Some products carry a unique nomenclature. For instance, a packer may classify onions into four product sizes: pre-pack, large medium, jumbo, and colossal.

Furthermore, buyers may find that some produce sold in their area is sized according to the approximate number of pieces per pound. For example, a "3 to 1" item size indicates that there are approximately three items per pound. Moreover, agribusiness has advanced to the point where several sizes exist within narrow product lines; for instance, Idaho potatoes come in 12 sizes, ranging from 4 to 18 ounces each.

Size of Container

Foodservice buyers prefer to purchase the size of container that is consistent with the needs of the operation. Generally, most fresh-produce items are packed in at least two, sometimes more, container sizes, and buyers usually find one of these sizes more suitable for their operation's needs. For instance, if buyers do not require a particular type of produce in large quantities, it would be appropriate for them to purchase it in small containers so that waste and spoilage are minimized.

As with many other types of foods and beverages, the fresh-produce product line has uniform container sizes. For instance, avocados typically come in a "flat," which is one layer of product, or a "lug," which consists of two layers of product. Some suppliers may be willing to "break" a case. However, chances are that you will need to pay for this added supplier service.

Type of Packaging Material

Several standardized types of packaging materials are available in the fresh-produce product line. However, these materials vary quite a bit in terms of quality and price. The AP price variation for some fresh produce is related solely to the quality of packaging; less expensive merchandise may be scantily packed, and more expensive products may be wrapped in high-quality packaging. Certain assurances come with the type of packaging material used. For instance, high-quality fiberboard costs more than thin brown-paper wrappings. With the former, the merchandise will be protected, whereas with the latter, you will probably lose some of it to damage during shipping and handling.

Packaging Procedure

Packing generally takes two forms: layered merchandise and slab-packed merchandise. With the layered style, the product is arranged nicely, usually between sheets of paper or cardboard. With the slab-packed style, the product is randomly placed into a container with no additional packaging. If it is important to preserve the edible yield of an item, buyers should probably purchase better packaging arrangements, or the savings associated with slab packing will be illusory.

For some products, buyers can request each item to be individually wrapped and layered in the case. They may also be able to purchase products that are layered in a "cell pack," which is a cardboard or plastic sheet with depressions in it; the items can sit in the depressions and not touch each other. Some apples are packed this way. These procedures are expensive, but if this process is necessary to preserve the appearance of the apples, probably no better way exists for buyers to accomplish their goal than to insist that cell packs are used.

Minimum Weight Per Case

Since so much fresh produce is purchased by the case, or by the container, the case weight can vary considerably. This has implications on the bid price because the weight will vary from case to case. Therefore, buyers should indicate the minimum acceptable weight during the bidding process and write it in the specifications. Also, fresh produce tends to "shrink," or dehydrate, while in transit. Thus, specifying a minimum weight enables buyers to receive the appropriate amount while simultaneously giving suppliers some shrink allowance. In some cases, buyers might also indicate a "decay allowance" on a fresh-produce product specification. For example, when buyers purchase ripe plums, slab packed, a few unusable ones are bound to be in the lot. Buyers and suppliers should agree on the number of bad pieces that will be acceptable.

Another dimension of this selection factor is the possibility that buyers might wish to indicate a weight range per case of produce they order. This gives suppliers a bit more flexibility and ensures that, once in a while, buyers might receive more weight, and, therefore, more usable servings of product than they would if they indicated only a minimum weight required on the specification.

Product Yield

Some of the fresh produce buyers purchase will be subject to a certain amount of "trim," or unavoidable waste. For example, when purchasing whole, fresh turnips, buyers should note on the product specification the minimum edible yield expected or, alternatively, the maximum amount of acceptable trim loss.

Point of Origin

If buyers shop around, they must be careful that they understand the differences in quality, texture, appearance, and taste that will accompany products from different areas of the world. If buyers do business with only one or two purveyors, generally these suppliers will try to provide fresh produce whose characteristics are somewhat consistent. But this may not be the case if buyers wish to deal with multiple purveyors.

Another dimension of this selection factor is the problem associated with noting on the menu the point of origin for certain menu offerings. For instance, if a hospitality operation indicates that it serves Idaho potatoes, the operator must be certain that its potatoes actually come from Idaho or it will be in violation of truth-in-menu legislation.

Color

Generally, when buyers specify the type or variety of merchandise desired, that in itself will indicate color. However, with the growing number of varieties of fruits and vegetables available, buyers may need to specify the color preferred. For example, peppers come in many colors, including green, red, purple, yellow, and orange.

Product Form

Buyers can purchase fresh produce in many forms. From the whole "fresh-off-the-vine" items, to the ready-to-serve products, to anything in between, buyers must determine which product best meets the requirements of their operations. When they purchase an item in anything other than its original form, varying degrees of value have been added: different degrees of economic "form" have given value to the item. These value-added products are more expensive, but in the long run may be more economical than whole products that require considerable labor and handling before service.[5] Today, many operators prefer purchasing value-added, "precut" fresh produce that has been subjected to additional cleaning, chopping, and so forth. For example, buyers can purchase peeled and sliced onions, potatoes, and similar items.

Degree of Ripeness

Buyers can purchase some fresh produce at varying stages of ripeness. Usually, buyers will purchase mature, fully ripened produce. However, immature, green produce can also be found. For some items, such as bananas and tomatoes, several stages of ripening can be ordered; these options may be quite suitable for an operation that wants to ripen some of its own fresh produce so that it is at its peak of quality when served.

Ripening Process Used

Buyers should be aware of the types of ripening processes. Some produce is naturally ripened on the plant. These items are a bit more expensive because of the difficulty of handling and the loss producers experience; they often end up with some rotting merchandise in their fields. The taste of the item, though, may be so desirable that buyers are willing to pay for it.

Some produce is ripened in a "ripening room." In this situation, the produce is picked when it is green and then is placed in a room, train car, or truck. Ethylene gas is then introduced into the "room." When some types of produce ripen naturally, they emit ethylene gas, so the introduction of this gas into the "room" speeds up the ripening process. Unfortunately, if the process is hastened too much, the fresh-produce item will not mature properly. Bananas, for example, will turn bright yellow, but the fruit under the skin will not have kept pace (i.e., it will not be very flavorful because it has not ripened fully even though the skin color suggests that it has).

Preservation Method

Fresh produce is not considered a potentially hazardous food; consequently, the federal government does not require any type of preservation. Nonetheless, storage conditions throughout the fresh-produce channel of distribution affect culinary quality and availability of produce. These conditions may also create unsanitary conditions that can cause a food-borne illness outbreak. However, this problem is typically due to contaminants present in the soil or water. The food itself is not usually the issue.

Some fresh produce items are refrigerated, while others can be left unrefrigerated. Buyers must note on their specification their refrigeration requirement, if any. Refrigeration is unnecessary for some items, such as bananas, potatoes, and most onions. However, they still should be kept in a cool environment so that they do not overripen or rot.

Refrigeration is expensive, so operations expect to pay more for refrigerated fresh produce. But unrefrigerated or uncooled product can rapidly deteriorate. As a result, the EP cost for such produce will usually be much greater than the EP cost associated with refrigerated produce.

If buyers are purchasing precut, convenience fresh produce, refrigeration is mandatory. The supplier must preserve the appropriate "pulp" temperature, otherwise, the produce will quickly deteriorate. For instance, chopped salad greens should be delivered at a pulp temperature of approximately 34°F to 36°F in order to preserve their culinary quality.

Waxing is another method used to preserve some produce. The wax prevents moisture loss and also contributes to the appearance of the produce. Mother Nature uses this process for such foods as peppers and cucumbers. Producers are also allowed to apply wax to some items, especially when they remove the natural wax when cleaning the product. Some items are traditionally waxed, and operators will receive them in this state unless they specify otherwise. Fresh-produce items most likely to be waxed by producers are apples, avocados, bell peppers, cantaloupes, cucumbers, eggplant, grapefruit, lemons, limes, melons, oranges, parsnips, passion fruit, peaches, pineapples, pumpkins, rutabagas, squash, sweet potatoes, tomatoes, and turnips.

Some fresh produce is preserved in controlled-atmosphere storage. The produce is put into a room, and the room then is sealed. Oxygen is removed, and a variety of other gases are introduced. The lack of oxygen reduces the rate of respiration, hence retarding spoilage. The produce held in atmosphere-controlled chambers will remain as is for a considerable period of time. Unfortunately, when this produce is removed from this environment, it deteriorates very rapidly. Generally, suppliers do not sell this type of merchandise to hospitality operators; rather, they sell it to the retail trade. But when buyers shop around, they should note on their specification that they prefer merchandise that has not been stored in a controlled environment.

Some of the fresh produce that comes to market has been chemically treated in order to preserve its shelf life and palatability. If buyers do not want this type of merchandise, they can opt to purchase fresh produce that has been "organically" grown—that is, grown without the use of synthetic chemicals and fertilizers.[6]

Organic food products have become so popular with consumers that in October of 2001, the USDA set up stringent guidelines and standards for products to be called "organic."[7] You can review these guidelines at www.ams.usda.gov/nop.

Buyers can also purchase fresh produce that has been grown in nutrient-rich water instead of chemically treated soil. This "hydroponic" fresh produce is especially popular with fine-dining establishments because they can grow it in the foodservice operation and serve it almost immediately after harvest. If, however, buyers purchase organic or hydroponic produce, they can expect to pay a premium AP price.

Trusting the Supplier

Specifying U.S. grades, packers' brands, and one particular supplier can make you a "house account" of the highest order, which may or may not be in accord with company policy. However, buyers building a trusting relationship with their suppliers may be the key to the success of their operations. This is because suppliers are linked to several crucial factors associated with operating an establishment. These may include: delivery schedules, seasonal changes, weather factors, the suppliers' buying capabilities, the transportation and storage facilities, and the speed with which the suppliers rotate products.

With fresh produce, and to a lesser extent with processed items, suppliers must assure buyers that they will get the quality they need and want, and that this will be as good on the table as it was in the field. If buyers do not work closely with a fresh produce supplier, they take some risks. For instance, just because you specify Sunkist and U.S. No. 1, there is no guarantee that you will receive them. The box might note U.S. No. 1, but it can easily have been repacked with lower-quality product.

Additionally, fresh-produce buying does not lend itself easily to bid buying. Buyers have many points to consider, and, in some parts of the United States, there are few suppliers to bid for their business. Consequently, this is an area in which they may wish to rely on the suppliers' highly developed expertise.

PURCHASING FRESH PRODUCE

The first step in purchasing fresh produce is to obtain *The PMA Fresh Produce Manual*. This unique publication contains very detailed specifications for many fresh-produce items. It also includes information on receiving, storing, and handling techniques.

The next step is to decide on the exact type of produce and quality wanted. Buyers may not be the ones to make this decision, unless they are user-buyers or owner-user-buyers. Once the decision on the type and quality of produce to be ordered has been made, the buyer should prepare specifications for each item. These specifications should be as complete as necessary and should include all pertinent information, especially if the buyer intends to engage in bid buying. (See the guidelines noted in Chapter 1. For examples of fresh-produce product specifications and a product specification outline, see Figures 4.4 and 4.5, respectively.)

Red Delicious apples
Used for fruit plate item
U.S. Fancy
Washington State
72 count
30- to 42-pound crate
Moisture-proof fiberboard
Layered arrangement, cell carton
Whole apples
Fresh, refrigerated
Fully ripened

Cauliflower, white
Used for side dish for all entrées
U.S. No. 1 (high)
12 count
18- to 25-pound carton
Moisture-proof fiberboard
Loose pack (slab pack)
Pretrimmed heads
Fresh, refrigerated
Fully ripened

Sweet Spanish, yellow globe onion
Used for onion rings
U.S. No. 1 (high)
Jumbo size
50-pound plastic mesh bag
Whole onions
Fresh, unrefrigerated
Fully ripened

Iceberg lettuce
Used for tossed salad
U.S. No. 1 (high)
10-pound poly bag
Loose pack (slab pack)
Chopped lettuce
Fresh, refrigerated
Fully ripened

FIGURE 4.4 An example of fresh-produce product specifications.

Intended use:

Exact name:

U.S. grade (or equivalent):

Packer's brand name (or equivalent):

Product size:

Size of container:

Type of packaging material:

Packaging procedure:

Minimum weight per case:

Product yield:

Point of origin:

Color:

Product form:

Degree of ripeness:

Ripening process used:

Preservation method:

FIGURE 4.5 An example of a product specification outline for fresh produce.

After buyers prepare the specifications, the next step is to consider the suppliers likely to satisfy their needs. Here, again, they can become a house account, use bid buying, or settle on some procedure in between. (Note that bid buying is not as prevalent in the fresh-produce trade as it is for processed foods and nonfoods.)

Over the years, several fresh-produce buying groups and trade associations have evolved. This, in turn, has made such products more readily available in the local markets. Buyers may also find several suppliers handling a few fresh-produce items as a sideline to their regular business of canned goods, frozen foods, and various nonfood items. Generally, though, it is difficult to find more than one or two consistently capable full-line, fresh-produce suppliers. Because of this, buyers may find it difficult to engage in bid buying.

An added attraction in the fresh-produce area is the independent farmers. These small businesspersons may occasionally want to sell buyers "farm-fresh produce." Some may sell on the roadside. Others may allow customers to come in and "pick their own." Still others may gather together one or two days a week in what is referred to as a "farmers' market." Over the years, farmers' markets have become more visible, numerous, and popular.

In some instances, the small, local farmer will be a wise choice. For example, as part of a college project, some students were required to compare an independent farmer's AP prices and supplier services a restaurant purchased to the same products and supplier services offered by other fresh-produce suppliers. The results of the study indicated that the restaurant's management was paying the farmer about 7 percent more per month. But it was the opinion of the class that the quality of the farmer's fresh produce was far superior. In addition, the farmer delivered daily. Unfortunately, he could not supply all the needs of this restaurant. But, for what he did supply, the class judged it to be the best overall value.

Reliable, independent farmers are, however, hard to find. Some of these "farmers" purchase their products from supermarket suppliers and then give the impression that the produce is home-grown. Another problem that might concern buyers is that neither the state or federal government is likely to inspect the farmer's facilities and products.

Adventurous operators might consider developing their own fresh produce gardens. Lately, several restaurant chefs have taken this unique step.[8] Gardens located on the premises ensure freshness. They can also be excellent marketing and promotional tools used to attract guests.

RECEIVING FRESH PRODUCE

Taking delivery of the correct amount of the specified quality is a major challenge. Some receivers may merely look at the box of lettuce, read "24 heads," and take it for granted that 24 heads of lettuce are in the carton. A skillful receiver resists the temptation to examine only the printing on the containers and cartons and to look at only the top layer of merchandise.

Before accepting a produce shipment, the receiver should conduct a visual inspection of the top layer, and check the weight of the entire carton. This inspection provides a quick and accurate idea of the quantity and quality of the shipment. The receiver should also conduct a random sampling of a proportion of containers. Here, a receiver carefully unpacks a box of produce to check the count, size, and quality throughout the carton. A good place for this inspection is in refrigerated quarters, such as a walk-in refrigerator, so that the produce does not become too warm, thus decreasing its shelf life.

Although receiving fresh produce can be a daunting task, we are not suggesting that receivers break open every carton to see whether it really contains full count. This undertaking can cause extensive damage to fragile fruit and vegetables, particularly if they are packaged in special protective films designed to extend their shelf life (see

Figures 4.6 and 4.7 for signs of acceptable and unacceptable quality in some fresh-produce items). However, buyers can make the task less stressful by establishing a partner-like relationship with vendors. If buyers have concerns about trusting their suppliers and receivers completely, it will be in their best interest to continually review their receiving practices very carefully. At the very least, they should verify that the produce is acceptable by checking to see that the quality is equal throughout and to make certain that no repacking has occurred.

After checking quality and quantity, receivers should check the prices and complete the appropriate accounting documents.

STORING FRESH PRODUCE

To prevent rapid quality degradation and loss of some nutritional value, it is imperative that hospitality operations properly store fresh fruits and vegetables. If buyers can arrange for frequent deliveries from their fresh-produce supplier and move the produce rapidly through production, they can be a little more flexible with their storage duties. But because many suppliers are not equipped to provide daily delivery, operators should plan for a suitable storage facility.

Fresh produce must be stored immediately at the proper temperature and humidity (see Figure 4.8 for requirements for some vegetables). Large foodservice operations usually have a separate cool area or refrigerator for these items. But any cool temperature is better than none. Hospitality operations must avoid all delays: most fresh produce deteriorates considerably when it is left at room temperature. In fact, most fully ripened produce becomes inedible quite quickly if it is held in the wrong storage environment. For example, fresh corn loses about 50 percent of its sugar in the first 24 hours after it is picked; proper refrigeration, however, can slow this deterioration.

To extend the shelf life of fresh produce, buyers must research the best possible storage environment for each fruit and vegetable. This includes knowing how the produce is packaged since some packages are designed to extend shelf life. Some fruits and vegetables, packed in a box, carton, or in cello wrap, are best stored by merely placing them in the refrigerator. On the other hand, some produce arrives in crates that are not designed to extend shelf life. Celery, for example, usually arrives in a crate and rapidly loses moisture and becomes limp if not repacked. Cello bags or plastic,

FIGURE 4.6 Signs of acceptable and unacceptable quality in some fresh fruit items. Reprinted with permission from *Applied Foodservice Sanitation Certification Coursebook,* 4th ed. Copyright 1992 by the National Restaurant Association Educational Foundation. All rights reserved.

	SIGNS OF GOOD QUALITY	SIGNS OF BAD QUALITY, SPOILAGE
Apples	Firmness; crispness; bright color	Softness; bruises. Irregularly shaped brown or tan areas do not usually affect quality.
Apricots	Bright, uniform color; plumpness	Dull color; shriveled appearance
Bananas	Firmness; brightness of color	Grayish or dull appearance (indicates exposure to cold and inability to ripen properly)
Blueberries	Dark blue color with silvery bloom	Moist berries
Cantaloupes (Muskmelons)	Stem should be gone; netting or veining should be coarse; skin should be yellow-gray or pale yellow	Bright yellow color; mold; large bruises
Cherries	Very dark color; plumpness	Dry stems; soft flesh; gray mold
Cranberries	Plumpness; firmness. Ripe cranberries should bounce.	Leaky berries
Grapefruit	Should be heavy for its size	Soft areas; dull color
Grapes	Should be firmly attached to stems. Bright color and plumpness are good signs.	Drying stems; leaking berries
Honeydew melon	Soft skin; faint aroma; yellowish white to creamy rind color	White or greenish color; bruises or water-soaked areas; cuts or punctures in rind
Lemons	Firmness; heaviness. Should have rich yellow color	Dull color; shriveled skin
Limes	Glossy skin; heavy weight	Dry skin; molds
Oranges	Firmness; heaviness; bright color	Dry skin; spongy texture; blue mold
Peaches	Slightly soft flesh	A pale tan spot (indicates beginning of decay); very hard or very soft flesh
Pears	Firmness	Dull skin; shriveling; spots on the sides
Pineapples	"Spike" at top should separate easily from flesh	Mold; large bruises; unpleasant odor; brown leaves
Plums	Fairly firm to slightly soft flesh	Leaking; brownish discoloration
Raspberries, Boysenberries	Stem caps should be absent; flesh should be plump and tender	Mushiness; wet spots on containers (sign of possible decay of berries)
Strawberries	Stem cap should be attached; berries should have rich red color	Gray mold; large uncolored areas
Tangerines	Bright orange or deep yellow color; loose skin	Punctured skin; mold
Watermelon	Smooth surface; creamy underside; bright red flesh	Stringy or mealy flesh (spoilage difficult to see on outside)

	SIGNS OF GOOD QUALITY	SIGNS OF POOR QUALITY, SPOILAGE
Artichokes	Plumpness; green scales; clinging leaves	Brown scales; grayish-black discoloration; mold
Asparagus	Closed tips; round spears	Spread-out tips; spears with ridges; spears that are not round
Beans (snap)	Firm, crisp pods	Extensive discoloration; tough pods
Beets	Firmness; roundness; deep red color	Gray mold; wilting; flabbiness
Brussels sprouts	Bright color; tight-fitting leaves	Loose, yellow-green outer leaves; ragged leaves (may indicate worm damage)
Cabbage	Firmness; heaviness for size	Wilted or decayed outer leaves. (Leaves should not separate easily from base.)
Carrots	Smoothness; firmness	Soft spots
Cauliflower	Clean, white curd; bright green leaves	Speckled curd; severe wilting; loose flower clusters
Celery	Firmness; crispness; smooth stems	Flabby leaves; brown-black interior discoloration
Cucumber	Green color; firmness	Yellowish color; softness
Eggplant	Uniform, dark purple color	Softness; irregular dark brown spots
Greens	Tender leaves free of blemishes	Yellow-green leaves; evidence of insect decay
Lettuce	Crisp leaves; bright color	Tip burn on edges of leaves. (Slight discoloration of outer leaves is not harmful.)
Mushrooms	White, creamy, or tan color on tops of caps	Dark color on underside of cap; withering veil
Onions	Hardness; firmness; small necks; papery outer scales	Wet or soft necks
Onions (green)	Crisp, green tops; white portion two to three inches in length	Yellowing; wilting
Peppers (green)	Glossy appearance; dark green color	Thin walls; cuts, punctures
Potatoes	Firmness; relative smoothness	Green rot or mold; large cuts; sprouts

FIGURE 4.7 Signs of acceptable and unacceptable quality in some fresh vegetable items. Reprinted with permission from *Applied Foodservice Sanitation Certification Coursebook*, 4th ed., Copyright 1992 by the National Restaurant Association Educational Foundation. All rights reserved.

	SIGNS OF GOOD QUALITY	SIGNS OF POOR QUALITY, SPOILAGE
Radishes	Plumpness; roundness; red color	Yellowing of tops (sign of aging); softness
Squash (summer)	Glossy skin	Dull appearance; tough surface
Squash (winter)	Hard rind	Mold; softness
Sweet potatoes	Bright skins	Wetness; shriveling; sunken and discolored areas on sides of potato. (Sweet potatoes are extremely susceptible to decay.)
Tomatoes	Smoothness; redness. (Tomatoes that are pink or slightly green will ripen in a warm place.)	Bruises; deep cracks around the stem scar
Watercress	Crispness; bright green color	Yellowing, wilting, decaying of leaves

FIGURE 4.7 (Continued)

USDA RECOMMENDED STORAGE REQUIREMENTS FOR VEGETABLES			
COMMODITY	STORAGE TEMPERATURE	RELATIVE HUMIDITY	MAXIMUM TOTAL STORAGE PERIOD*
Asparagus	32–36°F	95%	2–3 weeks
Broccoli	32–35	90–95	10–14 days
Carrots (topped)	32–35	90–95	4–5 months
Cauliflower	32–35	90–95	2–4 weeks
Celery	32–35	90–95	2–3 months
Lettuce	32–34	95	2–3 weeks
Onions, green (scallions)	32–35	90–95	—

*This maximum storage includes commercial storage of produce before it is delivered to your loading dock. If you intend holding fruits and vegetables in your walk-in for any length of time, consult your produce house or distributor to determine how long you can safely store produce that has already been in storage.

FIGURE 4.8 Recommended storage requirements for some fresh vegetables. Courtesy of *Restaurants & Institutions* magazine, a Cahners publication.

reusable, see-through tubs are excellent for repacking. Contrary to popular opinion, produce should not be washed before it is stored. Moisture enhances the growth of soft-rot microorganisms and invariably decreases the shelf life of the fruit or vegetable. Washing may also remove protective wax coating designed to extend the shelf life of vegetables such as green peppers and cucumbers. Products should be washed, if necessary, when they enter the production cycle.

Operators also should expect to extend fresh-produce shelf life if they practice proper fresh-produce handling techniques. The techniques are quite simple. The general rule is to handle the merchandise only when it is absolutely necessary. Employees should not pick it up, move it around, bend it, or bounce it because this will cause unnecessary bruising that will manifest itself in excessive spoilage and waste.

ISSUING FRESH PRODUCE

Fresh-produce purchases often bypass the central-storage facility and go directly to the food production department. If employees first move the fresh produce to a central-storage facility and issue it later to the food production department, they may want to issue these items as ready-to-go. That is, they may consider issuing cleaned, chopped onions instead of whole onions; sliced tomatoes instead of whole tomatoes; or topped, peeled, and cut carrots instead of whole carrots (see Figure 4.9 for preparation waste of some fresh fruit and vegetables). By doing this, they may be able to effect a labor cost savings or extract better value from higher-paid cooks and chefs.

Employees should follow proper stock rotation when issuing produce (see Figures 4.8 and 4.10 for storage times for some fresh-produce items). Since these items spoil quickly, make sure that the requisitioner takes no more than necessary. It may be wise to ask him or her to note the in-process inventory before asking for more stock.

IN-PROCESS INVENTORIES

Buyers will discover that control of purchasing, receiving, storing, and issuing fresh produce is easier than control of produce production and service. For instance, a great deal of supervision is required to ensure that salad greens do not sit out at room temperature too long.

Generally, once the fresh produce is issued to a user, buyers are relieved of their responsibility. But since this book is directed toward managers who need a more panoramic view of the hospitality industry, we offer commentary on how the production and service staffs use these purchased items.

	RAW WEIGHT	APPROXIMATE EDIBLE YIELD	APPROXIMATE WASTE
Apples	1 lb.	13 oz.	19%
Apricots	1 lb.	12 oz.	25%
Asparagus	1 lb.	9 oz.	44%
Avocado	1 lb.	12 oz.	25%
Bananas	1 lb.	13 oz.	19%
Beans, green	1 lb.	14 oz.	13%
Broccoli, whole head	1 lb.	10 oz.	38%
Brussels sprouts	1 lb.	16 oz.	0%
Cabbage	1 lb.	13 oz.	19%
Cantaloupe	1 lb.	11 oz.	31%
Carrots, no tops	1 lb.	12 oz.	25%
Cauliflower, trimmed	1 lb.	16 oz.	0%
Celery, whole stalk	1 lb.	12 oz.	25%
Cranberries	1 lb.	16 oz.	0%
Cucumbers	1 lb.	14 oz.	13%
Eggplant, whole	1 lb.	13 oz.	19%
Grapefruit	1 lb.	11 oz.	31%
Lemons	1 lb.	11 oz.	31%
Limes	1 lb.	11 oz.	31%
Lettuce, untrimmed	1 lb.	12 oz.	25%
Melon, honeydew	1 lb.	11 oz.	31%
Mushrooms	1 lb.	16 oz.	0%
Onions	1 lb.	14 oz.	13%
Oranges	1 lb.	11 oz.	31%
Pears	1 lb.	13 oz.	19%
Peppers, green	1 lb.	13 oz.	19%
Potatoes, sweet	1 lb.	11 oz.	31%
Potatoes, white	1 lb.	12 oz.	25%
Squash, summer	1 lb.	14 oz.	13%
Strawberries	1 lb.	14 oz.	13%
Tangerines	1 lb.	11 oz.	31%
Tomatoes	1 lb.	14 oz.	13%
Turnips	1 lb.	12 oz.	25%

FIGURE 4.9 The preparation waste of some fresh-produce items.

STORAGE TIMES FOR FRUITS AND VEGETABLES

APPLES, Fresh	Store in fruit or vegetable box three weeks to a month. Inspect daily to remove rotten fruit so that the balance will not be contaminated. Watch for blue mold or black rot.	CUCUMBERS	These are not sturdy and should be used within a week.
		EGGPLANT	Should not remain in storage more than a week.
APRICOTS	Easily stored for one to two weeks.	GARLIC	Can be kept for about two months at temperatures from 55 to 65 degrees. In a vegetable cooler at temperatures 32 to 36 degrees, garlic will last four months.
ASPARAGUS	Can be kept for two weeks but must be crated with the heels packed in moss.		
AVOCADO	May be kept in refrigerator one week after ripening.	GRAPEFRUIT	Will last six weeks at 32 to 36 degrees.
BANANAS	May be kept at 50 to 60 degrees and used within two to three days after ripening. Do not store in the cooler at any time.	GRAPES	White seedless or red Tokay grapes will keep for four weeks. Red Emperor, obtainable in late fall, will keep for two months.
BERRIES	All fresh berries can be kept for a week to 10 days. However, it is recommended that these be used as quickly as possible for best flavor.	KALE	In temperatures from 32 to 36 degrees, kale will remain in good condition for three weeks.
BROCCOLI	Can be stored for 8 to 10 days.	LEMONS	May be kept from one to two months at 50 to 60 degrees.
CABBAGE	EARLY VARIETY will keep about two weeks. LATE VARIETY is much sturdier—will last two months.	LETTUCE, Iceberg	If in good condition and inspected regularly, iceberg lettuce may be kept for four weeks. The leaves should not be removed until the lettuce is to be used, unless they have begun to rot. However, to obtain a maximum quality, lettuce should be used as soon as possible after arrival.
CANTALOUPE	Inspect daily for ripeness. When ripe, may be held in the cooler for one week.		
CARROTS	If in good condition, they may be kept in the storeroom for a few days. Under refrigeration, they will last three months.	LIMES	Will not last in storage over two weeks.
CAULIFLOWER	May be kept for two weeks if the leaves are not cut away. After the leaves are removed, it deteriorates rapidly.	MELONS	May be stored a maximum of three weeks. However, it is recommended that they be used as soon as the proper degree of softness is achieved.
CELERY	Should not be kept longer than a few days. If it is wilted, placing in water will freshen it.	MUSHROOMS	Fresh, should not be kept more than one or two days.
CORN	Corn is one of the most sensitive of vegetables and should be used within 24 hours after arrival.	ONIONS, Green	If kept under refrigeration, they will last a week or 10 days.
CRANBERRIES	May be stored in a vegetable box for as long as two months.	ONIONS, Yellow	If stored in a cool, dry place, unrefrigerated, they will last three months.

FIGURE 4.10 Storage times for some fresh-produce items. Reprinted from *Lodging* magazine.

ORANGES	Should be used within a week if possible. If necessary, they may be held in a reasonably good condition for a month or six weeks.	POTATOES, Sweet	These do not have the staying power of white potatoes. They require a cool place, 50 to 60 degrees, and should not be kept for more than a week. They will last for three or four weeks if the air is extremely dry.
PARSNIPS	Can be stored two to three months at 32 to 36 degrees.		
PARSLEY	A week is about the time limit for parsley. Keep it well iced.	PUMPKIN	Can last for a month at temperatures from 50 to 60 degrees. However, it is better to buy the canned variety.
PEACHES	Most varieties will last about a week; the yellow cling variety about two weeks. Peaches must be inspected and sorted each day.		
		RADISHES	Should not be kept longer than a week, and it is wiser to use them within a few days. The leaves should be removed as soon as possible.
PEARS	Summer or Bartlett variety—before ripening, they may be kept three weeks at 65 to 75 degrees. After ripening, they must be refrigerated and used within a few days. They require gentle handling to prevent bruising and must be sorted every 5 days. Bosc or Comice variety—may be kept six weeks before ripening if sorted weekly. After ripening, must be used within a few days. Winter Anjou or Winter Nelis variety—will keep eight to ten weeks before ripening.		
		RHUBARB	May be kept for a week, but since it loses flavor after a short time, it should be used as soon as possible.
		ROMAINE	Will generally last about 10 days.
		SPINACH	Must be properly iced to last any time at all. Even then, one week is its time limit.
		SQUASH, Summer	Will last only two weeks
		SQUASH, Winter	Can be held at 50 to 60 degrees for three months
		STRAWBERRIES	Must not be kept longer than 2 days and require a temperature from 32 to 36 degrees.
PEPPERS	May be held for about three weeks.		
PINEAPPLES	May be kept as long as two weeks on a ripening table at temperatures from 65 to 75 degrees. Once ripe, they should be used within 2 or 3 days.	TOMATOES	Should not be kept over a week after ripening. They require daily sorting for ripeness.
		TURNIPS	White—they will keep about 10 days or two weeks without refrigeration. Under light refrigeration, they will last three months.
PLUMS	Green Gage or red—will keep for two weeks but after ripening must be used within a few days.		
POTATOES, White	May last four months in cool, dry, well-ventilated place that is refrigerated. New potatoes, however, should not be kept for more than five or six weeks.	WATERMELON	May be held a week or 10 days, but no longer.

FIGURE 4.10 (Continued)

We have known several restaurants that undertook considerable expense to purchase fresh produce efficiently, only to see the savings disappear because an inexperienced manager did not, or could not, supervise its use. Many operations protect the AP price, or backdoor cost, very well, but the EP cost, or front door cost, does not often enjoy equal consideration. Figure 4.9 shows that considerable unavoidable loss is associated with fresh produce; the chances for excessive loss increase if proper supervision is absent.

KEY WORDS AND CONCEPTS

Agricultural Marketing Service (AMS)	Grading factors	Pulp temperature
Approximate waste percentages	Hydroponic fresh produce	Purchasing, receiving, storing, and issuing fresh produce
As-purchased price (AP price)	Independent farmer	Ready-to-serve produce
Broken case	In-process inventories	Reference books that can be used when preparing specifications
Cell pack	Intended use	
Color	Layered packaging	Ripening process used
Controlled atmosphere storage	Lug	Ripening room
	Minimum weight per case	Shelf life
Decay allowance	Organic fresh produce	Shrink allowance
Degree of ripeness	Perishable Agricultural Commodities Act (PACA)	Size of container
Difference between two levels of U.S. No. 1	Packaging material	Slab-packed
Edible-portion cost (EP cost)	Packaging procedure	Storage requirements for fresh produce
Ethylene gas	Packers' brands	Trim
Exact name	Point of origin	Trusting the supplier
Farmers' market	Precut fresh produce	Type of product
Field run	Preservation method	United States Department of Agriculture (USDA)
Flat	Produce Marketing Association (PMA)	U.S. grades
Form value	Product form	Value-added products
Genetically altered fresh produce	Product size	Variety of products
	Product yield	

REFERENCES

1. Patt Patterson, "A Hard Sell: Buying Produce Is No Day in the Park," *Nation's Restaurant News*, January 25, 1993, p. 23.

2. Thomas M. Burton, "Buying Fine Produce for Finicky Chefs Is No Bowl of Cherries," *The Wall Street Journal*, August 6, 1991, p. A1.

3. Kathleen Deveny, "America's Heartland Acquires Global Tastes," *The Wall Street Journal*, October 11, 1995, p. B1.

4. Brian O'Reilly, "Reaping a Biotech Blunder," *Fortune*, February 19, 2001, p. 156. See also: Scott McMurray, "New Calgene Tomato Might Have Tasted Just as Good Without Genetic Alteration," *The Wall Street Journal*, January 12, 1993, p. B1; Anonymous, "European Ruling Backs Banning of Biotech Crops," *The Wall Street Journal—Eastern Edition*, September 10, 2003, p. A22; Rick Charnes, "Genetically Altered Food: Myths and

Realities," *EarthSave International*, Retrieved on September 16, 2003 (Available http://www.earthsave.org/ge.htm).

5. Meg Major, "Produce Perspectives 2000," *Supermarket Business*, October 15, 2000, pp. 89–94. See also: Food Spectrum, "Retail Prepared Refrigerated Foods: The Market and Technologies Mini Study on Value-added Produce," Retrieved on September 16, 2003 (Available http://www.foodspectrum.com/value_added_produce.htm); Patt Patterson, "Fresh-Cut vs. Raw Produce: Where's the Value?" *Nation's Restaurant News*, September 13, 1993, p. 95; Elizabeth Schneider, "Veggies in Volume: Beating the Buffet and Banquet Blahs," *Food Arts*, November 1991, p. 76.

6. Betsy Block, "What You Need to Know About Organic Food," *Boston Globe*, March 29, 2000, p. E.1. See also: Charles Thienpont, "More Growers Plant Organic Crops—See Prices 50% Below Last Year," *FoodService Director*, April 15, 1990, p. 58; Diane Welland, "Chefs Consider Organic Produce," *Restaurants USA*, September 1991, p. 26; David Belman, "The Time Is Ripe for Organics," *Restaurants USA*, August 1995, p. 18.

7. Betsy Spethmann, "Planting the Seed," *Promo*, August 2002 15 (9), pp. 26–28. See also: Stephanie Salkin, "USDA Unveils Final Organics Rules," *ID: The Information Source for Managers & DSRS*, February 2001, 37(2), p. 20.

8. Ken Macqueen, "Kitchen Garden," *Maclean's*, July 1, 2003, 116(26/27), p. 77. See also: Amy Zuber, "On the Menu: Foodlife, Chicago," *Nation's Restaurant News*, September 4, 2000, p. 38; Kathy Blake, "New Herb Garden Spices Up Offerings at NYC's Lenox Hill Hospital," *Nation's Restaurant News*, August 10, 1998, 32(32), p. 78; Jennifer Batty, "Restaurants with Farms Start a Blooming Revolution," *Restaurants USA*, August 1992, p. 30.

QUESTIONS AND PROBLEMS

1. Name the U.S. grades for fresh produce.

2. What is the most frequently ordered grade of fresh produce in the hospitality industry?

3. A product specification for fresh lemons could include the following information:
   - (a)
   - (b)
   - (c)
   - (d)
   - (e)

4. Why does the foodservice operator normally not care for fresh produce that has been in controlled atmosphere storage?

5. Why should buyers note on their specifications for fresh produce the minimum weight per case?

6. Why is the point of origin for fresh produce very important?

7. What is an appropriate intended use for green tomatoes?

8. When would a buyer use a packer's brand in lieu of a U.S. grade when preparing a product specification for fresh tomatoes?

9. When would a buyer purchase fresh produce from a farmers' market?

10. Why might it be difficult to engage in bid buying when purchasing fresh produce? When would a buyer bid buy? Why? If possible, ask a school foodservice director to comment on your answer.

11. Outline the specific procedures a buyer would use to purchase, receive, store, and issue salad greens—lettuce, red cabbage, and carrots—and baking potatoes. Assume that these products will be used in a steak house. If possible, ask a steak house manager to comment on your answer.

12. Prepare product specifications for the products noted in Question 11.

13. Assume that you manage a cafeteria and have run out of salad greens at 7:30 P.M. on a Saturday night. What do you do? If possible, ask a cafeteria manager to comment on your answer.

14. Supplier A offers lettuce at $16.75 per case, and Supplier B offers it at $18.50 per case. The yield for Supplier A is 88 percent; for Supplier B, 94 percent. Supplier A expects a COD payment; Supplier B gives seven days' credit terms. Which supplier should a buyer purchase from? Why?

15. Why is it important to note the exact variety of fresh produce desired, instead of merely noting the type of item needed?

16. What is an appropriate intended use for U.S. No. 2 grade tomatoes?

17. What does the notation "3 to 1" indicate to a buyer?

18. Precut fresh produce usually carries an AP price that is much higher than that of raw fresh produce. Identify some of the reasons for this difference.

19. What are some of the methods fresh-produce suppliers can use to extend the shelf life of fruits and vegetables?

20. When would a buyer specify organically grown fresh produce on a product specification for fresh produce?

21. What is a "decay allowance"?

22. What critical information is missing from the following product specification for onions?
    Onions
    Used to make onion rings
    U.S. No. 1 Grade (or equivalent)
    Packed in 50-pound mesh bags

23. What is the advantage of purchasing genetically altered fresh tomatoes?

24. What is an appropriate intended use for "field run" fresh produce?

25. For which types of products is "pulp temperature" an important selection factor?

5

PROCESSED PRODUCE AND OTHER GROCERY ITEMS*

The Purpose of this Chapter

After reading this chapter, you should be able to:

- Identify management considerations surrounding the selection and procurement of processed produce and other grocery items.

- Identify the selection factors for processed produce and other grocery items, including government grades.

- Describe the process of purchasing, receiving, storing, and issuing processed produce and other grocery items.

INTRODUCTION

The purchasing procedures for convenience items, such as processed fruit and vegetables, and for other grocery items, such as spices, pastas, fats, and oils, are more routine than those required for fresh products. In general, the qualities are more predictable, and the as-purchased (AP) prices do not fluctuate so widely as those for fresh products.

To prevent the mistaken idea that this area of purchasing does not present difficulties, we must stress that purchasing processed items requires several management considerations. As is almost always the case, these considerations center on the

---

* Authored by Andrew Hale Feinstein and John M. Stefanelli.

determination of what a hospitality operation wants, what type of product is best suited for its needs, and which supplier can accommodate these needs.

MANAGEMENT CONSIDERATIONS

It is probably impractical to imagine any storeroom without a few cans of tomatoes on its shelves. Thus, the decision here is not an either/or proposition. It is more a question of which products should be fresh and which should be processed. In addition, some methods used to cook certain fruits and vegetables do not produce food that tastes substantially different from its processed counterpart. For example, a tomato sauce made with canned tomatoes may taste about the same as one made with fresh tomatoes. Finally, you can combine some fresh products with processed ones. For instance, a tomato and green bean casserole can be made with fresh tomatoes and canned or frozen beans, or with fresh beans and canned tomatoes.

Food processors process produce for many reasons in addition to preserving them. Food processors seek to smooth out seasonal fluctuations and to capture items at their peak of flavor while simultaneously adding value to the items. In doing so, food processors transfer some work from the foodservice kitchen to the food-processing plant. Thus, processing fruits and vegetables can be viewed as the procedure of extending the availability of perishable items.

One of the ironies about processed produce is that most items are processed to increase shelf life. When these items are lost due to mishandling in the foodservice operation, one of the main reasons why they were processed in the first place is defeated. Many processed products, once thawed, opened, or heated, have extremely short in-process shelf lives. In addition, reheating or reusing many of these items usually results in inferior finished products. Processed produce shares this problem with most convenience products.

Consequently, a major management decision involving processed produce items is whether to use them at all. (Hospitality operations usually have little choice for other grocery items, although some properties make their own pasta, render their own fat, and blend their own condiments.) Taking into consideration the current interest in "natural, whole foods," operators cannot take this decision lightly. Some make a point of reminding their patrons that all vegetables on their menus are cooked from the fresh state. Whether this approach has marketing value may be a matter of opinion.

Once hospitality operators realize that at least some processed produce and other grocery items will be used in preparing menu items, they face the question of which processing method to choose. For some products, they have little choice. For example, if buyers must purchase plain pasta, they must keep in mind that it is a dried product, though some fresh refrigerated and some precooked frozen pastas are available. They must also consider additional processing techniques, like pickling and other fermentation methods. Buyers almost exclusively purchase foods processed in these ways for the taste the processing imparts and not necessarily for convenience, AP price considerations, or other reasons. Also, some other preservation methods, such as adding chemical preservatives and refrigerating some soup bases, have become standard. Unless buyers specify otherwise, they will receive the product this way. The buyers' selection of a processing method, then, is affected by: (1) food quality, (2) AP price, and (3) the need for convenience. Although the standards of quality vary within each processing method, by and large, the processing method itself predetermines the taste, AP prices, and convenience.

If buyers opt for canned goods or shelf-stable products packed in aseptic packaging, they will receive the benefits of standardized packaging, longer shelf life, and less expensive storage costs. But the buyers also get a distinctive "canned" taste. For some items, such as tomato sauce, cans or aseptic packages may be the only choices. For others, such as white asparagus spears, buyers may have to settle for a can or a bottle.

When buyers choose frozen processing, they have the benefit of fresher flavor, or at least a taste as close as possible to natural flavor. Moreover, purveyors claim that only products picked at their peak of flavor are frozen. Some processed items usually are sold only in the frozen state. For example, corn on the cob and french fries are normally available only fresh or fresh-frozen.

Unfortunately, frozen-fruit and frozen-vegetable packaging is not quite as standardized as the cans used for produce packed that way. Buyers also take greater risks with frozen items: the chances of thawing and refreezing, of a freezer breakdown, and of freezer burn. The shelf life of many frozen items is not as long as that of canned and bottled items. The AP prices tend to be higher, and a higher storage cost is associated with frozen products.

One of the biggest difficulties with any frozen product is the possibility of thawing a larger amount than is needed. The excess cannot be refrozen without a considerable loss of quality. Usually, the item is then wasted entirely. But frozen products are just too costly to throw out; consequently, a hospitality operator may try to work them into the menu somehow at the risk of alienating customers.

When choosing dried products, buyers are obviously going to save on storage. In addition, if buyers care for the items properly, they will have a long shelf life. Also, since the food is lightweight and does not require refrigeration, its transportation costs remain low, which, in turn, reduces AP prices. On the other hand, the AP prices of many dried items may still be high because of the amount of time and energy used to process them. Unfortunately, buyers cannot purchase very many food items in a dried state. However, some processed items, including instant mashed potatoes, dried onion flakes, and dried spices, are usually sold only in a dried state.

Many dehydrated foods are expensive. For example, dried fruit requires very ripe fruit with a high concentration of natural sugar. These qualities are costly; however, they make such fruit particularly desirable. For example, dried pineapple rings used in making upside-down cakes probably have an AP price that exceeds that of the canned counterpart, but the taste is different—dried pineapple is extremely sweet and strong.

A major difficulty with some dried items is the need to reconstitute them. A mistake here, even a tiny one, can ruin the product. Another difficulty is the style of packaging. For instance, macaroni products come in all sorts of packaging materials and package sizes. Dried fruit is sometimes nicely layered on waxed paper and lined up neatly in a box. But it may also be slab-packed, or tossed in randomly and pushed together so that by the time buyers get it, some of it may be damaged.

In addition to deciding which processing method best suits the needs of the operation, management must make another major decision regarding processed products, which centers on the question of substitution. For example, a recipe for mixed vegetables could include some fresh product, to use leftovers; some frozen product, bought at bargain prices; and some dried product, to take advantage of the excess sweetness. But consider the problem of inertia: since these purchases do not usually represent a large percentage of the purchase dollar, few operators devote much effort to determining the least expensive recipe unless they have access to a computerized management information system (MIS). Although this area may not seem to offer a great deal of money-saving potential, some money can, nevertheless, be saved.

When buyers purchase processed food, they usually obtain what they want. They name it, and somebody will make it if the purchase volume is large enough. For instance, fats and oils can be manufactured almost according to individual specifications, but operators must pay for this service. Nevertheless, when buyers purchase these products, it is good to know that they can get what they need.

Some other management considerations involving processed food that occur intermittently are outlined below.

1.   Some buyers tend to neglect generally accepted purchasing principles when it comes to some processed produce and grocery items. This is probably because only a small amount of the total purchase dollar is involved, as the majority goes toward meat, fish, poultry, alcoholic beverages, and some desserts. For example, the temptation is strong for buyers to set the par stock for condiments and let it go at that. Manufacturers and suppliers who rely heavily on "pull strategies" for some of these items further foster this tendency. Some products, such as Heinz® ketchup

and A-1® steak sauce, that grace a dining room table seem almost traditional. To a lesser extent, other condiments, such as olives, pickles, and relishes, fall into this category.

2. The neglect mentioned earlier might also be nurtured by the cavalier attitude with which some employees approach inventory. For instance, some managers allow service personnel to bypass the normal issuing system when they need steak sauce, hot sauce, or similar condiments. In many small operations, service staff walk into the storeroom and take what they need. If a bottle or two spills or disappears, few supervisors get upset.

3. Numerous "impulse" purchases flood the market. For example, buyers can purchase devices to: drain near-empty catsup bottles, check the pressure in canned goods, and determine whether a product has been thawed and refrozen. These may be used once or twice and then tossed into the back of a drawer.

4. For one reason or another, several new products are introduced each year in grocery product lines. Of course, many food products are not really new, just new variations of existing foodstuffs. For example, buyers can find all sorts of new vegetable combinations and sauce variations. The same is true for rice and pasta concoctions. Taking the time to examine all these "new" ideas can finally force buyers to neglect other more important business.

5. Sometimes buying one processed item entails buying something else. For instance, if buyers purchase semolina flour to make their own pasta, they must also buy the pasta machine. Similarly, if they buy corn flour to make their own tortillas and taco shells, they may need a special basket to hold the shells in the fryer.

6. Processed foods present several "opportunity buys," such as introductory offers, quantity discounts, volume discounts, salvage buys,* and other hospitality operations' going-out-of-business sales. For operators who control a lot of purchase money, long-term contracts may also be available. These opportunities usually require a bit of extra analysis. Buyers must decide whether they are going to buy these items on a day-to-day basis or to succumb to a salesperson who comes in with a flamboyant special offer.

7. Buyers must decide which container size they should buy. Smaller packages have higher AP prices per unit, but they sometimes provide the best edible-portion (EP) cost. A related concern exists: Should buyers purchase individual, filled catsup

---

* Recall that these purchases may be outlawed by local health authorities. At any rate, these are questionable opportunities because buyers cannot be sure that the products have not been exposed to prolonged heat, chemicals, or other contamination.

bottles, or should they keep the empty bottles and refill them with catsup from a No. 10 can or some other bulk pack? The latter choice may entail some waste and labor, but it may also produce the best EP cost.

8.   A final major consideration relates to Point 6. Should buyers accept an offer that looks appealing but that would involve changing the form of the product they usually purchase? For instance, they may be offered a bargain in canned green beans, but they normally use the frozen form. Buyers should not take this temptation lightly. If profits are running a little low for the business, or if a particular buyer is just naturally conservative with money, it's surprising how big a few pennies can look. (This problem also arises with introductory offers or other types of "push strategies." Suppliers often try to switch a buyer from one item to another by temporarily manipulating the AP price.)

SELECTION FACTORS

Management, either alone or in conjunction with others, decides the quality, type, and style of food wanted for each processed item. During this decision-making process, the owner-manager should evaluate the selection factors in the following sections.

Intended Use

As always, owner-managers want to determine exactly the intended use of the item so that they will be able to prepare the appropriate, relevant specification. For example, canned fruit used in a recipe that has several other ingredients need not be as attractive as that used alone in a pie filling.

Exact Name

Confusing terminology clutters the market, especially the processed-produce market. For instance, buyers cannot simply order pickles; they must order Polish pickles, kosher pickles, sweet pickles, and so on. Similarly, they must specify canned Bartlett pear halves if that is what they desire, or extra virgin olive oil if they want olive oil with the lowest possible acidity.

The list of these designations can grow incredibly long, and more esoteric terms seem to exist in the area of processed produce and other grocery items than in most other areas. Nonetheless, buyers must become familiar with the market terminology.

The federal government has provided some assistance to buyers who are responsible for ordering many processed foods. Some standards set specific processing requirements, such as heating at very high temperatures in hermetically sealed containers. This is necessary to ensure the wholesomeness and safety of the finished product. Other standards are related to composition. For instance, any food that purports to be organic or to contain organically produced food ingredients—that is, the product label or labeling bears the term "organic" or makes any direct or indirect representation that the food is organic—must meet a particular set of standards. For example, canned organic vegetables must contain at least 90 percent organic ingredients, which are ingredients that are grown without added hormones, pesticides, herbicides, or synthetic fertilizers. If a can of soup, for instance, simply states "made with organic vegetables," then the percentage of organically produced ingredients in the soup must be stated on the label as "Contains _____ percent organic ingredients" (with the blank filled in with the actual total percentage of organically produced ingredients in the soup).

These types of standards, however, do not keep different companies from making distinctive recipes. For example, the United States Department of Agriculture (USDA) content requirement for beef stew specifies only the minimum percentage of beef (25 percent) that the stew must contain. The USDA requirement does not prevent a manufacturer from using its own combinations of other ingredients or increasing the amount of beef to make the product unique. As such, all brands of stew probably will taste somewhat different, which makes it risky for buyers to rely only on standards of identity when selecting processed produce and other grocery items, as well as some meat, dairy, fish, and poultry items.

Standards of identity are available for approximately 235 items if you care to use them. The USDA has established standards for meat and poultry products (see: www.ams.usda.gov), and the Food and Drug Administration (FDA) has set them for cocoa products; cereal, flour, and related products; macaroni and noodle products; bakery products; milk and cream products; cheese and cheese products; frozen desserts; sweeteners and table syrups; food flavorings; dressings for food; canned fruits and fruit juices; fruit butters, jellies, preserves, and related products; soda water; canned and frozen fish and shellfish; eggs and egg products; oleomargarine; nut products; canned and frozen vegetables; and tomato products.

For some items, buyers may be able to make do with only a standard of identity. For instance, although all types and brands of frozen orange juice are not the same, most of them are close. This is true of other frozen juices, too. Thus, buyers might be governed primarily by the AP price for these products.

U.S. Government Inspection and Grades (or Equivalent)

The USDA's Agricultural Marketing Service (AMS) Fruit and Vegetables Division and the FDA conduct mandatory inspections of processors' facilities. If any meat is incorporated in a product or if any items require egg breaking, the inspection falls under the jurisdiction of the USDA's Food Safety and Inspection Service (FSIS). Recall that continuous federal inspection or equivalent state inspection exists for any type of meat product that is sold. If no meat is involved, the foodstuffs are, nonetheless, inspected for wholesomeness, though less frequently. Of course, some state and local inspection may also be involved.

Government inspection is mandatory, but U.S. grading service is voluntary and food processors must pay for it. However, they have the option to pay for continuous inspection with or without a grade. Many buyers include federal-government grades in their specifications to ensure that the products are produced and packed under continuous government inspection. Understandably, bid buyers also seek out the relevant grading standards for their specifications.

Buyers can also specify that the products desired must carry the federal-government inspection shield. This shield indicates that the product was "packed under continuous inspection of the U.S. Dept. of Agriculture." The USDA provides this fee-based service to those primary sources who want inspection only and who do not want to purchase the federal-government grading service.

Federal grades have been established for canned, bottled, frozen, and dried produce and grocery items (see Figure 5.1).[1] Specific grading factors exist for different items. For instance, the grading factors for canned and bottled foods include the color, the uniformity of size and shape, the number of defects and blemishes, and the "character," which refers to the texture, tenderness, and aroma. Also, the quality of the packing

FIGURE 5.1 Federal grade and inspection stamps used for processed produce and other grocery items.
Source: United States Department of Agriculture.

medium—the water, brine, or syrup—may be important for some products. For some items, such as canned, whole tomatoes, during the evaluation process the grader considers the "drained weight," which is the servable weight that remains after the juice is removed.

Grading factors for frozen foods include the uniformity of size and shape, maturity, quality, color, and number of defects and blemishes.

Grading factors for dried foods include uniformity of size and shape, color, number blemishes and defects, moisture content,* and the way the products are packed—are they carefully layered or packed tightly together in a container, thus distorting their natural shape?

As with fresh produce, no single categorization of grading nomenclature exists for processed foods. As a matter of convenience, buyers could rely on the following grading categories for canned, bottled, and frozen items:

1. Grade A. The very best product with excellent color, uniform size, weight, and shape, and few blemishes.
2. Grade B. Slightly less perfect than Grade A.
3. Grade C. May contain some broken and uneven pieces, perhaps some odd-shaped pieces; the flavor usually falls below Grades A and B, and the color is not so attractive.

Again, for convenience, you could rely on the same nomenclature for dried foods:

1. Grade A. The most attractive and most flavorful product.
2. Grade B. Not quite so attractive as Grade A.
3. Grade C. More variations in taste and appearance, and usually broken pieces.

Although this convenient grading system for canned, bottled, frozen, and dried products might be ideal, it is not reality. Not all these types of foods use the grading system A, B, and C. For example, canned mushrooms, frozen apples, and several dried foods carry Grades A, B, and Substandard. Frozen apricots carry the Grades A, B, C, and Substandard. Sometimes you hear the terms "Fancy," "Choice," or "Extrastandard."

---

* No federal standard exists for moisture content. The usual dried item has at least 75 percent of its moisture removed, but this is not enough to make the produce last for an extended period. The term "sun-dried" implies high moisture residual, about 25 percent, whereas foods dehydrated in other ways usually contain 5 percent moisture.

These are alternate terms for U.S. grades that several people in the channel of distribution use. For example, many buyers use the following U.S. grade designations when purchasing canned fruit and canned vegetables:

CANNED FRUIT	CANNED VEGETABLES
Fancy	Fancy
Choice	Extrastandard
Standard	Standard

Another confusion in this area is that some processed items carry two grade designations, some carry three, and many carry four.

Still another difficulty: some products may display "Grade A," and not "U.S. Grade A" on the label. This is an indication that a federal inspector has not graded the products. Food processors are allowed to use "Grade" on their labels, as long as a nongovernmental graded item does not carry the "U.S." prefix. As a general rule, an item could carry types of grading nomenclature similar to those the federal government uses, even if the appropriate government agency did not grade the product. Since some discrepancy exists between these terms, to avoid any potential confusion, buyers should carefully note whether "U.S. Grade A" or "Grade A" is listed on the package label. (As we noted in Chapter 4, buyers usually can communicate with a purveyor and make their desires known.)

Packers' Brands (or Equivalent)

A bit of "pull strategy" is inherent in certain product lines. As such, buyers can sometimes be "coerced" into purchasing Heinz ketchup, A-1 steak sauce, Del Monte® relish, and so on.

In addition, many processed items come and go. So, if buyers want a particularly esoteric combination of fruit, they might find only one producer who handles it.

Also, since packaging can be unstandardized, buyers may seek out brands that meet their particular packaging requirements. (Recall that a brand normally implies more than just product quality.)

For some items, particularly something like frozen peas in cream sauce, a packer's brand may be the most important indication of quality and flavor. The quality of a fruit or vegetable varies from year to year and from place to place. The top-of-the-line brands make an effort to smooth out these annual fluctuations.

Packers' brands may be desirable, therefore, if only because subtle differences occur between, for example, tomato packers. After all, canned tomatoes can vary tremendously, not only in appearance, but in taste as well. Consequently, some buyers may be wary about trading Heinz for Del Monte simply because they detect a slightly different flavor.

A tremendous variety of brands is available to buyers. For instance, some companies package only the best-quality merchandise and will pack lower-quality products only if these items carry some other brand name. These firms refer to themselves as "premium" brands. Recall from Chapter 1 that some companies prepare several qualities under the same brand-name heading; that is, they carry several "packers' grades" (i.e., packers' brands) in their sales kits. For instance, buyers can sometimes purchase a specific producer's brand of carrots, but they will notice that different-colored labels exist for each quality of carrots that specific producer packs. These different-colored labels represent the packer's "grades" produced by the company.

Buyers can also opt for generic brands. These brands are not as plentiful in the wholesale-distribution channels as they are in retail grocery stores and supermarkets, probably because if buyers desire this type of quality, a packer's brand already exists to satisfy their needs. If they insist on purchasing generic brands, they probably will need to shop at the numerous warehouse wholesale clubs that cater to small businesses.

Generic brands can be very economical. Generally speaking, they typically are offered at very low AP prices for at least three reasons: (1) lower, or nonexistent, selling and advertising costs; (2) lower packaging costs; and (3) in some cases, lower quality. Keep in mind that lower quality does not necessarily imply lower nutritional value. Also, buyers are apt to receive a more uniform quality when ordering single-ingredient, generic-brand food. For instance, canned sliced peaches would tend to have better and more consistent quality than, say, mayonnaise, which includes several ingredients and involves relatively complicated processing.

Product Size

A very important consideration is the question of size, or count. For example, when buyers order pitted green olives in a No. 10 can, they should also indicate the olive size they want. They can do this by stating a specific count, which, in turn, implies a number, or count, of olives in a particular can. The higher the count, the higher the number of olives in the can and the smaller the olives. Sometimes, too, buyers can specify the approximate number of product pieces they would like in a can. They will, however, usually be limited in the count that they can have. Only a few choices are available (see Figure 5.2).

In lieu of specifying the count, buyers could use other marketing terms that essentially serve the same purpose. For instance, while it is true that buyers can indicate olive sizes by stating the count desired, they can also use such terms as "large," "extra

FIGURE 5.2 Typical product sizes for peach and pear halves. Courtesy of Canned Fruit Promotion Service.

large," "jumbo," or another appropriate marketing term to convey the necessary size information to their suppliers.

Buyers also may want to know how many cups they can get from a can or a frozen pack. Although these volume measurements appear on consumer products' package labels, buyers must be careful with volumes listed on commercial labels; they are sometimes misleading. For instance, on an instant mashed potato can label, there could be a reconstituted, or ready-to-serve, volume of 3 gallons stated—a volume that may be attainable only if you whip the potatoes long enough to incorporate a great deal of air.

Size of Container

Buyers must indicate on the specification the exact size of the container that they wish to buy (see Figure 5.3). To do this, they should determine whether the size of each package meets their needs. It costs more per ounce to buy dried oregano in a little bottle than in a much larger container. But if hospitality operations do not use much dried oregano, some product in the big container will go to waste. So the EP cost becomes the buyers' main consideration when they evaluate appropriate container sizes.

Type of Packaging Material

For frozen and dried products, especially frozen, the packaging materials are not nearly so standardized as those for cans and bottles. If buyers purchase large amounts of these products, such as an annual supply, they should examine the packaging very carefully. Will it hold up for a few weeks or a few months in the freezer? Also, especially for dried products, can moisture seep into the containers?

Another packaging consideration that emerges occasionally is ease of opening. Also, can the package be reclosed tightly enough to save the rest of the contents for later? Sometimes an item comes in an inconvenient package, which can lead to waste. In this case, the EP cost goes up. A more convenient package generally costs more. But as long as the EP cost is acceptable, it may be worth it.

We should address the need to use environmentally safe packaging whenever possible. When buyers purchase cans or bottles, these items may not be as convenient as, say, plastic pouches or aseptic containers. However, buyers can recycle them. This helps protect the environment and, in some parts of the United States, buyers may even

A Guide to Common Can Sizes

6 oz.	Approximately $\frac{3}{4}$ cup 6 fl. oz.	Used for frozen concentrated juices and individual servings of single strength juices.
8 oz.	Approximately 1 cup 8 oz. ($7\frac{3}{4}$ fl. oz.)	Used mainly in metropolitan areas for most fruits, vegetables and specialty items.
No. 1 (Picnic)	Approximately $1\frac{1}{4}$ cups $10\frac{1}{2}$ oz. ($9\frac{1}{2}$ fl. oz.)	Used for condensed soups, some fruits, vegetables, meat and fish products.
No. 300	Approximately $1\frac{3}{4}$ cups $15\frac{1}{2}$ oz. ($13\frac{1}{2}$ fl. oz.)	For specialty items, such as beans with pork, spaghetti, macaroni, chili con carne, date and nut bread—also a variety of fruits, including cranberry sauce and blueberries.
No. 303	Approximately 2 cups 1 lb. (15 fl. oz.)	Used extensively for vegetables: plus fruits, such as sweet and sour cherries, fruit cocktail, apple sauce.
No. 2	Approximately $2\frac{1}{2}$ cups 1 lb. 4 oz. (1 pt. 2 fl. oz.)	Used for vegetables, many fruits and juices.
No. $2\frac{1}{2}$	Approximately $3\frac{1}{2}$ cups 1 lb. 13 oz. (1 pt. 10 fl. oz.)	Used principally for fruits, such as peaches, pears, plums and fruit cocktail; plus vegetables, such as tomatoes, sauerkraut and pumpkin.
46 oz.	Approximately $5\frac{3}{4}$ cups 46 oz. (1 qt. 14 fl. oz.)	Used almost exclusively for juices, also for whole chicken.
No. 10	Approximately 12 cups 6 lbs. 9 oz. (3 qts.)	So called "institutional" or "restaurant" size container, for most fruits and vegetables. Stocked by some retail stores.

FIGURE 5.3 Average can sizes. Courtesy of American Can Company, Greenwich, CT.

be able to earn a small amount of income from recycling plants that purchase these materials.

Another packaging consideration is the issue of "personalized" packaging (see Figure 5.4). For some processed items, such as sugar packets, buyers can order packaging that contains the hospitality operation's logo or another form of advertisement.

FIGURE 5.4 An example of a personalized package. Courtesy of Lady Luck.

Sometimes these options will increase the AP price of the underlying food item; however, the advertising value may more than offset the added expense.

Packaging Procedure

This is an important consideration for some processed produce and other grocery items. As a general rule, packing usually has two styles: slab-packed merchandise and layered merchandise. Most processed produce and other grocery items are necessarily slab-packed—that is, they are poured into the container and the container is then sealed. Some products, though, such as dried fruit, can be slab-packed or neatly layered between sheets of paper or cardboard. Furthermore, some layered merchandise, such as frozen, double-baked, stuffed potatoes, might be individually wrapped.

As usual, the layered and/or individually wrapped products cost more, at least in terms of higher AP prices. However, because better packaging and a more careful packing style minimize product breakage and other forms of product loss, the resulting EP costs may be quite acceptable.

Drained Weight (Servable Weight)

Considering weights instead of volumes is usually a good idea. The weight of the contents of a can can vary. It is good practice for buyers to calculate "drained weight" (servable weight) when purchasing canned items; this is the weight of the product less its juice. To compute this figure, buyers drain a product in a specific sieve for a certain amount of time. Some packers' brands offer a great deal of fruit and little juice. Other brands could have more juice and less fruit. Buyers must be concerned with portions per can and drained weights—and with EP costs, to be sure.

Recall that food processors need to note on consumer package labels the serving size and number of servings in the package. Processors may eventually be required to list the weight of the fruit and, separately, the weight of the juice, water, or syrup. To a certain extent, buyers can estimate the amount of juice by the absence or presence of the words "heavy pack" or "solid pack." If these words do not appear, buyers should expect a lot of liquid—for example, about $5\frac{1}{2}$ ounces of liquid in a typical 16-ounce can of fruit. The term "solid pack" means no juice added; "heavy pack" means some juice added, but not much.

Best of all, however, buyers should measure the drained weight. It is more reliable than estimating weight by looking at a can's label.

When buyers compare the weights of two or more packers' brands of frozen products, they should thaw frozen fruit and cook frozen vegetables from the frozen state before weighing them. Buyers should also do the same with dried products: weigh them only after reconstituting them.

Buyers should be wary of purchasing anything after considering only the volume. The standard of fill protects buyers from a packer who fills a can only halfway and pretends to give a lot for the money. Remember, the possibility that an unscrupulous supplier may pump air into a product or lower the specific gravity of a product (by decreasing the product's density, he or she makes it lighter). Some possible examples here include ready-to-serve potato puffs. (Should buyers purchase them by count or by weight?) Similarly, tomato puree can vary in density; it can weigh about as much as an equal amount of water, which has a specific gravity of 1.00. Alternately, the puree can have a specific gravity of 1.06 or even a little more, and, thus, a greater weight.

Type of Processing

Buyers must indicate the type of processing desired. The type of processing implies certain flavor and texture characteristics, as well as specific product-preservation techniques. Generally, buyers will purchase canned, bottled, frozen, and/or dried merchandise.

Color

In some cases, buyers will need to specify a preferred color. Suppose, for instance, canned red apple rings are available but no green ones are. The same choices are available for bottled maraschino cherries. Generally, when buyers specify the exact name of the item wanted, they indicate the color required. If the exact name of a product does not include this notation, buyers must be sure to point it out elsewhere on the specification if it is relevant.

Product Form

At times, buyers will need to indicate the specialized form of the merchandise they want. Usually this is not necessary if they are purchasing common, ordinary processed produce and grocery items. But if buyers are purchasing, say, a particular vegetable casserole, they may want to note the amounts and types of vegetables desired. Alternately, buyers may find it necessary to point out that they want a minimum number of almonds in the frozen green beans almondine they want to procure for their establishment.

Packing Medium

For some products, several packing mediums are available. For example, for canned fruit, buyers can select fruit packed in water, in syrup, or with no added medium. They also can specify the syrup density desired by noting the minimum "Brix" level. Thick syrup has a higher Brix (i.e., more sugar) than light syrup. The federal government sets minimum Brix levels for some products, yet buyers may want higher levels. For instance, a higher Brix carries a higher AP price, but since fruits packed in heavy syrup do not break easily, the resulting EP cost may be quite satisfactory.

When purchasing vegetable products, buyers may be able to specify the type and amount of sauce desired. For instance, frozen broccoli could be packed with butter sauce, cheese sauce, or some other specialized sauce.

The Use of Additives and Preservatives

The concern some hospitality operators show regarding the issue of preservatives in canned, bottled, frozen, and dried items appears to be less than they exhibit regarding dairy and meat products. Operators could not get the products they need if they had to settle for "fresh" produce all the time. Moreover, they could not store fresh produce efficiently and would probably waste a great deal of it. But some processors are more discreet with their use of additives and preservatives than others.

If the thought of additives and preservatives bothers buyers, they may have difficulty weeding out the offending packers' brands. Furthermore, they will find "organic," "whole," and "natural" foods to be expensive. Why? Buyers have entered a market with few suppliers. Furthermore, shelf life is reduced, resulting in greater handling costs and subsequently greater spoilage loss.

Other Information That May Appear on a Package Label

The federal government requires a great deal of information to be displayed on consumer product labels. However, required label information aside, typical hospitality operators would be much more interested in other kinds of information that may or may not appear on a package label. Many operators would like to see packing dates, freshness dates (or shelf lives), and serving cost data noted on package labels. Many also would like to see the lot number on a package label. (Since the quality of product shifts, it would be nice to be able to reorder, for instance, corn of not only the same packer's brand but also from the same batch as the previously ordered corn.) School foodservice buyers seek out products that carry Child Nutrition (CN) labels, which indicate how the products conform to the nutritional requirements of the USDA. When food processors put these details on their package labels, buyers must expect to pay a bit for this added information value.

One-Stop-Shopping Opportunities

Small operations tend to prefer one-stop shopping for many items, including processed products. To capitalize in this area, some national corporations carry extensive lines of processed produce and other grocery items and compete directly with local and regional distributors. For instance, General Mills (www.generalmills.com/corporate/), Green Giant (www.greengiant.com/), Pillsbury (www.pillsbury.com), and McCormick (www.mccormick.com/index.cfm) produce, market, and distribute wide varieties of processed products. These companies normally offer only one level of product quality, and they usually control all aspects of production and distribution. As a result, their products are often referred to as "controlled brands."

Some national corporations carry extensive lines as well as extensive qualities of processed produce and other grocery items. For instance, U.S. Foodservice (www.usfoodservice.com)—a broadline distributor—distributes many types of products, as well as many different "packer's grades" (i.e., packer's brands) in most product lines.[2] U.S. Foodservice uses the name "Cross Valley Farms®" on their signature brand line of fruits and vegetables.

As the typical hospitality operation grows, it shows less and less of a tendency to use one-stop shopping for processed produce. Too many opportunity buys and long-term contracts at a good savings are available, but they can usually be exploited only if buyers shop around. Large firms usually have the time to do such shopping around. In addition, as its menu gets larger and incorporates more variety, an operation has less opportunity to satisfy its needs with the one-stop-shopping method.

However, one-stop shopping for processed produce, other grocery items, and meat products provides a subtle advantage. Shortages sometimes occur for these items, and being a good customer may ensure a buyer a continual supply. In fact, some suppliers "allocate" certain product lines; that is, they predetermine how much a buyer can purchase. This amount is referred to as the buyer's "allocation."

AP Price

The EP cost is the only relevant concern for a buyer. The EP cost includes not only the cost of the product, but, indirectly, the cost of the labor it takes to prepare and serve it, the cost of the energy needed to work with it, and other overhead expenses. Making these judgments is difficult. But a buyer must look beyond, for example, the drained weight of canned mushrooms.

Since an interminable number of varieties, styles, and packaging methods exist for processed foods, there are correspondingly different AP prices. It is not easy to tell whether it is advantageous to take a Grade C instead of a Grade B for a savings of, perhaps, 5 cents a can. In most cases, the trade-off here would be a matter of opinion. Since lower grades are perfectly acceptable in some recipes, lower qualities can save money without reducing a recipe's acceptability.

For similar items, a buyer usually pays similar AP prices. But some suppliers may give better quantity discounts and volume discounts than others. Typically, these discounts are quite lucrative. Therefore, bid buying can save money, but only if a buyer is willing to accept a large supply, put the cash up front, and make the buy at new pack times, that is, when packers process that year's products.

Unfortunately, there may not be enough bidders for a buyer's business, especially if the local supplier cannot find enough of a product to satisfy his or her requirements. But even if the buyer finds only one supplier, a large buy can be valuable: a substantial savings may result if he or she can store and protect a large supply. However, this has subtle disadvantages: if a buyer is "locked in" for a year, his or her menu is somewhat set. Also, if AP prices for similar items fall, the buyer cannot easily take advantage of them when his or her storage area is full.

Supplier Services

Normally, canned products require only nominal supplier service. Most of these items are not readily perishable, so buyers are not likely to be concerned with how quickly the supplier moves them. But this cannot be said of frozen items. If buyers doubt the capability of a supplier to maintain frozen products at 0°F or below, they should avoid this person. Frozen food costs more because it maintains a better culinary quality, but this quality rapidly deteriorates when storage temperatures are above zero or if they fluctuate; in addition, these problems reduce shelf life drastically.

Occasionally, buyers may want to obtain a stockless purchasing deal, in which they protect an AP price for six months to a year. This tactic rarely saves as much as traditional forward buying, in which buyers take delivery of a large amount. But now and then it can produce a reasonable saving.

Buyers would also like to receive reasonable "break points." They may not want to buy, for example, 100 cases before they get a price break or quantity discount. Fifty cases may be more acceptable. We have noticed that break points are somewhat standard among suppliers.

Sometimes an AP price shoots up dramatically. Buyers like to be warned about this beforehand, if possible. If buyers use bid buying exclusively, they may not receive that warning. If buyers are house accounts, chances are they will.

Together with fluctuating AP prices, which really are not so common with processed products as they are with fresh foods, comes the potential for shortages. A rainstorm can reduce the canned-peaches supply. Only certain buyers may be on the list of those slated to get some of that supply. Again, bid buyers may be left behind. They always take a chance that nobody will answer their solicitations for bids. House accounts may be better serviced.

Buyers should consider another supplier service: If they want a low grade, can they get it? Low-grade products, generally, are not in plentiful supply. Here again, house accounts find themselves in a better position than bid buyers.

Yet another service is the delivery schedule. Realistically, though, most buyers can live with one delivery every week, or even one every two weeks. If, however, buyers have only a limited storage area, they might want other arrangements.

Local Supplier or National Source?

Buyers can easily go to the primary source for a direct purchase and have the carload of canned tomatoes "drop-shipped" (which means that they will be delivered straight to the back door). Buyers can also buy direct and arrange for a local supplier to distribute the merchandise.

We have already expressed our views on buying directly and bypassing the local supplier. Still, buyers must do what they believe is best given their circumstances.

PURCHASING PROCESSED PRODUCE AND OTHER GROCERY ITEMS

A buyer's first step in purchasing processed produce and other grocery items is to obtain copies of reference materials that contain useful information that they can use to prepare specifications. Most suppliers publish several in-house materials, particularly individual brochures that detail their major product lines. They also have very informative Websites. These references usually include considerable product information, as well as detailed descriptions of grading factors used by U.S. government graders.

The buyer's next step is to determine precisely what he or she wants. This may not be an easy task since several management decisions are involved. Some items are traditionally purchased canned or bottled, some frozen, others dried. But the tradition does not hold true for all products.

Once buyers know what they want, they can make the actual purchase simple or difficult. It will be relatively simple if they buy on a day-to-day basis. If they become dissatisfied with a particular packer's brand, they can just switch to another brand. It is, however, a little more adventurous to enter the bid-buying route, especially when buyers seek to lock up a six-month or one-year supply.

If buyers enter into a long-term contract on a bid basis, they will need detailed specifications. They may find it beneficial to prepare these detailed specs even if they do not use bid buying; it is good discipline for buyers to put their ideas on paper before they actually commit to any type of purchase. (See Figure 5.5 for some example product specifications and Figure 5.6 for a sample product specification outline for processed produce and other grocery items.)

If buyers go the long-term route, which is usually available in this product area even for small operations, they will have a bit of work ahead of them. Before they sign a long-term contract, they should examine the bidders' products very carefully.

Pineapple slices Used for salad bar Dole® brand (or equivalent) 66 count No. 10 can Unsweetened clarified pineapple juice	White cake mix Used to prepare cupcakes Pillsbury® Food Service brand 5 pounds Cardboard box
Whole canned onions Used for plate granish U.S. Fancy 350 count No. 10 can Water pack	Confectioners' powdered sugar Used for baking C & H® brand 1-pound cardboard box Packed 24-pound per case

FIGURE 5.5 An example of processed produce and other grocery items product specifications.

Intended use:

Exact name:

U.S. grade (or equivalent):

Packer's brand name (or equivalent):

Product size:

Size of container:

Type of packaging material:

Packaging procedure:

Drained weight:

Type of processing:

Color:

Product form:

Packing medium:

FIGURE 5.6 An example of a product specification outline for processed produce and other grocey items.

A mistake in this area can be extremely costly, and unintentional mistakes happen easily. For example, buyers may examine three brands of canned tomatoes. They may perform all sorts of tests—comparing drained weights, looking for tomato skins, noting the clearness of the juice—but they may discover too late that one brand has slightly less acid, which may be unsuitable for some recipes. These tests are called "can-cutting tests." If competing salespersons or suppliers are in attendance, the testing is sometimes referred to as "holding court." Buyers usually complete a checklist for each brand and then compare each brand's scores (see Figures 5.7 and 5.8). If buyers are purchasing for large commissary production, a little oversight like insufficient acid in a recipe can significantly impair the culinary quality of the finished product.

Another problem can occur through purchasing errors. How do buyers return a year's supply of canned tomatoes to a supplier if they make an error? Buyers should consider a related problem: If they buy a year's supply and later notice that the product is not quite as good as what they contracted for, they must protect themselves by retaining a few unopened packages that were available for their cutting test just prior to signing the contract. Buyers then have on hand a standard of quality they can use to prove that they actually contracted for something better.

Buyers can consider other suggestions for cutting tests: (1) Always check frozen fruit after it has thawed; in particular, check the fruit's texture, which tends to suffer in

SAMPLE	1	2	3
Vendor/brand			
Drained weight			
Color			
Size			
Uniformity of size			
Defects			
Clearness of syrup			
Grade			
Flavor			
Case price			
Unit price			
Serving size			
EP cost/serving			

FIGURE 5.7 A checklist for canned sliced peaches.

the freezing process. (2) Be sure to cook frozen vegetables from the frozen state before testing them. (3) Check all canned goods immediately after opening them, especially their odor. Canned products are cooked during the canning process to kill harmful bacteria so that the products will stay wholesome and not deteriorate. Thus, they are ready to eat. (4) Conduct tests of dried and concentrated products, such as soup, on the reconstituted product.

Generally, then, buyers should test products after they have been prepared for customer service. In some instances, buyers may even want to prepare a full recipe with each one of the competing bidder's products and then perform their tests.

Unless buyers are invited to a supplier's headquarters for some type of product introduction or abbreviated cutting test, these tests can take time and effort. But it is normal practice in large organizations to spend a great deal of effort when considering a quantity buy.

Once buyers enter into a long-term contract, salespersons carrying similar products will still undoubtedly approach them. For instance, each competing supplier with its new brand of spaghetti sauce will bring it to a buyer's attention. Alternately, thanks to backdoor selling, a cook may urge the buyer to purchase a new type of soup base.

Actually, buyers will not see too many revolutionary products. But they may see subtle changes, such as different packaging. For instance, they may buy a new packer's

	CLING PEACHES	BARTLETT PEARS	FRUIT COCKTAIL
Color:			
Grade B:	Reasonably bright, possibly slight discoloration.	Reasonably uniform in color, may show tint of pink, appear translucent.	Fairly clear liquid and distinct color.
Grade C:	Reasonably bright, yellow-orange, possibly greenish-yellow.	May vary noticeably, appearance could be either white or brown.	
Texture/Character:			
Grade B:	Reasonably good texture, no more than 10% of fruit being mushy.	Reasonably tender, may possess moderate graininess.	Texture of fruits may vary, from firm to soft.
Grade C:	Fairly good texture.	Texture may vary noticeably, with soft or frayed edges.	
Defects:			
Grade B:	Reasonably free of pit material, not more than 5% crushed or broken.	Major defects may not exceed 10% of fruit and minor defects by 20%.	Refer to USDA guidelines for acceptable defects for each individual fruit in fruit cocktail.
Grade C:	Fairly free of pit material, less than 20% of fruit may be blemished.	Major defects may not exceed 20% of fruit and minor defects by 30%.	
Uniformity:			
Grade B:	Largest unit may not exceed smallest unit by more than 60%.	Largest unit may not exceed smallest unit by more than 75%.	No more than 20% of fruit may vary substantially in size.
Grade C:	Largest unit may not exceed smallest unit by more than 100%.	Largest unit may not exceed smallest unit by more than 100%.	
Smell:	No offensive odors should be present in fruit or liquid of any grade.		
Taste:	Distinct and fruity.	Similar to mature pears.	Ability to detect flavor of individual fruits.
Sample Size:	#10 can equals 30 halves or 100 slices.	#10 can equals 30 halves or 100 slices.	#10 can
Pack Ratio:	N/A	N/A	Peaches 30%–50% Pears 25%–45% Grapes 6%–20% Pineapple 48 pieces Cherries 24 halves

FIGURE 5.8 Guidelines for examining some canned fruit products.

Source: United States Department of Agriculture.

brand of soup base if it is similar to the one they currently use simply because it comes in a 30-pound pack instead of the 1-pound packs they usually buy. If buyers purchase the new product, their old supplier may come out with a 22-pound pack, so testing tends to be a continuing process.

A very complicated test arises when buyers want to evaluate, say, canned peas and frozen peas by using each in two versions of the same recipe, and when buyers want to try other combinations of canned, frozen, and dried. For instance, one stew recipe might be made three ways, with frozen onions, canned onions, and fresh onions. Buyers have to decide which meets their criteria the best.

We have not seen very many local small producers in the processed produce area. But suppose somebody in the family has won a state fair blue ribbon for his or her canned pears and now wants to sell them. We do not consider it a good idea to buy these items. Indeed, recall that most states prohibit the use of home-cooked products in foodservice operations because they come from unapproved sources.

Having struggled successfully with all these details, buyers will find the actual buying, at last, relatively easy. The ordering procedures themselves rarely present any burdensome difficulties.

RECEIVING PROCESSED PRODUCE AND OTHER GROCERY ITEMS

Generally accepted procedures for inspecting the quality of delivered processed produce and other grocery items have been established:

1. **Canned and Bottled Products.** Check the containers for any swelling, leaks, rust, dents, or broken seals. These characteristics, especially swelling, indicate contamination problems. Refuse damaged containers, as well as those that are dirty, greasy, or generally unkempt.

2. **Dried Products.** Check the condition of the containers. If the dried foods are visible, look for mold, broken pieces, and odd appearance.

3. **Frozen Products.** Check the condition of the container, looking for any indication of thawing and refreezing; stained packaging indicates this. Check the food temperature. Look for $-10°F$, but $0°F$ is acceptable. If the frozen foods are visible, check for "freezer burn," the excessive, often brownish dryness that occurs if food has not been protected properly in frozen storage.

Occasional problems occur with quantity checks. Most of these processed products come in cases, and when the cases are full, buyers may assume that these cases are full of exactly what they ordered. Repacking is rare, but it can happen. In any event, buyers should open at least some cases when they are receiving these items. This is particularly true when receiving a large quantity.

Carefully check incoming products against the invoice and a copy of the purchase order. Some packers' brands resemble one another. Also, be careful of supplier substitutions, which can occur from time to time, especially for these types of products.

Quality checks are not always easy, mainly because what you can actually see is limited. Rarely will a chef, for example, come to the receiving dock to check the quality of processed products, unless a large quantity is being delivered.

After checking the quality and quantity, check the prices, and complete the appropriate accounting documents.

STORING PROCESSED PRODUCE AND OTHER GROCERY ITEMS

Generally accepted procedures for storing processed products have also been established:

1. **Canned and Bottled Products.** Store these products in a dry area at approximately 50°F to 70°F. Avoid any wide fluctuation in temperature and humidity. Heat can be especially damaging. For instance, it robs spices of their flavor; it hastens the oxidation of frying fats, which means that these fats will not retain flavor or last as long in the french fryer; and it hastens the chemical changes in canned items, which means taste changes. Avoid dampness, too, which causes rust and attracts dust and dirt. In hot climates, consider refrigerating such items as spices and fats. Keep all canned and bottled products tightly covered, opening only what is necessary; otherwise, they will lose some shelf life.

2. **Dried Products.** Be especially careful of dampness as it can hasten the growth of mold, thereby ruining the products. Try to keep these products a little cooler than canned items so that insects are not attracted to them.

3. **Frozen Products.** If at all possible, store these products at −10°F or lower. This temperature will preserve the maximum flavor, especially if the other distribution channel members have maintained this temperature. Be careful not to damage any packages because that will eventually lead to freezer burn. Also, avoid fluctuations in temperature, which reduce shelf life drastically.

Perhaps the most unfortunate aspect of storing processed produce and other grocery items is that so many of them require slightly different storage environments. As a practical matter, we can hardly satisfy each requirement, so we try to reach a happy medium by paying especially close attention to proper stock rotation.

However, it is not always easy to ensure proper rotation because many employees grow complacent in their handling of processed items, thinking they will hold forever. Theoretically, canned, bottled, and dried food will last a long time, but no one would try to keep them very long. Nor will frozen food last forever; freezing merely slows, but does not eliminate, deterioration.

Other storage considerations include: (1) keeping the items off the floor, where they can attract dirt; and (2) when filling flour bins, and other such storage bins, trying not to mix the new flour with old. If possible, use bins that load from the top and unload from the bottom.

ISSUING PROCESSED PRODUCE AND OTHER GROCERY ITEMS

Hospitality operators often find a good deal of neglect for many processed items. Such products as individual containers of catsup, salt, and sugar—usually food that goes to the service personnel stations—are not always controlled closely. The best way to avoid this waste is to ensure that written stock requisitions exist for every item and that no requisitioner asks for more than necessary. Buyers can control a requisitioner by asking him or her to take note of the in-process inventory prior to asking for additional stock.

Buyers need to consider how they would issue half-cans or some similar amount. The fact is, if this problem arises frequently, they might be better off with smaller containers.

Finally, the EP cost is more vulnerable to attack in the area of in-process inventories. Here, as elsewhere, supervision is the key. Without effective and efficient supervision, it is futile to spend time and effort to save money in purchasing.

KEY WORDS AND CONCEPTS

Additives and preservatives	Can-cutting test	Exact name
Advantages and disadvantages of the various processing methods	Child Nutrition label (CN label)	Food and Drug Administration (FDA)
Agricultural Marketing Service (AMS)	Color	Food Safety and Inspection Service (FSIS)
Allocation	Common can sizes	Forward buying
Aseptic packaging	Controlled brands	Freezer burn
As-purchased price (AP price)	Drained weight	Freshness dates
Break points	Drop shipment	Generic brands
Brix	Edible-portion cost (EP cost)	Going-out-of-business sales
	Environmentally safe packaging	Grading factors

Heavy pack	Packers' brands (grades)	Shelf life
Holding court	Packing dates	Shelf-stable products
Impulse purchase	Packing medium	Single-ingredient, generic-brand food
In-process inventories	Personalized packaging	
Institutional can size	Premium brands	Size of container
Intended use	Product form	Slab-packed
Layered packaging	Product size	Solid pack
Lot number	Product-testing factors	Specific gravity
Management considerations when purchasing processed produce and other grocery items	Pull strategy	Standard of identity
	Purchasing, receiving, storing, and issuing processed produce and other grocery items	Stock rotation
		Sun-dried versus other drying methods
New pack time	Push strategy	Supplier services
One-stop shopping	Quantity discount	Type of processing
Opportunity buys	Recommended storage procedures	United States Department of Agriculture (USDA)
Organic, whole, natural foods		
Packaging material	Reference books that can be used when preparing specifications	U.S. grades
Packaging procedure		U.S. Grade A versus Grade A
Packed under continuous government inspection	Salvage buys	Volume discount
		Warehouse wholesale club

REFERENCES

1. For a detailed dicsussion of processed fruits and vegetables grades, see http://www.ams.usda.gov/fv/ppb.html. Retrieved October 2, 2003.

2. For a list of U.S. Foodservice's signature brands, see http://www.usfoodservice.com/products/signature/signature.html. Retrieved October 2, 2003.

QUESTIONS AND PROBLEMS

1. What are the U.S. grades for canned, bottled, frozen, and dried items?

2. Give an example of optional information a packer could note on a package label.

3. Assume that you own a small coffee shop and that you have no franchise affiliation. Your annual volume (open 24 hours, every day) is $825,000.00. You sell hamburgers, and you are currently using Heinz individual catsup bottles. You can save about 8 percent of your $438.00-a-year catsup expense (i.e., $0.08 \times \$438.00 = \$35.04$ per year) if you buy a different brand of bulk-pack catsup and plastic containers and fill these containers from the pack. What course do you recommend? If possible, ask a coffee shop manager or owner-manager to comment on your answer.

4. What is an appropriate intended use for canned peas?

5. Give an example of the pull strategy in the processed produce and other grocery items channel of distribution.

6. What is the primary difference between U.S. Grade B and U.S. Grade C products?

7. What are the grading factors for canned and bottled products?

8. What are the grading factors for frozen products?

9. What are the grading factors for dried products?

10. What critical information is missing from the following product specification for canned peach halves?
    Peach halves
    Packed in light syrup
    CODE brand, red label (or equivalent)
    Packed in No. 10 cans, 6 cans per case

11. Why is organic food more expensive than nonorganic, processed-produce products?

12. Note three examples of grocery products that you have seen in a restaurant operation that carry personalized packaging. What are some of the advantages and disadvantages of personalized packaging?

13. Assume that you normally purchase 1200 cases of canned peaches every three months. (You order once every three months.) The AP price per case is $8.75. Your supplier offers you a one-year supply for $8.60 per case, cash on delivery (COD). (You currently have 30 days in which to pay your invoices from this supplier.) Assume you are the purchasing director for a 400-room hotel that does excellent restaurant and banquet business. What course of action do you suggest? If possible, ask the manager of a comparable property to comment on your answer.

14. What is an appropriate intended use for dried apricots?

15. Assume that you are the purchasing director for a university food service with 15 campus housing buildings, 8 snack bars, a large dining commons, and an unpredictable banquet business. Currently, you operate a central commissary and a central distribution center. Outline the specific procedures you would use for the purchasing, receiving, storing, and issuing of canned peach halves. If possible, ask a university foodservice purchasing director to comment on your answer.

16. Your supplier calls to say that the price of tomato paste is due to rise soon and suggests that you purchase at least 2500 cases immediately. Assuming that you find your supplier completely trustworthy, what specifically should you consider before making your decision about this potential purchase?

17. What are the recommended storage temperatures for canned and bottled products and for frozen products?

18. Assume that you manage a steak house. You have been using individually wrapped half-ounce portions of catsup. One Saturday afternoon, you notice that you have very few of these packets left because the previous night's business was especially brisk. At first glance, you do not believe you can get a delivery from the commissary—the steak house is part of a national chain—until Monday morning. But you are open tonight until 9 P.M. and all day tomorrow, 11 A.M. to 9 P.M.

What course of action do you take? If possible, ask a steak house manager to comment on your answer.

19. Assume that you manage a college foodservice facility. You want to purchase your annual requirement of canned tomatoes. You have four brands of tomatoes from which to choose, three reasonably well-known brands and one generic brand. The AP prices vary only slightly among the three name brands, but the generic brand offers a 22 percent savings. Unfortunately, the generic brand contains mostly broken pieces and has a drained weight that is 15 percent less than the name-brand merchandise. In addition, the supplier warns you that the quality of the generic brand is not predictable from year to year. Which type of merchandise would you purchase? Why? If possible, ask a college foodservice manager to comment on your answer.

20. A product specification for frozen corn could include the following information:
    (a)
    (b)
    (c)
    (d)
    (e)

21. The choice of which food-processing method a buyer selects is usually affected by three major criteria. Identify these three criteria.

22. Why would a canned tomato puree with a high specific gravity usually be more expensive than one with a lower specific gravity?

23. What is the difference between the designations "U.S. Grade A" and "Grade A"?

24. What is an appropriate intended use for frozen asparagus spears?

25. Define or explain the following terms:

(a)	AMS	(i)	New pack time
(b)	CN label	(j)	Can-cutting test
(c)	Standard of identity	(k)	Holding court
(d)	Stock rotation	(l)	Freezer burn
(e)	Break point	(m)	Solid pack
(f)	Drained weight	(n)	Premium brand
(g)	Heavy pack	(o)	Generic brand
(h)	Specific gravity	(p)	Packer's grade

6

DAIRY PRODUCTS⋆

The Purpose of this Chapter

After reading this chapter, you should be able to:

- Explain the selection factors for dairy products, including government grades.
- Describe the process of purchasing, receiving, storing, and issuing dairy products.

INTRODUCTION

Purchasing dairy products can be an arduous task. Foodservice buyers are inundated with numerous varieties and forms of these products, including milk, cheeses, and frozen concoctions. Whether the product is fresh, aged, dried, or fermented, the savvy buyer is knowledgeable in the distinguishing factors that characterize each product. The most noted component of dairy products is butterfat. The butterfat content affects quality, including flavor and mouthfeel. Because it is such a key component, the amount of butterfat in a dairy product has a direct correlation with the as-purchased (AP) price.

Buyers may ask what the difference is between one whole-milk brand and another? There should be little, since most dairies use fairly standardized management techniques. But dairies may vary somewhat in their quality control programs and their processing methods, so that, perhaps, a taste comparison between dairies is not a

---

⋆ Authored by Andrew Hale Feinstein and John M. Stefanelli.

wasted effort. The type of feed, seasonal variation, and the stage of lactation of the dairy herds can also cause slight flavor variations.

Chances are that even though little difference in flavor may exist between one brand of whole milk and another, the same cannot be said of other dairy products. The taste of many of these products, especially cheese, tends to be unique to each producer. Once buyers settle on a particular brand—for example, a specific cheese—especially if the cheese is served alone, they may find it difficult to discontinue it in favor of another brand. If they do, their customers probably will notice any change. Thus, for these items, there is a good possibility that buyers will inevitably become a house account with the supplier of the chosen brand.

As mentioned, butterfat is an important component in dairy products, and it is considered an expensive fat. Since butterfat content is almost directly related to the AP price of a dairy item, many products have been manufactured in which the butterfat has been either reduced or replaced. These types of substitutions can be cost-effective. For that reason, the owner-manager should evaluate the types of substitution possibilities. Will it be butter or margarine; half-and-half or nondairy coffee cream; and natural cheese or cheese food made with vegetable fat? The type of foodservice establishment will help answer these questions on the number and types of allowable substitution possibilities. For example, margarine chips are not a suitable substitute for butter pats in a gourmet dinner house.

Buyers face another series of substitution issues: one dairy item may be substituted for another in food-production recipes. More possibilities exist for dairy products than for other ingredients because any time an item contains fat, at least one substitution is a possibility: for example, yogurt for sour cream, skim milk for whole milk, and pasteurized process cheese for natural cheese. These factors can complicate the decision-making process for buyers. For example, they do not have to use sour cream on the baked potato; they have at least two alternatives: a cultured dressing, which is like a low-fat sour cream, or an imitation nondairy product.

Whatever the decision, if hospitality operations use a suitable dairy substitute product, the possible flavor and nutrition alterations in the final product must be addressed. To a certain extent, it is a matter of opinion whether these flavors are different. Whenever operators combine two or more ingredients in a recipe, the possibility of recipe change always exists. So it is to the firm's benefit to experiment (see Figure 6.1, which notes some dairy product substitutions).

Proper representation of substitute products is critical. Truth-in-menu legislation in some parts of the country prohibits misrepresentation—plus it is unethical. In addition, operators must be careful not to serve one product and imply that it is another; for example, they cannot serve half-and-half and imply that it is cream, a richer product that has more butterfat and is more expensive.

1 cup butter	1 cup margarine
	$7/_8$ to 1 cup hydrogenated fat plus $1/_2$ teaspoon salt
	$7/_8$ cup lard plus $1/_2$ teaspoon salt
	$7/_8$ cup rendered fat plus $1/_2$ teaspoon salt
1 cup coffee cream (20 percent)	3 tablespoons butter plus about $7/_8$ cup milk
1 cup heavy cream (40 percent)	$1/_3$ cup butter plus about $3/_4$ cup milk
1 cup whole milk	1 cup reconstituted nonfat dry milk plus $2^1/_2$ teaspoons butter or margarine
	$1/_2$ cup evaporated milk plus $1/_2$ cup water
	$1/_4$ cup sifted dry whole milk powder plus $7/_8$ cup water
1 cup milk	3 tablespoons sifted nonfat dry milk powder plus 1 cup water
	6 tablespoons sifted nonfat dry milk crystals plus 1 cup water
1 cup buttermilk or sour milk	1 tablespoon vinegar or lemon juice plus enough sweet milk to make 1 cup (let stand 5 minutes)
	$1^3/_4$ teaspoons cream of tartar plus 1 cup sweet milk

FIGURE 6.1 Some dairy product substitutions.

Once operators decide what dairy products they want, they must then determine the exact products and supplier(s) they would like to use. The typical foodservice operation can use one supplier for its dairy products—a sort of built-in one-stop buying strategy. Alternately, a buyer can take the time to evaluate the wide variety of suppliers available for each type of dairy product.

Small operators may prefer one-stop shopping for the majority of their dairy products. But a single supplier may not have everything needed. Consequently, the decision sometimes involves a choice between one or two suppliers, who may not always have exactly what the firm wants and who make fewer deliveries, versus several suppliers who carry what the firm needs and who make more frequent deliveries.

This decision is not easy to make. On the one hand, dairy products do not usually represent a great deal of the purchase dollar. Hence, hospitality operators could argue that little potential gain is associated with evaluating every available brand and supplier. Conversely, the substitution possibilities can be lucrative. A single sure way of an operation knowing whether it has examined all the substitution possibilities is to plow through every available brand and supplier. To complicate matters, many new products come and go. In some cases, it is preferable to procure the desired dairy products from one or two suppliers. But bear in mind that in many instances, a more careful search of the possibilities may pay for itself.

SELECTION FACTORS

Management personnel usually determine the varieties and qualities of dairy products they want to include on the menu. They may or may not work in concert with other company personnel in making these decisions. Regardless, they usually consider many of the selection factors outlined below.

Intended Use

As always, buyers want to determine exactly the intended use of the item so that they will be able to prepare the appropriate, relevant specification. For instance, if a cheese is needed primarily for flavor and only secondarily for appearance, the specification should reflect this.

Exact Name

It is very important for buyers to note the exact, specific name of the item they want. The majority of dairy products carry a standard of identity established by the federal government. This standard is based primarily on the minimum amount of butterfat content. For some products, the standards of identity also prescribe minimum or maximum amounts of milk solids allowed (see Figure 6.2 for some dairy products' legally defined minimum fat contents).

If an operation can use these minimum governmental standards, it may find bid buying the easiest procedure. Assuming that everything else is equal, the buyer could write the specifications with just a few words, for example, "vanilla-flavored ice cream." So, assuming that every bidder meets the required standard, a buyer could save money when one supplier bids lower than the rest.

U.S. Government Grades (or Equivalent)

The Agricultural Marketing Service (AMS) of the United States Department of Agriculture's (USDA's) Poultry and Dairy Division has set federal grading standards for poultry, eggs, and dairy products. U.S. grades, however, do not exist for every type of dairy product.[1] But milk, which is the base for all natural dairy products, usually is graded. As is true with most foods, the grading of milk and milk products is voluntary; however, many states require milk to be graded by the federal government (see Figure 6.3).

ITEM	MINIMUM PERCENTAGE FAT
Cheddar cheese	30.5%
Cottage cheese, creamed	4.0%
Cottage cheese, dry curd	0.5% (Maximum)
Cottage cheese, low fat	0.5% to 2.0%
Cream cheese	33.0%
Ice cream	10.0%
Ice milk	2.0% to 7.0%
Milk, evaporated	7.9%
Milk, low fat	1.0% to 2.0%
Milk, skim	0.1% to 0.5%
Milk, whole	3.25%
Mozzarella cheese	18.0% to 21.6%
Mozzarella cheese, part skim	12.0% to 18.0%
Neufchatel cheese	20.0% to 33.0%
Pasteurized process American cheese	26.8%
Pasteurized process American cheese food	23.0%
Pasteurized process American cheese spread	20.0%
Ricotta cheese	11.0%
Ricotta cheese, part skim	6.0% to 11.0%
Sour cream	18.0%
Whipping cream	30.0%

FIGURE 6.2 Some dairy products' legally defined minimum fat contents.

Like most foods high in protein, milk is a good medium for harmful bacteria. Consequently, most states and local municipalities have stringent health codes covering milk production. As such, milk must be produced and bottled under government-prescribed conditions. The U.S. Public Health Service's Milk Ordinance and Code (termed the Pasteurized Milk Ordinance [PMO]) contains provisions covering such activities as the approved care and feeding of dairy cows, the handling of the milk, the pasteurization* requirement, and the holding temperature of the milk.[2]

Because of these safety controls, dairies have little influence regarding milk production. They do, however, have the option of homogenization. This is the dividing of

---

* Heating the milk to kill pathogens (disease-causing but not spoilage bacteria). Milk that is not pasteurized is referred to as "raw milk." Hospitality operators usually are not allowed to serve raw milk to their customers.

FIGURE 6.3 Federal grade and inspection stamps used for dairy products.
Source: United States Department of Agriculture.

the butterfat globules so that they stay suspended in the milk and do not rise to the top. The dairies can also dictate what to do with their milk: sell it to ice cream makers, sell it to dry-milk producers, market it to households, and so on.

Fluid milk grades are based primarily on the finished product's bacterial count. There are two federal-government grading designations for fluid milk:

1.   **Grade A.** This is the milk the government considers to be fluid milk, to be sold in retail stores and delivered to consumers (see Figure 6.4).

2.   **Manufacturing Grade.** This milk is sometimes called Grade B. More bacteria are allowed in Manufacturing Grade milk than in Grade A; this milk is used for manufacturing milk products, such as butter, cheese, and ice cream.

Some persons discuss a third grade of milk, a certified grade. This refers to milk that has very little bacteria and can be used for infants and sick persons; technically, "certified" is not classified as a grade, but some buyers treat it as such.

Some spoilage bacteria always exist in pasteurized milk, but they are harmless. In addition to the number of bacteria, the grader also considers the milk's odor, taste, and appearance.

Milk can be fortified with vitamins A and D, and some states allow other types of nutrient additives.

As we noted previously, few dairy products are graded; this is primarily because the fluid milk used to produce them is usually graded and produced under continuous government inspection. Furthermore, the federal government has not established grading standards for most dairy products. In addition to fluid milk, U.S. grading standards have been determined for dry milk; Cheddar, Swiss, Colby, and Monterey Jack cheeses; and butter.

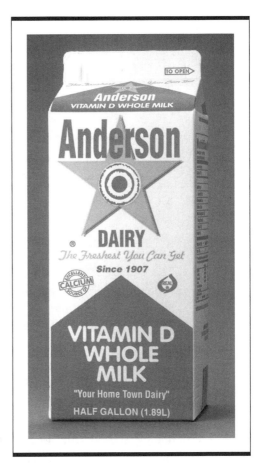

FIGURE 6.4 An example of a retail carton of milk. Courtesy of Anderson Dairy, Inc.

The federal grades for dry nonfat milk are:

1. Extra

2. Standard

The federal grades for dry whole milk are:

1. Premium

2. Extra

3. Standard

The grading factors for dry milk that the grader evaluates include the product's color, odor, flavor, bacterial counts, how scorched the milk is, how lumpy it is, how well it will go into solution, and how much moisture it contains.

The federal grades for Cheddar, Swiss, Colby, and Monterey Jack cheeses are:

1.   AA

2.   A

3.   B

4.   C

The grading factors for cheeses that the grader evaluates include the product's color, odor, appearance, flavor, texture, finish, and plasticity (i.e., body).

The federal grades for butter are:

1.   AA

2.   A

3.   B

The grading factors for butter that the grader evaluates include the product's flavor, odor, freshness, texture, and plasticity.

Some states use their own grading systems. For example, Wisconsin imposes grades for cheese. Also, the "U.S. Grade A" designation on a package label indicates that the federal government graded the dairy product, whereas the notation "Grade A" on the package label signifies that the dairy product meets specific criteria that a state, county, and/or local government agency established.

Additional terminology sometimes appears in the dairy products market, but this terminology does not necessarily represent governmental grades. Instead, this terminology tends to comprise terms and designations that, for one reason or another, have become popular. For example, ice cream carries several designations. It can be called "premium," "regular," or "competitive" (premium has 15 to 18 percent butterfat, regular has about 12 percent, and competitive has 10 percent—see Figure 6.5). Alternately, ice cream might be called "French," which means that eggs have been used as a thickening agent.

Another type of terminology especially prevalent on dairy product package labels is dating information. In some states, "pull" dates must be listed on dairy products' package labels. These dates tell supermarket managers, suppliers, and consumers the last day that the products can be sold. If states do not require pull dates to be listed, they usually require some sort of coded dates, or "blind" dates, to be noted on the package labels.

Typically, foodservice buyers opt for Grade A milk and a comparable quality for all the other dairy products. Since several dairy products differ in taste as one goes from one supplier to the next, it is not difficult to understand why U.S. Grades, and even local government grades, are not the major selection criterion.

Packers' Brands (or Equivalent)

The difference in taste between one supplier and another can be remarkable for cheese, yogurt, ice cream, sherbet, and dry milk. It is, in fact, amazing how different the taste can be between two brands of apparently equal merchandise.

Consequently, brand names tend to become important to buyers. Foodservice managers cannot be easily persuaded to drop their current ice cream for competing products. Of course, the bid buyer, or the buyer who has the time, occasionally checks out different brands of dairy products. New dairy items enter the market periodically, and some of them may be deemed to be good substitutes. Some of these new products may even save hospitality operators a bit of money.

Product Size

A few dairy products require size designations. For instance, butter could be ordered in 1-pound prints, 50-pound slabs, or one or more "chip" sizes. Cheese slices usually come in two or more sizes; for example, you might order a 1-ounce size for the cheeseburger platter and a 2-ounce size for the grilled cheese sandwich plate.

Size of Container

Dairy products are sold in various package sizes. Because of the highly perishable nature of dairy items, experience shows that the size of the container is very important. Buyers should purchase only the necessary amount in order to minimize leftovers and reduce waste.

Naturally, the smaller the package size unit—for example, half-pint milk containers in contrast to half-gallons—the higher the AP price. The edible-portion (EP) cost could, however, be lower. For instance, bartenders use cream in some drinks. A small package could carry a premium AP price, but if the cream drink volume is low, bartenders may waste cream if you use large containers. In this case, the EP cost would jump to an unacceptable level.

Not every supplier carries the package sizes buyers want. For example, a buyer may be satisfied with a particular brand but find that the supplier does not stock that brand in the individual portion packs the buyer desires.

Type of Packaging Material

Generally, dairy products packaging materials are quite standardized throughout the hospitality industry. One of the major reasons for this is that dairy regulations usually specify minimum packaging requirements that protect the culinary quality and wholesomeness of the products.

This standardization does not mean, however, that all dairy products are packaged alike. Buyers usually can select from a wide variety of packaging materials. There are plastic, fiberboard, metal, glass, and aseptic containers. Typically, two or more choices are available for many products.

Custom-packaging options may also be available for some dairy products. For instance, some dairies will include an operation's name and/or logo on individual half-pint containers of milk, individually wrapped butter chips, and single-serve creamers. Of course, the buyer must be prepared to pay a bit more in exchange for this added value.

Packaging Procedure

This can be an important consideration, especially for the single-serve dairy product items many restaurant operators purchase. For instance, buyers can purchase butter chips that are layered in a 5-pound container and separated by pieces of waxed paper. Alternately, buyers can obtain individually wrapped butter chips (which, by the way, the local health district may require in order to protect the wholesomeness and cleanliness of the butter).

As mentioned several times, the layered and/or individually wrapped products will carry premium AP prices. But the end result—that is, the EP costs—may be quite acceptable if buyers purchase premium packaging and packaging procedures that tend to protect the shelf life of the merchandise.

Product Yield

For some dairy products, buyers may need to indicate the maximum waste they will accept (or the minimum yield acceptable). For instance, they might need to indicate whether they would accept rind on the cheese they want to buy. Similarly, buyers should note on the specification that, for example, they will not accept more than two broken cheese slices per hundred.

Product Form

For some dairy items, buyers may need to note the exact form of the product. For example, they might need to note sliced, whole, grated, shredded, or crumbled cheese. Similarly, they might need to note whipped butter, if applicable, instead of just butter.

Preservation Method

Most dairy items are kept under continuous refrigeration. Although refrigeration is not required for some items, such as certain cheeses, if buyers want these types of items kept under refrigerated conditions, they need to note this on the specification.

Some dairy items are frozen. The obvious ones are ice creams and frozen yogurts. However, some suppliers freeze the cheeses and butter they sell. So if buyers do not want frozen dairy items, they may have to specify this for some items that they purchase.

A few dairy items are traditionally canned. Evaporated milk, sweetened condensed milk, and canned whole milk are usually marketed in metal containers. Whole milk also comes in aseptic packages and can be kept at room temperature for months. This "shelf-stable" product is pasteurized using "ultra-high temperatures" (UHT), and its taste is very similar to fresh, refrigerated, whole, fluid milk and coffee creamers. This technique is sometimes referred to as "ultra-pasteurized" (UP). Although individual UP creamers are used extensively in foodservice operations, the whole-milk product has yet to gain widespread popularity in the United States.

When considering preservation methods, wise buyers also take the time to specify the maximum pull date allowed at time of delivery. If stored correctly, dairy products will remain safe to consume for a few days after the pull date; however, their culinary quality could be compromised to the point where these products should not be served to guests. Furthermore, the local health district may not allow the use of out-dated products.

Butterfat Content

In general, as the butterfat increases, so does the AP price, but more butterfat also makes for a better product. Moreover, producers tend to treat dairy products with a high butterfat content with more respect. For example, a premium ice cream typically contains high-quality flavorings—fresh fruits rather than fruit syrups, for example.

If buyers are satisfied with the amount of butterfat mandated by the federal government's standard of identity, they can ignore this selection factor. However, if they want a product that is more or less "creamy," they must note this requirement on the specification.

Milk Solids Content

The federal government also mandates the maximum amount of nonfat, dried milk solids that some dairy products can have. If these standards are acceptable, this selection factor is irrelevant. However, if buyers desire fewer solids than the maximum allowed, they must indicate their exact requirement on the specification.

Overrun

The amount of air in a frozen dairy product is referred to as "overrun." Most people in the foodservice industry consider overrun to be the amount of air incorporated into any type of dairy product. Some dairy products contain a good deal of air. When a chef whips butterfat, he or she incorporates air. Also, butterfat holds the air for quite a while, and even longer when the product is frozen or contains some added emulsifiers.

The air content is crucial to the flavor of such items as ready whipped cream in an aerosol can and ice cream.

Whipped cream is usually sold by the number of ounces in the can. But it can also be sold by volume, which is a typical measure of quantity for many types of dairy products. If buyers start to compare AP prices on the basis of volume, they must keep in mind that air costs nothing. So they could be buying more volume but less solid product.

The federal government standard of identity for ice cream dictates that 1 gallon must weigh at least $4\frac{1}{2}$ pounds and contain at least 1.6 pounds of total food solids. Therefore, for this type of product, buyers are protected to some extent. But this is not the case for such items as whipped topping. For these types of products, buyers must be ever mindful of the exact value of the purchase.

Chemical Additives

Because milk is a food for babies, it has been kept natural for many decades. However, a few dairy products contain chemical additives that stabilize, emulsify, and preserve them, and at times, the dairy industry has been unjustly criticized for this. It is easy to assume that all dairy items include several chemical additives. But this is just not the case. The products that typically contain chemicals are nondairy items. All things considered, dairy products in their natural form, processed or relatively unprocessed (pasteurized), have significantly fewer chemical additives than other processed foods.

Untreated Cows

In this age of biotechnology, dairies are able to treat their herds with synthetic and naturally occuring hormones—such as bST (bovine somatotropin) and rBGH or rBST (Recombinant Bovine Growth Hormone)—designed to increase milk production. Some buyers, though, may not want to purchase products made from this type of milk.[3] If so, they need to note on the specification that they will accept only products coming from cows that have not been treated this way.

How the Product Is Processed

Dairy processing methods usually fall under government inspection. But these inspections ensure only wholesomeness, not flavor, convenience, or packaging.

The type of processing can be very important for some dairy products. For example, all Swiss cheeses are a bit different. Although they all meet a minimum standard of identity, substantial differences in aging methods and aging times can exist. The packer's brand usually indicates these processes.

Buyers may want to know whether or not the process is "natural." For example, some cottage cheeses contain an absolute minimum of additives; others may contain extra acid, such as phosphoric acid to set the curd, and artificial flavorings. If buyers want all of the dairy items they purchase to be natural, they will have to search out the appropriate brand.

FIGURE 6.6 A nondairy item. Courtesy of Sysco Corporation.

Nondairy Products

Many operators may use nondairy items for several reasons, including: (1) AP prices may be lower; (2) nondairy products, being less perishable, save on storage costs and reduce waste; and (3) weight watchers and those people who cannot tolerate lactose (milk sugar) may represent a clientele worth accommodating (see Figure 6.6).

Unfortunately, most imitation items contain some chemical additives and usually are not nutritionally equivalent to the products they imitate; consequently, some customers may refuse to use them. The fat substitutes being marketed today do not impress many nutritionists, who doubt very much that these products will make people healthier or slimmer. Furthermore, nondairy products may not work in some recipes that call for dairy ingredients.

AP Price

In a free-market area, some opportunities to reduce the AP price may exist. In a controlled state, you must legally pay at least the minimum AP price.

Most states have a variety of price control and credit control policies; that is, the local governments regulate the price and the type and amount of credit a dairy can extend to its customers. Periodically, someone starts a drive to eliminate the local government and the federal government's power in this area. But the price and credit controls seem to weather these attacks very well.

Dairy products provide few quantity buy opportunities, although there may be money-saving long-term contracts for some items if it is legally possible.

Also, some dairy products, especially cheese, are imported. Import taxes tend to add up, thereby increasing AP prices.

One-Stop Shopping

Most foodservice operators like one-stop dairy shopping because of the standing order they can bargain for and some dairy suppliers provide. If buyers can get a supplier to bring his or her current stock of dairy items just up to par, this supplier service is an added plus. However, the supplier then controls the inventory level. But since dairies often provide frequent deliveries, this reduces the amount of inventory of perishable items hospitality operations must carry.

One-stop shopping tends to entail a higher AP price. To receive the convenience we noted, however, operators may be willing to pay more.

PURCHASING DAIRY PRODUCTS

Since most dairy products are highly perishable, the buyers' first purchasing steps are to determine precisely what they want and then to determine the delivery schedule they think will be appropriate. Buyers prefer daily delivery; however, they should negotiate for and follow any supplier service or purchasing tactic that helps control the quality of these items.

Preparing elaborate specifications for dairy products usually is not necessary unless the buyer is a bid buyer or if he or she expects to enter into a long-term contractual arrangement with the supplier. As always, it might be good discipline for a buyer to reduce his or her ideas to detailed written specifications before committing to purchasing any item. (See Figure 6.7 for some example product specifications and Figure 6.8 for an example product specification outline for dairy products.)

If a hospitality operation uses the ordinary types of dairy products and is located in a noncontrolled state, bid buying might be profitable. But as a percentage of total purchase dollars, these savings are liable to be small. However, if operators can live with some variations, their savings can add up through the optimal selection of many suppliers and packers' brands.

Butter Used for customer service U.S. Grade AA Butter chips 90 count Layered arrangement, easily separated 5-pound box Waxed, moisture-proof, vapor-proof Refrigerated	Nondairy coffee whitener, liquid Used for customer service House brand $3/8$-ounce portion, single serve 400 servings per case Loose pack (slab pack) Moisture-proof carton Unrefrigerated
Bleu cheese Used for tossed salad Frigo® brand Crumbles 5-pound poly bag 4 bags per case Moisture-proof, vapor-proof Frozen	Half-and-half Used for customer service U.S. Grade A $3/8$-ounce portion, single serve 400 servings per case Loose pack (slab pack) Moisture-proof carton Refrigerated

FIGURE 6.7 An example of dairy product specifications.

If buyers can live with variations and/or purchase a large quantity of dairy products, they should take the time to evaluate some of the substitution possibilities. Several products are capable of providing comparable culinary quality in a recipe, and a few minutes of cost calculation might indicate that a particular recipe is much less expensive to produce with one of these instead of with a comparable one. For instance, if buyers use large quantities of fresh milk to prepare breads, a switch to dry milk may yield a comparable-quality finished product at a lower EP cost.

Sometimes independent farmers seek to do business with hospitality operations. It might be best to avoid them no matter how "natural" their products may seem or how good the deals they offer appear to be. The products might not come under the rigid quality control standards the federal and state governments have established.

RECEIVING DAIRY PRODUCTS

When receivers get dairy products, they should take the time to carefully examine them for dirt, broken containers, and faulty wrapping. Milk cartons can get dented, and cheese wrappings sometimes crack or split. Since these products deteriorate

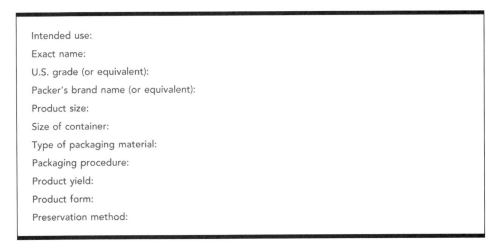

Intended use:

Exact name:

U.S. grade (or equivalent):

Packer's brand name (or equivalent):

Product size:

Size of container:

Type of packaging material:

Packaging procedure:

Product yield:

Product form:

Preservation method:

FIGURE 6.8 An example of a product specification outline for dairy products.

quickly, receivers should be reluctant to accept anything that does not look clean and properly packaged.

Receivers also must check to see that they receive everything that was ordered. This check can be difficult for at least two reasons. First, with so many dairy items on one invoice, either the receivers or the suppliers may miss something. Second, some dairy items are delivered on a standing-order basis. This is typical with ice cream and sherbet, and it may allow the delivery agent to work alone stocking an operation's freezers. When leaving, the agent may present an invoice for the receiver to sign. If the receiver is busy, he or she may not thoroughly check what has been delivered. Alternately, an unscrupulous delivery agent may tell a receiver that a container that already was in the dairy box was delivered that day.

A related problem centers on supplier substitutions. For example, a supplier may be out of Roquefort cheese and send bleu cheese instead, not wanting to see the operation serve guests without at least a similar item. But some substitutes in the dairy line do not always match well, especially cheeses.

Although receivers do not often take the time to make planned but random taste tests, they should. This may not be necessary if buyers purchase a proprietary brand; they will then have some assurance of quality.

Since most dairy items are perishable, receivers might consider moving everything into a refrigerated area before they make their inspection. After checking qualities and quantities, receivers should check the invoice arithmetic and complete the appropriate accounting procedures.

STORING DAIRY PRODUCTS

Most dairy products should be stored in a refrigerator or freezer as soon as possible. Dried, canned, and bottled items can go to the storeroom, as can, possibly, some nondairy products.

If chefs are going to serve a certain dairy product—for example, if they plan on presenting a cheese platter on that evening's menu—they should bring the cheese to the correct serving temperature by leaving it out at room temperature before serving. If they are going to serve the cheese later in the week, it should be refrigerated, since room temperatures cause most cheeses to age. This will have a detrimental effect on the cheese by causing it to quickly change in odor, flavor, and appearance.

Most dairy items readily pick up odors. Therefore, maintaining a separate dairy refrigerator is recommended. If that is impossible, hospitality operators should keep dairy products tightly covered, in a segregated area in the common refrigerator, and away from odorous foods.

As much as possible, operators should also keep dairy products, particularly cheeses, in their original packaging. When they store these items, they should try not to nick or cut the packaging. This is easy to do and hastens spoilage and waste.

When storing dairy products, operators should take a bit of extra time to ensure that they rotate the products on the shelves properly. They cannot take a chance that a customer will get sour milk. It is not easy to tell whether the food is rotated properly unless they take the extra time to check the pull dates many dairies put on their products. Dairy products are not like lettuce: if a head of lettuce is bad, you know it, but whole milk in individual half-pints is harder to monitor.

ISSUING DAIRY PRODUCTS

Hospitality operators should issue older dairy products first. Many of these items, especially ice cream, go straight from receiving to a production area. If they are issued from a central storeroom, make sure that the requisitioner receives the correct product. For example, if the requisitioner wants milk for a cake recipe, make sure that he or she gets the appropriate dry milk.

Since most dairy products deteriorate rapidly, hospitality operators should try not to handle them any more than necessary. Also, they should make sure that requisitioners do not take more than they need for any one particular work shift or job. Operators might consider asking the requisitioner to make a note of the in-process inventory before asking for more stock.

IN-PROCESS INVENTORIES

Dairy products fall victim to spoilage, waste, and pilferage whenever they stay in-process for any extended period. They spoil because butter, cheese slices, and coffee cream, for example, are often left at room temperature too long. Also, employees waste dairy products by failing to empty milk containers and cans of whipped cream completely. Pilferage is particularly common. For instance, employees may help themselves to a quick glass of milk once in a while.

As usual, supervision is the key. It helps both to head off waste and pilferage and to prevent, as well, such embarrassing situations as a customer tasting curdled coffee cream or rancid butter.

KEY WORDS AND CONCEPTS

Agricultural Marketing Service (AMS)	Milk solids content	Raw milk
As-purchased price (AP price)	Nondairy products	Recombinant bovine growth hormone (rBGH or rBST)
Blind dates	One-stop shopping	Shelf-stable products
Bovine somatotropin (bST)	Overrun	Size of container
Butterfat content	Packaging material	Spoilage bacteria
Chemical additives	Packaging procedure	Standard of identity
Custom packaging	Packers' brands	Substitution possibilities
Disease-causing bacteria	Pasteurization	Synthetic hormones
Edible-portion cost (EP cost)	Pasteurized Milk Ordinance (PMO)	Truth-in-menu legislation
Exact name	Preservation method	Ultra-high temperature (UHT) pasteurization
Fortified milk	Price and credit controls	Ultra-pasteurized (UP)
Grading factors	Processing method	U.S. grades
Homogenization	Product form	U.S. Grade A versus Grade A
In-process inventories	Product size	U.S. Public Health Service's Milk Ordinance and Code
Intended use	Product yield	
Lactose	Pull dates	Vitamins A and D
Manufacturing grade	Purchasing, receiving, storing, and issuing dairy products	

REFERENCES

1. For a detailed dicsussion of dairy grades, see http://www.ams.usda.gov/dairy. Retrieved October 2, 2003.

2. See, for example, Anonymous, "Milk Safety References," U. S. Food and Drug Administration Center for Food Safety and Applied

Nutrition, Retrieved October 3, 2003 (Available http://www.cfsan.fda.gov/~ear/prime.html).

3. See, for example, Anonymous, "POSILAC® bovine somatotropin," Monsanto Dairy Website, Retrieved October 3, 2003 (Available http://www.monsantodairy.com/). See also: Deana Grobe and Robin Douthitt, "Consumer Risk Per-ception Profiles Regarding Recombinant Bovine Growth Hormone (rbGH)," *Journal of Consumer Affairs*, Winter 1999, 33(2), pp. 254–276; David Smith and Robert Skalnik, "Biotechnology in the Agricultural Sector: A Challenge to Consumer Welfare," *International Journal of Consumer Studies*, September 2003, 27(4), pp. 277–283.

QUESTIONS AND PROBLEMS

1. What are the U.S. grades for fresh fluid milk?

2. What are the U.S. grades for butter?

3. What is the minimum weight of a gallon of ice cream?

4. What is the minimum butterfat content for ice cream?

5. What is an appropriate intended use for nondairy creamer?

6. What is an appropriate intended use for margarine?

7. What is an appropriate intended use for dry nonfat milk?

8. Assume that you operate the food service in a minimum security prison. You serve approximately 500 inmates and 120 civilian staff members a day, three meals and various snacks. You have a severely tight food budget. Outline the specific procedures you might use to purchase, receive, store, and issue whole, fluid milk for use in cooking and as a beverage. Note: The prisoners are your workers. If possible, ask a prison foodservice official to comment on your answer.

9. What is the primary purpose of pasteurization?

10. What is the primary purpose of homogenization?

11. What are the U.S. grades for dry nonfat milk?

12. What are the grading factors for Cheddar, Swiss, Colby, and Monterey Jack cheese U.S. grades?

13. What critical information is missing from the following product specification for milk?
    Milk, fluid
    U.S. Grade A
    Used for cooking and baking
    Bulk container

14. What is the primary difference between premium ice cream and competitive ice cream?

15. What is an appropriate intended use for low fat milk?

16. Explain why dairy products should not be stored with fresh produce.

17. A product specification for an ice cream bar could include the following information:
    (a)
    (b)

(c)

(d)

(e)

18. One-stop dairy product shopping is especially popular among small operators. Why do you think this is the case? What advantages are there? What disadvantages are there?

19. Why do most government jurisdictions prohibit restaurants from serving raw milk?

20. Define or explain the following terms:

(a)	Pull dates	(i)	Custom packaging
(b)	Certified milk	(j)	Product form
(c)	Fortified milk	(k)	Aseptic container
(d)	Lactose	(l)	Nondairy products
(e)	Minimum butterfat content	(m)	Product yield
(f)	Overrun	(n)	Standard of identity
(g)	Controlled AP prices	(o)	U.S. Grade A versus Grade A
(h)	UHT pasteurization		

EGGS*

The Purpose of this Chapter

After reading this chapter, you should be able to:

- Explain the selection factors for eggs, including government grades.
- Describe the process of purchasing, receiving, storing, and issuing eggs.

INTRODUCTION

The most important egg purchasing considerations center on determining what hospitality operations want and which type of product is best suited to their needs. Purchasing fresh shell eggs is a relatively easy task. Buying processed eggs, whether frozen, dried, or imitation eggs, or preprepared egg products, can be more of a challenge.

SELECTION FACTORS

As with all products, management personnel, either alone or in cooperation with others, normally decide in advance the quality of eggs the operation needs. They use the following selection factors to evaluate the standards of egg quality and, to a certain extent, the egg suppliers.

---

* Authored by Andrew Hale Feinstein and John M. Stefanelli.

Intended Use

As always, buyers want to determine exactly the intended use of an item, so that they will be able to prepare the appropriate, relevant specification. For instance, an egg product may be needed primarily for flavor and only secondarily for appearance. If so, the specification should reflect this.

Exact Name

Generally, this selection factor causes very little difficulty. Fresh eggs are chicken eggs, so if buyers use the term "eggs," they will receive chicken eggs (see Figure 7.1). The term "fresh shell eggs" refers to eggs that are fewer than 30 days old. Conversely, "storage eggs" are shell eggs older than 30 days.[1]

Storage eggs are rarely utilized in the foodservice industry, as egg production has been stabilized throughout the year and most shell eggs are now distributed within a few days. Shell eggs kept under refrigeration or in a controlled atmosphere for long periods of time are now considered unacceptable. However, if an operator's specification notes

FIGURE 7.1 A flat of chicken eggs. Courtesy of the American Egg Board.

"shell eggs" or "eggs" instead of "fresh shell eggs," he or she may be disappointed with the purchase.

The term "egg products" refers to eggs that have been removed from their shells for processing at facilities called "breaker plants." Whole eggs, whites, yolks, and various blends—with or without added ingredients—that are processed and pasteurized, are basic types of egg products (see Figure 7.2). These products are available in liquid, frozen, and dried forms. Because of the different combinations of types and forms, buyers need to carefully consider their selection factors when ordering processed egg products. Buyers must be absolutely certain that they indicate the exact name of the desired item to prevent receiving an unacceptable product.

U.S. Government Inspection and Grades (or Equivalent)

The Egg Products Inspection Act (EPIA) of 1970 requires the United States Department of Agriculture (USDA) to ensure that egg products are safe, wholesome, unadulterated, and accurately labeled, for the protection of the health and welfare of consumers (www.fsis.usda.gov/OPPDE/rdad/Acts/epia_toc.htm). The original impetus for this act was driven by states' concerns regarding the contamination risk present with eggs and

FIGURE 7.2 Processed egg products.
Courtesy of Michael Foods, Inc.

egg products. Under the EPIA, breaker plants must submit to continuous government inspection. (Conversely, federal shell egg inspection is a voluntary program and is under the jurisdiction of the Food and Drug Administration [FDA].)

In most states, an egg producer must submit to either a federal-government inspection or a state-operated inspection. These inspectors also examine the condition of the laying hens, their environment, and their feed. The USDA's Food Safety and Inspection Service (FSIS) employs inspectors to examine egg products at official breaker plants. However, federal and state agencies often work together to perform inspections.

Unlike the FSIS, the USDA's grading service is voluntary. Therefore, shell egg processing plants can elect whether or not to purchase the egg quality grading program the USDA's Agricultural Marketing Service (AMS) administers. The USDA grade shield can be placed on the carton when eggs are graded for quality and checked for weight (size) under the supervision of a trained USDA grader.

The shell eggs produced under continuous federal-government inspection as mandated by certain states for monitoring the compliance with quality standards, grades, and weights can bear the USDA grade shield. States that elect not to purchase the USDA grading service but that use their own agencies to monitor egg packers for compliance can allow the cartons to bear such a term as "Grade A" without the USDA shield.

Three federal-government consumer grades for fresh shell eggs have become familiar quality guidelines (see Figure 7.3).

1. **U.S. Grade AA.** These eggs have whites that are thick and firm; yolks that are high, round, and practically free from defects; and clean, unbroken shells. Because these eggs are the top quality produced, only the freshest products earn this grade. The grade is very hard to obtain because once an egg is about a week old, its quality deteriorates to Grade A.

2. **U.S. Grade A.** These eggs have characteristics of Grade AA eggs except that the whites are "reasonably" firm. This is indicative of slightly lower quality. A fresh egg older than about a week usually falls into this grade category. This product's egg white and egg yolk are not quite as firm as those found in Grade AA merchandise. However, both Grade A and Grade AA eggs are generally suitable for all finished menu products where appearance is important.

3. **U.S. Grade B.** These eggs have whites that may be thinner and yolks that may be wider and flatter than eggs of higher grades. The yolks are also susceptible to breaking under even the slightest pressure. The shells must be unbroken but may show slight stains. Therefore, because of these appearance problems, Grade B merchandise is suitable only for finished menu products in which eggs are used as an ingredient or as scrambled eggs (or omelettes).

FIGURE 7.3 Federal grade and inspection stamps used for eggs.
Source: United States Department of Agriculture.

Within these grade categories, an egg can be rated "high," "medium," or "low." As such, nine possible grades exist.

Fresh eggs are graded mainly on interior and exterior quality factors. Interior quality, such as the firmness of the egg yolk and egg white, is determined primarily by freshness. The exterior quality factors, such as shape, cleanliness, and soundness, are determined by the age of the hen, the feed eaten, and the general management of the flock and of the egg-laying facilities. The fresher an egg, the better it is, assuming that the laying hen is the right age (from 6 months to $1\frac{1}{2}$ years old), is eating a proper diet, and is living in an appropriate environment.

The grader uses a process called "candling" to check the interior quality of fresh eggs. This involves passing the egg over a light source, which reveals the yolk (a dead center implies freshness), the size of the air space (which gets larger as the egg becomes older), impurities, and cracks. The grader may also crack open randomly selected eggs to determine the height and firmness of the egg white. He or she might also evaluate the condition of the shell, especially if the laying hens are relatively old. Older hens produce eggs that have rougher and thinner shells.

The use of U.S. grades for fresh-egg purchasing is widespread, even though a wide tolerance exists between egg grades. A major advantage of graded shell eggs is that they have been produced under continuous government inspection. It is a good idea for buyers to insist on this type of inspection because shell eggs are on the FDA's list of potentially hazardous foods. Fresh shell eggs, even those that appear to be sound, can be contaminated with salmonella bacteria by the hens on rare occasions. Continuous government inspection, as well as buyers insisting on constant refrigeration at 45°F, should help mitigate this problem.

Foodservice buyers usually purchase Grade A eggs. Grade AA eggs are difficult to obtain. In addition, Grade A eggs normally suffice, particularly for fried and scrambled eggs and omelettes.

Packers' Brands (or Equivalent)

Some supermarkets pack their own eggs, so these eggs carry a packer's brand. But foodservice buyers do not show a great deal of brand loyalty for fresh shell eggs. The federal grade is the most common quality indicator.

However, packers' brands can be extremely important when buyers purchase processed and convenience forms of eggs, such as frozen, refrigerated liquid, and dried egg products. Processed egg products, while produced under continuous government inspection, do not have established grading standards. As a result, in almost every instance, brand names are the only reliable indication of quality.

Product Size

Buyers are normally interested in the size and uniformity of the shell eggs they purchase. The U.S. government helps in this area because graded eggs must meet quality and size standards. If a federal inspector grades the eggs, the producers are required to indicate the egg size somewhere on the container. Buyers should choose the size most useful and economical. Shell eggs are available in six sizes (see Figure 7.4).

Peewee eggs, sometimes called "pullet eggs," come from younger hens, usually at the beginning of their laying life. Jumbo eggs come from relatively older hens, usually at the end of their laying life. These extreme sizes are not common in the fresh-egg trade. If buyers want these sizes, normally they must visit a farmer with very young and very old laying hens.

When choosing eggs to be fried, scrambled, poached, or prepared in omelettes, buyers usually prefer the large size. These eggs have a very acceptable appearance and show up well on the plate. Also, those who purchase fresh eggs for use in cake, batter, or

SIZE OR WEIGHT CLASS	MINIMUM NET WEIGHT PER DOZEN (OUNCES)
Jumbo	30
Extra large	27
Large	24
Medium	21
Small	18
Peewee	15

FIGURE 7.4 U.S. weight classes for shell eggs.
Source: United States Department of Agriculture.

drink recipes tend to choose large eggs because most quantity recipes, as well as recipes found in most cookbooks, assume a 2-ounce egg.

Buyers who keep close track of the as-purchased (AP) prices for each size might be able to save a few pennies when purchasing fresh shell eggs. For instance, in a recipe calling for eggs by weight, if large eggs cost $1.20 per dozen and medium eggs cost $1.00 per dozen, buyers could determine the best price per ounce in the following manner:

$1.20 per dozen ÷ 24 ounces per dozen = $0.05 per ounce
$1.00 per dozen ÷ 21 ounces per dozen = $0.0476 per ounce

So, medium eggs represent a slightly better buy.

We do not normally see buyers use this procedure; it is probably more helpful to homemakers than to commercial buyers, who are often restricted to the large-size egg. But if buyers want to purchase a large volume of eggs for use in various recipes and the labor is available to process them, such a procedure might save money.

Size of Container

For shell eggs, the normal package size is a 15-dozen or 30-dozen case. At times, buyers might order one or more "flats"; a flat contains $2^1/_2$ dozen fresh eggs. It is unusual to purchase eggs in 1-dozen or $1^1/_2$-dozen containers.

Processed egg products offer a bit more variety in package sizes, so buyers must be prepared to describe the exact packaging characteristics that best suit their needs.

Type of Packaging Material

For fresh shell eggs, the packaging is standardized. Buyers probably will not need to consider this selection factor if they order only fresh eggs. The same may not be true with processed egg products, since a considerable variation in packaging quality can exist. Buyers must be concerned with this selection factor because improperly packaged products can support tremendous bacterial multiplication. This is especially true if hospitality operators must store the processed egg products on their premises for a reasonably long period of time.

Usually, frozen egg products are packaged in moisture-proof, vapor-proof containers. In most instances, a processed product, such as frozen, precooked scrambled eggs, normally is packaged in a heavy plastic pouch; these pouches are sometimes referred to as "Cryovac® bags," or "Cryovac packaging." (Cryovac is a brand name. It is the company that developed the "vacuum shrink-wrap" technology that enables food processors to store products for considerable lengths of time.)

Some frozen items are also packed in metal or plastic containers. For instance, frozen eggs, frozen egg yolks, and frozen egg whites are typically packaged in 30-pound plastic containers, tankers, or poly-lined drums.

Dried egg products—which are becoming less frequently used in the hospitality industry—are, generally, sold in 6-ounce pouches, and 3-pound and 25-pound poly packs, which are airtight, plastic-lined bags. Some of these products are also packaged in metal or plastic containers.

Packaging Procedure

The packaging process is not a concern for most egg products that the typical hospitality buyer purchases. The types of packaging and packaging procedures for fresh shell eggs are very standardized, as they are for most processed products.

In a few instances, buyers may need to specify a desired packaging procedure. For example, if they want to purchase frozen, precooked, plain omelettes, they could purchase them individually wrapped and stacked neatly in the case, or in a layered arrangement, where they are separated by sheets of waxed paper. In some cases, then, it is possible for buyers to specify a combination of inner wrapping and outer wrapping that meets their needs.

Color

The breed of the hen determines the color of an egg. White-feathered chickens, such as the Leghorn, White Rock, and Cornish chickens, lay white eggs. Chickens with red feathers, such as the Rhode Island Red, New Hampshire, and Plymouth Rock chickens, lay brown eggs. Although no differences in flavor and nutrition have been proved, in some parts of the United States, such as New England, consumers request brown eggs. Consumers may perceive some psychological difference in shell color. So, in addition to a potential price difference between brown- and white-shell eggs, buyers must keep this customer preference in mind.

Product Form

In some instances, buyers may wish to purchase one or more convenience egg products. For example, they may want to buy precooked, refrigerated, whole, peeled eggs and use them to garnish their salad bar offerings. Alternately, buyers might wish to purchase cheese-stuffed, frozen, precooked omelettes. As always, the added value of the desired form will increase the AP prices, but the ultimate edible-portion (EP) costs may be quite affordable.

Preservation Method

Buyers should know how fresh and processed egg products are preserved, so that they can make additional judgments concerning the items' quality. The most common preservation methods are discussed in the following paragraphs:

Refrigeration This is the most common preservation method for fresh shell eggs. As fresh eggs get older, they lose quality: moisture dissipates, the white gets thinner, and the yolk becomes weaker. Refrigeration is the best deterrent to this quality loss. The FDA recommends that state and local health districts require fresh shell eggs to be received and stored at 45°F or less in order to minimize food-borne illnesses that can result if the eggs are contaminated with small amounts of salmonella bacteria. However, no federal law requiring egg refrigeration exists. Wise buyers do not jump to the conclusion that the fresh shell eggs they have purchased have been kept under constant refrigeration.

Oil Spraying or Dipping When a hen lays an egg, it puts a protective coating on the outside of the egg. At the plant, federal-government regulations require that USDA-graded eggs be carefully washed and sanitized using a special detergent, which removes this coating. To counteract this, usually the egg is coated by the supplier with a tasteless, natural mineral oil. This helps to preserve egg freshness. Oil spraying is not quite as effective as refrigeration, though it is a reasonably effective alternative.

Overwrapping To retard moisture loss and the tendency for its yolk and white to thin, an egg can be doubly wrapped in heavy plastic film. If wrapped correctly, this method is superior to oiling eggs.

Controlled-Atmosphere Storage Some producers hold eggs in an oxygen-free environment. The oxygen is removed and replaced by carbon dioxide. As an egg ages, it loses moisture and carbon dioxide. The carbon dioxide in the environment acts as a counter-pressure, thereby preventing the loss of carbon dioxide and keeping the egg white firm. This is an expensive method, but producers who wish or need to hold shell eggs for a while use it.

As a general rule, suppliers do not sell controlled-environment eggs to foodservice operators. These eggs are intended for supermarket and grocery store distribution. As with fresh produce, once eggs are removed from the controlled atmosphere, their quality deteriorates very rapidly; hospitality operators cannot tolerate this problem.

The Processing Method When buyers purchase processed egg products, they are actually purchasing eggs that have been preserved in a manner other than in the shell. Among the most common forms of processed eggs are frozen whole eggs, frozen egg yolks, and frozen egg whites. Large bakeries normally use these products. Some labor savings are associated with these products since no one in the operation has to process them.

But problems are associated with these egg products, too. First, the freezing process must be conducted using optimal conditions and techniques. Frozen yolks will become rubbery if sugar or glycerine is not added to them before freezing and if they are not frozen correctly. Second, the thawing process for frozen eggs should be done in a controlled refrigerated temperature. (Operators should allow three to five days in the refrigerator for thawing a 30-pound can.) Unsupervised employees may not be patient enough to wait out this relatively long thawing procedure. Also, if frozen eggs are thawed at room temperature, harmful bacteria can multiply rapidly.

Other familiar forms of processed eggs used in food-service operations and large bakeries are dried eggs, dried egg yolks, and dried egg whites. Since reconstitution may be problematic, dried eggs are not normally used for scrambling. They tend to work better in recipes that call for eggs because cooks can measure them easily. The only possible problem associated with dried eggs, other than their reconstitution, is that some of the product may be scorched during the drying process. This is much less of a problem today due to improved processing technology. Spray drying, or spraying the liquid egg mixture into a heated environment, is a superior method. Freeze-drying, or freezing liquid eggs and going from the frozen state directly to the dried state, is also successful.

Buyers can purchase other types of processed egg products. For instance, they can order frozen deviled eggs and frozen, cooked scrambled eggs. These items are expensive because of the convenience they offer, but the potential labor savings might more than offset the high AP price.

Trust the Supplier

The problem with buying eggs is not so much selecting the quality: buyers will rarely go wrong if they stipulate U.S. grades for fresh shell eggs or if they settle on a particularly desirable brand name for a processed egg product. Buyers must be concerned with choosing the right supplier. Since suppliers abound in the egg trade, the choice can be wide. Buyers must make sure that a supplier can make adequate deliveries and moves only fresh product. Also, if buyers expect that fresh eggs be refrigerated, they may need to check that the supplier maintains the specific storage environment.

In addition to supplier services, buyers must consider purchasing options. Bid buying and becoming a house account are two alternatives. Both have their advantages and disadvantages, and buyers must make their choice.

PURCHASING EGGS

As usual, the first step in egg purchasing is for hospitality operators to determine precisely what they want. As noted earlier, fresh shell eggs present few problems: the most widely used quality is U.S. Grade A, and the normal size is large. However, considerably more combinations of qualities and sizes are available. As a result, management, either alone or in conjunction with other key employees, must determine the efficacy and usefulness of these combinations as they relate to a particular operation.

The qualities and styles of processed egg products are not so easily chosen. If buyers purchase these items, their best bet is to consider brand names. No federal quality standards exist for processed egg products (although buyers could specify that the processed product be prepared with fresh eggs of a certain U.S. grade); the convenience and reliability of the packer's brand name eventually become the overriding factors.

Once buyers make their decision, they need to prepare a complete specification that includes all pertinent information, whether or not they use it in bid buying. If

Fresh shell eggs
Used for fried, poached, scrambled eggs
U.S. Grade AA
Large
30 dozen per shipping case; 20 cases
Cartons labeled with an expiration date
    not to exceed 28 days from date of
    packaging
Eggs delivered within 7 days of official
    grading
White shell

Scrambled egg mix
Used for scrambled eggs on buffet
Fresh Start® brand
2-pound carton
Moisture-proof, vapor-proof carton
6 cartons per case
Refrigerated liquid

Meringue powder
Used to prepare dessert topping bakery
    products
R & H® brand
6-pound container
Plastic, resealable container
Unrefrigerated

Frozen, whole, shelled eggs
Used to prepare bakery products
McAnally® brand
30-pound container
Metal can

FIGURE 7.5 An example of egg product specifications.

nothing else, the discipline of preparing this document will help to ensure that buyers have considered all of the relevant factors and are, indeed, purchasing the egg product that suits their needs (see Figure 7.5 for some sample product specifications and Figure 7.6 for an example of a product specification outline for egg products).

After determining what they need and when they need it, buyers must evaluate potential suppliers. Buyers will find a reasonable number of potential suppliers in the fresh-egg trade, which is good news for those who like the bid-buying strategy. The biggest problem in supplier selection probably involves determining which supplier provides the freshest eggs. A "U.S. Grade A" designation on the box is no guarantee of the quality of the eggs inside.

The processed-egg trade does not offer as many suppliers or brands. If buyers want dried eggs, the limited number of purveyors stocking them may be a surprise. Furthermore, if buyers want reduced-cholesterol eggs, they may find even fewer suppliers. Similarly, not too many producers carry such specialty items as frozen deviled eggs.

Intended use:

Exact name:

U.S. grade (or equivalent):

Packer's brand name (or equivalent):

Product size:

Size of container:

Type of packaging material:

Packaging procedure:

Color:

Product form:

Preservation method:

FIGURE 7.6 An example of a product specification outline for egg products.

As we found with fresh produce, independent farmers are very active in the fresh-egg trade. But buyers should be wary of them. Keep in mind that their flocks may be too young or too old, and the hens may not be well managed. Also, independent farmers can purchase eggs from someone else and then resell them, implying that the eggs are from their own farms. Moreover, government inspectors may not check these farmers.

One potential advantage of buying from independent farmers is that they may be able to get shell eggs to buyers one or two days after laying. When purchasing from a typical supply house, buyers can expect to receive eggs that are almost a week old. But if the farmer does not provide refrigerated storage, the older egg may be a better buy because an unrefrigerated, 2-day-old egg has less quality than a refrigerated 7-day-old egg.

Another potential advantage might be the willingness of the independent farmer to bargain for an AP price based on an agreed-upon markup of the wholesale egg market price. This might represent a reasonable saving. Ordinarily, most suppliers are not eager to use this type of pricing technique except for their large customers.

Before purchasing any egg product, buyers should take some time to evaluate the substitution possibilities. Several processed items substitute nicely for fresh eggs, and vice versa. If buyers purchase a lot of eggs, it may pay for them to have recipes printed several ways to include various forms of eggs and egg substitutions. For example, cake recipes might be written to incorporate fresh eggs, dried eggs, or frozen eggs. If the AP prices vary favorably, a few minutes of cost calculation might signal that one of these recipes is demonstrably more economical than the others.

RECEIVING EGGS

When receiving fresh shell eggs, receivers should take the time to examine them carefully for cracks, dirt, and lack of uniformity. They should also check the temperature to see if the eggs are at 45°F or lower without being frozen. In addition, receivers should make sure that all of the eggs are there. Weighing the containers might be the easiest quantity check.

It is difficult for receiving personnel to determine the age of fresh eggs. The American Egg Board recommends that receivers randomly break a few eggs and inspect them to determine if they meet the guidelines of their given grade.[2] The delivery agents may think that the receivers have lost their mind if they expect the agents to wait around while they conduct this ritual. But a little skepticism never hurt a buyer.

Processed eggs usually require other sorts of inspection. Assuming that all of the products ordered have been delivered, receivers then need to assess the quality of the items. This is not very easy. Receivers can check frozen egg products to see whether any crystallization has occurred; this is an indication of refreezing. They can also use a temperature probe to test frozen egg products.

Furthermore, receivers are able to check the can pressure of any canned dried egg products; devices on the market make this quick, simple test possible. An abnormally high pressure could indicate that a can's contents are contaminated.

If receivers do not take or make the time to check egg quality, or if they do not have the time, they must trust their supplier and delivery agent. Because processed-egg quality is always difficult to determine, we suppose that a certain degree of trust is inherent here in any case.

After checking the quality and quantity, receivers should check the prices and complete the appropriate accounting procedures.

STORING EGGS

Fresh eggs should be refrigerated as soon as possible at 45°F or below, but do not freeze them. Eggs stored at 45°F or below will retain their quality for weeks. In addition, since these items pick up odors quickly, they should be kept in their original containers. Some large operations maintain a dairy refrigerator to keep fresh eggs away from particularly odorous products, such as onions, fish, cabbage, and apples (see Figure 7.7). Processed eggs also require a specific storage environment, either frozen or dry storage, suggested by the form in which they come.

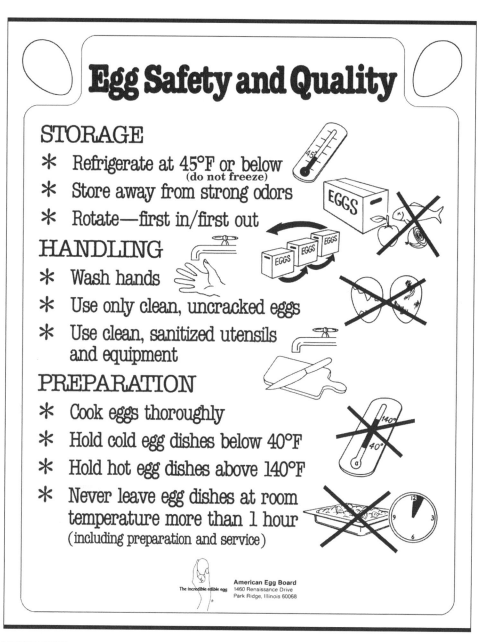

Egg Safety and Quality

STORAGE

* Refrigerate at 45°F or below
  (do not freeze)
* Store away from strong odors
* Rotate—first in/first out

HANDLING

* Wash hands
* Use only clean, uncracked eggs
* Use clean, sanitized utensils
  and equipment

PREPARATION

* Cook eggs thoroughly
* Hold cold egg dishes below 40°F
* Hold hot egg dishes above 140°F
* Never leave egg dishes at room
  temperature more than 1 hour
  (including preparation and service)

The incredible edible egg

American Egg Board
1460 Renaissance Drive
Park Ridge, Illinois 60068

FIGURE 7.7 An American Egg Board egg safety and quality flyer. Courtesy of the American Egg Board.

ISSUING EGGS

Hospitality operators should properly rotate stock, so that the oldest items are issued first. This may be accomplished by dating containers as they are received. In some cases, the egg purchases go straight into production. If operators issue eggs from a central storeroom, they must make sure, for example, that the requisitioner who wants eggs for a cake recipe gets the frozen or dried eggs, if applicable.

Because fresh shell eggs deteriorate rapidly once they leave refrigeration, operators must ensure that the requisitioner takes no more than he or she needs for any one particular work shift or job. They might consider asking the requisitioner to take note of the in-process inventory before asking for additional stock.

IN-PROCESS INVENTORIES

The benefit of good purchasing effectiveness can be immediately offset if hospitality operators do not control in-process inventories. For example, if breakfast cooks leave fresh shell eggs out at room temperature all day, the egg quality could drop a grade. The same is true of processed eggs. As always, supervision is the key. Without it, there is little sense in buyers taking care to purchase the proper items for the production staff. The best a supervisor can do is insist that all eggs be kept in the recommended environment at all times and removed from this environment only when necessary.

KEY WORDS AND CONCEPTS

As-purchased (AP price)	Food Safety and Inspection Service (FSIS)	Preservation method
Breaker plants	Fresh-egg flat	Processed eggs
Candling	Fresh shell eggs	Processing method
Color	Grading factors	Product form
Controlled-atmosphere storage	Independent farmer	Product size
Cost per ounce of fresh shell eggs	In-process inventories	Purchasing, receiving, storing, and issuing eggs
Cryovac	Intended use	Shrink-wrap
Edible-portion cost (EP cost)	Large egg, most typical size purchased	Size of container
Egg Products Inspection Act (EPIA)	Packaging material	Storage eggs
Exact name	Packaging procedure	United States Department of Agriculture (USDA)
	Packers' brands	U.S. grades
	Potentially hazardous food	

REFERENCES

1. American Egg Board, *Eggcyclopedia*, 3rd ed. (Park Ridge, IL: American Egg Board, 1994).

2. American Egg Board, *The Incredible Edible Egg: A Natural for Any Foodservice Operation* (Park Ridge, IL: American Egg Board, 2003). See also:

American Egg Board, *Egg Handling and Care Guide*, 2nd ed. (Park Ridge, IL: American Egg Board, 2000); for more information about eggs, visit http://www.aeb.org.

QUESTIONS AND PROBLEMS

1. What are the U.S. grades for fresh shell eggs?

2. List the sizes for fresh shell eggs.

3. What procedure can receivers follow to determine the freshness of shell eggs?

4. What will be the AP price per ounce of a large shell egg at $1.25 per dozen?

5. Assume that you are the manager of an employee food service. Outline the specific procedures you would use for the purchasing, receiving, storing, and issuing of fresh shell eggs that will be used for 3-minute eggs. If possible, ask an employee foodservice manager to comment on your answer.

6. Assume that your buffet brunch has another hour to go and that you have just run out of fresh eggs. You were preparing omelettes and scrambled eggs for use on the buffet line. You have a couple of cans of dried eggs and one can of frozen eggs in storage. Can you use these processed eggs to tide you over? If not, what do you suggest? If possible, ask a foodservice manager to comment on your answer.

7. An independent farmer calls on your country club to solicit its fresh-egg business. The following offer is made: daily delivery, eggs no more than a day old, an AP price 2 cents higher than the AP prices other suppliers charge. What do you suggest? (*Hint:* Eggs that are too fresh should not be used for hard-boiled eggs because the shells may be difficult to peel; otherwise, we know of no problems with very fresh eggs.) If possible, ask a country club manager to comment on your answer.

8. Assume that you are the kitchen supervisor for a resort hotel. Your Sunday brunch normally includes scrambled eggs. Your cooks have been preparing them in a steamer and serving them in a chafing dish; customers then help themselves. The quality of this product is not as good as you would like it to be, but the alternative of scrambling a few eggs at a time to order is not viable. You could purchase frozen scrambled eggs packed in 5-pound Cryovac bags. These eggs need only to be steam-heated for 20 minutes. Their quality, in your opinion, is superb. But the AP price is very high—approximately three times the price of fresh eggs. What do you suggest? If possible, ask a resort hotel's food and beverage director or kitchen supervisor to comment on your answer.

9. What is an appropriate intended use for frozen whole eggs?

10. What is an appropriate intended use for dried egg whites?

11. A product specification for dried eggs could include the following information:
    (a)
    (b)
    (c)
    (d)
    (e)

12. A foodservice buyer normally specifies the large-size fresh egg because:
    (a)
    (b)

13. What are the two typical package sizes for fresh shell eggs?

14. Fresh shell eggs can be preserved in the following ways:
    (a)
    (b)
    (c)

15. What happens to a fresh shell egg as it becomes older?

16. What is the preferred preservation method for fresh shell eggs?

17. When is a fresh shell egg at its highest quality?

18. A "flat" contains _____ dozen fresh shell eggs.

19. What are the primary grading factors for fresh shell egg grades?

20. What is the primary indication of quality of a processed egg product?

21. What is the difference between the designations "U.S. Grade A" and "Grade A"?

22. When would buyers purchase reduced-cholesterol egg products?

23. A product specification for frozen omelettes could include the following information:
    (a)
    (b)
    (c)
    (d)
    (e)

24. What is an appropriate intended use for U.S. Grade B shell eggs?

25. What critical information is missing from the following product specification for a frozen egg mix?
    Frozen egg mix
    Used for low-fat entrées
    EggBeaters® brand

8

POULTRY*

The Purpose of this Chapter

After reading this chapter, you should be able to:

- Explain the selection factors for poultry, including government grades.
- Describe the process of purchasing, receiving, storing, and issuing poultry.

INTRODUCTION

"Poultry" is a term applied to all domesticated birds used for food. Poultry is not an especially difficult item to purchase, unless you are in the market for certain types of processed items. Generally, raw poultry is still considered to be a "commodity," which means the typical buyer does not perceive a great deal of difference between one frying chicken and another.

The poultry that food services typically buy are chicken, turkey, and duckling. On occasion, some might also purchase goose, squab, and Cornish hen (see Figure 8.1). For the most part, a specific class of bird is raised the same way all over the country. For example, frying chickens are raised in about eight weeks. They consume a relatively standardized diet—standardized, at least, according to nutritional needs—and

---

* Authored by Andrew Hale Feinstein and John M. Stefanelli.

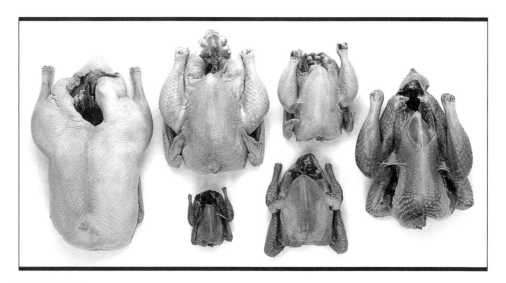

FIGURE 8.1 Clockwise from left: duckling, free-range chicken, poussin, guinea fowl, squab, quail. Copyright © 2003, John Wiley & Sons, Inc. This material is used by permission of John Wiley & Sons, Inc.

are slaughtered, cleaned, and packed with similar production line techniques. In short, raising any bird these days is a standard, scientific undertaking. Hospitality operators find very few small-scale producers, although a few independent farmers here and there may seek their business.

As with other products, the operators' major problem is deciding exactly what they want. If they want fresh or fresh-frozen poultry, they will have several suppliers from which to choose. Also, unless the buyers include packers' brands of fresh poultry, the as-purchased (AP) price will be about the same among the suppliers, provided, of course, that these suppliers offer the same quality and supplier services.

If buyers want other types of processed poultry, they obviously face the question of the degree of convenience they would like built into the products. They can usually purchase whole, dressed birds (it is not easy to purchase live birds today); cut-up birds; and precooked, prebreaded, presliced, and prerolled poultry. Numerous processed poultry products, as well as some imitation items, are available.

Only a few food processors undertake some types of processing. For instance, buyers can purchase cut-up frying chickens from a variety of sources. But they can purchase canned, cooked whole chickens from only a few suppliers.

In general, buyers will encounter little difficulty when purchasing poultry items. Numerous suppliers exist, and numerous styles of poultry items are available in the market.

SELECTION FACTORS

As with all products, the owner-manager normally decides the quality, type, and style of poultry products desired. Either alone or in cooperation with other employees, the owner-manager usually evaluates the following selection factors when determining the desired standards of quality and, to a certain degree, the supplier for poultry items.

Intended Use

As always, buyers want to determine exactly the intended use of the item so that they will be able to prepare the appropriate, relevant specification. For example, poultry used for soup will differ from that needed for a deep-fried menu item.

Buyers may want to select a poultry item they can use for two or three purposes. This approach is not ordinarily recommended because each item has one best use. However, poultry has a short shelf life, especially if it is fresh. Consequently, fresh poultry should be turned quickly. One way for buyers to do this is to use the poultry they purchase in several different dishes.

If buyers purchase a lot of poultry, they should examine the substitution possibilities because bargains may await them. They might, for instance, substitute turkey rolls for turkey breasts or for whole turkeys. They could also substitute canned poultry for fresh; purchase precooked, chopped chicken pieces if the intended use is for chicken salad; and use precooked sliced turkey breast instead of cooking their own.

Substitutions can disrupt the production and service functions, however. In addition, buyers must, of course, keep track of the distinctive culinary differences between these items. The culinary quality varies for at least two reasons: (1) for different menu items, food processors use poultry of different ages; and (2) the processing method itself could rob or add favorable qualities.

A major issue with any processed product centers on the substitution possibilities and on how much convenience buyers want built into it. Whatever type of processing they want, they have a generous number of suppliers from which to choose, which gives them additional flexibility. Poultry products encourage bid buying.

Exact Name

This is an important consideration because the federal government has established standards of identity for many poultry items. For example, some fresh products are standardized according to the birds' sex and/or age at the time of slaughter (see Figure 8.2).

CHICKEN

Young (tender) birds

Broiler/fryer—9 to 12 weeks old; $1^1/_2$ to $3^1/_2$ pounds; either sex

Roaster—3 to 5 months old; $3^1/_2$ to 6 pounds; either sex

Capon—less than 8 months old; 6 to 10 pounds; desexed male bird

Cornish game hen—5 to 7 weeks old; 1 to $1^1/_2$ pounds; immature bird

Old (less tender) birds

Stewing hen—more than 10 months old; 3 to 7 pounds; mature female bird

Stag—more than 10 months old; 3 to 7 pounds; mature male bird

TURKEY

Fryer/roaster—less than 16 weeks old; 4 to 8 pounds; either sex

Young hen—5 to 7 months old; 8 to 14 pounds; female bird

Young tom—5 to 7 months old; over 12 pounds; male bird

Yearling hen—under 15 months old; up to 30 pounds; mature female bird

Yearling tom—under 15 months old; up to 30 pounds; mature male bird

DUCK

Duckling—under 8 weeks old; under 4 pounds; either sex

Duck—over 16 weeks old; 4 to 6 pounds; either sex

FIGURE 8.2 Definitions of some poultry products.

Age can affect the intended use of the poultry. Poultry to be cooked with dry heat should be young if buyers want a tender product. (As poultry ages, it becomes less tender. However, it also develops more fat, which carries flavor.) If chefs want to simmer a chicken for chicken soup, buyers would probably opt for an old bird. It will have more flavor, and moist heat will ensure tenderness. As a bonus, an old bird tends to have a high conversion weight, that is, a high edible yield. If a bird gets too old, though, much of the weight begins to collect in the abdominal fat. This fat can be collected and used for, perhaps, a roux for cream of chicken soup. But the yield of cooked meat per pound of raw chicken may be less than expected because of the extra fat.

Sex is not particularly important for young birds, but in older birds, the differences in taste, texture, and yield diverge dramatically between the sexes. Females tend to be tastier, juicier, and have higher conversion weights than males. So, if buyers purchase mature poultry, they should take note of the birds' sex, especially if they are bid buying.

As with other product lines, buyers occasionally encounter market terminology that defines very specifically what a poultry product is. For instance, "free-range" chickens are allowed to roam free instead of spending their lives in cages. "Kosher" chickens, prepared according to Jewish dietary laws, are also allowed to roam free, are usually a strong breed, tend to be a little older at the time of slaughter in order to promote flavor development, and are free of hormones and other chemical and artificial ingredients. Since these types of poultry products are usually much more expensive than those raised in the traditional way, buyers must not use this terminology carelessly.

If buyers purchase processed poultry products, merely noting the exact name may not be enough because even though standards of identity exist for such products as chicken pot pies, the producers of these items need to meet only some minimum standard. Also, even though buyers may not be averse to a producer's particular formula, they must keep in mind that chicken pot pies can come with several onion varieties and potato varieties. If buyers are dealing with fresh poultry, either whole birds or standardized parts, using the exact name is normally adequate because these fresh items are more consistent among producers. But the same cannot be said for processed products. These items can and will vary significantly among producers, so overreliance on standards of identity for them can be a bit chancy.

U.S. Government Grades (or Equivalent)

Poultry inspection became mandatory with the 1957 Poultry Products Inspection Act. This law applies to all raw poultry sold in interstate commerce, as well as to such processed products as canned and frozen items.

Some states conduct their own poultry inspection programs, and the 1968 Wholesome Poultry Products Act requires state programs to be at least equal to the federal inspection program. Poultry inspected under a state program, however, can be sold only within that state. Any poultry product transported across state lines or exported to another country must be produced under continuous federal inspection. In states that do not conduct inspection programs, all plants are required to be under continuous federal-government inspection (see Figures 8.3 and 8.4).

The Food Safety and Quality Service (FSQS) of the United States Department of Agriculture (USDA) performs federal inspection for wholesomeness and federal grading (see Figure 8.4a and 8.4b). Assuming that the product is wholesome, a poultry producer can elect to purchase the grading service.

Some states leave poultry producers no choice: they must have their products federally graded after they are inspected. As a practical matter, most poultry product

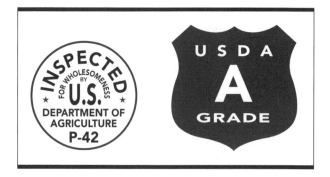

FIGURE 8.3 Federal grade and inspection stamps used for poultry products.
Source: United States Department of Agriculture.

specifications contain a U.S. grade designation. Consequently, producers have little choice in the matter.

Federal inspectors grade poultry according to several grading factors. Inspectors, or graders, consider: (1) conformation (Does the bird have good form?); (2) fleshing (Does the bird have a well-developed covering of flesh?); (3) fat covering (Does any flesh show through the skin—that is, is it a "thin-skinned" bird?); and (4) other factors (Does the bird have any bruises, excessive pin feathers left after cleaning, broken bones, missing parts, or discoloration?).

Several poultry grades exist. The consumer grades are as follows:

1. **Grade A.** This is the top poultry quality produced. It indicates a full-fleshed bird that is well finished and has an attractive appearance.

2. **Grade B.** This bird usually has some dressing defects, such as a torn skin. Also, the bird is, generally, less attractive. For example, it might be slightly lacking in fleshing, and the breast bone may be very visible.

3. **Grade C.** This bird resembles a Grade B bird, but it lacks even more in appearance. Also, parts of its carcass might be missing.

In addition to consumer grades, the federal government offers a "procurement" grading system, which consists of two procurement grades: I and II. These grades are intended for use by noncommercial, "institutional" food services and are based almost entirely on the amount of edible yield the poultry products contain. The appearance of the birds is deemphasized.

Some state and local markets use the following three commercial grades: Extra, Standard, and No Grade. They are similar to procurement grades, although the tolerances between these grades are wider than those of the consumer and procurement grades.

(a)

(b)

FIGURE 8.4 (*a*) and (*b*) FSQS inspecting the wholesomeness of poultry.
Source: United States Department of Agriculture.

A specification for poultry products usually contains some grade reference. This is especially true when buyers purchase fresh or fresh-frozen whole birds or parts. If buyers purchase processed products, such as prebreaded, precooked chicken patties, they are more apt to rely on a packer's brand name to specify the desired quality.

The use of U.S. grades for poultry, especially the consumer grades, is very popular in the foodservice industry. Buyers usually opt for U.S. Grade A products when appearance is very important. For instance, chefs would most often prepare a fried-chicken entrée with the highest-quality raw products. Lower grades of poultry are usually sold to food processors. These grades are often referred to as "manufacturing grades" because they are not usually intended for foodservice operations unless they undergo some sort of processing that alters their appearance and culinary quality. When appearance is not as important to a hospitality operation, such as when poultry is used to prepare chicken salads, turkey casseroles, or pot pies in the operation's kitchen, these lesser grades may be adequate.

More information on grading regulations and standards of poultry is available online. Buyers can find regulations governing the grading of poultry and rabbits (7 CFR Part 70) at www.ams.usda.gov/poultry/regulations. In addition, they can find U.S. classes, standards, and grades for poultry (AMS 70.200 *et seq.*) at www.ams.usda.gov/poultry/standards.

Packers' Brands (or Equivalent)

For fresh and fresh-frozen poultry, brand loyalty rarely comes into play. Buyers seem to rely a little more on specific brands of raw turkey or duck. But this does not seem to be the case with raw chicken, which is usually viewed as a basic commodity.

Some manufacturers attempt to take poultry out of the commodity class and instill brand loyalty in consumers, who have several brands from which to choose. For instance, Perdue®, Tyson®, and Foster Farms® are some of the brand labels buyers can specify for fresh chicken.

Some value, as well as a higher AP price, is associated with proprietary chicken brand names. For example, producers generally slow down the assembly line to use a "soft-scald" procedure, which removes feathers at a lower temperature, thereby significantly increasing the tenderness of the birds. Some producers use a "chill pack" preservation procedure for their finished products. This maintains the chickens' temperature at about 28°F to 29°F (they freeze at about 27°F to 28°F), which extends the products' shelf life without freezing them. Furthermore, these brand name items are usually produced in exceptionally clean environments. Very high levels of sanitation will increase the products' shelf life because it is directly related to the numbers of bacteria found on the skin of the birds.

The brand name campaign seems aimed primarily at homemakers, however. Hospitality operations that list a lot of poultry signature items on their menus strive to ensure that their customers identify such poultry with the operations that prepare it, and not with a particular packer's brand name. Generally, packers' brands are important to hospitality buyers only when they purchase processed poultry products.

Product Size

When purchasing raw poultry products, buyers usually cannot specify an exact product size. Instead, they must indicate the acceptable weight range. For example, they need to indicate a weight range for whole birds and, to some extent, for raw poultry parts, such as chicken thighs and turkey breasts.

As a general rule, the larger the bird, the higher its edible yield. Buyers might find it helpful—and interesting—to know that, for example, a turkey's bone structure stops developing when it reaches about 20 pounds. Turkeys that weigh more than 20 pounds have more fat and a bit more meat on the same-size skeleton as "thinner" birds. If the operations' intended use allows, their buyers should purchase large birds because these may be the most economical choice.

Buyers also must note the product size of any processed products that their operations need. Fortunately, they normally can specify an exact size for these items and do not have to rely on weight ranges. Many sizes are available for such products as precooked, breaded chicken breasts; chicken patties; and turkey pot pies.

Product Yield

For some poultry products, such as frozen, breaded chicken patties, buyers may want to indicate the maximum yield, or the minimum trim, expected. For example, they may want to note on the specification that they will accept no more than two broken pieces in a 48-piece container of chicken patties.

Size of Container

As always, buyers must indicate the package size that they prefer. If necessary, they would need to note the size of any inner packs. For instance, buyers may want a 30-pound case of frozen chicken chow mein, with six 5-pound plastic pouches per case.

Type of Packaging Material

A reasonable variety of packaging quality is available, so buyers would do well to consider this selection factor when preparing their specifications. The variety is considerable for fresh product—for instance, fresh poultry may arrive at your back door wrapped in butcher paper; packed in cardboard or wooden crates; packed in shrink wrap, or packed in large, reusable plastic containers that the suppliers will pick up when they deliver the next shipment. The relatively brief shelf life of fresh items makes it necessary for buyers to reject packaging that would do anything to shorten a product's shelf life drastically.

When buyers purchase several types of processed products, they will experience a greater variety of packaging qualities. Normally, processed items used in the hospitality industry are packed in moisture-proof, vapor-proof materials that are designed to withstand freezer temperatures. Buyers must be certain that the items they purchase in this manner are packaged properly so that they can avoid any unnecessary loss of product.

Packaging Procedure

Like most products, poultry is packaged in many different ways. For instance, most raw, refrigerated items are slab-packed, while their frozen counterparts are usually layered. Whole birds, typically, are individually wrapped if they are frozen, whereas fresh, refrigerated birds may not be; in fact, fresh birds may be slab-packed with crushed ice covering them, which is sometimes referred to as an "ice pack procedure."

Fresh or frozen boneless poultry products are sometimes packaged in "cello packs." Products packed this way usually come in a 5-pound box that has six cello-wrapped portions, each of which contains two to four poultry pieces.

Fresh birds or bird parts can also come in "gas-flushed packs." In this arrangement, the poultry is placed in plastic bags, and the air is "flushed" out of the bag and replaced with carbon dioxide. This packaging procedure, which is a type of controlled-atmosphere packaging, is done primarily to extend the fresh poultry's shelf life.

Some fresh, refrigerated products can be purchased in a "marinade pack." Here, suppliers pack individual poultry parts, for example, chicken wings, in a reusable plastic tub and pour a specific marinade solution over them. The chicken wings then absorb the required flavor as they journey through the channel of distribution, so that by the time they are delivered to the hospitality operation's back door, they can be put directly into production.

Most processed poultry products that the typical hospitality operation purchases are frozen and layered. These items are usually referred to as "individually quick frozen"

(IQF). This designation indicates that the products are flash-frozen and layered in the case. In some instances, they may even be individually wrapped.

IQF is sometimes referred to as a "snap pack" or "shatter pack." When you remove a layer of product and drop it onto a counter top, the pieces should snap apart cleanly. If they don't, it's likely the product has been refrozen.

Product Form

One of the amazing aspects of the poultry trade is the seemingly endless number of products available and the forms in which buyers can purchase them. At one extreme are the raw products, while at the other extreme several artificial meat items, such as "ham" and "hot dogs," are produced with poultry.

If buyers purchase whole birds, they will have the option of receiving them whole or they can request that the birds be cut into a specific number of pieces. Buyers also may be able to specify a particular cutting pattern, although the standards of identity the federal government has established address this issue fairly well; hence, it is often unnecessary to dwell on this aspect.

When purchasing whole birds, buyers can opt to have the variety meats included or excluded. These are the organ meats, such as the liver and heart. Because they are not usually included with whole birds intended for use by foodservice establishments, if buyers want them, they normally will need to indicate this desire on the specification. Ordinarily, if foodservice operators need to purchase variety meats, they will buy them separately; for example, chicken livers are normally purchased separately, packed in a 5-pound Cryovac® bag.

Sometimes raw poultry, especially raw-poultry parts, includes a bit of cutting and trimming that adds to the AP price of the items, but enhances their convenience. For example, buyers can order boneless, skinless chicken breasts. This convenience, while initially a costly alternative, could, in the long run, result in the most economic edible-portion (EP) costs.

If buyers need to purchase a good deal of highly processed poultry products, such as precooked turkey rolls or frozen, preprepared chicken chow mein, and are not satisfied with the federal-government minimum standards of identity, they will need to do something to indicate the particular type of formula they want. For example, they may find it necessary to specify in great detail a chicken patty's proportion of white and dark meat, the amount and type of breading, and so forth. Buyers might be able to utilize a packer's brand, one that resembles what they want, but this convenience may not always be available.

Preservation Method

Most poultry purchased for use in the hospitality industry is preserved in one of two ways: refrigerated or frozen. Many refrigerated products are packed at chill pack temperatures. If suppliers do not provide the chill pack alternative, usually they will provide the ice pack method, which tends to accomplish the same effect as the chill pack—namely, the reduction of the storage temperature to just above freezing. Both chill packs and ice packs maintain temperatures of about 28°F to 29°F.

Foodservice operations that strive for a good poultry reputation usually purchase fresh, ice-packed, or chill-packed poultry. For instance, if fried chicken is an operation's signature item, it is very unlikely that the firm will use a frozen item because it can cause several problems. These items can, for example, very easily thaw just a bit during the receiving cycle and become freezer-burned when they refreeze in storage. In addition, frozen poultry products get a red tinge around the bones when they are cooked. Furthermore, these items lose flavor and moisture when they are thawed too long before cooking.

Some fresh poultry (as well as fresh meat, fish, and produce) may be preserved with irradiation. Irradiation removes almost all traces of harmful bacteria in meat and fish, and spoilage bacteria in fresh produce. However, many critics maintain that nutrients are lost during the irradiation process and that not enough information is known about the safety of this procedure. As a result, many food services are not eager to embrace this technology.

Generally, buyers purchase processed poultry products in the frozen state. Some canned products exist, but the typical hospitality operation seldom uses them. For instance, buyers could purchase either frozen or canned chicken noodle soup. The canned item is usually less expensive, but many foodservice operators opt for the frozen variety because the culinary quality is superior.

AP Price

Raw-poultry products offer little spread in AP price from one supplier to another. The distance that a finished poultry item must travel to get to a hospitality operation's back door, does, however, make a difference. For example, Colorado free-range chicken usually costs more than a similar item raised closer to home. But given the same style of poultry, the same quality, and the same supplier services, the AP prices are pretty much identical.

In addition, AP prices for raw products tend to be more predictable because farmers can produce poultry much more easily and quickly than many other foods. For instance, it takes about two years to bring a steer from birth to the dining room table. With a frying chicken, this process takes only a few weeks. Also, fresh-poultry AP prices reflect the standardized, scientific management used in raising poultry. For instance, suppliers use pretty much the same feeding formulas and environments so that the poultry looks the same, tastes the same, and has the same conversion weight. Since AP prices vary little, the EP costs, theoretically, should also be similar between competing suppliers' raw products.

AP prices vary quite a bit for processed items, however, and this requires buyers to estimate EP costs as well as customer acceptance of these processed products. So, if buyers purchase a lot of chicken, they should consider entering into long-term contracts or, at least, using the hedging technique. Buyers may be able to maintain an AP price that they can live with by hedging in the commodity futures market. In addition, quantity buys offer reasonably good savings.

Trust the Supplier

No matter what type of poultry buyers purchase—fresh or processed—they can find several potential suppliers in the marketplace. Hospitality operations do not need to become house accounts unless they want to.

The AP prices of raw poultry do not vary significantly among suppliers. Consequently, bid buying these items may not be as profitable as it is for processed products, unless, of course, buyers can get some additional service. For instance, if they purchase fresh poultry, they will want to keep it at a temperature of at least about 30°F to 35°F. Also, they might want the supplier or processor to cut up the poultry and put it into a marinade. Alternately, they might want a whole chicken cut into eight, nine, or ten parts, depending on the intended use. Not every supplier can provide these services. But if two or three have these capabilities, buyers might consider bidding out their business once in a while.

In our experience, buyers tend, eventually, to become fresh-poultry house accounts; that is, they tend to purchase from one trusted supplier. This product requires at least a 30°F to 35°F temperature environment, as well as one that is very sanitary. We have known buyers whose only evaluation of a fresh-poultry supplier centered on the odor emanating from his or her poultry storage refrigerator. To these buyers, a clean refrigerator meant a reliable supplier.

Several suppliers sell various processed poultry items. The exact one buyers want, though, may be available from only one purveyor.

If buyers are very particular, they may find that they have to cast their lot with one supplier. If they are satisfied with a variety of choices, bid buying and extra negotiating might generate rewards.

PURCHASING POULTRY

As always, the first step in purchasing is for buyers to decide on the quality and type of product they want. For fresh poultry, and even for some processed poultry, they may find few suppliers who can provide exactly what they want, particularly if they have special requirements. As we have noted, though, usually enough potential suppliers exist to satisfy bid buyers.

Buyers normally use U.S. Grade A quality for poultry, especially when appearance is important. Of course, they can buy several combinations when other grades are available. However, buyers cannot always assume that Grade B is available for every item, since most producers strive to achieve the Grade A.

Once buyers know what they want, it is usually beneficial for them to prepare a complete specification. This should include all pertinent information, whether or not they use it in bid buying, simply because the discipline of preparing this written document helps to ensure that buyers have considered all relevant factors and are, indeed, purchasing the right quality and quantity (see Figure 8.5 for some example product specifications, and Figure 8.6 for an example product specification outline for poultry products).

Broiler/fryer, raw
Used for fried chicken lunch entrée
U.S. Grade A
Quarter chicken parts, cut from
    whole birds weighing between
    $2^1/_2$ to $3^1/_4$ lb. dressed weight
No variety meats
Ice packed in reusable plastic tubs
Approximately 30 lb. per tub
	Boneless chicken breast, raw
	Used for dinner entrée
	Tyson® brand
	4-oz. portions
	48, 4-oz. portions packed per case
	Moisture-proof, vapor-proof case
	    with plastic "cell-pack" inserts;
	    products layered in cell packs
	Frozen

Chicken base Used to prepare soups and sauces Minor's® brand 16-oz. resealable plastic containers 12 containers packed per case Refrigerated	Turkey breast, raw Used for sandwiches U.S. Grade A Bone in, skin on Under 8 lbs. Wrapped in Cryovac® (or equivalent) Refrigerated

FIGURE 8.5 An example of poultry product specifications.

Intended use:

Exact name:

U.S. grade (or equivalent):

Packer's brand name (or equivalent):

Product size:

Product yield:

Size of container:

Type of packaging material:

Packaging procedure:

Product form:

Preservation method:

FIGURE 8.6 An example of a product specification outline for poultry products.

After preparing the specs, buyers must evaluate potential suppliers. Keep in mind that for raw poultry, the important consideration probably is supplier services, since quality and AP prices usually vary only a little. Buyers need to consider such matters as freshness, delivery capabilities, temperature control, and plant appearance. Buyers usually can find many processed poultry product suppliers, but not too many if they have strict requirements. For example, if buyers want turkey or chicken cold cuts, they will not find many suppliers.

As in the fresh-produce and fresh-egg trades, independent farmers can probably supply poultry. We do not recommend this, however, unless a farmer's products and plant are under continuous inspection.

RECEIVING POULTRY

First, when accepting poultry, receivers must make the customary quality and quantity checks. Because harmful bacteria can multiply rapidly on poultry, especially at room temperatures, many hospitality operations receive and inspect poultry in refrigerated storage. The delivery agent and the receiver go straight to this area.

The raw-poultry quality check is not difficult. The grade shield is usually displayed prominently on the carton, and with whole poultry, on the wing of each bird. But receivers must be careful that the boxes have not been repacked. They can never be quite sure of what is in the boxes, regardless of the grade noted on the carton. A trustworthy supplier is the best insurance.

The quality check can cause some trouble if receivers are concerned with the age of the poultry. For instance, buyers might purchase hens to get the flavorful meat. But how do receivers know whether the hens are as old as they should be? Receivers can look at their size and amount of abdominal fat. To the trained eye, this check is routine. But some receiving agents lack this skill.

Processed products usually require other types of quality checks. Receivers must perform the normal checks of frozen products—looking for proper temperature, signs of thawing and refreezing, and inadequate packaging—and of canned products—looking for leaks, rust, and swollen cans.

Once receivers are satisfied with the quality, they must check quantity. Normally, buyers purchase poultry by the pound or by the bird. Also, some poultry comes packed in ice, which tends to make weighing difficult. Receivers might have to weigh enough birds to see whether they are within the weight range that the buyers have specified. As such, they may have to dig around in the ice a little or weigh the poultry with the ice on it. It is preferable, however, to temporarily remove the ice before weighing the poultry. Another option is to weigh the packer's brand, prepackaged, chill pack chicken to compare it with the weight stated on the label.

Receiving agents also need to check the types of parts, variety meats, and processed items they get. Sometimes, buyers order legs and receivers get wings, buyers order chicken livers and receivers get gizzards, or buyers order chicken franks and receivers get turkey franks. These are usually honest mistakes, but they can ruin a production schedule.

Receiving agents can streamline the poultry-receiving process by using the USDA's Acceptance Service. This service is popular among large foodservice operators for meat and poultry items. Remember also that buyers can hire an inspector to help them write specifications. Also, under the Acceptance Service, the federal inspector, or the state counterpart, will accept or reject the product according to what the buyers specify.

Finally, after making quality and quantity checks, receivers must check the prices and complete the appropriate accounting procedures.

STORING POULTRY

Hospitality operators should store fresh and frozen poultry immediately and at the proper temperatures and humidity in the environment its form suggests.

Fresh poultry has a short shelf life. Following proper storage practices can extend this shelf life from three to four days to up to a week. For example, if chickens are received in an ice pack, they should be stored as is, but in such a way that any melted ice runs out of the storage package and does not soak into the birds. As the ice melts, operators should add more ice, but usually not more often than every other day. When receivers get poultry items packed without ice, they can increase the shelf life by placing them in a perforated pan, layering in ice, and refrigerating them. This maintains the temperature at approximately 28°F to 29°F.

Some operations may wish to marinate their poultry to lengthen the shelf life; this imparts a distinctive flavor as well. They can also extend poultry's shelf life by precooking it, though this could hamper their standard production schedule. Whatever storage procedures they follow, operators should not handle poultry any more than is absolutely necessary. If improperly handled, it will become contaminated.

Since poultry is expensive, operators might consider keeping a perpetual inventory of it. To do this, they need to enter the appropriate information into their inventory management system.

ISSUING POULTRY

If hospitality operators use a perpetual inventory system, they must deduct the quantity issued. If applicable, they will need to make the following decision: Should they issue the item as is, or should they issue it as ready-to-go (for example, cut-up and breaded, or precut)? Recall that any choice involves several advantages and disadvantages.

Operators should follow proper stock rotation guidelines when issuing these items. Also, since these products, especially fresh ones, deteriorate rapidly and are expensive, operators must make sure that the requisitioner does not take more than is absolutely necessary. If they can, they should force this person to note the in-process inventory before asking for more poultry.

IN-PROCESS INVENTORIES

Depending on the type of poultry product, several degrees of waste are possible. For example, it is easy for a chef to burn a breaded poultry item on the outside while failing to cook it thoroughly on the inside. Similarly, if a chef is carving a whole roast turkey, a lot of usable meat can stick to the bones. Also, leaving a roast turkey under a glow lamp too long can make a once beautiful roast turkey collapse into charred rubble.

If hospitality operators use several types of poultry, they need to keep them straight. For example, a cook might unknowingly use the chopped, cooked chicken slated for chicken salad in the soup.

Probably the biggest consideration with in-process poultry inventory is the sanitation problem. Staphylococcus and salmonella bacteria should not be present on cooked poultry products. However, when poultry items are contaminated after cooking, usually by a human handler or by being placed on a contaminated surface, bacteria grow very quickly at warm temperatures (40°F to 120°F). For example, a finished chicken à la king kept in a warm instead of a hot steam table for four or five hours can become sufficiently contaminated to cause an outbreak of food-borne illness.

KEY WORDS AND CONCEPTS

Age of bird at time of slaughter

As-purchased price (AP price)

Cello pack

Chill pack

Commodity

Contamination problems

Conversion weight

Edible-portion cost (EP cost)

Exact name

Food Safety and Quality Service (FSQS)

Free-range chicken

Gas-flushed pack

Grading factors

Ice pack

Independent farmer

Individually quick frozen (IQF)

In-process inventories

Intended use

Irradiation

Kosher chicken

Manufacturing grade

Marinade pack

Material used in processed products

Number of pieces per bird

Packaging procedure

Packers' brands

Poultry Products Inspection Act

Poultry used for more than one menu item

Preservation method

Procurement grades

Product form

Product size

Product yield

Purchasing, receiving, storing, and issuing poultry products

Sex of bird

Shatter pack

Shelf life

Signature item

Size of container

Snap pack

Standard of identity

State grades

Substitution possibilities

Trust the supplier

Type of packaging material

United States Department of Agriculture (USDA) Acceptance Service

U.S. grades

Variety meats

Weight range

Wholesome Poultry Products Act

QUESTIONS AND PROBLEMS

1. What are the U.S. consumer grades for poultry?

2. Why is the age of a bird at the time of slaughter an important selection factor?

3. Why do hen turkeys generally have a higher AP price than tom turkeys?

4. What is an appropriate intended use for a broiler-fryer?

5. The primary grading factor for the poultry grades "Procurement I" and "Procurement II" is: _____.

6. What are lower-quality poultry products generally used for?

7. Outline the specific procedures hospitality operators should use for purchasing, receiving, storing, and issuing frozen, prebreaded broiler-fryer parts. Assume that these parts will be used in a school

food service or in a hospital food service. If possible, ask a school foodservice director or a hospital dietitian to comment on your answer.

8. Which type of poultry product would buyers purchase if they were planning to prepare chicken and dumplings and wanted to use fresh chicken? Why?

9. When hospitality operators serve turkey and dressing, they could use fresh turkey or a processed turkey product, such as a turkey roll. What are the potential advantages and disadvantages of using the fresh product? What are the potential advantages and disadvantages of using the processed product?

10. What is another name for the term "conversion weight"?

11. What is the primary difference between the poultry grades U.S. Grade A and U.S. Grade B?

12. Why is the weight range of a fresh, whole bird an important selection factor?

13. What critical information is missing from the following product specification for sliced, cooked chicken breast?
    Sliced, cooked chicken breast
    Used for deli sandwiches
    Country Pride® brand (or equivalent)
    Packed in Cryovac bags
    Refrigerated

14. Describe the necessary storage conditions for fresh and processed poultry.

15. Prepare a product specification for the product mentioned in Question 8.

16. Why are free-range and kosher chickens more expensive than chickens raised in the typical way?

17. Why are frozen chicken parts, such as breasts and thighs, unacceptable to many foodservice operators?

18. What method can hospitality operators use to extend the shelf life of fresh poultry?

19. Explain why chicken is referred to as a "commodity item."

20. Assume that you manage a school food service. You serve lunch only—5000 lunches per day, five days a week. A poultry purveyor calls to tell you that he is going out of business. He has about 7500 pounds of frozen, cut-up broiler-fryers. He will sell you this stock for 50 percent of the current AP price. You have to let him know your decision tomorrow. What do you do? If possible, ask a school foodservice manager to comment on your answer.

21. A product specification for turkey franks could include the following information:
    (a)
    (b)
    (c)
    (d)
    (e)

22. What are the major advantages of the chill pack procedure?

23. When would buyers substitute a processed chicken patty for a boneless, skinless chicken breast?

24. Prepare a product specification for the following poultry products:
    (a)   Chicken wing
    (b)   Turkey
    (c)   Chicken egg roll
    (d)   Duckling
    (e)   Chicken patty

25. What is the difference between the ice pack procedure and the marinade pack procedure?

9

FISH*

The Purpose of this Chapter

After reading this chapter, you should be able to:

- Explain the selection factors for fish, including government grades.
- Describe the process of purchasing, receiving, storing, and issuing fish.

INTRODUCTION

Buying fresh fish can be one of the most frustrating jobs in all of purchasing. Processed—that is, canned, salted, and frozen—fish is easier to buy. With fresh items, not only will buyers find very few suppliers, but they may also have to take whatever fresh fish is available. If hospitality operations want fresh fish on the menu, they might have to offer whatever their supplier has in stock. At times, however, the supplier may have nothing.

Obtaining a wide variety of fresh fish is very difficult unless operators are willing to deal with all potential suppliers. It also is very difficult for operators to maintain consistent culinary quality unless they have a working relationship with all of the suppliers. It is not unusual for a foodservice operator to purchase only one type of fresh

---

* Authored by Andrew Hale Feinstein and John M. Stefanelli.

fish from as many seafood suppliers as possible in order to obtain a steady supply and consistent quality.

One nice feature of fish is that operators can usually get by with processed products, unless they want to advertise fresh items. If they insist on fresh fish, it is not particularly difficult to buy something fresh; it is just that, as we said earlier, they have few suppliers from which to choose.

Moreover, fresh fish can cause other challenges. Unless operators are close to a major transportation hub, the "fresh" fish can be in tired condition; its as-purchased (AP) prices can fluctuate, which can force operators to price their menu almost every day; once they get the fish, production employees may not be able to handle it properly unless they are highly skilled in this area; and some choice items, like Dover sole, Maine lobster, and Alaska king crab, cannot always be purchased fresh. So, unless operators can obtain a reasonably steady supply, they might want to reconsider any decision to feature fresh fish on the menu.

Some companies take the guesswork and the difficulty out of the selection and procurement of fresh fish by "growing their own," or by purchasing products that suppliers grow under controlled conditions. For instance, large restaurant companies that own commissaries can practice "aquaculture," or fish farming, which serves to produce fish items of consistent size and culinary quality. Any foodservice operation, though, can purchase farm-raised fish from suppliers who provide this option.

Aquaculture is not a new development; in fact, its roots go back to China, where the Chinese people have farmed fish since before the birth of Christ. However, the procedures have become very popular only in the last few years.

Many buyers are fond of farm-raised fish because it eliminates a lot of the risk from the fish-purchasing process. It also ensures stable quality and a consistent supply.

Each year sees an increase in the amount of farm-raised fish purchased in the United States. According to the National Restaurant Association (NRA), farm-raised catfish is the largest segment of the aquaculture industry. Other popular farm-raised fish are trout, tilapia, salmon, oysters, mussels, clams, scallops, abalone, crawfish, and shrimp. While aquaculture, technically, can be expanded to include other fish species, such as halibut and flounder, the technology needed to do it effectively and efficiently is still being refined.

Fresh-fish buyers must be very knowledgeable about the fish products they purchase. These items are not very standardized, and, unlike other fresh products, such as fresh produce and fresh meats, few market guidelines are available for most items. So operators who are responsible for purchasing fresh-fish products can never know too much about them.

SELECTION FACTORS

M anagement must determine the varieties and qualities of fish they want on the menu. Remember, they usually make this type of decision in cooperation with other company personnel. Management usually considers several of the following selection factors.

Intended Use

As always, hospitality operators want to precisely determine the intended use of an item so that they will be able to prepare the appropriate, relevant specification. For example, a whitefish that will be broiled needs a more attractive appearance than a whitefish that will be used in breaded fish patties.

Exact Name

Many varieties of fish exist in the world. In the United States, more than 200 varieties are sold. Obviously, then, buyers must be especially careful and indicate the precise name of the item they want (see Figure 9.1 for some of the more popular varieties of fish product used in the foodservice industry).

Unfortunately, even if buyers indicate an exact name, they could receive an unwanted item because the fish industry is fond of renaming fish. For example, on the East Coast of the United States, the name "lemon sole" refers to a particular size of flounder, while on the West Coast and in Europe, it refers to other fish species. A similar problem occurs with the name "snapper." It sometimes seems that almost every fish under the sun is called snapper.

The federal government actually encourages renaming fish because it would like to see the public eat products that are quite good, yet suffer from an image problem because they have unappealing names. Many perfectly delicious and nutritious fish species abound primarily because they are protected by repugnant names. Lately, though, several "trash" fish have become more popular.

Renaming fish is also done as an attempt to increase its marketability and profitability. For example, we recall that years ago a fish product called "slimeheads" or "Australian perch" was on the market. There was little demand for it. However, once Australians renamed it "orange roughy," sales skyrocketed. And, we might add, so did its price.

Renaming fish is a particularly sensitive issue in the hospitality industry. Receivers must be careful to get only those products that buyers actually order. If customers are

SALTWATER SPECIES	FRESHWATER SPECIES
Cod	Catfish
Flounder	Lake perch
Haddock	Lake trout
Halibut	Pike
Mackerel	Rainbow trout
Mahi Mahi (Dolphin)	Smelt
Monkfish	Tilapia
Ocean catfish	Whitefish
Ocean perch	
Orange roughy	SHELLFISH SPECIES
Pollock	Abalone
Salmon	Clam
Sea bass	Crab
Sea trout	Crawfish
Snapper	Lobster
Sole	Mussel
Swordfish	Oyster
Tuna (Ahi, Bluefin, Yellowfin, etc.)	Scallop
Turbot	Shrimp
Whiting	

(a)

(b)

FIGURE 9.1 (*a*) Varieties of fish. (*b*) Varieties of fish that are popular in the foodservice industry. Top row from left: tilapia, blue runner. Middle row: jack, mahi mahi. Bottom row: trigger fish, two red mullets.

Source: Gisslen, *Professional Cooking*, 4th ed. Copyright © 1999, John Wiley & Sons, Inc. This material used by permission of John Wiley & Sons, Inc.

adventurous and accept more exotic species, then buyers will order them. If not, buyers must be certain to select only those suppliers who will help them achieve their purchase objectives.

In addition to the exact name of the product, operators must be careful to use, where applicable, the appropriate market terminology in their particular area that further identifies what they want. While some market terms are common, each growing and harvesting area tends to adopt its own peculiar nomenclature to identify some items (see Figure 9.2 for some of the most common marketing terms).

A few fish products meet federal-government standards of identity. For instance, a "lightly breaded" shrimp product must contain at least 65 percent shrimp, whereas a "breaded" shrimp product must contain at least 50 percent shrimp. As with all standards

---

Block—a solid cube of raw fish; usually skinless; normally weighs about 10 to 20 pounds.

Breaded/battered—fish product coated with a seasoned crumb or batter mixture.

Butterfly fillet—two small fillets held together by a small, thin piece of skin.

Chunk—cross section of a large, dressed fish. It contains the cross section of the backbone. It is similar to a bone-in beef pot roast.

Drawn—whole fish that has been eviscerated, i.e., the entrails have been removed.

Dressed—a completely clean fish; can be cooked as is or processed into steaks, fillets, portions, etc.

Fillet—boneless fish, cut away from the backbone.

Fin fish—fish that has fins and a backbone. There are "fat" fin fish and "lean" fin fish. There are saltwater and freshwater species.

Green, headless—usually refers to a raw, unprocessed shrimp.

Peeled and deveined (P&D)—a shrimp without its shell or black vein.

Portion—a piece of fish cut from a block of fish. It is similar to a fillet, but it does not meet the exact definition of the fillet.

Shellfish—fish products that are completely or partially covered by a shell. There are crustaceans, whose shells are soft (e.g., shrimp), and mollusks, which have hard shells (e.g., oysters).

Shucked—fish that has been removed from its shell. Normally used when ordering shell-less mollusks.

Steak—a cross section of a large fish that has been cut from a dressed fish carcass.

Stick—a small piece of fish usually cut from a fish block.

Whole (round)—fish right out of the water; nothing has been done to it.

FIGURE 9.2 (a) Common marketing terms for fish products.

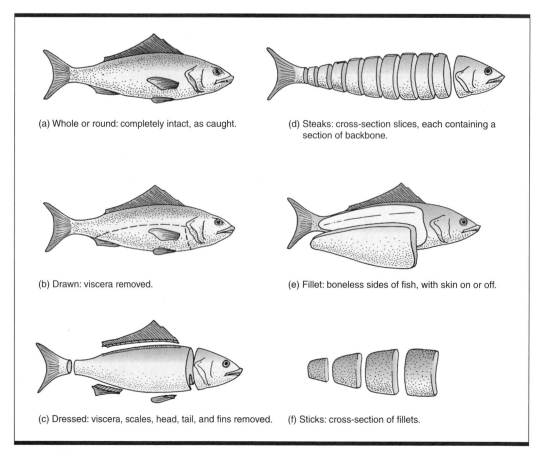

(a) Whole or round: completely intact, as caught.

(d) Steaks: cross-section slices, each containing a section of backbone.

(b) Drawn: viscera removed.

(e) Fillet: boneless sides of fish, with skin on or off.

(c) Dressed: viscera, scales, head, tail, and fins removed.

(f) Sticks: cross-section of fillets.

FIGURE 9.2 (b) Market forms of fish.

Source: Gisslen, *Professional Cooking,* 5th ed. Copyright © 2003, John Wiley & Sons, Inc. This material is used by permission of John Wiley & Sons, Inc.

of identity, though, the figures represent minimal requirements. Furthermore, in the fish trade, only a handful of processed products are subject to the standard of identity regulations. Consequently, wise buyers usually do not rely on standards of identity when preparing specifications for fish products.

U.S. Government Grades (or Equivalent)

The U.S. Department of Commerce's (USDC) National Marine Fisheries Service publishes grade standards and offers grading services for fishery products similar to those that the United States Department of Agriculture (USDA) established for other foods.

The USDC's grading program also provides for official inspection of the edibility and wholesomeness of fishery products, and many of its grade standards specify the amount of fish component required in a processed product.

The U.S. government grades few fish items. The products for which the federal government has established grading standards are processed products, such as breaded and/or precooked items.

Fish grades are based on several grading factors. Normally, the grader evaluates the fish's appearance, odor, size, uniformity, color, defects, flavor, and texture, as well as its point of origin. The federal grades for fish products are:

1. **Grade A.** This is the best quality produced. The appearance and culinary quality are superior. Grade A products have a uniform appearance and are practically devoid of blemishes or other defects.

2. **Grade B.** This is good quality and, generally, is suitable for many foodservice applications. Grade B items have significantly more blemishes and/or defects than Grade A products.

3. **Grade C.** This grade resembles Grade B, but it is lacking in appearance. It is suitable only for finished menu items, such as soups and casseroles, where appearance is not critical.

When buyers use U.S. grades as one of their selection factors, normally, only the U.S. Grade A designation is specified (see Figure 9.3). Lower grades usually do not carry a federal grade shield; rather, these manufacturing grades are left unmarked and are usually sold to food processors who will process them into several types of convenience fish products.

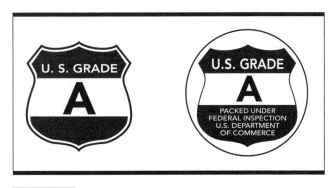

FIGURE 9.3 Federal grade stamps used for fish.
Source: United States Department of Agriculture.

Packed Under Federal Inspection Seal

Fish products are not subject to mandatory, continuous, federal-government inspection. The Food and Drug Administration (FDA) provides periodic inspections for all food-fabricating plants, monitors imported and interstate fish shipments, and requires fish processors to adopt the Hazard Analysis Critical Control Point (HACCP) system to increase food safety. Also, a cooperative agreement exists between federal and state agencies that monitors the farm beds where oysters, clams, and mussels are raised. However, these inspections fall far short of those the USDA uses to monitor meat-packing, poultry-packing, and egg-breaking plants.

Fish are cold-blooded animals; hence, the diseases afflicting them supposedly do not threaten the humans who eat them. However, fish can be exposed to many toxins, bacteria, and parasites that can be harmful to humans. Even though properly prepared and served fish causes no more health problems than meat products and fewer problems than poultry items, many fish buyers seem to be reluctant to purchase fish that has not been produced under continuous government inspection.

One way for buyers to ensure this type of inspection is to demand that all fish products they purchase carry a U.S. grade designation. Fish products that are produced and graded under the USDC inspection program may carry the USDC "Federal Inspection" mark or the U.S. grade shield. Unfortunately, grading designations, as mentioned earlier, are available only for a few fish items.

Another problem is that only fish produced in the United States can carry the federal government's grade or inspection shield; according to the NRA, about two-thirds of the fish consumed in the United States comes from approximately 120 other countries.

An alternative to requiring a U.S. grade designation or federal inspection is to purchase only those fish products that are produced under the continuous inspection of a state or local government agency. However, these agencies normally do not provide an extensive array of inspection services, and the ones they do provide usually are not very comprehensive.

The only sure way to obtain fish items that are produced under continuous government inspection is to demand that any fish product purchased carry the Packed Under Federal Inspection (PUFI) seal (see Figure 9.4). This seal indicates that the product is clean, safe, and wholesome, and that it has been produced in an establishment that meets all of the sanitary guidelines of the USDC's National Marine Fisheries Service. The product is not graded for quality, but it meets the acceptable commercial quality standards the federal inspection agency has set.

FIGURE 9.4 This seal signifies that the fish product is clean, safe, and wholesome and has been produced in an acceptable establishment with appropriate equipment under the supervision of federal inspectors.
Source: United States Department of Commerce.

Specifying that all of the fish buyers purchase must carry the PUFI seal probably will significantly reduce the potential number of suppliers who can bid for their business. Continuous fish inspection is a voluntary program, and not many fish-processing plants participate. Furthermore, time-consuming inspection is not easily adaptable to many fish suppliers who sell fresh fish products; in many instances, the products are caught in the morning and sent via air express to a foodservice operation, where they will be used on the menu that evening.

Packers' Brands (or Equivalent)

Because most fresh fish does not carry a brand name, this criterion is not a useful selection factor for fresh items. But packers' brands for processed fish items, especially canned fish, abound. In fact, a brand name, as well as the reputation of the item's producer and distributing suppliers, is very important to many buyers. For example, we probably could safely assume that StarKist® and Chicken of the Sea® brands instill a good deal of confidence in the marketplace.

In some instances, a brand name may be the only guide to seafood consistency. In addition, the type of processed product you want may be available from only one company. For example, not too many firms are producing marinated baby sardines.

Few beginning purchasers are familiar with the brand names in the seafood area. As a result, these brands may not be very useful until a buyer has studied the brands and experimented with them for awhile.

Some brand-name, processed fish products carry the U.S. Grade A designation. Items like fish sticks and raw breaded shrimp usually carry this type of grading mark. The brand is indicative of a certain culinary quality, and the grade shield assures wholesomeness.

Product Size

As usual, size information is a very important selection factor. Fish products come in so many sizes that a specification without this type of information is seriously deficient.

Some shellfish items, such as crab legs and lobster tails, are usually sized by count. For these two products, the count is based on 10 pounds. For example, if lobster tails are sized "10/12," a 10-pound lot comprises approximately 10 to 12 pieces.

For many fresh fish products, buyers may be able to specify a weight range only. For example, if buyers are purchasing large salmon fillets, the supplier may be unable to accommodate an exact size.

If buyers are purchasing whole-fish products, they will also need to settle for a weight range. For instance, they cannot specify an exact size for whole lobsters. They must be satisfied with one of the five traditional sizes available: "chickens" (approximately 1 pound), "quarters" (1 to $1\frac{1}{2}$ pounds), "selects" ($1\frac{1}{2}$ to $2\frac{1}{2}$ pounds), "jumbos" ($2\frac{1}{2}$ to 5 pounds), and "monsters" (more than 5 pounds).

For processed products, buyers are normally able to indicate the exact desired weight per item. For instance, when purchasing breaded fish sticks, they usually can choose among several available sizes.

The sizing system is an informal procedure that has developed over the years. No federal-government standards deal with this issue. Furthermore, many producers attach their own type of size designation to their items. When this happens, buyers will need to determine the exact nomenclature so that their specifications are adequate.

Product Yield

Buyers may need to indicate the minimum yield they will accept, or the maximum trim they will allow, for the fish products they receive. For example, buyers could note on their specification that they will accept no more than, perhaps, 2 percent broken fish sticks in every 20-pound case purchased. Alternately, the buyer will accept no more than 2 percent dead oysters for each barrel purchased.

Size of Container

When considering the size of the individual fish products buyers purchase, they must also give some thought to the size of the container they would prefer. Container sizes vary sufficiently to satisfy most buyer preferences. Generally speaking, the size and type of packages available for fish products are very similar to those used to package poultry products.

Type of Packaging Material

Fresh fish often is delivered in reusable plastic tubs or styrofoam containers. The products normally also are packed in crushed ice—that is, fresh product is often "ice packed."

Processed fish items usually are packaged in cans, bottles, or moisture-proof, vapor-proof materials designed to withstand freezer temperatures.

Live-in-shell fish items usually are packaged in moisture-proof materials. These products also may be packed in seaweed, or some other similar material designed to prevent dehydration. Usually, these products are not packed in ice or in fresh water because these packing media reduce their shelf lives.

As mentioned, fish products are generally packaged in the same type and variety of materials used to package poultry products. Buyers, therefore, have several choices.

Packaging Procedure

Once again, we note the similarity between poultry and fish products. As with poultry, fish items are slab-packed, layered, chill-packed, ice-packed, cello-packed, and individually quick frozen (IQF). Some items are also available in a marinade pack; for example, buyers may be able to purchase Cajun-seasoned sole fillets.

Fresh shellfish are, typically, slab-packed. Fresh fin fish normally are ice-packed in order to preserve their culinary quality and extend the shipment's shelf life. Some fresh fin fish are placed in "modified-atmosphere packaging" (MAP), which is a type of controlled-atmosphere packaging that involves chilling freshly harvested fish; placing it in plastic wrap; and pumping a combination of carbon dioxide, oxygen, and nitrogen into the wrap. The shelf life of fresh fish packaged this way can be as long as three weeks.

Fresh-frozen fish products are usually trimmed, cut, IQF, and packaged on board a fish-factory ship. These items are often cello-wrapped and placed in moisture-proof, vapor-proof containers.

Processed frozen-fish items are most often processed, IQF, layered on plastic or waxed sheets or plastic "cell packs," and placed in moisture-proof, vapor-proof containers. As with similarly processed frozen poultry products, this type of packaging procedure is sometimes referred to as a "snap pack" or "shatter pack."

Product Form

Fish is processed into many forms. Foodservice operations that strive for a high-quality seafood reputation normally need to use dressed fresh fish and, if further processing is necessary, to perform these tasks in the operation's kitchen. High-quality processed convenience items, though, are usually available. For example, breaded/battered, portion-control products are very popular, as are portion- control stuffed, marinated, and other preseasoned fish products.

Many unique types of convenience fish products are available. For example, "flaked and re-formed" "shrimp" items, which consist of odd scraps of shrimp material shaped to look like whole shrimp, are available. Also, several imitation fish products, such as imitation shrimp and seafood salad, are made with a fish-based paste called "surimi." Surimi is also used to produce various "meat" products—such as imitation frankfurters and bacon bits. While these items do not necessarily provide the same health benefits as the real products, some of these low-cost substitutes are very attractive to budget-conscious restaurant operators.

If buyers purchase a good deal of processed fish, they may need to rely exclusively on packers' brands as an indication of quality and other desired product characteristics. If buyers do not use brand name identification and are purchasing, say, a frozen-fish patty, they need to note on the specification the types of fish and other food matter that must be used to process these items. Buyers also would need to note the proportion of these materials desired. If they do not note these characteristics and do not wish to specify a particular packer's brand name, they cannot expect to maintain quality control. While federal standards of identity exist for some fish products, hospitality operators cannot rely on them exclusively because they represent only minimum guidelines to which commercial fish processors must adhere.

Usually, the major issues that need to be resolved when contemplating the purchase of processed fish products are: (1) what are the substitution possibilities, and (2) what degree of convenience do you want?

Several substitution possibilities exist. Processed fish products come in many forms. For example, buyers can substitute fillets for steaks, butterfly shrimp for headless shrimp, and imitation crab for real crab. These alternatives are limited because, as buyers go from one item to the next, the culinary quality changes. Managers must not risk alienating their customers.

The degree of convenience desired usually is related to the labor skill, equipment, and utensils available to produce menu items, and the size of the kitchen and storage facilities. Today, the economics of the foodservice industry tend to favor the use of many convenience products, and so long as the culinary quality is acceptable, managers will seriously consider using them.

Whatever the degree of processing buyers desire, they will, typically, have more than one supplier from which to choose. Fresh fish may be scarce, but buyers usually have two or more brands of processed fish from which to choose. This makes it easier for buyers who like to shop around and bid buy. It also makes it convenient for hospitality operators who want to move fish items around on their menus. This gives buyers a greal deal of flexibility.

Preservation Method

Fish is preserved in many ways: frozen, dried, smoked, refrigerated, ice-packed, cello-packed, chill-packed, live, live-in-shell, and canned.

The operation that offers fish signature menu items prefers live, live-in-shell, and/or ice-packed or chill-packed dressed fresh fish. If fresh product is unavailable, the frozen item is normally the preferred alternative, because at least it ensures a steady, year-round supply.

For some products, canned is the preferred choice. For example, snails and sardines are usually purchased in cans or bottles. Indeed, Americans buy more canned fish than any other type, fresh or processed.

Packing Medium

In some instances, buyers need to indicate the specific packing medium desired. For example, canned tuna is packed with water or with several varieties of oil. The same is true for other canned fish products. Since the packing medium significantly affects a product's culinary quality, buyers must be careful to include this selection factor on the specification.

Point of Origin

A lobster is not simply a lobster. If a lobster comes from the Gulf of Mexico, it is not the same type as the lobster that comes from Australia. This is true of any other type of fish, fresh or processed: the area it comes from influences its distinctive character, flavor, and texture. Hence, buyers sometimes carefully specify the origin of the fish they buy, especially if they purchase a great deal of fresh fish.

Many foodservice operators note on their menus the points of origin for some menu offerings. This seems to be a very popular practice for fish products. For instance, it is quite common to see menus that advertise Lake Superior whitefish, Alaskan salmon and crab legs, Australian lobster tails, and Chilean sea bass (see Figure 9.5). This menu nomenclature forces buyers to purchase fish products that originate from these locales. Substitute items cheat the customer. Furthermore, in some parts of the United States, any substitutes would violate truth-in-menu legislation.

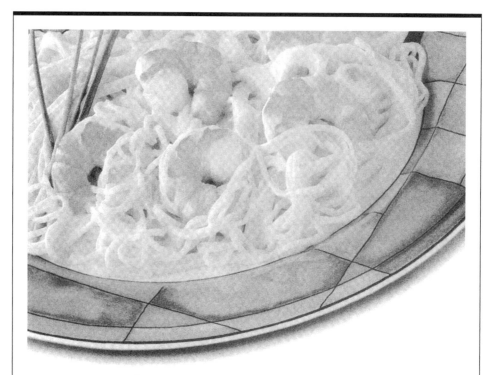

There is No Equal to Mexican Shrimp From Ocean Garden

Mexican Shrimp from Ocean Garden is the standard by which all other shrimp is measured. For consistently high quality, great taste, firm texture and customer satisfaction, there is no equal.

OCEAN GARDEN PRODUCTS, INC.
1-800-4-SHRIMP
For recipe please visit www.oceangarden.com

OCEAN GARDEN® COMPASS® PRIDE OF MEXICO® MARK® OCEAN SHELL® OCEAN GLO® ASIAN GARDEN®
FOR FURTHER INFORMATION CIRCLE 44 ON READER SERVICE CARD.

FIGURE 9.5 An example of point of origin. Courtesy of and © Ocean Garden Products, Inc.

Trust the Supplier

When buyers use a great deal of fresh fish, they may have to cast their lot with one primary supplier, assuming that they are satisfied with his or her capability. Together, a buyer and a supplier take what they can get from several sources, usually dictated by season, from the various seafood-producing areas of the world.

For instance, if a buyer purchases fresh, live-in-shell lobsters, the typical scenario may go something like this. The buyers will either go through a local supplier or purchase directly from a supplier on the East Coast, and order what is wanted for the next two or three days. The lobsters will be shipped FedEx air freight. At the airport in the buyer's city or a nearby city, the lobsters will be put into a local FedEx van, and brought right to the buyer's back door. The buyer will pay the East Coast fish supplier the going rate for that day, usually on a cost-plus basis. Alternately, the buyer may pay a local supplier, and he or she will then take care of all the details. Scenarios like this often take place for other types of fresh fish as well. But with other types, buyers may need to accept what is available.

The cost-plus purchasing procedure will influence an operator's menu pricing procedures. When buyers purchase fresh-fish products, they may wish to set a different menu price every day or every week. In the live-lobster example, a typical price-setting strategy is as follows: Pay for the lobsters. Throw out any dead ones, perhaps two out of 24. Divide the total as-purchased (AP) price plus any freight cost by 22. Add a markup to cover labor, overhead, and profit, and use this figure as the live-lobster menu price for today or for however long it takes to sell this batch of lobsters.

It is a good idea for buyers to deal consistently with trusted suppliers when they purchase a great deal of fresh fish. The items are not standardized, the quality is variable, the supply is erratic, and the prices change continuously. Buyers need the supplier's expertise. Also, they must be confident that he or she will charge the correct market price for the items and not try to take advantage of the buyers' inability to track the fresh-seafood market on a daily basis. Furthermore, the suppliers must ensure that the products they distribute are wholesome and safe for human consumption. Fresh fish are very difficult items to move successfully through the distribution channels; several contamination opportunities exist. The competent fresh-fish supplier deserves the buyer's respect.

When buyers purchase processed fish, they do not need to become a house account. Processed-fish buying, like most other processed-food purchasing, lends itself nicely to bid buying. Buyers learn, in time, the qualities associated with particular brand names. They can also learn to specify the type of processing they want and the area from which the fish should come.

Bear in mind, too, that there is a "new pack time" for canned and frozen fish, similar to the new pack time for canned fruits and vegetables. If hospitality operators have the money and the expertise, and large firms often have both, bid buying for a six-month or one-year supply can be rewarding.

PURCHASING FISH

The first step in purchasing fish is to acquire some of the indispensable reference materials available. For instance, fish buyers might want to have a copy of *The New Fresh Seafood Buyer's Guide: A Manual for Distribution, Restaurants, and Retailers* (New York: Kluwer, 1991—although it is now out of print). They should consult some of the leading trade journals, such as *The Seafood Leader*; the March/April annual buyer's guide issue is especially useful. Buyers should also consider subscribing to the *Seafood Price-Current*. This report contains twice-weekly market prices for many fish products from various regions of the United States. The Web also has a large amount of current seafood information. For instance, Seafood.com provides daily price information and news regarding the seafood industry.

The next step is for buyers to contact the FDA Office of Seafood Safety and ask for a list of approved interstate fish suppliers operating in their area. The local health district can provide a list of suppliers who operate only in their local market.

Next, buyers and other management personnel need to decide the exact type of product and quality they want. Once they determine what they want to include on the menu and what quality they want, the buyers should take the time to prepare a specification for each product. However, we do not think that specs are very valuable for fresh fish since buyers usually have to take what is available or go someplace else. They might prepare a statement of quality and give it to their suppliers, who then might call them when something meeting their standard comes in. A statement of quality might include the area the fish is to come from, as well as its size, form, and preservation method. If the fish is grown under controlled conditions, buyers might be able to specify its feed and nurturing techniques.

If buyers purchase processed fish, they should prepare detailed specifications so that they compile ideas as to exactly what they want. The specs must include all pertinent information, especially if buyers will use them in bid buying (see Figure 9.6 for some examples of product specifications and Figure 9.7 for an example product specification outline for fish products).

If buyers purchase processed fish in quantity, it might be worth a little trouble for them to shop around. For instance, purchasing a lot of frozen fish, such as a six-month

Tuna, solid white, albacore
Used to prepare tuna salad
Chicken of the Sea® brand
Water pack
66.5-ounce can
6 cans per case
Moisture-proof case

Australian lobster tails
Used for dinner entrée
U.S. Grade A (or equivalent)
16/20 count
25-pound moisture-proof, vapor-
    proof container
Layered pack
Frozen

Seafood Newburg
Used for banquet service
Overhill® brand
6-ounce individual portion pack
48 portions per case
Packaged in moisture-proof, vapor-
    proof material
To be reconstituted in its package
Frozen

Clam juice (ocean)
Used for beverage service
Nugget® brand
46-ounce can
12 cans per case
Moisture-proof case

FIGURE 9.6 Examples of fish product specifications.

supply, can save buyers a bit of money, as long as the storage costs are reasonable. Many frozen-fish suppliers will set up a stockless purchase plan for you or will provide the same type of supplier service for a slight carrying charge.

Another interesting aspect of buying processed fish is the reasonable spread in the AP prices between one brand name and another. Naturally, when the AP price is lower, buyers usually suspect a lower quality. Although this conclusion is typically true when buyers purchase other processed products, especially canned fruits and vegetables, it is not always the case with processed fish. For example, buyers will see a big difference in AP price between canned dark-meat tuna and white-meat tuna. Some people care very little about this color differential. Another example is the processed fish stick: a little more cod and a little less haddock may yield a fine-tasting product at a more economical AP price. Similarly, different formulas for fish patties, fish stews, and other processed entrées can yield substantially different AP prices while maintaining acceptable quality standards.

Intended use:

Exact name:

U.S. grade (or equivalent):

PUFI seal:

Packer's brand name (or equivalent):

Product size:

Product yield:

Size of container:

Type of packaging material:

Packaging procedure:

Product form:

Preservation method:

Packing medium:

Point of origin:

FIGURE 9.7 An example of a product specification outline for fish products.

Minimum-order requirements may trouble buyers. If they buy fresh fish, but only a little, they might find that a high minimum-order requirement and freight cost hinder their desire to serve a high-quality fish entrée.

Sometimes, buyers may be tempted to purchase fish from a neighbor who has just returned from a fishing trip. No mandatory federal-inspection requirement exists for the neighbor, and the fish may be perfectly good. The state or local health district, however, may have some regulations prohibiting the sale of this fish. We believe that buyers should avoid this practice since they never really know where the fish came from or how it has been handled.

Since fish availability can be somewhat unpredictable, buyers can sometimes find real seafood bargains. For example, a supplier may have some merchandise on a "move list." Perhaps more shrimp has suddenly shown up on the market and its AP price has gone down, or the tuna industry has a larger promotional effort. Before buyers jump at these bargains, though, they need to consider four factors: (1) Can their employees handle the item properly if it is a new item to them? (2) Do they have the proper equipment to prepare and serve the new item? (3) If the item is currently on their menu, should they drop its menu price? (4) Should they use this bargain as a loss leader on their menu? That is, would they be willing, for instance, to serve bargain shrimp in their lounge at a very inexpensive menu price in order to attract customers who supposedly will then order more profitable merchandise from them during a subsequent visit to the restaurant? This type of promotion may establish a trend whereby buyers would be forced to continue it well after the AP price of shrimp soars.

Management and buyers must decide what fish they want. Then they must engage the supplier that can handle their needs. For processed fish, bid buying may be the answer, but this is not always possible when purchasing fresh merchandise.

RECEIVING FISH

When a fish shipment is delivered, the receivers' first step is to check its quality. When they examine fresh fin fish or fresh, shucked shellfish, the product should have a mild, not fishy, scent. Its flesh should be firm, and it should spring back when slight pressure is applied. The product should be slime-free. Gills should be bright pink or red. If the head is attached, the eyes should be clear and bright. To ensure quality and maximum shelf life, the product should be ice-packed or chill-packed.

Sometimes, suppliers will send "slacked-out" fish, that is, thawed fish, instead of the fresh product ordered. Usually, this product looks a little dry, or it has "ice spots," which are dried areas on its flesh.

When examining live fish, receivers should see a product that is very active. They do not want live fish that seem sluggish or tired. Also, the product should be heavy for its size.

Live-in-shell crustaceans, upon examination, should also be very active and feel heavy for their size.

When receivers evaluate live-in-shell mollusks, the shells should be closed, or they should at least close when tapped with fingers. Open shells indicate that the products are dead and are past their peak of culinary quality. Live-in-shell also should feel heavy for their size.

Frozen fish should be frozen solid. They should be packaged in moisture-proof, vapor-proof material. No signs of thawing or refreezing, such as crystallization, dryness, items stuck together, or water damage on the carton, should be visible. Also, if applicable, the amount of "glaze" on the items—a protective coating of ice on the frozen items that the producer adds to prevent dehydration—should not be excessive.

Canned merchandise should show no signs of rust, dents, dirt, or swelling. Canned fish is especially dangerous if it is contaminated, so the receiving agent can never be too careful when evaluating the containers.

One problem receivers have when checking the quality of fresh fish is knowing whether or not they have gotten the right species—the item they ordered. Many fish products look similar to the untrained eye. For example, it is not easy to distinguish bay scallops from shark meat cubes, red snapper from Pacific rockfish, or cod fillets from haddock fillets.

Here is a corollary problem: What represents good quality to one nose may be offensive to another. We have seen receivers try to send back fresh-fish items because these items did not "look right" or "smell right." They had to be convinced that the slipperiness or ocean aroma was natural.

Another problem with receiving fresh fish arises when receiving agents do need to return it. Suppose that the deliveryperson made a mistake. If he or she takes the fish back, chances are that the product will go bad before the supplier can resell it. Also, operators are left without a fresh-fish item they may need that very night. Consequently, they might keep it and use it if the quality is acceptable in spite of its being the wrong variety. Receivers then would complete a Request for Credit memorandum or make some mutually agreeable settlement with the supplier.

An unfortunate problem with some processed fish is that it is hard to tell whether they will be acceptable once they are prepared for customer service. For instance, chefs may not know whether a frozen lobster tail is bad until they cook it and it falls apart. With a fresh lobster tail, experienced chefs can usually judge its quality before cooking it.

After satisfying their nose and eyes, receiving agents should go on to weigh, preferably without the ice, or count the merchandise and then get it into the proper storage environment as quickly as possible. Fresh fish will deteriorate right before their eyes if they fail to keep it refrigerated. Whatever receivers do, they should not let fresh and frozen fish stay on the receiving platform any longer than is absolutely necessary. Some companies put scales and checking equipment in the walk-in refrigerator to receive fish, poultry, and meat. Others roll the scale into the refrigerator temporarily.

If receivers are getting a shipment of shellfish from a supplier on the FDA's Interstate Certified Shellfish Shippers List, a tag in the container will note the number of the bed where the shellfish were grown and harvested. Operators are required to keep this tag on file for 90 days if the products are fresh and two years if they are frozen. If an outbreak of food-borne illness is traced to the shellfish, health officials must be able to determine the lot number of the offending products so that any as-yet-unused parts of that batch can be removed from the channel of distribution.

In some areas, other tags may accompany a fish shipment. For instance, some parts of the United States do not allow the importation of fish from other areas unless they carry identifying tags. These tags are usually issued to indicate that the products are acceptable to the local Fish and Game office and the local health district, and that they have not been purchased from unapproved local sources.

After doing the quality and quantity checks, receivers must verify the prices and complete the appropriate accounting documents.

STORING FISH

Fresh fin fish and fresh, shucked shellfish should be maintained at approximately 32°F and at no less than 65 percent relative humidity. These products are best stored on a bed of crushed ice and covered with waxed paper to prevent dehydration. If hospitality operators cannot use crushed ice, they should wrap the products tightly in plastic wrap, aluminum foil, or some other suitable container and store them in the coldest part of the refrigerator. The maximum shelf life for these items is about two days.

Operators must store live fin fish in tanks specifically designed to hold the particular types of products they are purchasing. Operators can also store live-in-shell fish in the appropriate water tanks, although they could be kept in their original containers and covered with damp cloths. Generally, these shellfish items should not be stored in fresh water or crushed ice. This is especially true for mollusks because the fresh water can kill them.

Hospitality operators should store frozen fish at or below 0°F. Ideally, these fish should be stored at −10°F to −15°F because these temperatures are conducive to maximum shelf life, which is about three months. If the products have the correct amount of glaze on them, they could maintain acceptable culinary quality for up to one year.

Operators should keep canned or bottled fish products in a dry storeroom. The ideal temperature of this storeroom would be 50°F, though 70°F is acceptable. The storeroom's relative humidity should not exceed 60 percent.

ISSUING FISH

Most fresh fish goes directly into production. But if fish enters an issue-controlled storage area, hospitality operators will need to prepare issue documents when it goes to production. Just as when operators issue most items, they have a choice: Should they issue the product as is, or should they issue it as ready to go? For example, oysters can be shucked ahead of time. Similarly, escargot can be preprepared, and someone might portion a fresh snapper, put it in pie pans, and season it, so that all the cook has to do is pop it into the oven.

Operators should try to avoid the temptation of prepreparing the snapper and then storing the portions in a freezer. On the one hand, this practice allows the cook to take the preprepared snapper out of the in-process freezer and put it in the oven. This is convenient; it also reduces leftovers, which usually go straight into the garbage can. On the other hand, some of the flavor is lost when fish are frozen this way. Besides, if operations have fresh fish on their menu, they have to risk some loss; they will also have to price these menu items so as to take into account these probable losses.

Of course, it is absolutely essential for operators to follow proper stock rotation when issuing fish products. Also, since these items deteriorate rapidly and are expensive, operators must make sure that requisitioners do not take more than absolutely necessary. Management should ask them to note the in-process inventory before requisitioning more fish.

IN-PROCESS INVENTORIES

A great deal of risk—spoilage, especially—is associated with fish, particularly fresh fish. This risk increases dramatically when fish is not handled properly at this step. Merely listing fish on the menu probably places hospitality operators in a high category."

Operators must try to reduce losses in the in-process inventories; typically, however, some loss occurs, particularly if several fresh fish items appear on the menu. But two rules are paramount: (1) Employees should not handle the product needlessly because this spreads bacteria and hastens the deterioration of fish quality. (2) Do not preprepare any more fish than chefs can use during the shift. Conservative prepreparing can cause production problems later on if the operation gets an unexpected rush, but it is necessary if the firm wants to avoid excessive leftovers.

KEY WORDS AND CONCEPTS

Aquaculture	Food and Drug Administration Office of Seafood Safety (FDA Office of Seafood Safety)	Intended use
As-purchased price (AP price)		Interstate Certified Shellfish Shippers List
Cello pack	Glaze	Live
Chill pack	Grading factors	Live-in-shell
Crustacean	Hazard Analysis Critical Control Point system (HACCP system)	Loss leader
Exact name		Lot number
Fin fish		Manufacturing grade
Fish and Game office requirements	Health district requirements	Marinade pack
	Ice pack	Marketing terms for fish products
Fish "frankfurter"	Ice spots	
Flaked and re-formed fish products	Individually quick frozen (IQF)	Modified-atmosphere packaging (MAP)
	In-process inventories	

Mollusk	Product form	Statement of quality
Move list	Product size	Stockless purchase plan
National Marine Fisheries Service	Product yield	Substitution possibilities
New pack time	Purchasing, receiving, storing, and issuing fish products	Surimi
Packaging procedure	Renaming fish	Tagged fish
Packed Under Federal Inspection (PUFI) seal	Shatter pack	Trash fish
Packers' brands	Shelf life	Trust the supplier
Packing medium	Shellfish	Truth-in-menu legislation
Point of origin	Shucked fish	Type of packaging material
Popular varieties of fish products	Size of container	United States Department of Agriculture (USDA)
Preservation method	Slacked out	United States Department of Commerce (USDC)
Processed fish	Snap pack	U.S. grades
	Standard of identity	Voluntary inspection

QUESTIONS AND PROBLEMS

1. List the U.S. grades for fish.

2. A product specification for frozen, breaded shrimp could include the following information:
   (a)
   (b)
   (c)
   (d)
   (e)

3. Briefly describe the necessary storage environments for fresh-fish products.

4. The deliveryperson arrives at 5 P.M. with 20 pounds of fresh whitefish. The buyer ordered 20 pounds of fresh snapper. The supplier was out of snapper and sent whitefish. He tried to call earlier about it, but could not get through. What should the buyer do in this situation? If possible, ask a foodservice manager to comment on your answer.

5. Outline the specific procedures hospitality operators should use for purchasing, receiving, storing, and issuing frozen lobster tails. Assume that the tails will be used in a steak house for a steak and lobster tail entrée. If possible, ask a steak house manager to comment on your answer.

6. What is an appropriate intended use for canned, dark-meat tuna?

7. What does the acronym PUFI stand for?

8. What are the differences between fin fish and shellfish?

9. What are the two main types of shellfish?

10. What is the difference between a fish fillet and a fish portion?

11. KWG Enterprises sells a frozen, breaded shrimp, eight to a pound, for $7.69 per pound. Its fresh, raw shrimp, 14 to a pound, sell for $6.24 per pound. A restaurant normally sells about 24 orders of fried shrimp each day. Which product should its buyer purchase? Why? If possible, ask a foodservice manager to comment on your answer.

12. Why is the point of origin of a fish product an important selection factor?

13. What does the term "14/16 crab legs" indicate to a foodservice buyer?

14. What is the primary difference between "round" fish and "drawn" fish?

15. A supplier calls to tell a buyer that he has just gotten a good buy on frozen ocean perch. The hospitality operation has never used this item before on its menu, but, with the low AP price the supplier quoted, the buyer is tempted. She has to buy 500 pounds, but this does not seem too troublesome. Assume that you operate a family-style restaurant. What would you do in this situation? Why? If possible, ask a family-style-restaurant manager to comment on your answer.

16. What critical information is missing from the following product specification for shrimp?
       Whole, raw, headless shrimp
       Used for shrimp cocktail
       U.S. Grade A (or equivalent)
       Packed in 5-pound laminated cardboard containers

17. List the grading factors for fish products.

18. Under what conditions would buyers purchase an imitation fish product?

19. What is the purpose of the tagging system that is in effect for shellfish products?

20. When would buyers purchase a flaked and re-formed fish product?

10

MEAT*

The Purpose of this Chapter

After reading this chapter, you should be able to:

- Identify the management considerations surrounding the selection and procurement of meat.
- Explain the selection factors for meat, including government grades.
- Describe the process of purchasing, receiving, storing, and issuing meat.

INTRODUCTION

Meat represents a major portion of the foodservice purchase dollar. Consequently, buyers tend to be especially careful when making meat-purchasing decisions. Although the meat industry and the federal government provide buying guidelines, purchasing meat is a time-consuming experience.

TYPES OF MEAT ITEMS PURCHASED

Many foodservice operations purchase some type of beef, veal, pork, or lamb item in addition to many types of processed meats, such as cold cuts, sausages, ham, and bacon. Operations use preprepared meat entrées, such as beef stew, and other less familiar products to a lesser extent (see Figure 10.1 for some of the more popular types of meat products the foodservice industry uses).

---

* Authored by Andrew Hale Feinstein and John M. Stefanelli.

BEEF

Brisket
Bottom sirloin butt
Butt steak
Chicken fried steak
Chuck
Cubed steak
Eye of round
Ground beef
Inside round
Outside round
Porterhouse steak
Rib
Ribeye roll
Ribeye steak
Short loin
Skirt steak
Strip loin steak
T-bone steak
Tenderloin
Tenderloin steak
Top sirloin butt
Top sirloin steak

CURED MEAT AND SAUSAGE

Bacon
Bologna
Bratwurst
Breakfast sausage
Corned beef
Frankfurter
Ham
Italian sausage
Knockwurst
Luncheon meat
Pepperoni
Polish sausage
Salami

LAMB

Breast
Hotel rack
Leg
Loin
Loin chop
Rib chop
Shoulder

PORK

Back rib
Boston butt
Center-cut chop
Cutlet
Fresh ham
Ground pork
Loin
Loin chop
Rib chop
Spare rib
Tenderloin

VARIETY MEAT

Calf liver
Steer liver
Sweetbread

VEAL

Breast
Cubed steak
Cutlet
Hotel rack
Ground veal
Leg
Loin chop
Rib chop
Shank
Tenderloin

FIGURE 10.1 Types of meat products that are popular in the foodservice industry.

MANAGEMENT CONSIDERATIONS

As a rule, deficient supervision is not an issue in the purchasing, receiving, storing, and issuing of meat, nor is it an issue in production. But management finds that deciding on the quality they want and the cuts they prefer is not easy. They must seek suppliers who can provide what they want on a continuous basis. Managers who are responsible for the ordering, expediting, and receiving of meat products can attest that this is no simple task.

Some of the major managerial meat-purchasing decisions are discussed in the following sections.

Should Management Offer Meat on the Menu?

Or should you minimize the amount of meat items that you offer? The question usually is, how tied do you want to be to meat? (In other words, how much of your image do you want to be associated with it?) Although some operations, such as vegetarian restaurants, exclude meat on the menu, realistically, most operations offer meat entrées. As such, it is difficult to alter an image of the establishment when the concept is tied to meat. This is evident in theme restaurants, such as a steakhouse, where meat represents the "signature item." Consequently, you must stay with your specialty, regardless of availability or increases in the as-purchased (AP) price.

Given the high cost of meats in relation to that of other foods, the restaurateur's goal is to utilize every edible morsel. As such, meat may appear as menu items in discreet ways. What is intended to be served once sometimes blossoms into an unwieldy number of meat-related menu items: the luncheon chef's special created from the preceding night's leftovers, a meat loaf or beef stew prepared from the trimmings from a large piece of meat, and new menu selections based on customer request.

Alternatives

Many alternatives for conventional meat items exist. Fish and poultry are excellent substitutes. Also, hospitality operators can take a chance on exotic types of meat dishes, such as buffalo, bison, and venison. Additionally, they can experiment with different grades of meat. For instance, operators can use a lower meat grade from an older, tougher, but more flavorful animal when a moist cooking method is employed. Otherwise, they can purchase lower grades of meat and tenderize them chemically or mechanically. Meat recipes that contain several ingredients make good candidates for some lower-cost substitutions. Meat loaves, stews, and casseroles can be manipulated by including or excluding certain fats, meat qualities, and fillers. Also, meat alternatives, such as soybean extender, can be incorporated into certain recipes.

Although the list of meat alternatives is seemingly endless, operators always take a chance when they make substitutions. This is especially true when customers are used to one item and taste a substitute product that is unfamiliar, or when the replacements are visible to consumers, such as when a low-quality steak that has been mechanically or chemically tenderized is substituted for a naturally tender steak of higher quality.

The Quality Desired

As always, the quality hospitality operators need reflects the intended use of the product and the image they wish to project. Spending considerable time determining the quality of the product they want does not guarantee that the desired item is readily available.

One reason for this is that differences exist from animal to animal and, since many meat items come fresh, quality variations result. To minimize this phenomenon, most livestock farmers have standardized the care and feeding of their animals. In addition, this variability is generally less of a problem with processed meats. For example, pork that is cured for ham and bacon offers a great deal of predictability.

A second reason buyers cannot always find desired items is that meat of the highest- (or lowest-) quality grade may be in short supply. Because this can also thwart the operators' plans, buyers may need to rethink their strategy.

A third challenge relates to bid buying. Obtaining consistent quality should be a concern when purchasing in this manner. When bids are accepted, they often cover a long period. Changes in quality may become evident over time. The problems associated with finding consistent quality and a continuous supply of meat may drive buyers to stay with one supplier.

Type of Processing

While buyers can still purchase a "side of beef," which is half of an animal carcass, most buy fresh meat items that are butchered to some extent. Although many dining establishments prefer only fresh cuts of meats, numerous foodservice operations rely on frozen meat products in order to operate efficiently. Also, when it comes to some processed-meat items, such as bacon, ham, and bologna, most operations would never consider purchasing fresh meat and processing it in-house. Ultimately, buyers must decide on what type and amount of processing is most economical for the meat's intended use.

A key factor used in making this decision is the AP price; usually, it has a direct correlation with the degree of fabrication. A considerable spread may exist between the AP price per pound for "portion-cut" meat that needs nothing more than some cooking or heating, such as ready-to-cook sirloin steak, and the AP price per pound for "wholesale cut" meat, such as a whole loin, which can be butchered into sirloin steaks.

An even greater AP-price-per-pound spread may exist between a portion-cut steak and a side of beef.

Many restaurants find it more economical to let suppliers wield the cutting tools because it is too expensive to devote a great deal of space to a butcher shop. They prefer to concentrate on devoting as much space as possible to a revenue-generating dining room and lounge.

Still, some foodservice operations perform a great deal of fabrication in-house. They may have determined that in-house processing is more economical or that it enables them to have better control of the quality. Some may even use this feature in their marketing efforts. In any case, the hospitality industry generally thinks that in-house meat fabrication presents four major problems:

1.   This practice requires additional labor hours and skilled laborers, which may lead to increased operating expenses and recruiting cost.

2.   This practice tends to increase the level of pilferage; it is easy to steal items that are being preprepared, usually in a remote area of the kitchen.

3.   Avoidable waste also tends to increase whenever hospitality operators engage in major production efforts.

4.   Meat production creates a considerable amount of "working" dirt and waste products that must be removed continuously in order to prevent contamination. As such, an operation's sanitation needs increase substantially. The typical restaurant kitchen does not have the equipment, time, and skill needed to perform these duties adequately.

Conversely, some disadvantages exist with portion-cut meats. These drawbacks include:

1.   A premium may be charged for uniform weight, shape, and thickness. (This uniformity, however, may be very useful to relatively inexperienced cooks.)

2.   Pilferage may increase with portion cuts of meat; they are easy to steal, especially since they usually come in convenient packages.

3.   Sometimes, a case of precut steaks contains one or two steaks of demonstrably lower quality. Can the receiver catch this problem or will the cook notice the difference?

When purchasing in large quantities, some buyers may find it advantageous to choose frozen-meat items. With today's processing technology, freezing does not harm the products. Proper freezing, thawing, and cooking methods can make any quality differences in taste or texture almost imperceptible. Quality problems are usually due to improper handling.

Another option buyers can consider is convenience foods. Frozen-meat entrées, including fajitas, baked and sliced meat loaf, and stuffed peppers, may not be popular in some hospitality operations, but college, hospital, and other institutional food services use these products readily. These preprepared, portion-control items are sometimes used for large banquets and employee meals, too.

Reducing the AP Price and the EP Cost

Meat purchasers can find all sorts of ways to reduce the AP price while keeping the edible-portion (EP) cost, profit margins, and dollar profits acceptable. Some opt for substitutions, such as replacing meat with fish, offering casseroles instead of sandwich steaks, and widening the menu to include lower-cost products. This latter tactic, though, reduces both AP prices and menu prices. The food cost as a percentage of menu sales prices may decrease, but the sales revenue may also decline and cut into the number of dollars left to cover labor, overhead, and profits.

Buyers can also substitute meat of lower quality and tenderize it; purchase some "formed-meat" products, which are less expensive cuts of meat that have been "flaked," then re-formed and sliced to resemble a steak; or add soybean extenders, as long as chefs follow the legal requirements. (For example, soybean-extended ground beef cannot be called hamburger.)

Furthermore, hospitality operations can reduce the AP price and the EP cost by shrinking portion sizes. Alternately, instead of including a baked potato with the steak, they can charge extra for it. (This reduces the EP cost of the steak dinner, not the EP cost of the steak itself.) Yet another option is to consider serving coleslaw as a substitute for tossed salad. Remember, these substitution strategies always carry a certain amount of risk.

One way to protect the AP price and EP cost is to enter into a long-term contract with the supplier. This arrangement can help operators retain their standards of quality while considerably reducing the risk of both blurring their image and annoying their customers—a risk that may accompany the other cost-cutting strategies just noted. Buyers may realize a saving if they contract for, perhaps, a six-month supply of beef. Of course, they must have the wherewithal to procure the huge quantities that are necessary to interest a supplier. Consequently, only large chains regularly use this method. Not everyone in the hospitality industry thinks that firms can save money this way. For instance, the daily cash price for meat may drop considerably at any time during the contractual period.

The hedging procedure may be a viable way of maintaining a relatively stable AP price. Large foodservice companies that have the resources and skills needed to practice this procedure may save money in the long run.

As a practical matter, small operators may have to make do on a day-to-day basis. If they try to spend too much time concentrating on the AP price, they may neglect the EP cost. It may be better for small operators to concentrate on the EP cost, especially on ways of reducing it that represent little or no risk. For instance, they should ensure that there is minimal waste in production and service, minimal shrinkage, zero pilferage, and so forth.

Every once in a while, buyers may find a bargain. Unfortunately, most bargains come from new suppliers in the form of temporary introductory offers. In addition, bargains may force buyers to purchase a new convenience entrée at a special AP price or to take fresh-frozen meat in lieu of fresh-refrigerated meat.

Normal quantity buy opportunities do appear. But, aside from items on a move list, fewer meat bargains are available in comparison to those offered in other product areas. Furthermore, buyers do not shift suppliers or meat specifications quickly. Meat is a major purchase, so meat buyers practice discretion more widely.

SELECTION FACTORS

As with all products, management decides on the quality and style of meat desired. Then, either alone or in concert with other personnel, management evaluates several of the following selection factors when determining standards of quality desired and suppliers.

Intended Use

As always, buyers want to determine exactly what the intended use of an item is so that they will be able to prepare the appropriate, relevant specification. For instance, bacon used on the breakfast menu should be cut differently from bacon that will be cooked, crumbled, and used as a salad topping.

Exact Name

As with all other products, it is very important for hospitality operators to note the exact name of the meat items they want so they do not receive products that will not suit their needs. To some extent, identifying the exact product that they prefer is a bit easier in this channel of distribution than it is in some other product lines. Over the years, a great deal of standardization has evolved, which the meat industry, meat users, and the United States Department of Agriculture (USDA) have spurred.

The federal government has set several standards of identity for meat products. For instance, if buyers specify that they want hamburger, they will get a mixture that is 70 percent lean meat and 30 percent fat. Of course, as with all standards of identity, a producer is free to improve upon the government definition. In this example, the producer can use beef from just about any part of the animal. So, if buyers expect to receive a specific type of meat in their hamburger, they must include this information in their specification.

Using standards of identity, therefore, is a bit risky unless buyers include additional appropriate information on the specification. They will be able to reduce this amount of information considerably if they utilize in the specification the Institutional Meat Purchase Specifications (IMPS) numbering system for meat items. These numbers take the place of part of a meat specification. For example, if buyers order a 1112 ribeye steak, they will get a particular style and trim.

The IMPS numbers evolved from a cooperative effort by the National Association of Meat Purveyors (NAMP [now called the North American Meat Processors Association]), the National Live Stock and Meat Board, foodservice purchasing agents, and the USDA's Agricultural Marketing Service (AMS) Livestock Division. These numbers are included in *The Meat Buyers Guide (MBG)*, published by the NAMP. In addition to the numbers, a description and a picture of each item are included. This is very desirable to buyers who can then "order by the numbers" and be assured of receiving the exact cut of meat they want.

The IMPS numbers, sometimes referred to as the IMPS/NAMP numbers or the *MBG* numbers, provide a considerable degree of convenience. Typical buyers would never think of preparing meat product specifications without first consulting this major reference book. IMPS numbers are indexed according to product category. The first digit of the number refers to the type of product; the remaining digits indicate a specific cut and trim (see Figure 10.2 for the IMPS numbers for some meat cuts and Figure 10.3 for the *MBG* Table of Contents).

If, for some reason, buyers do not wish to use the IMPS numbers when preparing specifications, they must at least be able to indicate the exact cut of meat they want or the exact type of processed item they need. Fresh meat comes in four basic cuts: (1) the whole carcass; (2) a side, essentially half a carcass; (3) a wholesale (primal) cut; and (4) a retail cut. Numerous other cutting terms, such as hotel-sliced bacon, Spencer-cut prime rib of beef, and square-cut chuck, are involved, too. Buyers are responsible for becoming familiar with these cuts and related terminology, especially if they decide to forgo the use of IMPS numbers.

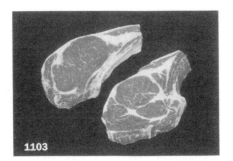

 1103

Beef Rib, Rib Steak, Bone In

Bone in rib steaks may be prepared from any bone in rib item. The *latissimus dorsi, infraspinatus* and *trapezius* muscles above the blade bone and the *subscapularis* and *rhomboideus* muscles below it including the blade bone, related cartilage, feather bones, chine bones and backstrap shall be excluded. The short ribs shall be excluded at a point which is no more than 3.0 inches (7.5 cm) from the ventral edge of the *longissimus dorsi* muscle.

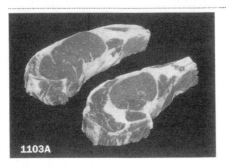

 1103A

Beef Rib, Rib Steak, Boneless

This boneless rib steak item is prepared as described in Item No. 1103, except, in addition, all other bones, cartilages and the intercostal meat shall be excluded.

 1112A

Beef Rib, Ribeye Steak, Lip-On, Boneless

Boneless ribeye steaks, lip-on shall be prepared from a rib item meeting the end requirements of Item No. 112A. The lip shall be cut on the short rib side with a straight cut which is ventral to, but no more than 2.0 inches (5.0 cm) from the *longissimus dorsi,* leaving the lip firmly attached.

 1112B

Beef Rib, Ribeye Steak, Lip-On, Short-Cut, Boneless

This item is as described in Item No. 1112A, except in this item the lip shall be cut on the short rib side ventral to, but no more than 1.0 inch (2.5 cm) from the *longissimus dorsi.*

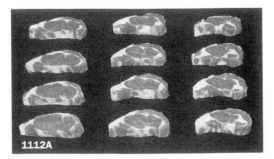

FIGURE 10.2 IMPS numbers for some meat products. Reprinted from *The Meat Buyers Guide,* Author and Publisher: The North American Meat Processors Association. Copyright 1997, Third Printing—May 2003. All rights reserved. Visit www.namp.com for further production information.

1112

Beef Rib, Ribeye Roll Steak, Boneless

Boneless ribeye roll steaks shall be prepared from any boneless ribeye roll item. Any lip, if present on the product being used to prepare this item, shall be excluded so as to expose the natural seam immediately ventral to the *longissimus dorsi* muscle.

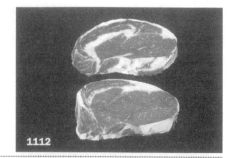

1112

1114D

Beef Chuck, Shoulder Clod, Top Blade Steak, Boneless (for Braising)

Boneless top blade steaks shall be prepared from product meeting the end requirements of Item No. 114D.

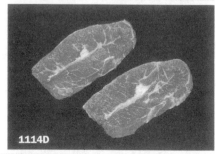

1114D

1116D

Beef Chuck, Chuck Eye Roll Steak, Boneless (for Braising)

Boneless chuck eye roll steaks shall be prepared from product meeting the end requirements of Item No. 116D.

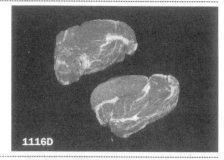

1116D

1121D

Beef Plate, Inside Skirt Steak, Boneless

The boneless steaks shall be prepared from product meeting the end requirements of Item No. 121D.

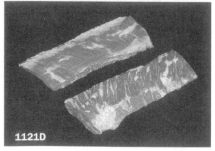

1121D

FIGURE 10.2 (Continued)

CONTENTS

Table of

FIGURE 10.3 *The Meat Buyers Guide* Table of Contents. Reprinted from *The Meat Buyers Guide*, Author and Publisher: The North American Meat Processors Association. Copyright 1997, Third Printing—May 2003. All rights reserved. Visit www.namp.com for further product information.

In addition to cuts of meat, buyers need to be knowledgeable about other commonly used trade terms. Two major ones are "variety meats" (sometimes referred to as "edible by-products"), which refer to such organs as the liver, tongue, and heart, and "sausages," which are preserved, usually dried or salted, chunked or chopped meat and spices shaped into tubes. Some sausages have skins; some do not. Some are cooked; some are not.

The federal government can modify the existing list of terms via additions or deletions. It can also include definitions of the terms used in its grading practices and grading standards. Buyers should stay abreast of these changes if they purchase a great deal of meat.

If buyers need a great deal of processed meat products, they must be concerned with the exact name, specific form, and culinary quality of these items. Standards of identity exist for many products; for example, a minimum formula exists for preprepared beef stew. Generally, though, these identity formulas are not very useful as a major selection factor for processed-meat products.

U.S. Government Inspection and Grades (or Equivalent)

The inspection of meat for wholesomeness has been mandatory since the passage of the Federal Meat Inspection Act in 1907. This law applies to all raw meat sold in interstate commerce, as well as meat exported to other countries. Processed products, such as sausages, frozen dinners, canned meats, and soups made with meat, must also be inspected. An exception is rabbit meat; it is not required to be federally inspected. However, the rabbit industry has a voluntary program that requires rabbit meat packers to pay for inspection.

Federal inspection falls under the jurisdiction of the USDA's Food Safety and Inspection Service (FSIS) (www.fsis.usda.gov). The principal inspection system the FSIS uses for most meat-slaughtering and -processing facilities is the Hazard Analysis Critical Control Point (HACCP). The food industry instituted HACCP to ensure food safety and nonfood safety conditions exist in all critical stages of food handling by reducing and eliminating defects that pass through traditional inspection. FSIS inspectors conduct online carcass inspection and verification inspection to make certain that plants are meeting the FSIS's performance standards for food safety or nonfood safety defects (www.fsis.usda.gov/OA/haccp).

States with companies that sell meats solely through intrastate commerce channels have the discretion of conducting their own meat-inspection programs. Under the 1967 Wholesome Meat Act, these state inspection programs are required to be at least equal to the federal inspection programs.

The federal inspection program begins with the approval of plans for a slaughtering or processing plant to ensure that the facilities, equipment, and procedures can adequately provide for safe and sanitary operations. Facilities and equipment in plants must be easy to clean and to keep clean. The floor plan, water supply, waste disposal methods, and lighting must be approved for each plant facility. Each day before operations begin, the inspector checks the plant and continues the inspection throughout the day to ensure that sanitary conditions are maintained. If, at any time, he or she finds that the equipment is not properly cleaned or an unsanitary condition is present, slaughtering or processing operations are stopped until corrective steps have been taken.

The inspection of animals is done both before and after slaughtering. Before slaughter, USDA inspectors examine all livestock for signs of disease, and any animal appearing sick undergoes a special examination. No dead or dying animal is allowed to enter the slaughtering plant. After slaughter, the inspectors examine each carcass and its internal organs for signs of disease or contamination that would make all or part of the carcass unfit as human food (see Figure 10.4). To ensure uniformity in the inspection process, veterinary supervisors regularly monitor the inspectors' procedures and work.

Meat that passes the rigorous USDA inspection is marked with a federal-inspection stamp (see Figure 10.5). This stamp indicates the number of the meat-processing plant where the meat was slaughtered and packed. The stamp does not appear on all meat cuts; usually, it is visible only on wholesale cuts of meat. Some retail cuts, however, may include remnants of an inspection stamp unless it is completely removed during the cutting and trimming process.

The inspection program the USDA provides is the most trusted inspection program available. The program the United States military uses to inspect its meat products before use in troop feeding is comparable to the USDA's procedures. Although several types of inspection programs exist, they are not the same as USDA inspection for wholesomeness. For instance, religious inspections are performed in some meat plants. These inspections certify only that the meat items satisfy the religious codes, not that they have met a certain standard of quality.

In addition to its meat inspections, the federal government also prepares guidelines on such topics as humane slaughter techniques, animal husbandry, and transportation techniques.

In addition to mandatory inspection, the USDA offers voluntary grading programs. It has long recognized the importance of a uniform system of grading slaughter animals to facilitate the production, marketing, and distribution of livestock and meats. These grading programs also provide an objective evaluation of the culinary quality and edible yield of fresh meat items. The initial U.S. standards for grades of beef were

FIGURE 10.4 Federal inspector checking a beef carcass. Courtesy of National Cattlemen's Beef Association.

formulated in 1916, with the official standards adopted in 1928. Today, these services are conducted under the regulations of the Agricultural Marketing Act of 1946. The AMS Live Stock & Feed Division publishes the grades (see Figure 10.6).

Quality grading systems exist for beef, lamb, pork, and veal; voluntary yield grading systems are available for beef and lamb. While no formal yield grading system exists for pork, these items' quality grades are based primarily on yield; other culinary characteristics play a lesser role. Veal is not graded for yield.

Because grading services are not mandated, they must be purchased. Nonetheless, purchasing such services is popular with many manufacturers, distribution intermediaries, and retailers since it is quite common for buyers to use government grades as one selection factor when they purchase meat. However, some cattle breeders and meat packers feel that grading is too capricious and inconsistent, and some people are not happy with the changes and proposed changes (usually done for health and

FIGURE 10.5 Federal inspection stamps used for meat products. The number appearing in the stamp identifies the meat plant where the meat was processed and inspected.

Source: United States Department of Agriculture.

nutrition concerns) in the grading systems that have occurred over the years. Nevertheless, approximately 50 percent of the meat sold in the United States is graded for quality. Despite the problems associated with federal grades, those mentioned above among others, buyers have not been deterred from using those systems. Buyers may not rely so heavily on the grades as they once did, but most meat specifications contain some reference to a U.S. grade.

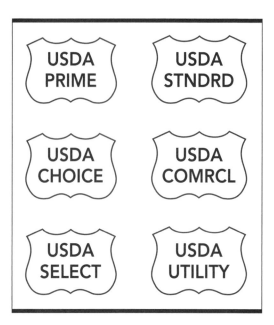

Quality grades for slaughter cattle are based on an evaluation of factors related to the palatability of the meat. Federal graders primarily evaluate beef quality by the amount and distribution of "finish," that is, the degree of fatness; the firmness of muscling; and the physical characteristics of the animal that are associated with maturity. Other grading factors the federal grader evaluates include: the age of the animal at the time of slaughter; the sex of the animal; the color of the flesh; the amount of external finish; the shape and form of the carcass; and the number of defects and blemishes. Foodservice buyers consider the amount of "marbling," or the little streams of fat that run through meat, present in the flesh to be a primary factor.

U.S. quality grades for beef also are subject to several limiting rules. The USDA grading system strictly defines some, while others grant some discretion to the federal grader. The most severe limiting rule, and the one that normally causes a great deal of anxiety among meat producers, is the regulation associated with the beef animals' "maturity class." These animals are divided into five maturity classes:

Class A: Age at the time of slaughter is 9 to 30 months.
Class B: Age at the time of slaughter is 30 to 42 months.
Classes C, D, E: Age at the time of slaughter is greater than 42 months.

It is the grader's responsibility to determine an animal's physiological age at the time of slaughter. He or she does this by examining the color and texture of the lean

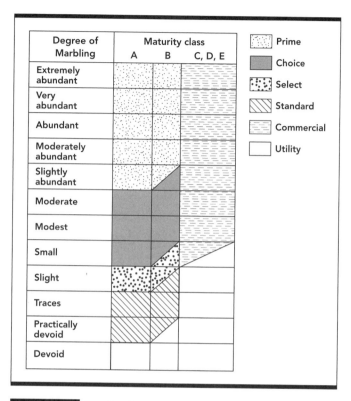

Degree of Marbling	Maturity class			Prime
	A	B	C, D, E	Choice
Extremely abundant				Select
Very abundant				Standard
Abundant				Commercial
Moderately abundant				Utility
Slightly abundant				
Moderate				
Modest				
Small				
Slight				
Traces				
Practically devoid				
Devoid				

FIGURE 10.7 How beef quality grades are determined.

meat, the condition of the bones, and the amount of hardening, or ossification, of the cartilage.

Beef animals' quality grades are affected by their maturity class. For instance, regardless of the quality present in Class C cattle, it cannot be graded Prime. In general, any beef animal older than 42 months at time of slaughter cannot receive the highest-quality grade regardless of the score it received on the other factors the grader considers (see Figure 10.7).

The federal quality grades for beef are:

1. **Prime.** This is the best product available. Tender and very juicy, Prime beef contains 8 to 10 percent fat. Usually, the animal has been grain-fed for at least 180 days in order to develop the exceptionally large amount of firm, white fat. The meat is extremely flavorful.

2. **Choice.** Choice grade beef contains at least 5 percent fat. Three levels exist: high, medium, and low. High Choice is similar to Prime, although the animal has

been grain-fed for only about 150 days. Medium Choice indicates that the animal has been grain-fed about 120 days; Low Choice, for about 90 days. Foodservice operators normally purchase High and Medium Choice; Low Choice is typically sold through supermarkets and grocery stores.

3. **Select.** This beef is a very lean product, containing 4 percent fat and sometimes referred to as "grass-fed beef." The fat on this product usually is not very white, nor very firm. This grade is popular in supermarkets. It is a low-cost item and is more healthful than higher-quality grades, but it lacks flavor.

4. **Standard.** While similar to Select, Standard beef is even less juicy and tender. It has a very mild flavor. This product also tends to be referred to as "grass-fed beef."

5. **Commercial.** Beef from older cattle, Commercial grade is especially lacking in tenderness. Usually, dairy cows receive this quality grade. Because of the animals' age at the time of slaughter, some of this meat may be quite flavorful.

6, 7, 8. **Utility, Cutter, and Canner.** No age limitation exists for these quality grades. Old breeding stock (older cows and bulls) is usually classified into one of these three quality grade categories. Generally, these products are not available as fresh meat. Rather, these manufacturing grades are intended for use by commercial food processors.

Lamb quality grades are based primarily on the color, texture, and firmness of the flesh; quality and firmness of the finish; the proportion of meat to bone; and the amount and quality of the "feathering," which is the fat streaking in the ribs and the fat streaking in the inside flank muscles.

The federal quality grades for lamb are:

1. Prime

2. Choice

3. Good

4. Utility

If a foodservice operation offers fresh lamb on the menu, the Prime and Choice quality grades are normally used. Lower-graded products are not intended for use as fresh meat items. As is typically the case in the hospitality industry, commercial food processors primarily use manufacturing grades to prepare convenience food items.

Pork quality grades are based almost exclusively on yield. The most important consideration is the amount of finish, especially as it relates to color, firmness, and texture. Feathering is also an important consideration. Grain-fed pork make better-quality products, which are far superior to those animals that are given other types of feeds.

The federal quality grades for pork are:

1. No. 1
2. No. 2
3. No. 3
4. No. 4
5. Utility

Most pork is used by commercial food processors to fabricate a variety of convenience and processed items, such as ham and bacon. As a result of this and the fact that most pork is separated into small cuts before it leaves the meat-packing plant, very few pork carcasses are graded for quality. Meat packers and purchasing agents rely on the various packers' brands (i.e., packers' "grades") that are available.

If fresh pork is included on the menu, the typical foodservice operation necessarily must use the No. 1 or No. 2 quality grade, or equivalent packers' brands. Lower-quality items will shrink too much during the cooking process, resulting in an unacceptable finished menu item.

Veal quality grades are based on the color, texture, and firmness of the flesh; the proportion of meat to bone; the quality and firmness of the finish; and the amount and quality of the feathering. High-quality veal will have a pink color and smooth flesh.

The federal quality grades for veal are:

1. Prime
2. Choice
3. Good
4. Standard
5. Utility
6. Cull

The Prime and Choice quality grades are intended for use as fresh products. The lower-quality products tend to be tough and, therefore, are more commonly used to fabricate convenience items.

For more information on U.S. government inspection and grades, see www.ams.usda.gov/lsg/ls-mg.

Product Yield

The federal government provides a voluntary yield grading service for beef and lamb. Yield grades are numbered 1 through 5, with Yield Grade 1 representing the highest yield of cuts and Yield Grade 5, the lowest. Grading criteria are based on the following factors: (1) thickness of fat over ribeye; (2) area of ribeye; (3) percent of kidney, pelvic, and heart fat; and (4) carcass weight. Some limiting rules apply. For instance, USDA Prime beef cannot earn a Yield Grade of 1 or 2 because of its high percentage of fat, which reduces its edible yield.

Prime beef has a great deal of marbling—and, hence, flavor—and the amount of muscle is considerably reduced for this quality grade. For similar reasons, USDA Choice beef also cannot earn a Yield Grade 1. So, to a certain degree, quality and yield grades are interrelated.

Buyers who purchase beef sides, quarters, or wholesale cuts want to specify a desired yield grade. But with such large cuts of beef, an exact yield grade cannot always be specified. Instead, a yield range must be denoted. For example, a buyer may purchase a USDA Commercial beef brisket, with a U.S. Yield Grade 1–2, or its equivalent (see Figures 10.8 and 10.9).

Buyers who purchase only retail cuts can use the federal-government standards to specify an exact yield grade. For example, a specification for sirloin strip steak can state that it must be cut from a beef carcass that carries U.S. Yield Grade 1, or its equivalent.

If buyers use the IMPS system and/or the USDA yield grades, they can be assured of standardized edible yields for the fresh products they purchase. If, however, they do not utilize these systems, they will need to indicate on the specification the minimum yield, or the maximum trim, they will accept.

FIGURE 10.8 A federal yield grade stamp used to indicate the yield of a carcass.
Source: United States Department of Agriculture.

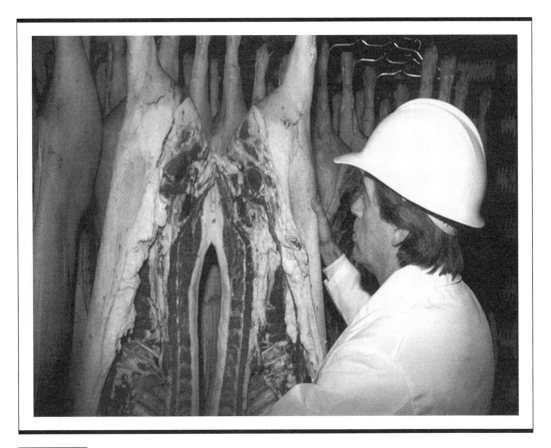

FIGURE 10.9 Federal inspector checking the quality of beef carcasses.

Source: United States Department of Agriculture.

No formal yield grading system is available for pork and veal. Recall, though, that the U.S. quality grades for pork are based primarily on yield. The pork grades are sometimes referred to as "yield standards." The U.S. No. 1 pork grade represents the leanest, meatiest product, whereas U.S. No. 4 product has about twice as much fat and one-third less muscle.

Packers' Brands (or Equivalent)

Many meat producers have developed their own grading procedures for branding their fresh meat. As mentioned earlier, this branding system is sometimes referred to as "packers' grades." For instance, Armour sells "Star Deluxe," "Star," "Quality," and

"Banquet." A significant number of foodservice buyers find this procedure quite acceptable. Although packers' grades are similar to federal grades, they may not reflect an objective viewpoint.

Some meat producers sell "certified organic" products. A qualifying entity must have: (1) standards for what constitutes an agricultural product that is "organically" produced, and (2) a system for ensuring that the product meets those standards. Once the associated animal husbandry and production methods meet USDA standards, the meat items can be labeled as "certified organic by [the certifying entity]" (see www.fsis.usda.gov/OPPDE/larc/Certified_Organic and www.fsis.usda.gov/OPPDE/larc/Organic_Claims for more information on this topic).

Any type of processed meat, other than fresh-frozen or portion-cut, is often purchased on the strength of a brand name. Most pork products, particularly those that are cured, are purchased this way. In almost every instance, these brand names are the only indication of quality. We have seen few specifications for cold cuts that did not rely on a brand name or its equivalent. (It is possible to detail the material used in these items. For a product like breakfast sausage, it would be easy to specify the amount of pork, fat, seasonings, water, and type of grind. For most other types of products in this area, it would not be this simple.)

Packers' brands are sometimes used when purchasing portion-cut meats. Buyers should be concerned with uniformity. Although portion-cut items have a standardized weight, not all packers necessarily supply the same shape. Second, the packaging can differ drastically; steaks may be individually wrapped and neatly stacked, or tossed together in a box. In addition, equal weights are not necessarily an indicator of uniform appearance; some items may be sloppily cut, with excess nicks and tears.

Other forms of brand name identification exist in the meat channel of distribution. For instance, the SYSCO brand Supreme Angus Beef products include only young beef of predominantly Angus breed that fall within the upper two-thirds of the USDA Choice grade. Also, if a meat packer's products meet the quality standards that the popular Certified Angus Beef Program mandates, they are allowed to carry the certification seal. The program, developed by the American Angus Association, stipulates that a beef product must have at least modest marbling, be in the youngest maturity class, and qualify for U.S. Yield Grade 1, 2, or 3.

Product Size

Invariably, buyers must indicate the size of the particular piece of meat they order. This task is made easy because the *MBG* notes weight ranges for wholesale cuts of meat, as well as standardized portion sizes for retail cuts. For example, all large cuts are categorized into four weight ranges: A, B, C, and D. Small retail cuts can be purchased in several sizes. The IMPS regulations stipulate that portion cuts specified between 6 to 11 ounces must be accurate to within $1/2$ ounce; those specified between 12 to 17 ounces must have a tolerance of $3/4$ ounce; and those specified 18 ounces must be accurate to within 1 ounce.

If buyers purchase processed convenience items, such as frozen, preprepared, stuffed peppers, a size indication may also be required. Usually, a particular packer's brand carries only one size, so if buyers consistently indicate the same brand on their specification, they will not need to specify product size information.

Size of Container

Container sizes are standardized in the meat channel of distribution. Processed products come in package sizes that normally range from about 5 pounds to more than 50 pounds. For instance, frozen, preprepared beef stew may be packed in a 5-pound Cryovac® bag, packed six bags to a case. Alternately, the stew may be packed in a 5-pound, oven-ready, foil tray, which is sometimes referred to as a "steam-table" pack, packed six trays to a case.

Fresh, portion-cut meat products are usually packed in 10-pound cases. Some, such as ground beef, normally come in 5-pound and 10-pound Cryovac bags. Fresh, wholesale cuts of meat are typically packed in a container that is big enough to accommodate them. For example, if buyers order beef rib, IMPS number 109, weight range C (18 to 22 pounds each), the items will usually be packed three to a case. In such instances, buyers would not specify the size of the container if an accurate size could not be determined. Another option is to specify a "catch weight" (approximate weight) of 60 pounds per case.

If buyers purchase canned or bottled merchandise, they must specify the appropriate can number or volume designation. For instance, they can purchase canned chili in a No. 10 can and beef jerky in a 1-gallon jar.

Type of Packaging Material

Packaging materials play an integral role in maintaining product quality by minimizing the degree of product deterioration. As a result, meat buyers should specify materials that optimize product quality in accordance with intended use. The typical packaging used in the industry comprises moisture- and vapor-proof materials. Cryovac plastic, or its equivalent, is particularly popular because it is ideal when a meat packer wants to "shrink wrap" the product; this involves packing the meat in plastic and pulling a vacuum through the parcel so that air is removed and the wrapping collapses to fit snugly around the product.

Packaging quality has a significant effect on a meat product's AP price. Most suppliers adhere to standardized materials. However, a supplier who seeks to undercut his or her competitor could easily do so by using inferior packaging materials. This is false economy because buyers must be concerned with protecting meat properly, especially if it is or will be frozen.

Packaging Procedure

The packaging procedure for meat items parallels those procedures used in the poultry and fish channels of distribution. With the exception of the ice pack method, all other procedures are available from at least one supplier. The choice of packaging procedures varies. Large cuts of meat are necessarily slab-packed in containers that are big enough to hold them. Portion cuts are usually layered, but if buyers prefer, the packer will wrap each portion cut individually and then layer them in the case. Bacon comes in a "shingle pack," which is the way it normally appears in a supermarket; a "layout pack," in which several bacon strips are placed on oven paper and the cook can then conveniently place a layer of this bacon on a sheet pan and pop it into the oven; or a "bulk pack," which is a type of slab-packing procedure. Not every supplier offers a wide range of packaging options because this practice adds to the cost of doing business. However, most suppliers will accommodate buyers' needs if they are willing to pay for these services.

Product Form

Once again, we note the usefulness of the *MBG*. If buyers use the IMPS numbers when specifying meat cuts, and sellers follow them, the meat will be cut and trimmed according to the standards that exist for those items. If buyers are purchasing processed items, they can rely, to some extent, on the *MBG* because it contains IMPS numbers for several convenience items. However, many more items are available in the meat channel of distribution that are not noted in this reference book. Buyers must be very careful to indicate the exact product desired; usually, the best way for buyers to ensure that they obtain a suitable convenience product is to rely on a packer's brand name.

Preservation Method

Most meat products that foodservice buyers purchase are preserved in one of two ways: refrigerated or frozen. Canned, dehydrated, and pickled products are also available. For example, buyers can purchase canned soups and canned chili products; however, many operators tend to favor the frozen varieties.

Meat is also preserved by curing and/or smoking. Curing is accomplished when the meat is subjected to a combination of salt, sugar, sodium nitrite, and other ingredients. Smoking preserves the meat and, in most instances, cooks it as well. Many cured items are also smoked. Foodservice buyers purchase a good deal of these products. The primary factors used for the selection of cured and smoked meats are the unique flavor, texture, and aroma that these preservation methods create. Usually, these products are refrigerated when delivered to a hospitality operation, though many of them could be frozen. For instance, bacon, which is a cured and smoked product, may be refrigerated or frozen.

The hospitality industry has witnessed a great deal of controversy concerning the use of nitrites. Sodium nitrite combines with certain amino acids to form nitrosamine, a carcinogenic substance. Nitrites continue to be used, though in lesser amounts than before, because of their superior preservation qualities and because they can control the growth of *Clostridium botulinum*, the deadly bacterium that causes botulism food poisoning. Nitrites also are responsible for the characteristic color and flavor of cured meat products.

If operators use cured and/or smoked products, they must ensure consistent culinary quality by specifying very clearly the types of products they desire. Ordinarily, the only way to obtain consistency is to specify a particular packer's brand. The many combinations of curing and/or smoking procedures that can be used almost force buyers to select one desired packer's brand for each item purchased. Product substitutions are inadvisable because customers would notice them very quickly.

Tenderization Procedure

If buyers purchase meat to be used for steaks and chops—meat that will be broiled or fried—it has to have a certain degree of tenderness. High-quality meats come with a good measure of natural tenderness. In other cases, it may be necessary for someone in the channel of distribution to introduce a bit of "artificial" tenderization. And, usually, a primary source or an intermediary contributes this.

The natural tenderization process is referred to as "aging" the meat. Beef and lamb can be aged. Pork and veal usually are not. One of several aging methods may be employed. The first is called "dry aging." This method tenderizes the meat and adds flavor. Only high-quality grades of meat can be dry aged successfully. Although a very old animal would be flavorful, no amount of aging would tenderize it. Conversely, young meat that is aged goes through a rushed maturation process. In fact, a couple of weeks of dry aging may produce as much flavor as an extra year of life. USDA Prime beef and USDA Choice beef (High Choice and Medium Choice, not Low Choice) are good candidates for dry aging.

Dry aging is expensive. This method requires the meat to be held for about 14 days in carefully controlled temperature and humidity. As a result, the meat loses moisture. So it weighs less after aging, which forces up its AP price. Also, dry aging requires an additional investment in facilities and inventories, which forces up the AP price even more. Buyers should never assume that the meat they purchase has been dry aged. They must request this expensive procedure—and must be prepared to pay for it.

In the late 1970s, the Food and Drug Administration (FDA) approved a new aging process for beef that involves spraying meat with a mold. (Mold naturally forms during dry aging; this new process simply speeds it along.) This process claims to accomplish in 48 hours what used to take two weeks or more to accomplish under the conventional dry-aging method.

Another type of aging done in the trade is called "wet aging," also referred to as "Cryovac aging" (www.cryovac.com). This method is less expensive than the conventional dry-aging process and causes no weight loss. This method involves wrapping the meat cuts in heavy plastic vacuum packs, sealing them tightly, and keeping them refrigerated for about 10 to 14 days. The wrapped meat can be in transit, aging, while it is trucked to the restaurant. Unfortunately, wet aging causes very little flavor development. Also, wet-aged meat seems to be much drier than dry-aged meat if it is cooked past the medium state. The wet-aging process is the most common form of aging available. If buyers do not specify a tenderization procedure, they can expect that the fresh meat they purchase will undergo this wet-aging process.

The biggest disadvantage of using aged meat is that it quickly cooks to the well-done state. However, if hospitality operators are going to roast large wholesale cuts, they can do a bit of aging themselves. By cooking these items in a slow oven, say, at between 200°F and 225°F, they actually simulate aging. They do not add much flavor, but at these temperatures the meat tenderizes somewhat while it cooks. Whatever the method used, aging usually provides a good meat product.

Some people are under the impression that they can purchase unaged meat, which is sometimes referred to as "green" meat, and successfully age the product themselves in the refrigerator or freezer. Certainly, purchasing green meat is tempting because the AP price would be significantly lower than that of a properly aged product. Unfortunately, meat will not age in the typical refrigerator. Meat also will not age if it is frozen or once it is cooked. It will age properly only if its storage environment has the required temperature and humidity.

Another type of tenderization procedure that can be used on beef animals is a process called "beef electrification." This was introduced in 1978. It consists of subjecting a beef carcass to three 15-minute, 600-volt electric shocks. This process allows meat to be aged for only about two-thirds of the normal aging procedure. In addition, the electrification process not only reduces the aging time, but also increases tenderization by about 50 percent.

Meat can also be "chemically" or "mechanically" treated to tenderize it. Chemical tenderizing involves adding an enzyme to the meat that changes the protein structure. Meat packers can inject enzymes into live animals just before slaughter. They also can give postmortem injections. Alternately, after dressing the meat and, usually, cutting it into no more than $1/2$-inch-thick pieces, packers can dip the meat into an enzyme solution and allow it to remain there for about 30 minutes. This dipping method, however, prevents the enzyme from penetrating the muscles too deeply.

Restaurants may use the dipping tenderization procedure in their own kitchens when they offer a low-priced steak dinner. The steaks probably came from a low-quality animal. The operations want it tender, though, and chemical aging is a way of achieving this.

These chemical procedures are not as popular today as they once were in the hospitality industry. A major problem concerns the possibility that the enzymes used will continue to attack the muscles and connective tissues of the meat if the meat product is kept at a temperature range of approximately 120°F to 140°F. This could easily result in a product that is very mushy and, hence, unacceptable to customers.

Mechanical methods, such as grinding and cubing, alter the shape of the product, but not so much the taste. Tenderizing via physical techniques may be preferable to chemical means. However, if not applied properly, mechanical methods can ruin the product.

Mechanical tenderization can be accomplished using the "needling" method. This procedure involves submitting a wholesale cut to a machine with several tiny needles. The needles penetrate the meat, tenderizing it without altering its shape. Chefs have to look carefully to see the needle marks, and, once the meat is cooked, they are not visible. This method can be used on boneless or bone-in wholesale cuts. Most often, operators use it on wholesale cuts that will be served as steaks or roasts.

Inexpensive, tough pieces of meat can also be tenderized by the "comminuting," or flaking and re-forming, process. These pieces are flaked, not ground, and then pressed together to resemble, for example, a loin of beef. "Steaks" are then cut from this "loin."

Hospitality operators can accomplish mechanical tenderization themselves; they can even buy an expensive needling machine if they wish. For that matter, they can age their own meat and apply a dipping chemical bath. The question is: Who can provide these services less expensively—the buyer or the supplier?

Usually, the tenderization question arises only when buyers purchase fresh beef products. If veal is tenderized at all, the mechanical method is used. More typically, veal is roasted very slowly. Also, operators rarely experience a tenderness problem with pork. Lamb is aged, but only about half as long as beef.

Point of Origin

Occasionally, a foodservice operator notes on the menu the point of origin for a meat entrée. For example, guests may see "Iowa Corn-Fed Beef," "West Virginia Ham," "Wisconsin Veal," or "Belgian Blue Cattle" (a rare, costly, imported breed that is exceptionally low in fat and calories) listed on a menu. If hospitality operations want to use this form of advertising, they must purchase the appropriate product or else they will be violating any relevant truth-in-menu legislation.

Inspection?

Some buyers may be concerned with the various chemicals and additives that suppliers can use to enhance meat production. If buyers are worried, they should seek out meat producers who do not use these methods. Alternately, buyers can hire private inspectors, such as the USDA's Acceptance Service, to ensure that meat products meet their standards. The USDA also has an inspection program for which buyers can contract that guarantees that the meat they purchase is free of pesticides and pesticide residues.

As noted earlier, rabbit meat is not required to be federally inspected for wholesomeness. If inspection of this item is important to buyers, they need to indicate this on their specification.

If buyers purchase meat that comes from another country, its inspection for wholesomeness may not be as demanding as the one U.S. meat must undergo. For instance, many U.S. foodservice operations purchase large quantities of beef from other countries. The federal government must inspect these meat products before they are allowed to enter the United States, but the inspection is hampered somewhat because some exporting countries use additives and chemicals that U.S. regulations do not cover. As far as we can determine, this practice has not caused any health hazards. But, again, if buyers are very concerned with product safety, they should indicate on the specification that they want considerably more inspection than is normally provided.

Imitation Meat Products

Several imitation meat products are available. For instance, hospitality operators can use soybean and oat bran as meat extenders. Alternately, they can use these products to create such items as "bacon bits." Many meat producers sell "ham," "hot dogs," and other similar items that are made with chicken, fish, and/or turkey.

Imitation meat products seem to be very popular in the institutional segment of the foodservice industry. However, there is no reason why they cannot enjoy success in any type of foodservice operation. If operators introduce them as a new "alternative" menu item, they may sell briskly.

Some foodservice operators like imitation meat products because they can manipulate the fat content. This generally translates into a lower AP price; however, a low-fat product may sometimes be more expensive than the traditional item. Imitation meats may even have positive health implications. Reducing the fat content and/or substituting polyunsaturated oils can lead to an imitation meat product that is lower in calories and cholesterol compared to the real item.

One-Stop-Shopping Opportunity

Not every meat supplier carries all the meat products buyers need, especially if they occasionally purchase some unusual items or convenience entrées. In general, the more processing hospitality operations want, the more suppliers they need. One-stop-shopping opportunities are not the rule in meat buying unless the shopping list contains only the common items. Buyers must ask themselves a question: "Should I tailor my menu around one or two suppliers, or should I write the menu and take my chances with a lot of suppliers?" Buyers should not take this issue lightly. On one hand, buyers like to make deals and bid buy because of the potential savings. But, on the other hand, they do not enjoy taking risks with signature items.

AP Price

Since meat represents such a large part of the foodservice purchase dollar, buyers are mindful of AP prices. AP prices vary for meat items, types of packaging, and any other value-added feature in a rather predictable way. However, some buyers are concerned about how the meat industry sets contract prices. In some cases, large meat contracts are prepared in such a way that the eventual AP price is not known until the day the hospitality operation takes delivery. On that day, the price reported on the "Green

Sheet," or in some comparable market pricing report, may be the one the buyer must pay the supplier.

The Green Sheet is the trade nickname for the *HRI Meat Price Report*. It is a weekly guide to current AP prices buyers are paying to U.S. meat producers for beef, lamb, pork, veal, and poultry, as well as for several types of processed-meat items. While it is true that with this type of pricing mechanism, the eventual AP price may be lower than expected, it is also true that it could be much higher. This is a risk some buyers do not want to take—or it may be a risk that some foodservice firms forbid their buyers to take.

One way to keep AP prices down is to bid buy among acceptable suppliers. The time involved, as well as the inconvenience, may be worthwhile. Good savings can accrue by this method. But switching suppliers indiscriminately, especially for signature items, can be risky. Different delivery times, supplier capabilities, and product form may be trivial concerns for other items, but they are usually crucial for meat.

If buyers have adequate cash reserves, they may do well with quantity buys once or twice a year for fresh meat and many processed-meat items. On the cash market, purchasing day to day, buyers take their chances. While a good deal represents potentially great savings, a miscalculation represents potentially serious losses. Several companies keep statistics on both AP prices and the availability of meat supplies. By tracking the supply and demand over time, buyers can develop models to predict the optimal times to purchase meat in large quantity. (Recall that not everyone agrees that large contracts, the futures market, and other such strategies are profitable ventures.)

PURCHASING MEAT

A buyer's first step in purchasing meat is obtaining a copy of the *MBG*. This unique publication is an indispensable reference source that every meat buyer should have. Other good meat reference books are available, but the *MBG* is the only source that addresses the purchasing function exclusively.

The buyer's next step is to determine precisely what meat his or her operation needs. As noted earlier, fresh meats are usually selected on the basis of U.S. grades and IMPS numbers, while processed convenience items are typically selected on the basis of packers' brands. Supplier selection may be based on numerous criteria, including availability, reliability, accountability, shipping, and price.

After the buyer determines the meat products the operation requires, it is always wise to prepare specifications for each item whether or not he or she uses them in bid buying (see Figure 10.10 for example meat specifications, and Figure 10.11 for an example of a product specification outline for meat products).

New York strip steak
Used for dinner entrée
IMPS number 1180
USDA Choice (High Choice)
Cut from USDA Yield Grade 2 carcass
Dry aged 14 to 21 days
12-ounce portion cut
Individually wrapped in plastic film
Layered pack
10- to 12-pound case
Refrigerated

Flank steak
Used for London broil entrée
IMPS number 193
Sipco® brand
Weight range B (1 to 2 pounds)
Packed 8 pieces per Cryovac® bag
Packed 6 bags per case
Case weight, approximately 70
    pounds (catch weight)
Refrigerated

Beef base
Used to prepare soups and sauces
LeGout® brand
16-ounce resealable plastic containers
Packed 12 containers per case
Refrigerated

Vegetable beef soup
Used for lunch appetizer
Campbell's® brand
51-ounce can
Packed 12 cans per case
Unrefrigerated

FIGURE 10.10 Examples of meat product specifications.

Intended use:

Exact name:

U.S. grade (or equivalent):

Product yield:

Packer's brand name (or equivalent):

Product size:

Size of container:

Type of packaging material:

Packaging procedure:

Product form:

Preservation method:

Tenderization procedure:

Point of origin:

FIGURE 10.11 An example of a product specification outline for meat products.

Once the buyer writes the specifications, he or she must evaluate potential suppliers, determine order sizes, fix order times, and so on. Many suppliers work in the fresh- and processed-meat areas. Most parts of the United States are a bid buyer's paradise. Purchasing meat, therefore, can be as easy or as difficult as the buyer wants to make it. In fact, as long as the buyer avoids obscure meat items, he or she can often find one-stop-shopping opportunities. Conversely, the buyer can shop around or practice trade relations.

Even though meat buying is a bid buyer's dream, most buyers approach it cautiously. We may see some long-term contract bidding, perhaps three or six months, but we rarely see indiscriminate shopping around. Most meat buyers seem to be concerned with supplier services, especially dependability. A good reputation helps meat suppliers tremendously. After all, their customers must have their signature items. Stockouts are intolerable to buyers; they must have the right item at the right time.

Naturally, cautious buyer attitudes make it difficult for new meat suppliers to establish themselves. In theory, no difference in items should exist for the same grade and cut of meat. But suppliers' item handling, delivery service, and dependability tend to overshadow this fact. Consequently, new suppliers must resort either to offering low AP prices, at least on an introductory basis, or to offering exceptionally attractive supplier services.

The quality and style of processed-meat items are not easily decided. So packers' brands and suppliers' capabilities tend to weigh heavily in these decisions. The owner-manager, either alone or in consultation with others, determines the requirements. Of course, if the hospitality operation needs esoteric meats, it will have more trouble finding suppliers. Packers' brands are important guides, but the brand the owner-manager wants may not be available in the local community, or, more likely, only one supplier will stock that brand. We have often noticed such items available on a cost-plus basis only.

Of course, the independent farmer is ever-present. Some independent farmers sell fresh meat, but many sell products like homemade sausage. We suggest staying away from all uninspected meat.

Before purchasing any meat item, and usually before or during the writing of specifications, buyers should take the time to evaluate the multitude of substitution possibilities.

Meat buying does not have to be difficult, but it certainly is not easy, especially if the buyer is responsible for procuring a wide variety of meat products. In general, the minimum knowledge the buyer needs can be summarized as follows: the different types of meat, the U.S. grades, the appropriate brand names, the various cuts, and the intended uses for the meat items (see Figures 10.12 through 10.15).

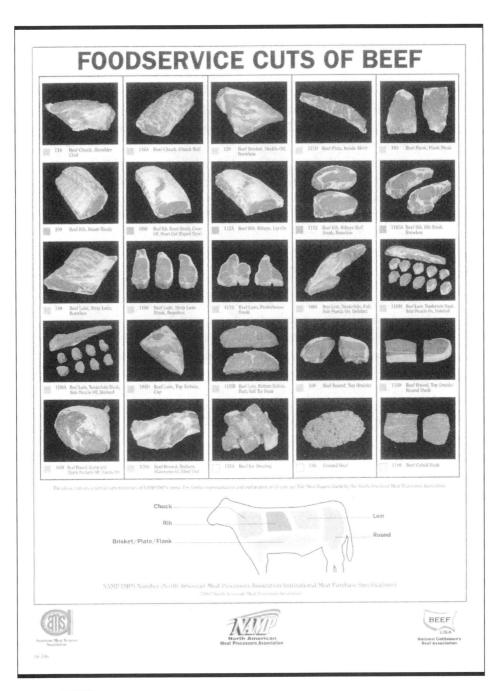

FIGURE 10.12 Beef chart. Courtesy of National Cattlemen's Beef Association.

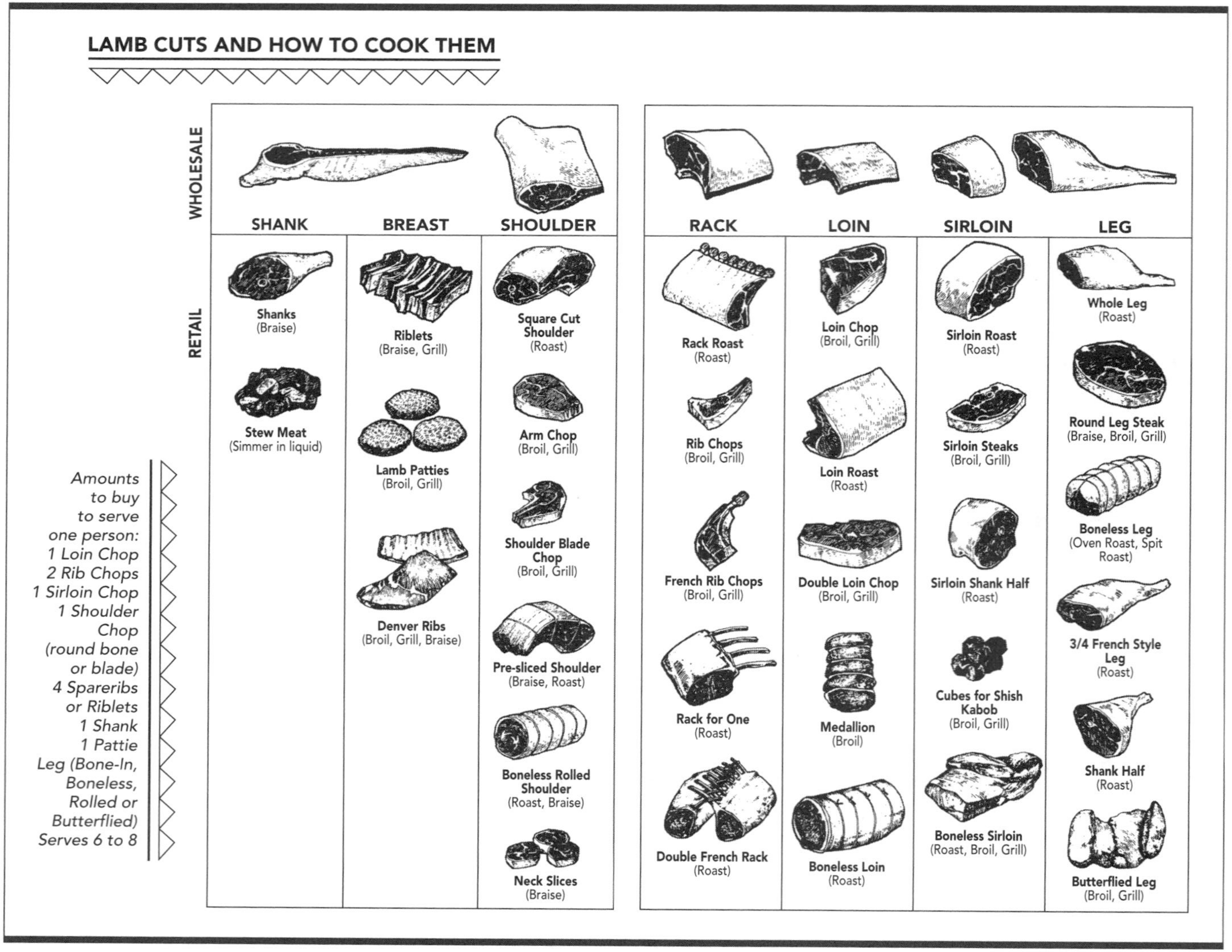

FIGURE 10.13 Lamb cuts and how to cook them. Courtesy of American Sheep Industry Association/American Lamb Council.

FIGURE 10.14 Retail cuts of pork. Courtesy of National Pork Producers Council.

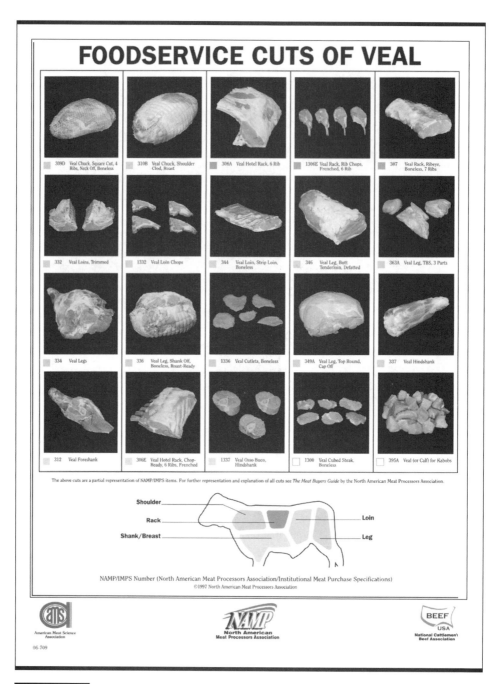

FOODSERVICE CUTS OF VEAL

309D Veal Chuck, Square Cut, 4 Ribs, Neck Off, Boneless	310B Veal Chuck, Shoulder Clod, Roast	306A Veal Hotel Rack, 6 Rib	1306E Veal Rack, Rib Chops, Frenched, 6 Rib	307 Veal Rack, Ribeye, Boneless, 7 Ribs
332 Veal Loins, Trimmed	1332 Veal Loin Chops	344 Veal Loin, Strip Loin, Boneless	346 Veal Leg, Butt Tenderloin, Defatted	363A Veal Leg, TBS, 3 Parts
334 Veal Legs	336 Veal Leg, Shank Off, Boneless, Roast-Ready	1336 Veal Cutlets, Boneless	349A Veal Leg, Top Round, Cap Off	337 Veal Hindshank
312 Veal Foreshank	306E Veal Hotel Rack, Chop-Ready, 6 Ribs, Frenched	1337 Veal Osso Buco, Hindshank	1300 Veal Cubed Steak, Boneless	395A Veal (or Calf) for Kabobs

The above cuts are a partial representation of NAMP/IMPS items. For further representation and explanation of all cuts see *The Meat Buyers Guide* by the North American Meat Processors Association.

Shoulder
Rack
Shank/Breast
Loin
Leg

NAMP/IMPS Number (North American Meat Processors Association/Institutional Meat Purchase Specifications)
©1997 North American Meat Processors Association

American Meat Science Association

North American Meat Processors Association

BEEF USA
National Cattlemen's Beef Association

06-709

FIGURE 10.15 Veal chart. Courtesy of National Cattlemen's Beef Association.

RECEIVING MEAT

When meat products reach a hospitality operation, many owner-managers insist that all inspection be done in the walk-in refrigerator. This practice minimizes spoilage opportunities, but it may be too cautious for some. However, it drives home the fact that an operation cannot take chances with meat—it is too expensive to treat carelessly.

The chef, or someone else in the operation who is also knowledgeable about meat quality, should handle the quality check. This receiver must be able to determine that what the buyer ordered matches what is being received. For instance, suppose a purchase order is for top round, High Choice, but the item received is bottom round, High Choice. This mistake could lead to a stockout and potential disgruntled customers at the dinner hour.

The receiver should check the condition of the meat. For example, all meats have a characteristic color. Fresh beef is a bright, cherry red. If it is not, it could be old. Also, it could be vacuum packed, such as the Cryovac-packaged meat, so that it is not exposed to oxygen to give it the bright, cherry red color, or "bloom," as it is sometimes called. If the characteristic color does not appear to be correct, the receiver should double-check the meat.

Odor is another sign of bad meat. If meat has an unpleasant odor, refuse it. (Fresh pork is difficult to check for odor because it deteriorates from the inside out, not the outside in.)

If meat products have a slimy appearance and are slimy to the touch, the receiver should refuse them. This slime consists of spoilage bacteria.

The receiver should also check the packaging. It should be appropriate for the item purchased. Proper packaging is essential for reducing loss in quality, especially for frozen meats. For instance, frozen meats whose packaging is split should be refused.

The receiver should also consider the temperature of the item. It should be about 40°F, minimally, for refrigerated meat, and 0°F, minimally, for frozen meat. Other processed items—those that are preserved with chemicals, like some sausages—could be less chilled. But a receiver should not make a habit of accepting even these products when warm because this could imply a certain degree of general carelessness on the part of the supplier or a willingness on the part of the operation to relax standards.

Making these kinds of checks on fresh meat and some processed items is not difficult. However, checking frozen meat can be problematical because detecting spoilage in frozen items is more difficult.

Next, the receiver should check quantities. He or she should not rely completely on what is printed on the meat packages. Repacking is not impossible for an unscrupulous

FIGURE 10.16 Federal stamps used to indicate that a meat product meets the buyer's specification under USDA Acceptance purchasing.

Source: United States Department of Agriculture.

supplier or a larcenous employee. The receiver should look for such specifications as weight, count, and sizes. If all of the meat the buyer ordered is in one container, the receiver should separate and weigh the contents individually. He or she must also make sure to deduct the weight of the carton and the packaging.

After checking the quality and quantity, the receiver should move to the prices. As might be expected, meat requires a strong emphasis on record keeping. The receiver should devote a reasonable amount of attention to recording the deliveries. Receiving sheets, bin cards, and meat tags are all methods that can be used to protect expensive meat items. Before a purchase, most buyers express considerable concern with suppliers regarding the actions needed to correct any potential problems. And, if a supplier is out of stock on the buyers' regular orders, he or she often quickly adjusts any agreements in the policies regarding returns, credit terms, and substitute meat items to maintain the buyers' loyalty.

Hospitality operators can streamline the meat-receiving process by using the USDA's Acceptance Service. This practice is popular in meat purchasing, even though it is expensive. Using this service, a hospitality operator can pay a meat expert to write meat specifications and to ensure that the specified products are actually delivered. Acceptance buying is sometimes referred to in the meat trade as "certified buying" or "certification" (see Figure 10.16).

A related program the USDA offers that may also interest meat buyers is called the "Product Examination Service." With this service, a federal inspector examines the purchased meat while it is in transit. The primary purpose of this service is to ensure that product quality does not deteriorate during shipment.

STORING MEAT

Mandatory government inspection ensures that most meat is very clean when it is delivered. The onus is on the foodservice operation to receive and store meat in the correct environment. Whether fresh or frozen, meat must be kept clean and cold. In addition, the stock must be rotated properly.

Meat products are susceptible to bacterial contamination; keeping them clean and sanitary is a big challenge. A dirty storage refrigerator can contaminate good meat. Therefore, it is important for hospitality operations to perform the necessary housekeeping chores in order to minimize contamination.

Operators should store fresh meat in a meat refrigerator apart from cooked-meat items, at a temperature of 35°F to 40°F. If this is not possible, operators should designate a segregated area of the refrigerator for meat storage. To prevent contamination from raw-meat drippings, operators should place cooked items above the raw meats. Also, they should not wrap fresh meat too tightly or stack it too tightly. Both of these practices tend to cut down on the beneficial cold-air circulation around the pieces of meat.

Operators should store frozen products at −10°F or lower. If they must freeze some chilled meat, they must be careful to wrap it correctly and to store it in a freezer no longer than suggested. (It is not a good idea to freeze fresh meat because the typical hospitality operation's freezer is designed to hold frozen foods, not to freeze fresh products.)

ISSUING MEAT

Operators should properly rotate the stock so that the oldest items are issued first. Meat is rarely received and sent straight to production; an employee typically needs a stock requisition to get it. More control of meat items tends to exist at the requisition stage. In many cases, operators add to a perpetual inventory when the meat is delivered. Also, the stock requisition is commonplace. The requisitioner should return any unused meat at the end of his or her shift. The meat consumed should be consistent with the guest checks, that is, the number of meat items sold during that shift. Managing these controls may be somewhat time-consuming, but the practice is quite common in hospitality operations for meat and other expensive items.

By all means, operators must make sure that the requisitioner gets the right item and the right quantity, and must make sure that the in-process inventory does not get so large that it encourages waste or pilferage.

IN-PROCESS INVENTORIES

Surprisingly, in-process meats cause relatively little trouble. These items receive the bulk of the supervisory efforts; moreover, the penalties for pilferage and waste are normally quite severe.

Some employees will make "mistakes." For example, they may burn a steak, accidentally on purpose, and give it to a friend or eat it themselves. Operators can reduce the number of these "errors" by demanding that the mistake, along with the rest of the leftover meat, be turned in to the storeroom at the end of the shift so that it can be accounted for at that time.

Meat also provides some opportunity for shortchanging customers. For example, a server might slice the beef a little thin and keep the extra few ounces handy to trade for a few ounces of gin that the bartender saved in a similar fashion. As is usually the case, effective supervision is the best answer.

KEY WORDS AND CONCEPTS

Advantages and disadvantages of portion-cut meat

Agricultural Marketing Act

Agricultural Marketing Service Livestock Division (AMS Livestock Division)

American Angus Association

As-purchased price (AP price)

Beef electrification

Botulism

Bulk pack

Carcass

Catch weight

Certified Angus Beef Program

Certified buying

Chemical tenderization

Comminuting process

Cryovac aging

Curing

Dry aging

Edible by-product

Edible-portion cost (EP cost)

Exact name

Feathering

Federal Meat Inspection Act

Finish

Flaked and reformed meat products

Food Safety and Inspection Service (FSIS)

Grading factors

Grass-fed beef

Green meat

Green Sheet

Has meat been subjected to United States Department of Agriculture (USDA), or equivalent, inspection?

Hedging

Imitation meat products

In-process inventories

Institutional Meat Purchase Specifications/North American Association of Meat Processors numbers (IMPS/NAMP) numbers

Institutional Meat Purchase Specifications numbers (IMPS numbers)

Intended use

Layout pack

Limiting rule

Long-term contract

Management considerations when purchasing meat

Manufacturing grade

Marbling

Maturity class

Meat Buyers Guide (MBG)

Meat Buyers Guide numbers (*MBG* numbers)

Mechanical tenderization

National Live Stock and Meat Board

Needling procedure

North American Association of Meat Processors (NAMP)

One-stop shopping

Packaging procedure

Packers' brands

Point of origin

Popular types of meat products

Portion-cut meat

Preservation method

Primal cut

Product form

Product size

Product yield

Purchasing, receiving, storing, and issuing meat products

Reluctance to change meat suppliers

Retail cut

Sausage

Shingle pack

Shrink-wrap

Side

Signature item

Size of container

Smoking

Sodium nitrite

Standard of identity

Standardized cut

Steam-table pack

Substitution possibilities

SYSCO Brand Supreme Angus Beef	Service (USDA Acceptance Service)	U.S. quality grades
Tenderization procedure	United States Department of Agriculture Agricultural Marketing Service (USDA AMS)	U.S. yield grades
Truth-in-menu legislation		Variety meat
Type of packaging material		Voluntary inspection for rabbit meat
United States Department of Agriculture (USDA)	United States Department of Agriculture Product Examination Service (USDA Product Examination Service)	Weight range
		Wet aging
United States Department of Agriculture Acceptance		Wholesale cut
		Wholesale Meat Act

QUESTIONS AND PROBLEMS

1. On what are the quality grades for beef primarily based?

2. All meat must be inspected during production in the United States except ＿＿＿＿＿＿ meat.

3. What are the USDA quality grades for beef?

4. What are the USDA quality grades for pork?

5. What are the USDA quality grades for veal?

6. What are the USDA quality grades for lamb?

7. Assume that you manage a high-check-average, full-service club, with annual food and beverage sales of $1.8 million. Your normal purchase order size of T-bone steaks, per week, is approximately 1200 pounds. The current AP price is $7.80 per pound, which will probably hold steady for the next six months. Your current supplier is a longtime good friend, and you are his biggest account. He has carried you during lean times in the past, and you have never had any problems with him. He is dependable and delivers twice a week, in the morning. Down the street is the ABC Corporation's central distribution center, whose management is after your T-bone steak business. Their deal comprises a six-month contract; an AP price of $7.55 per pound; the same quality, cut, yield, and packaging of the T-bones; and afternoon deliveries twice a week. What do you suggest?

8. What are lower-quality grades of meat typically used for?

9. What is the primary reason food buyers use the IMPS numbering system when preparing meat specifications?

10. You notice that your delivery of 500 portion-cut steaks is not up to your usual standard; each weighs 10 ounces instead of the normal 12. It is Friday at 4 P.M., and these steaks are for that weekend. Your meat supplier is usually closed on weekends. What would you do? If possible, ask the owner of a specialty restaurant to discuss this problem with you.

11. On what are USDA yield grades primarily based?

12. A product specification for fresh pork chops could include the following information:
     (a)
     (b)

(c)

(d)

(e)

13. It is painfully evident to you that meat prices are continually rising, and no relief is in sight. But this inflationary spiral has not been enough to encourage you to purchase large quantities and invest in freezer storage facilities. Just recently, your meat supplier has indicated that he intends to go out of business. This supplier calls you and asks whether you wish to purchase his large inventory of meat products at distress prices. Obviously, you are overwhelmed at the possibility of tremendous cost savings.

    (a) What must you know before making an intelligent decision concerning the purchase of this huge inventory?

    (b) Assume that you wish to purchase this inventory. Must you also purchase a freezer? Why or why not?

    (c) Under what conditions would it be advisable to purchase a freezer in this situation?

14. Briefly describe the maturity classes for beef.

15. What is the primary purpose of using the USDA's Product Examination Service?

16. What critical information is missing from the following product specification for lamb loin chops?

    LAMB, LOIN CHOPS

    USED FOR DINNER ENTRÉE

    IMPS NUMBER 1232

    6-OUNCE PORTION CUT

    PACKED IN 10- TO 12-POUND CONTAINERS

17. Assume that you manage a full-service country club. Wesson Brothers, purveyors of fine meats, has been your main source of meat supply for a considerable period. Recently, Wesson proposed to its customers a rather interesting "trade-off." Wesson would like to get out of the transportation business and is seriously considering allowing all its customers to pick up their own orders; in return for their picking up their orders, Wesson will reduce their AP price by 8 percent. Wesson is asking its customers for their opinions. If at least 50 percent of Wesson's existing customers perform the transportation function for themselves, the company will expect all other customers to do likewise.

    (a) Would you like to perform the transportation function? Why or why not?

    (b) If Wesson introduces this proposed policy, how much would you want the AP price of your purchases to decrease? Is an 8 percent reduction enough? Do you think the EP cost of the proposed method would equal that of the current method? Why or why not?

    (c) Assume that Wesson introduces the policy, that you do not wish to perform the transportation function, and that Wesson is an exclusive distributor for certain products you must have. What do you suggest? If possible, ask a country club manager to comment on your answer.

18. Assume that you manage a hospital food service. Currently, it costs about $52 per day to feed one patient. This figure includes food, labor, and direct operating supplies. You expect meat prices to increase by approximately 7 percent the following year, but the hospital administration will not increase your food budget. You must make do with the $52 amount all next year. Currently, you are serving meat at least two times a day on the average. You do not want to reduce this frequency, but you must cut meat costs somewhere. You ask the food buyer for some suggestions. What do you think he or she would propose? Why? If possible, ask a hospital foodservice director to comment on your answer.

19. List some differences between dry aging and wet aging.

20. Give an appropriate intended use for hamburger that has been extended with soybean.

21. Why are meat buyers reluctant to shop indiscriminately for their meat items?

22. The AP price of lean hamburger is $1.89 per pound. The AP price of regular hamburger is $1.09 per pound. The lean meat shrinks 10 percent when cooked; the regular meat shrinks 30 percent.
    (a) At what AP price must the lean hamburger sell at to make it equal in value to the regular hamburger?
    (b) Assume that the EP cost of both the lean and regular meat is equal. What other specific considerations should you examine before purchasing either the lean or the regular?

23. Prepare a product specification for the following meat products:
    (a) Veal cutlet
    (b) Skirt steak
    (c) Prepared chili with beans
    (d) Breakfast sausage
    (e) Ham

24. When would a buyer purchase an imitation meat product?

25. When would a buyer purchase a flaked and re-formed meat product?

26. What benefit would a restaurant owner gain if he or she listed on the menu the point of origin for the meat offerings?

27. Briefly describe one type of meat product you might purchase that would be a good candidate for the needling tenderization procedure.

28. What are the quality grading factors for veal and lamb?

29. Define or explain briefly the following terms:

(a)	Variety meat	(i)	Certified buying
(b)	Wholesale cut	(j)	Layout pack
(c)	Retail cut	(k)	Bloom
(d)	Marbling	(l)	Beef electrification
(e)	Product Examination Service	(m)	Green Sheet
(f)	NAMP	(n)	Shrink wrap
(g)	Curing	(o)	Manufacturing grade
(h)	Feathering		

YIELD, SIZES AND CAPACITIES, AND MEASUREMENT CONVERSION TABLES*

Dry Herbs and Spices

ITEM NAME	NUMBER OF TABLE-SPOONS PER OUNCE	NUMBER OF OUNCES PER TABLE-SPOON	NUMBER OF OUNCES PER CUP	NUMBER EACH PER OUNCE	NUMBER EACH PER TABLE-SPOON
Achiote (Annato) Powder	3.08	0.325	5.2		
Allspice, Ground	4.92	0.203	3.25		
Anise Seed, Whole	5.00	0.200	3.20		
Basil, Ground	5.70	0.175	2.81		
Basil, Whole Leaf	11.40	0.088	1.40		
Bay Leaf, Whole				130	
Bay Leaves, Ground	4.21	0.238	3.8		
Caraway Seed, Whole	4.23	0.236	3.78		
Cardamom, Ground	4.88	0.205	3.28		
Cayenne Pepper	5.30	0.189	3.02		
Celery Salt	1.95	0.513	8.21		
Celery Seed, Whole	3.90	0.256	4.10		
Chervil, Whole	13.00	0.077	1.23		
Chile Flakes, Green	5.90	0.169	2.71		
Chile Flakes, Red	5.90	0.169	2.71		

(continued)

---

* Authored by Francis T. Lynch.

Dry Herbs and Spices (Continued)

ITEM NAME	NUMBER OF TABLE-SPOONS PER OUNCE	NUMBER OF OUNCES PER TABLE-SPOON	NUMBER OF OUNCES PER CUP	NUMBER EACH PER OUNCE	NUMBER EACH PER TABLE-SPOON
Chile Pods, Cascabel				9	
Chile Pods, California and New Mexico				4	
Chile Pods, de Arbol				52	
Chile Pods, Guajillo				5	
Chile Pods, Japones (Japanese)				81	
Chile Pods, Morita				9	
Chile Pods, Pasilla				2	
Chile Pods, Pequin				511	
Chile Powder	4.25	0.235	3.76		
Chile Powder (Ancho)	3.87	0.258	4.13		
Chile Powder (Chipotle)	3.35	0.298	4.77		
Chinese Five-Spice	4.25	0.235	3.76		
Chives, Chopped	56.00	0.018	0.29		
Cilantro	36.00	0.028	0.44		
Cinnamon, Ground	4.00	0.250	4.00		
Cinnamon, Whole Sticks, 5" long				3	
Cloves, Ground	4.30	0.233	3.72		
Cloves, Whole	5.33	0.188	3.00		50
Coriander Seed, Ground	4.58	0.218	3.49		
Coriander Seed, Whole	5.68	0.176	2.82		
Cream of Tarter	2.46	0.407	6.50		
Cumin Seed, Whole	4.72	0.212	3.39		
Cumin, Ground	4.80	0.208	3.33		
Curry Powder	4.50	0.222	3.56		
Dashi No Moto (No MSG)	2.32	0.431	6.9		
Dashi No Moto (With MSG)	2.50	0.400	6.4		
Dill Seed, Whole	4.50	0.222	3.56		
Dill Weed	9.50	0.105	1.68		
Epazote	10.06	0.099	1.59		

Dry Herbs and Spices (Continued)

ITEM NAME	NUMBER OF TABLE-SPOONS PER OUNCE	NUMBER OF OUNCES PER TABLE-SPOON	NUMBER OF OUNCES PER CUP	NUMBER EACH PER OUNCE	NUMBER EACH PER TABLE-SPOON
Fennel Seed, Whole	4.20	0.238	3.81		
Fenugreek Seed, Whole	2.55	0.392	6.27		
Garlic Powder	4.32	0.231	3.70		
Garlic Salt	2.00	0.500	8.00		
Garlic, Granulated	2.66	0.376	6.02		
Ginger, Ground	4.20	0.238	3.81		
Hibiscus Flowers				40	
Lavender Flowers	22.54	0.044	0.71		
Mace, Ground	5.25	0.190	3.05		
Marjoram, Ground	5.93	0.169	2.70		
Marjoram, Whole Leaf	10.60	0.094	1.51		
Mint, Whole Leaf	34.00	0.029	0.47		
Monosodium Glutamate (MSG)	2.86	0.350	5.6		
Mustard Seed, Whole	2.50	0.400	6.40		
Mustard, Ground (powder)	5.16	0.194	3.10		
Nori (Seaweed) Sheets 8" x 8.5"				10	
Nutmeg, Ground	4.25	0.235	3.76		
Onion Powder	4.32	0.231	3.70		
Oregano, Ground	5.70	0.175	2.81		
Oregano, Whole Leaf	10.00	0.100	1.60		
Paprika, Ground	4.10	0.244	3.90		
Parsley Flakes, Whole	22.00	0.045	0.73		
Pepper, Black Whole	4.00	0.250	4.00		175
Pepper, Black, Coarse-Cut	4.32	0.231	3.70		
Pepper, Black, Cracked	4.00	0.250	4.00		
Pepper, Black, Table Grind	4.20	0.238	3.81		
Pepper, Red, Crushed (flakes)	5.90	0.169	2.71		

(continued)

Dry Herbs and Spices (Continued)

ITEM NAME	NUMBER OF TABLE-SPOONS PER OUNCE	NUMBER OF OUNCES PER TABLE-SPOON	NUMBER OF OUNCES PER CUP	NUMBER EACH PER OUNCE	NUMBER EACH PER TABLE-SPOON
Pepper, Szechuan, Whole	8.27	0.121	1.93		
Pepper, White, Ground	3.55	0.282	4.51		
Pepper, White, Whole	4.00	0.250	4.00		
Poppy Seed, Whole	3.20	0.313	5.00		
Poultry Seasoning	7.66	0.131	2.09		
Pumpkin Pie Seasoning Mix	5.06	0.198	3.16		
Rose Blossoms	24.24	0.041	0.66		
Rosemary, Ground	5.70	0.175	2.81		
Rosemary, Whole Leaf	11.85	0.084	1.35		
Saffron, Whole	13.50	0.074	1.19		
Salt, Hawaiian, White	1.79	0.559	8.95		
Sage, Rubbed	11.00	0.091	1.45		
Salt, Kosher, (Diamond Crystal)	3.40	0.294	4.7		
Salt, Kosher, (Morton Coarse)	1.87	0.534	8.55		
Salt, Kosher Flake	1.70	0.588	9.41		
Salt, Red Hawaiian, Medium Grain	1.77	0.566	9.05		
Salt, Regular	1.55	0.645	10.32		
Salt, Seasoning	1.95	0.513	8.21		
Savory, Ground	6.45	0.155	2.48		
Sesame Seed, Whole	3.00	0.333	5.33		
Tarragon, Ground	5.90	0.169	2.71		
Tarragon, Whole Leaf	13.00	0.077	1.23		
Thyme, Ground	6.60	0.152	2.42		
Thyme, Whole Leaf	10.00	0.100	1.60		
Turmeric, Powder	3.75	0.267	4.27		
Wasabi Powder	5.90	0.169	2.71		

Fresh Herbs

ITEM NAME	OUNCES PER BUNCH OR PER AP UNIT	GARNISH LEAVES OR SPRIGS PER BUNCH	GARNISH LEAVES OR SPRIGS PER AP OUNCE	OUNCES OF STEMLESS LEAF PER BUNCH	WEIGHT YIELD PERCENT: STEMLESS LEAF PER BUNCH	OUNCE WEIGHT OF 1 TABLE-SPOON CHOPPED	YIELD: TABLE-SPOONS OF CHOPPED LEAF PER PURCHASED OUNCE	OUNCE WEIGHT OF 1 CUP, CHOPPED
Basil, Sweet	2.5	59	23.6	1.4	56.00%	0.088	6.4	1.408
Bay Leaves	0.6	68	113	0.48	80.00%	0.113	7.1	1.803
Chives, 6" Lengths	1	115	115	0.95	95.00%	0.095	10	1.52
Cilantro	2.8	93	33	1.3	46.43%	0.093	5	1.486
Dill Weed	4.5	105	23	2	44.44%	0.112	4	1.785
Marjoram	1	38	38	0.76	76.00%	0.069	11	1.105
Mint	3.35	80	24	1.4	41.79%	0.108	3.88	1.724
Oregano	1	40	40	0.78	78.00%	0.065	12	1.04
Parsley, Curly	3.4	75	22	1.8	52.94%	0.080	6.62	1.28
Parsley, Italian	5.7	91	16	2.3	40.35%	0.113	3.51	1.8
Rosemary	1	22	22	0.8	80.00%	0.150	5.33	2.4
Sage, Green	1	68	68	0.6	60.00%	0.075	8	1.2

(continued)

273

Fresh Herbs (Continued)

ITEM NAME	OUNCES PER BUNCH OR PER AP UNIT	GARNISH LEAVES OR SPRIGS PER BUNCH	GARNISH LEAVES OR SPRIGS PER AP OUNCE	OUNCES OF STEMLESS LEAF PER BUNCH	WEIGHT YIELD PERCENT: STEMLESS LEAF PER BUNCH	OUNCE WEIGHT OF 1 TABLE-SPOON CHOPPED	YIELD: TABLE-SPOONS OF CHOPPED LEAF PER PURCHASED OUNCE	OUNCE WEIGHT OF 1 CUP, CHOPPED
Tarragon	1	48	48	0.8	80.00%	0.114	7	1.828
Thyme	1	43	43	0.65	65.00%	0.100	6.5	1.6
Watercress	6.1	25	4.1	1.65	27.05%	0.092	2.95	1.47

Notes

1. Ginger yields 70% when peeled.

2. These measurements are based on herbs of normal commercial size and quality with respect to their size, maturity, freshness, moisture, and conformation.

3. Leaves for garnish are large and attractive.

4. Stemless leaf yield includes the garnish leaves plus remaining good leaves.

5. Leaves were stripped from stems before chopping.

6. Chopped leaves were cut "chiffonade" then cross-cut and chopped a bit more.

7. Volume measures of chopped leaves were tapped down but not pressed down hard.

8. The "Yield: Tablespoons of Chopped Leaf per Purchased Ounce'" column was obtained by physically measuring (in cups) the total yield of the purchased amount after stemming and chopping, then multiplying that amount by 16 and dividing the answer by the ounces purchased.

Vegetables

ITEM NAME	AP UNIT	NUMBER OF MEASURES PER AP UNIT	MEASURE PER AP UNITS	TRIMMED/ CLEANED OUNCE WEIGHT OR COUNT	YIELD PERCENT	TRIMMED/ CLEANED OUNCE WEIGHT PER CUP	NUMBER OF TRIMMED/ CLEANED CUPS PER AP UNIT
Alfalfa Sprouts	bag	16	ounce	15.25	95.3%	1.70	8.971
Artichokes #18 (1 Pound Each)	case	18	each	17.5 each	97.2%		
Artichokes #36 (8 Ounce Each)	case	36	each	35 each	97.2%		
Artichokes, Baby	pound	9	each	9 each			
Artichoke Hearts, Marinated, Drained	jar	60	ounce	43	71.67%		
Artichoke Hearts, Marinated, Drained	jar	60	ounce	80 each			
Arugula, Young Leaves, Cleaned	bag	5.15	ounce	5	97.09%	0.6	8.333
Arugula, Young Leaves, Cleaned, Chopped	bag	5.15	ounce	5	97.09%	0.8	6.250
Asparagus, Jumbo Trimmed	pound	16	ounce	9.52	59.5%	4.50	2.110
Asparagus, Jumbo Whole	pound	9	each	8.5 each	94.4%		

(continued)

275

Vegetables (Continued)

ITEM NAME	AP UNIT	NUMBER OF MEASURES PER AP UNIT	MEASURE PER AP UNITS	TRIMMED/ CLEANED OUNCE WEIGHT OR COUNT	YIELD PERCENT	TRIMMED/ CLEANED OUNCE WEIGHT PER CUP	NUMBER OF TRIMMED/ CLEANED CUPS PER AP UNIT
Asparagus, Standard Trimmed	pound	16	ounce	9.13	57.1%	4.75	1.920
Asparagus, Standard Whole	pound	16	each	14.5 each	90.6%		
Asparagus, Thin Export Trimmed	pound	16	ounce	9.02	56.4%	5.00	1.800
Asparagus, Thin Export Whole	pound	38	each	34 each	89.5%		
Avocado, Whole	each	7	ounce	5.5	78.6%		
Avocados, 1/2" Dice	each	7	ounce	5.5	78.6%	5.40	1.019
Avocados, Puree	each	7	ounce	5.5	78.6%	8.10	0.679
Bamboo Shoot, Sliced from Whole Piece	pound	16	ounce	15.4	96.25%		
Bamboo Shoot, Sliced from Whole, Cored	pound	16	ounce	11.55	72.19%	4.5	2.567
Bean Sprouts (Mung)	pound	16	ounce	15.5	96.9%	3.20	4.844
Beans, Green	pound	16	ounce	14.1	88.1%		
Beans, Green, 1" cut	pound	16	ounce	14.1	88.1%	3.90	3.615
Beets, Whole 2" Diameter	pound	16	ounce	10.56	66.0%		
Belgian Endive	head	4	ounce	3.3	82.5%		
Belgian Endive Leaves	head	1	each	12 leaves			

Bitter Melon, Sliced	15.15	ounce	12.25	80.86%	3.20	3.828
Bok Choy, Baby	12	pound	11	91.7%		
Bok Choy, Regular	24	head	21	87.5%		
Broccoli, Bunch, Whole	21.5	each	13.5	62.8%		
Broccoli, Chinese	13.7	bunch	12.8	93.43%	4.30	2.977
Broccoli, Florets—chopped	21.5	each	13.5	62.8%	3.10	4.355
Broccoli, Florets	21.5	each	13.5	62.8%	2.50	5.400
Brussel Sprouts	16	pound	14.2	88.8%	3.20	4.438
Brussel Sprouts, Medium, Each	20	each	20			
Cabbage, Green/Red	40	head	32	80.0%		
Cabbage, Green/Red—Chopped	40	head	32	80.0%	2.50	12.800
Cabbage, Green/Red—Shredded	40	head	32	80.0%	3.30	9.697
Cabbage, Napa	34	head	29.9	87.9%		
Cabbage, Napa, Shredded 1/4"	34	head	29.9	87.9%	2.47	12.105
Cabbage, Savoy	36	head	29.8	82.8%		
Cabbage, Savoy, Shredded 1/4"	36	head	29.8	82.8%	2.50	11.920
Capers, Nonpareil, Drained	7	jar	4.7	67.14%	4.7	1.000
Cardoon, Large	35	head	18	51.4%		

(continued)

Vegetables (Continued)

ITEM NAME	AP UNIT	NUMBER OF MEASURES PER AP UNIT	MEASURE PER AP UNITS	TRIMMED/ CLEANED OUNCE WEIGHT OR COUNT	YIELD PERCENT	TRIMMED/ CLEANED OUNCE WEIGHT PER CUP	NUMBER OF TRIMMED/ CLEANED CUPS PER AP UNIT
Carrots (Table-Medium)	pound	16	ounce	13	81.3%		
Carrots, Baby, Cut and Peeled	pound	16	ounce	66 each			
Carrots, Chopped	pound	16	ounce	13	81.3%	4.90	2.653
Carrots, Diced 1/3" – 1/2"	pound	16	ounce	13	81.3%	5.00	2.600
Carrots, Grated	pound	16	ounce	13	81.3%	3.90	3.333
Carrots, Ground	pound	16	ounce	13	81.3%	5.33	2.439
Carrots, Sliced 1/4" – 1/6"	pound	16	ounce	13	81.3%	4.20	3.095
Carrots, Whole, Baby	pound	16	ounce	35 each	99.0%	6.20	2.550
Carrots, Whole, Jumbo	each	5.5	ounce	4.6	83.6%		
Carrots, Whole, Medium	each	4.1	ounce	3.3	81.3%		
Cauliflower	head	30	ounce	18	60.0%		
Cauliflower, Cut 1" Florets	head	30	ounce	18	60.0%	4.70	3.830
Celeriac (Celery Root)	head	19	ounce	11.5	60.5%		
Celeriac, Julienne	head	19	ounce	11.5	60.5%	3.00	3.833
Celery	bunch	32	ounce	22	68.8%		
Celery, Diced 1/2"– 1/3"	bunch	32	ounce	22	68.8%	4.00	5.500

Item	Unit						
Celery, Stalks per Bunch	bunch	1	each	10 each			
Celery, Tops per Bunch	bunch	32	ounce	9 ounce Top	28.0%		
Chard, Swiss	bunch	14	ounce	12.75	91.1%		
Chard, Swiss, Chopped	bunch	14	ounce	12.75	91.1%	2.30	5.543
Cocktail Onions, Large, Drained	jar	16.9	ounce	9.6	56.80%	4.95	1.939
Cocktail Onions, Large, Drained, Cup Count	jar	16.9	ounce				
Cocktail Onions, Large, Drained, Jar Count	jar	16.9	ounce				
Collard Greens	bunch	12	ounce	7.8	65.0%		
Collard Greens, Chopped	bunch	12	ounce	7.8	65.0%	1.90	4.105
Corn Cob, Fresh Niblets	whole	1	each	5	29.0%	5.75	0.870
Cranberry Beans in Pods	pound	16	ounce	8.55	53.44%	5.05	1.693
Cucumber	each	10	ounce	9.5	95.0%		
Cucumber, Sliced	each	10	ounce	9.5	95.0%	4.80	1.979
Cucumber, Peeled, Seeded, Sliced	each	10	ounce	5.5	55.0%	4.40	2.159
Cucumber, Peeled, Seeded, Diced	each	10	ounce	5.5	55.0%	5.30	1.792

(continued)

Vegetables (Continued)

ITEM NAME	AP UNIT	NUMBER OF MEASURES PER AP UNIT	MEASURE PER AP UNITS	TRIMMED/ CLEANED OUNCE WEIGHT OR COUNT	YIELD PERCENT	TRIMMED/ CLEANED OUNCE WEIGHT PER CUP	NUMBER OF TRIMMED/ CLEANED CUPS PER AP UNIT
Cucumber, Whole, English	each	16	ounce	15.7	98.1%		
Cucumber, Whole, English, Sliced	each	16	ounce	15.7	98.1%	4.50	3.489
Edamame, Hulled (Fresh Soybeans)	carton	19.6	ounce	17.6	90.0%	6.33	2.78
Eggplant, Japanese, Sliced	bag	9.3	ounce	8.4	90.32%	2.68	3.140
Eggplant, Peeled	each	19	ounce	16	84.2%		
Eggplant, Peeled, Cubed	each	19	ounce	16	84.2%	2.90	5.517
Eggplant, Thai, Quartered	bag	7.45	ounce	9 each	90.60%	4.20	1.770
Fennel with 6" Stem	head	14	ounce	13	92.9%		
Fennel, Stemmed	head	14	ounce	7.8	55.7%		
Fennel, Stemmed, Sliced 1/2"	head	14	ounce	7.8	55.7%	3.90	2.000
Fiddlehead Ferns	pound	16	ounce	150 each			
Garlic	head	2.1	ounce	1.85	83.0%		

Item							
Garlic, Chopped—Fresh	head	2.1	ounce	1.85	88.1%	4.80	0.402
Garlic, Cloves per Head	head	2.1		12 cloves			
Garlic Cloves, Already Peeled	jar	16	ounce	15.8	98.75%	5.15	3.068
Garlic Cloves, Peeled and Roasted	jar	16	ounce	15.7	98.13%	6.05	2.595
Garlic, Chopped, in Oil	jar	16	ounce	15.6	97.50%	9.2	1.696
Garlic, Elephant	each	7.55	ounce	7.05	93.38%		
Grape Leaves, Bottled	jar	8	ounce	42 each			
Haricots Verts	bag	14.4	ounce	13.05	90.63%		
Haricots Verts, 1" Lengths	bag	14.4	ounce	13.05	90.63%	3.35	3.896
Hojas (Dry Corn Husks)	bag	8	ounce	64 each			
Horseradish Root, Peeled	pound	16	ounce	11.6	72.50%		
Horseradish Root, Peeled, Shredded	pound	16	ounce	11.6	72.50%	2.5	4.640
Jerusalem Artichoke	pound	16	ounce	11	68.8%		
Jerusalem Artichoke, Sliced	pound	16	ounce	11	68.8%	5.30	2.075
Jicama	pound	16	ounce	13	81.3%		
Kale, Flowering—Leaves	bunch	12	ounce	24 leaves			
Kale, Green	bunch	20	ounce	12	60.0%		
Kohlrabi, Leaves on	pound	16	ounce	7.5	46.9%		

(continued)

Vegetables (Continued)

ITEM NAME	AP UNIT	NUMBER OF MEASURES PER AP UNIT	MEASURE PER AP UNITS	TRIMMED/CLEANED OUNCE WEIGHT OR COUNT	YIELD PERCENT	TRIMMED/CLEANED OUNCE WEIGHT PER CUP	NUMBER OF TRIMMED/CLEANED CUPS PER AP UNIT
Kale, Green, Chopped	bunch	20	ounce	12	60.0%	2.36	5.085
Kohlrabi, Sliced	pound	16	ounce	7.5	46.9%	4.70	1.596
Leeks	pound	16	ounce	7	43.8%		
Leeks, Cross-sliced 1/4"	pound	16	ounce	7	43.8%	3.10	2.258
Lemon Grass, Minced	pound	16	ounce	10.4	65.00%	3.05	3.410
Lemon Grass, Trimmed	pound	16	ounce	10.4	65.00%		
Lemon Grass, Whole	pound	16	ounce	6 each	100.00%		
Lettuce, Bibb, Leaves	head	6	ounce	9 leaves			
Lettuce, Butter/Bibb 5"	head	6	ounce	4.8	80.0%		
Lettuce, Butter/Bibb, Chopped	head	6	ounce	4.8	80.0%	1.95	2.462
Lettuce, Greenleaf	head	16	ounce	13	81.3%		
Lettuce, Greenleaf, Chopped	head	16	ounce	13	81.3%	1.95	6.667
Lettuce, Greenleaf, Leaves	head	16	ounce	12 leaves			
Lettuce, Iceberg	head	26	ounce	19	73.1%		
Lettuce, Iceberg, Chopped	head	26	ounce	19	73.1%	1.95	9.744
Lettuce, Redleaf	head	14	ounce	10.5	75.0%		

Item	Unit	Qty	Measure	Value	%		
Lettuce, Redleaf, Chopped	head	14	ounce	10.5	75.0%	1.95	5.385
Lettuce, Redleaf, Leaves	head	14	ounce	12 leaves			
Lettuce, Romaine	head	24	ounce	18	75.0%		
Lettuce, Romaine, Chopped	head	24	ounce	18	75.0%	2.00	9.000
Lettuce, Romaine, Leaves	head	24	ounce	16 leaves			
Lotus Root	each	5	ounce	4	80.0%		
Malanga, Peeled	pound	16	ounce	11.77	73.56%		
Malanga, Peeled, Thinly Sliced	pound	16	ounce	11.77	73.56%	4.5	2.616
Mushrooms, Crimini, Stemmed	pound	16	ounce	11.89	74.31%		
Mushrooms, Crimini, Stems Tipped	pound	16	ounce	14.3	89.38%		
Mushrooms, Crimini, w/ Stems, Sliced	pound	16	ounce	11.89	74.31%	2.2	5.405
Mushrooms, Large, Each	pound	16	ounce	12 each			
Mushrooms, Medium, Each	pound	16	ounce	22 each			
Mushrooms, Morels, Dry	bag	16	ounce	416 each		0.7	

(continued)

Vegetables (Continued)

ITEM NAME	AP UNIT	NUMBER OF MEASURES PER AP UNIT	MEASURE PER AP UNITS	TRIMMED/ CLEANED OUNCE WEIGHT OR COUNT	YIELD PERCENT	TRIMMED/ CLEANED OUNCE WEIGHT PER CUP	NUMBER OF TRIMMED/ CLEANED CUPS PER AP UNIT
Mushrooms, Morels, Fresh	pound	16	ounce	72 each	85.00%	3.80	4.200
Mushrooms, Oyster, Sliced	basket	5.8	ounce	5.5	94.83%	2.10	2.619
Mushrooms, Oyster, Whole	basket	5.8	ounce	5.5	94.83%	1.90	2.895
Mushrooms, Porcini, Dry	bag	7.5	ounce	7.4	98.67%	0.94	7.893
Mushrooms, Portobello 2"	pound	16	ounce	15 each			
Mushrooms, Shiitake	pound	16	ounce	46 each			
Mushrooms, Shiitake, Culled, Stem Trimmed	pound	16	ounce	12.86	80.38%		
Mushrooms, Shiitake, Dried	package	2.6	ounce	2.5	96.15%	1.2	2.083
Mushrooms, Shiitake, Dried	package	2.6	ounce	25 each			
Mushrooms, Small, Each	pound	16	ounce	45 each			
Mushrooms, White	pound	16	ounce	15 ounce	93.8%		
Mushrooms, White, Sliced	pound	16	ounce	15 ounce	93.8%	2.50	6.000

Mushrooms, White, Whole	pound	16	ounce	15 ounce	93.8%	3.40	4.412
Mustard Greens	bunch	12	ounce	9	75.0%		
Mustard Greens, Chopped	bunch	12	ounce	9	75.0%	2.50	3.600
Nopales (Cactus Leaves)	bag	22.1	ounce	19.85	89.82%		
Olives, Kalamata, Pitted	pound	16	ounce	13.5	84.4%	3.50	3.857
Olives, Kalamata, with Pits	pound	16	ounce	46 each			
Okra, Trimmed	pound	16	ounce	184 each	98.00%	5.4	2.963
Okra, Trimmed, Sliced 1/2"	pound	16	ounce	92 each			
Okra, Whole	pound	16	ounce	13.5	84.4%		
Onions, Boiling, Peeled	pound	16	ounce	14.4	90.00%		
Onions, Boiling, Peeled	pound	16	ounce	10 each			
Onions, Bulb	pound	16	ounce	14.5	90.6%		
Onions, Bulb, 1/2" Diced	pound	16	ounce	14.5	90.6%	3.90	3.718
Onions, Bulb, 1/4" Diced	pound	16	ounce	14.5	90.6%	4.45	3.750
Onions, Bulb, Sliced	pound	16	ounce	14.5	90.6%	3.00	4.833
Onions, Dehydrated, Chopped	pound	16	ounce	16	100%	3.00	5.330
Onions, Each Large	each	13.7	ounce	12.5	91.2%		

(continued)

Vegetables (Continued)

ITEM NAME	AP UNIT	NUMBER OF MEASURES PER AP UNIT	MEASURE PER AP UNITS	TRIMMED/ CLEANED OUNCE WEIGHT OR COUNT	YIELD PERCENT	TRIMMED/ CLEANED OUNCE WEIGHT PER CUP	NUMBER OF TRIMMED/ CLEANED CUPS PER AP UNIT
Onions, Each Small	each	7.8	ounce	6.9	88.5%		
Onions, Green	bunch	3.5	ounce	2.9	82.9%		
Onions, Green, Chopped	bunch	3.5	ounce	2.9	82.9%	2.00	1.450
Onions, Green, Each	bunch	3.5	ounce	7 each			
Onions, Pearl	basket	11	ounce	9.3	84.5%		
Parsnips	pound	16	ounce	13.5	84.4%		
Parsnips, Sliced	pound	16	ounce	13.5	84.4%	4.70	2.872
Peas, Snap	pound	16	ounce	15	93.8%	2.20	6.818
Peas, Snap, Chopped	pound	16	ounce	15	93.8%	3.50	4.286
Peas, Snap, Whole, Each	pound	16	ounce	90 each			
Peas, Snow, Whole, Each	pound	16	ounce	120 each			
Peppers, All, Brunoise	pound	16	ounce	8	50.0%	7.00	1.143
Peppers, Anaheim 7"	pound	16	ounce	8 each			
Peppers, Anaheim, Seeded, Diced	pound	16	ounce	13.85	86.56%	3.80	3.645
Peppers, Green Bells 7 Ounce	pound	16	ounce	13	81.3%		
Peppers, Green, Chopped	pound	16	ounce	13	81.3%	5.20	2.500

Item							
Peppers, Green, Sliced	pound	16	ounce	13	81.3%	3.20	4.063
Peppers, Habañero, 1/6" Rings	pound	16	ounce	14.04	87.75%	2.10	6.686
Peppers, Habañero, Chopped	pound	16	ounce	14.04	87.75%	3.85	3.647
Peppers, Habañero, Stemmed	pound	16	ounce	14.04	87.75%		
Peppers, Habañero, Whole	pound	16	ounce	52 each			
Peppers, Jalapeño Rings, Canned, Drained	can	12	ounce	6.618	55.15%	4.55	1.455
Peppers, Jalapeño Rings, Canned, Undrained	can	12	ounce	12	100.00%	8.25	1.455
Peppers, Jalapeño, Chopped w/ Seeds	pound	16	ounce	14.96	93.50%	3.75	3.989
Peppers, Jalapeño, Chopped, Seedless	pound	16	ounce	14.96	93.50%	3.9	3.836
Peppers, Jalapeño, Sliced w/ Seeds	pound	16	ounce	14.96	93.50%	2.8	5.343
Peppers, Jalapeño, Sliced, Seedless	pound	16	ounce	14.96	93.50%	3.4	4.400
Peppers, Jalapeño, Stemmed	pound	16	ounce	14.96	93.50%		

(continued)

Vegetables (Continued)

ITEM NAME	AP UNIT	NUMBER OF MEASURES PER AP UNIT	MEASURE PER AP UNITS	TRIMMED/ CLEANED OUNCE WEIGHT OR COUNT	YIELD PERCENT	TRIMMED/ CLEANED OUNCE WEIGHT PER CUP	NUMBER OF TRIMMED/ CLEANED CUPS PER AP UNIT
Peppers, Jalapeño Whole	pound	16	ounce	26 each			
Peppers, Pasilla, Cored, Seeded	pound	16	ounce	12.16	76.00%		
Peppers, Pasilla, Diced	pound	16	ounce	12.16	76.00%	3.8	3.200
Peppers, Pasilla Whole	pound	16	ounce	5 each			
Peppers, Red Bells 10 Ounce	pound	16	ounce	13.5	84.4%		
Peppers, Red, Chopped	pound	16	ounce	13.5	84.4%	4.50	3.000
Peppers, Red, Julienne	pound	16	ounce	13.5	84.4%	3.60	3.750
Peppers, Red, Sliced	pound	16	ounce	13.5	84.4%	3.30	4.091
Peppers, Serrano, Chopped Fine w/ Seeds	pound	16	ounce	13.88	86.75%	3.8	3.653
Peppers, Serrano, Stemmed	pound	16	ounce	13.88	86.75%		
Peppers, Serrano Whole	pound	16	ounce	106 each			

Peppers, Yellow-Hot Whole	pound	16	ounce	32 each			
Potato, Baker-Russet	pound	16	ounce	12.5	78.1%	5.00	2.500
Potato, Peeled, Diced	pound	16	ounce	12.5	78.1%	6.00	2.083
Potato, Peeled, Shredded	pound	16	ounce	12.5	78.1%		
Potato, Peeled, Sliced	pound	16	ounce	12.5	78.1%	5.20	2.404
Potatoes, Dehyd. Flakes	pound	16	ounce	16	100.0%	3.20	5.000
Potatoes, Dehyd. Granules	pound	16	ounce	16	100.0%	7.10	2.250
Pumpkins, Miniature (6.8 oz. each)	pound	16	ounce	2.35 each			
Pumpkin, Whole	each	96	ounce	60.5	63.0%		
Purslane	pound	16	ounce	12	75.0%	1.50	8.000
Purslane, Stemmed	pound	16	ounce	8.87	55.44%	1.4	6.336
Radicchio	head	8.7	ounce	8	92.0%	3.00	2.667
Radish, Daikon	each	16	ounce	14	87.5%		
Radish, Red, Large 1 3/8"	bunch			12 each			
Radish, Red, Sliced—Trimmed	bunch					4.10	1.000
Ramps (Wild Leeks)	pound	16	ounce	45 each	92.0%		
Rhubarb	pound	16	ounce	4 stalks			
Rhubarb, Cubed	pound	16	ounce	14.7	92.0%	3.80	3.868
Rutabaga	pound	16	ounce	13	81.3%		

(continued)

Vegetables (Continued)

ITEM NAME	AP UNIT	NUMBER OF MEASURES PER AP UNIT	MEASURE PER AP UNITS	TRIMMED/ CLEANED OUNCE WEIGHT OR COUNT	YIELD PERCENT	TRIMMED/ CLEANED OUNCE WEIGHT PER CUP	NUMBER OF TRIMMED/ CLEANED CUPS PER AP UNIT
Rutabaga, Brunoise	pound	16	ounce	13	81.3%	4.40	2.955
Rutabaga, Diced 1/4"	pound	16	ounce	13	81.3%	4.50	2.889
Rutabaga, Julienne	pound	16	ounce	13	81.3%	3.70	3.514
Salsify (Oyster Plant)	pound	16	ounce	14	87.5%		2.788
Shallots	pound	16	ounce	14.5	90.6%	5.20	2.788
Sorrel Leaves, Stemmed	pound	16	ounce	10.096	63.10%		
Sorrel Leaves, Stemmed, Chopped	pound	16	ounce	10.096	63.10%	1.1	9.178
Soy Bean Curd (Tofu), Firm, 3/4" Cubes	Package	12	ounce	11.8	98.33%	5.5	2.145
Soybean Sprouts	pound	16	ounce	15.5	96.9%	2.50	6.200
Spinach	pound	16	ounce	10.5	65.6%		
Squash, Acorn	pound	16	ounce	12.1	75.6%		
Squash, Acorn, Cubed	pound	16	ounce	12.1	75.6%	4.60	2.630
Squash, Banana	each	116.8	ounce	88.77	76.00%		
Squash, Butternut	pound	16	ounce	13.5	84.4%		
Squash, Butternut, Cubed	pound	16	ounce	13.5	84.4%	4.60	2.935
Squash, Chayote, Sliced	bag	24.8	ounce	19.75	79.64%	4.70	4.202
Squash, Crookneck	pound	16	ounce	15.6	97.5%		

Squash, Crookneck, Sliced	pound	16	ounce	15.6	97.5%	3.90	4.000
Squash, Hubbard	pound	16	ounce	11.4	71.3%		
Squash, Hubbard, Cubed	pound	16	ounce	11.4	71.3%	4.10	2.780
Squash, Kabocha, Seeded	pound	16	ounce	14.7	91.88%		
Squash, Kabocha, Seeded, Peeled	pound	16	ounce	11.67	72.94%		
Squash, Patty Pan (Summer)	pound	16.4	ounce	15.5	94.51%		
Squash, Patty Pan, Sliced	pound	16.4	ounce	15.5	94.51%	2.95	5.254
Squash, Spaghetti	pound	16	ounce	11	68.8%		
Squash, Summer	pound	16	ounce	15.2	95.0%		
Squash, Summer, Sliced	pound	16	ounce	15.2	95.0%	3.90	3.897
Squash, Zucchini	pound	16	ounce	15	93.8%	3.95	3.797
Squash, Zucchini, Chopped	pound	16	ounce	15	93.8%		
Squash, Zucchini, Sliced	pound	16	ounce	15	93.8%	3.80	3.947
Sweet Potato	pound	16	ounce	12	75.0%		
Sweet Potato, Cubed	pound	16	ounce	12	75.0%	4.70	2.553
Taro Root	each	22.25	ounce	19.05	85.62%		
Taro Root, 1" Dice	each	22.25	ounce	19.05	85.62%	4.20	4.536
Taro Root, 1/4" Dice	each	22.25	ounce	19.05	85.62%	3.90	4.885

(continued)

291

Vegetables (Continued)

ITEM NAME	AP UNIT	NUMBER OF MEASURES PER AP UNIT	MEASURE PER AP UNITS	TRIMMED/ CLEANED OUNCE WEIGHT OR COUNT	YIELD PERCENT	TRIMMED/ CLEANED OUNCE WEIGHT PER CUP	NUMBER OF TRIMMED/ CLEANED CUPS PER AP UNIT
Taro Root, Shredded	each	22.25	ounce	19.05	85.62%	2.75	6.927
Tomatillos	pound	16	ounce	14	87.5%		
Tomatillos, 1.5" Diameter	pound	16	ounce	10 each			
Tomato, Cherry Large	pound	16	ounce	22 each	95.0%	5.00	
Tomato, Cherry Small	pound	16	ounce	66 each	95.0%	5.13	
Tomato, Grape	pint	11.2	ounce	82 each	98.00%	4.85	2.310
Tomato, Sun-Dried—Dry	pound	16	ounce	16	100.0%	2.00	8.000
Tomatoes, Sun-dried, Julienne, Drained	jar	16	ounce	12.327	77.04%	5.3	2.326
Tomatoes, Sun-dried, Julienne, in Oil	jar	16	ounce	16	100.00%	7.35	2.177
Tomatoes, (5x6) Cored	pound	16	ounce	15.75	98.4%		
Tomatoes, (5x6) per Pound	pound	16	ounce	3 each			
Tomatoes, Cored and Peeled	pound	16	ounce	14.76	92.3%		
Tomatoes, Peeled, Seeded, Chopped	pound	16	ounce	12.55	78.4%	5.90	2.127
Tomatoes, Roma	pound	16	ounce	15	93.8%		
Tomatoes, Roma, Diced	pound	16	ounce	15	93.8%	5.70	2.632

Tomatoes, Roma, Sliced	pound	16	15	93.8%	4.20	3.571
Turnips	pound	16	13	81.3%		
Turnips, Dried 1/4"	pound	16	13	81.3%	4.50	2.889
Turnips, Julienne	pound	16	13	81.3%	3.70	3.514
Wasabi Root, Peeled, Grated Very Fine	pound	16	11.73	73.31%	8.8	1.333
Water Chestnuts	pound	16	11.5	71.9%		
Water Chestnuts, Sliced	pound	16	11.5	71.9%	4.40	2.614
Yam, Whole, Peeled	pound	16	14	87.5%		
Yucca, Peeled	pound	16	12.53	78.31%		
Yucca, Peeled, 1/3" Dice	pound	16	12.53	78.31%	4.6	2.724

Notes

1. Asparagus: Whole asparagus items show the yield after "culling" (discarding unservable stalks) but before trimming off the stalk ends. Trimming the ends yields about 63% of the whole asparagus. So, trimmed asparagus item yields reflect both the culling and stalk-end trimming, whereas the "whole" asparagus item yields reflect just that amount remaining after culling, but not trimming.

2. Beans, Green: Green beans are the Blue Lake variety. There are 66 each 4–6 inch beans per pound, AP. Strings and trim: 12%.

3. Carrots: Breakdowns of carrots: diced, grated, and so on are based on medium table-size carrots.

4. Cucumber: Ends are trimmed to obtain the yield percentage.

5. Lettuce, Bibb, Leaves: All lettuce leaf values are large leaves, exclusive of heart leaves.

6. Onion, Pearl: There are about 30 trimmed, peeled pearl onions per pint, and a pint of cleaned pearl onions weighs 9 ounces.

7. Peppers: Brunoise: Interior membrane of peppers have been sliced off as part of the cutting.

8. Potato: Russet, peeled and eyed.

9. Tomatoes, Roma: Trim is from removing the ends.

Fruit

ITEM NAME	AP UNIT	NUMBER OF MEASURES PER AP UNIT	MEASURE PER AP UNIT	TRIMMED/ CLEANED AVOIRDUPOIS OUNCE WEIGHT OR COUNT PER AP UNIT	YIELD PERCENT	TRIMMED/ CLEANED AVOIRDUPOIS OUNCE WEIGHT PER CUP	NUMBER OF TRIMMED/ CLEANED CUPS PER AP UNIT
Applesauce	#10 Can	111	ounce	110	99.0%	8.6	12.80
Apples, Fiji, 88 count	pound	16	ounce	11.91	74.44%		
Apples, Fiji, 88 count, peeled, cored, diced	pound	16	ounce	11.91	74.44%	4.3	2.770
Apples, Fiji, 88 count, peeled, cored, sliced	pound	16	ounce	11.91	74.44%	4.1	2.905
Apples, Golden Delicious, 80 count	pound	16	ounce	14.85	92.81%		
Apples, Golden Delicious, 80 count, peeled, cored, diced	pound	16	ounce	14.85	92.81%	3.92	3.788
Apples, Golden Delicious, 80 count, peeled, cored, sliced	pound	16	ounce	14.85	92.81%	3.73	3.981
Apples, Granny Smith, 88 count	pound	16	ounce	11.82	73.88%		
Apples, Granny Smith, 88 count, peeled, cored, diced	pound	16	ounce	11.82	73.88%	3.88	3.046

Apples, Granny Smith, 88 count, peeled, cored, sliced	pound	16	11.82	ounce	73.88%	3.73	3.169
Apples, Macintosh, 88 count	pound	16	11.56	ounce	72.25%		
Apples, Macintosh, 88 count, peeled, cored, diced	pound	16	11.56	ounce	72.25%	4	2.890
Apples, Macintosh, 88 count, peeled, cored, sliced	pound	16	11.56	ounce	72.25%	3.88	2.979
Apples, Red Delicious, 80 count	pound	16	14.38	ounce	89.88%		
Apples, Red Delicious, 80 count, peeled, cored, diced	pound	16	14.38	ounce	89.88%	4.3	3.344
Apples, Red Delicious, 80 count, peeled, cored, sliced	pound	16	14.38	ounce	89.88%	4.2	3.424
Apricots	pound	16	14.7	ounce	91.9%		
Apricots, Halves	pound	16	14.7	ounce	91.9%	5.46	2.69
Apricots, Halves, Dry	pound	16	16	ounce	100.0%	4.58	3.49
Banana, Chips, Dry	pound	16	16	ounce	100.0%	2.3	6.96
Bananas	pound	16	10.6	ounce	66.3%		
Bananas, Dwarf (finger)	bunch	15.2	11 each	ounce			
Bananas, Dwarf (finger), peeled	bunch	15.2	10.58	ounce	69.61%		

(continued)

Fruit (Continued)

ITEM NAME	AP UNIT	NUMBER OF MEASURES PER AP UNIT	MEASURE PER AP UNIT	TRIMMED/ CLEANED AVOIRDUPOIS OUNCE WEIGHT OR COUNT PER AP UNIT	YIELD PERCENT	TRIMMED/ CLEANED AVOIRDUPOIS OUNCE WEIGHT PER CUP	NUMBER OF TRIMMED/ CLEANED CUPS PER AP UNIT
Bananas, Dwarf (finger), peeled, sliced	bunch	15.2	ounce	10.58	69.61%	4.55	2.325
Bananas, Sliced	pound	16	ounce	10.6	66.3%	5.29	2.00
Blackberries	pound	16	ounce	15.2	95.0%	5	3.04
Blackberries, Frozen	pound	16	ounce	16	100.0%	5.3	3.02
Blueberries	pound	16	ounce	14.3	89.4%	5.1	2.80
Blueberries, Dried	pound	16	ounce	16	100.0%	4.5	3.56
Blueberries, Frozen	pound	16	ounce	16	100.0%	5.46	2.93
Boysenberries	pound	16	ounce	15.2	95.0%	5	3.20
Boysenberries, Frozen	pound	16	ounce	16	100.0%	5.3	3.02
Breadfruit	pound	16	ounce	12.48	78.0%	7.75	1.61
Carambola (Star Fruit)	pound	16	ounce	15	93.8%	7.3	1.14
Cherimoya	pound	16	ounce	8.3	51.9%	7.3	1.14
Cherries, Dried	pound	16	ounce	16	100.0%	5.2	3.08
Cherries, Sweet	pound	16	ounce	14	87.5%	5.11	2.74
Cranberries, Dried	pound	16	ounce	16	100.0%	5	3.20
Cranberries, Whole	pound	16	ounce	15.4	96.3%	3.35	4.60
Currants, Dry Zante	pound	16	ounce	16	100.0%	5.07	3.16
Currants, Red and White	pound	16	ounce	15.5	96.9%	3.95	3.92
Dates	pound	16	ounce	14.4	90.0%		
Dates, Pitted, Chopped	pound	16	ounce	14.4	90.0%	5.08	2.83

Item	Unit	Qty	Unit	Measure	Percent		
Dates, Pitted, Whole	pound	16	ounce	14.4	90.0%	4.4	3.27
Feijoa	pound	16	ounce	12	75.0%		
Figs	pound	16	ounce	15.8	98.8%		2.28
Figs, Dried	pound	16	ounce	16	100.0%	7.02	0.95
Grapefruit	each	13.3	ounce	7	52.6%	7.4	4.63
Grapes, Green Seedless	pound	16	ounce	15	93.8%	3.24	4.27
Grapes, Red Seedless	pound	16	ounce	14.3	89.4%	3.35	2.21
Guava	pound	16	ounce	12.8	80.0%	5.8	
Jackfruit	pound	16	ounce	5.5	34.4%		
Jackfruit, Sliced	pound	16	ounce	5.5	34.4%	5.67	0.97
Key Lime Juice Yield of 1 Pound	pound	16	ounce	6.3	39.38%		
Key Limes	pound	16	ounce	15 each			
Kiwi Fruit	pound	16	ounce	13.5	84.4%		
Kiwi Fruit, Sliced	pound	16	ounce	13.5	84.4%	6.2	2.18
Kumquats by Cup	cup	1	cup	16 each		5	
Kumquats by Pound	pound	16	ounce	15	93.8%		
Lemon Juice Yield 1 Each	each	4	ounce	1.65	41.3%	8.3	0.20
Lemon Juice Yield 1 Pound	pound	16	ounce	6.62	41.4%	8.3	0.80
Lemons 165 Count Whole	pound	16	ounce	4 each	49.0%		
Lemons 165, Flesh	each	4	ounce	1.96	49.0%		
Lime Juice Yield of 1 Each	each	3.4	ounce	1.44	42.4%	8.3	0.17
Lime Juice Yield of 1 Pound	pound	16	ounce	6.76	42.3%	8.3	0.81

(continued)

Fruit (Continued)

ITEM NAME	AP UNIT	NUMBER OF MEASURES PER AP UNIT	MEASURE PER AP UNIT	TRIMMED/CLEANED AVOIRDUPOIS OUNCE WEIGHT OR COUNT PER AP UNIT	YIELD PERCENT	TRIMMED/CLEANED AVOIRDUPOIS OUNCE WEIGHT PER CUP	NUMBER OF TRIMMED/CLEANED CUPS PER AP UNIT
Limes 2" Diameter	pound	16	ounce	4.7 each			
Litchi Nut	pound	16	ounce	9.6	60.0%		
Loquats	pound	16	ounce	10.4	65.0%		
Mango	pound	16	ounce	11	68.8%		
Mango, Sliced	pound	16	ounce	11	68.8%	5.82	1.89
Melon, Cantaloupe	pound	16	ounce	9.3	58.1%		
Melon, Cantaloupe, Ball	pound	16	ounce	6	37.5%	6.25	0.96
Cantaloupe, Sweet Tuscan style, 3/4" cubed	pound	16	ounce	10.16	63.50%	5.05	2.012
Melon, Cantaloupe, Cubed	pound	16	ounce	9.3	58.1%	5.65	1.65
Melon, Casaba	pound	16	ounce	9.5	59.4%		
Melon, Casaba, Cubed	pound	16	ounce	9.5	59.4%	6	1.58
Melon, Crenshaw	pound	16	ounce	10.6	66.3%		
Melon, Crenshaw, Cubed	pound	16	ounce	10.6	66.3%	5.8	1.83
Melon, Honeydew	pound	16	ounce	9.2	57.5%		
Melon, Honeydew, Cubed	pound	16	ounce	9.2	57.5%	5.9	1.56
Melon, Watermelon	pound	16	ounce	7.9	49.4%		

Melon, Watermelon, Cubed	pound	16	7.9	49.4%	5.36	1.47
Nectarines	pound	16	12	75.0%	5.6	2.14
Nectarines, Sliced	pound	16	12	75.0%	8.3	0.72
Orange Juice, Yield 1 (Pound) 72 Count	pound	16	6.01	37.6%	8.3	0.39
Orange Juice, Yield of #1, 72-count	each	8.5	3.2	37.6%		
Oranges, Navel 72 Count	pound	16	10	62.5%		
Oranges, Sections No Membrane	pound	16	5.5	34.4%	6.1	0.90
Oranges, Valencia 72 Count	pound	16	10.5	65.6%		
Papaya	pound	16	10.7	66.9%		
Papaya, Cubed	pound	16	10.7	66.9%	5.3	2.02
Papaya, Mexican (Large), cubed	pound	16	11.98	74.88%	5.47	2.190
Passion Fruit	pound	16	10.2	63.8%		
Paw Paw, peeled, seeded	pound	16	5.94	37.13%	8.46	0.702
Peaches, Peeled, Seeded	pound	16	12.5	78.1%		
Peaches, Sliced	pound	16	12.5	78.1%	6	2.08
Peaches, Slices—Dry	pound	16	16	100.0%	4	4.00
Pears, Asian, 1/2" diced	pound	16	11.92	74.50%	4.75	2.509
Pears, Asian, 1/4" slices	pound	16	11.92	74.50%	4.55	2.620
Pears, Bosc, 100 count	pound	16	14.24	89.00%		

(continued)

Fruit (Continued)

ITEM NAME	AP UNIT	NUMBER OF MEASURES PER AP UNIT	MEASURE PER AP UNIT	TRIMMED/ CLEANED AVOIRDUPOIS OUNCE WEIGHT OR COUNT PER AP UNIT	YIELD PERCENT	TRIMMED/ CLEANED AVOIRDUPOIS OUNCE WEIGHT PER CUP	NUMBER OF TRIMMED/ CLEANED CUPS PER AP UNIT
Pears, Bosc, 100 count, peeled, cored, diced	pound	16	ounce	14.24	89.00%	5.75	2.477
Pears, Bosc, 100 count, peeled, cored, sliced	pound	16	ounce	14.24	89.00%	4.98	2.859
Pears, Comice, 90 count	pound	16	ounce	12.61	78.81%	5.73	2.201
Pears, Comice, 90 count, peeled, cored, diced	pound	16	ounce	12.61	78.81%		
Pears, Comice, 90 count, peeled, cored, sliced	pound	16	ounce	12.61	78.81%	4.9	2.573
Pears, d'Anjou, 90 count	pound	16	ounce	13.05	81.56%		
Pears, d'Anjou, 90 count, peeled, cored, diced	pound	16	ounce	13.05	81.56%	5.1	2.559
Pears, d'Anjou, 90 count, peeled, cored, sliced	pound	16	ounce	13.05	81.56%	4.9	2.663
Persimmon	pound	16	ounce	13.1	81.9%		
Pineapple	pound	16	ounce	7.75	48.4%		
Pineapple, Cubed	pound	16	ounce	7.75	48.4%	5.9	1.31
Plantains	pound	16	ounce	11.5	71.9%		
Plums	pound	16	ounce	14.4	90.0%		

Item						
Plums, Sliced	pound	16	14.4	ounce	90.0%	5.8	2.48
Pomegranates	pound	16	8	ounce	50.0%		
Prickly Pear Fruit	pound	16	11	ounce	68.8%		
Prunes, Pitted	pound	16	16	ounce	100.0%	4.5	3.56
Quince	pound	16	9.75	ounce	60.9%		
Raisins, Not Packed	pound	16	16	ounce	100.0%	5.1	3.14
Raisins, Packed Down	pound	16	16	ounce	100.0%	5.8	2.76
Raspberries	pound	16	15.3	ounce	95.6%	4.3	3.56
Sapotes	pound	16	11.35	ounce	70.9%		
Soursop	pound	16	10.7	ounce	66.9%		
Strawberries	pound	16	14.7	ounce	91.9%		
Strawberries, Halves	pound	16	14.7	ounce	91.9%	5.36	2.74
Strawberries, Pureed	pound	16	14.7	ounce	91.9%	7.9	1.86
Strawberries, Sliced	pound	16	14.7	ounce	91.9%	5.85	2.51
Strawberries, Whole, Medium	pound	16	24 whole	ounce			
Tamarind, fresh, peeled with seeds	pound	16	11.81	ounce	73.81%		
Tamarind, fresh, peeled, seedless	pound	16	5.66	ounce	35.38%		
Tangelo, Mineola	pound	16	10.677	ounce	66.73%		
Tangelo, Mineola, count	pound	16	1.74 each	ounce			

(continued)

Fruit (Continued)

ITEM NAME	AP UNIT	NUMBER OF MEASURES PER AP UNIT	MEASURE PER AP UNIT	TRIMMED/ CLEANED AVOIRDUPOIS OUNCE WEIGHT OR COUNT PER AP UNIT	YIELD PERCENT	TRIMMED/ CLEANED AVOIRDUPOIS OUNCE WEIGHT PER CUP	NUMBER OF TRIMMED/ CLEANED CUPS PER AP UNIT
Tangerine	pound	16	ounce	11.5	71.9%		
Tangerine Sections	pound	16	ounce	11.5	71.9%	6	1.92

Notes

1. Apricots: These are pitted to obtain the yield.

2. Lemons: One pound of raw lemons yields 6.62 avoirdupois ounces juice (or 6.35 fluid ounces). There are usually 10 sections in a lemon.

3. Limes: One pound of limes yields 6.76 avoirdupois ounces juice (or 6.49 fluid ounces). There are usually 10 sections in a lime.

4. Melon Balls: Balling reduces melon yields from 55–60% to 35–40%. Use waste if possible. One cup equals 13 standard balls.

5. Oranges: There are 10 to 12 sections in an orange (10 of full size). One pound of 72-count oranges yields 6.01 avoirdupois ounces juice (or 5.76 fluid ounces).

6. Citrus Juices—Purchasing Formulas: The ounce measures shown are ounces by weight. Fluid ounces can be substituted, but remember that a weighed ounce is about 95% of a fluid ounce of juice. To convert a weight ounce (avoirdupois) to a fluid ounce, multiply the number of weighed ounces by .96 to arrive at its fluid-ounce equivalent. When working with fluid ounces, use the cups purchasing formula for more efficiency in determining how much raw citrus to buy for an as-served volume of juice.

Canned Foods (in Number 10 Cans)

ITEM	USDA RECOMMENDED DRAINED WEIGHT MINIMUM IN OUNCES PER #10 CAN
Apples, Sliced	96
Apricots, Halves or Slices in Heavy Syrup	62
Apricots, Halves or Slices, Light Syrup, Juice, or Water	64
Apricots, Whole, Peeled in Heavy Syrup	60.4
Apricots, Whole, Peeled in Light Syrup, Juice, or Water	62
Apricots, Whole, Unpeeled in Heavy Syrup	60
Apricots, Whole, Unpeeled in Light Syrup, Juice, or Water	61.5
Asparagus, Center Cuts and Tips	60.2
Beans, Garbanzo	68
Beans, Green or Wax, 1.5" Cut	60
Beans, Green or Wax, French Cut	59
Beans, Green or Wax, Mixed or Short Cut	63
Beans, Green or Wax, Whole	57.5
Beans, Kidney	68
Beans, Lima, Fresh	72
Beans, Pinto	68
Beets, Diced, 3/8"	72
Beets, Julienne	68
Beets, Sliced Medium, 1/4"	68
Beets, Whole, Size 1 to 3	69
Beets, Whole, Size 4 to 6	68
Blackberries, Heavy Pack in Light Syrup or Water	74
Blackberries, Regular Pack in Heavy Syrup	62
Blueberries	55
Blueberries, Heavy Pack in Light Syrup, or Water	70
Boysenberry, Regular Pack in Heavy Syrup	55
Carrots, Diced, 3/8"	72
Carrots, Julienne	68
Carrots, Sliced	68
Carrots, Whole	68
Cherries, Red, Tart, Pitted in Syrup	70.2
Cherries, Red, Tart, Pitted in Water, or Juice	72
Cherries, Sweet, Dark and Light, Pitted in Heavy Syrup	66.5
Cherries, Sweet, Unpitted in Light Syrup, Juice, or Water	70
Corn, Whole Kernel, Grade A	70

(continued)

Canned Foods (in Number 10 Cans) (Continued)

ITEM	USDA RECOMMENDED DRAINED WEIGHT MINIMUM IN OUNCES PER #10 CAN
Corn, Whole Kernel, Grades B & C	72
Figs, Kadota, 70 Count or Less	63
Figs, Kadota, 71 Count or More	66
Fruit Cocktail (in all Packing Media)	71.15
Fruits for Salad (in All Media)	64.5
Grapes, Light, Seedless (in All Packing Media)	62
Hominy, Whole, Style I	72
Loganberry, Regular Pack in Light Syrup, or Water	60
Mushrooms, Stems and Pieces	61
Olives, Ripe, Chopped	90
Olives, Ripe, Pitted, Jumbo, Colossal, Super-Colossal	49
Olives, Ripe, Pitted, Small, Medium, Large, Extra-Large	51
Olives, Ripe, Whole, Jumbo, Colossal, Super Colossal	64
Olives, Ripe, Whole, Small, Medium, Large, Extra-Large	66
Olives, Sliced, Wedges, Quartered	55
Onions, Whole, 100 to 199 Count (Small)	63
Onions, Whole, 200 Count (Tiny)	64
Onions, Whole, 80 to 99 Count (Medium)	60
Peaches, Clingstone, Diced, All Media	70
Peaches, Clingstone, Halves, 23 Count or Less in Heavy Syrup	65
Peaches, Clingstone, Halves, 23 Count or Less in Light Syrup, Juice, or Water	67
Peaches, Clingstone, Halves, 24 Count or More in Heavy Syrup	66.5
Peaches, Clingstone, Halves, 24 Count in Light Syrup, Juice, or Water	68.5
Peaches, Clingstone, Heavy Pack, All Media	76
Peaches, Clingstone, Quarters, Pieces (Irregular) in Heavy Syrup	66.5
Peaches, Clingstone, Quarters, Pieces (Irregular) in Light Syrup, Juice, or Water	68.5
Peaches, Clingstone, Slices in Heavy Syrup	66.5
Peaches, Clingstone, Slices in Light Syrup, Juice, or Water	68.5
Peaches, Clingstone, Solid Pack, Unsweetened	92

Canned Foods (in Number 10 Cans) (Continued)

ITEM	USDA RECOMMENDED DRAINED WEIGHT MINIMUM IN OUNCES PER #10 CAN
Peaches, Freestone, Halves, 23 Count or Less, Heavy Syrup, Light Syrup, Juice, or Water	61.5
Peaches, Freestone, Halves, 24 Count or More, Heavy Syrup, Light Syrup, Juice, or Water	62.5
Peaches, Freestone, Quarters, Mixed Pieces of Irregular Shape in All Media	64.5
Peaches, Freestone, Slices in Heavy Syrup, Light Syrup, Juice, or Water	61
Pears, Diced in All Media	67
Pears, Halves, 25 Count or Less in All Media	62.7
Pears, Halves, 25 Count or More in All Media	64.1
Pears, Slices or Quarters in All Media	65.5
Peas and Carrots (Diced Carrots)	71
Peas and Carrots (Sliced Carrots)	70
Peas, Field and Blackeye Peas	72
Peas, Grade A	70
Peas, Grade B	72
Pimentos, Diced, Chopped	74
Pimentos, Pieces	74
Pimentos, Sliced	71.7
Pimentos, Whole—Halves	70.7
Pimentos, Whole—Pieces	72.2
Pineapple Chunks in All Media	65.75
Pineapple Cubes in All Media	71.25
Pineapple Tidbits in All Media	65.75
Pineapple, Crushed in Syrup (Minimum is 63% of weight by content)	69.3
Pineapple, Crushed, Solid Pack (Minimum is 78% of weight by content)	85.8
Pineapple, Sliced, in All Media	61.5
Plums, Purple, Halves in All Media	60.2
Plums, Purple, Whole in All Media	54.7
Potatoes, Sweet	73
Potatoes, White, Diced	76
Potatoes, White, Sliced	75

(continued)

Canned Foods (in Number 10 Cans) (Continued)

ITEM	USDA RECOMMENDED DRAINED WEIGHT MINIMUM IN OUNCES PER #10 CAN
Potatoes, White, Whole	74
Prunes, Heavy Pack in All Media	110
Prunes, Regular Pack in All Media	70
Raspberries, Grade C in Water	60
Raspberries, Red, Grade A or B in Syrup	53
Sauerkraut	80
Spinach	58.4
Tomatoes, Whole, Peeled, Grade B	63.5
Tomatoes, Whole, Peeled, Grade C	54.7

Canned Foods Weight-to-Volume

PRODUCT	TOTAL OUNCES PER #10 CAN	NET OR DRAINED WEIGHT IN OUNCES	DRAINED WEIGHT YIELD PERCENTAGE	OUNCES PER SINGLE CUP	OUNCES PER QUART	OUNCES PER HALF-GALLON
Applesauce	111	110	99.10%	8.6	34.4	68.8
Apples, Sliced, Dry Pack in Juice	97	95.8	98.76%	5.65	24.6	53.25
Baked Beans with Brown Sugar and Bacon	112.7	112.7	100%*	9.15	37.75	76.4
Beets, Diced (3/8"), in Water	107.6	72.4	67.29%	5.3	21.8	48.25
Beets, Sliced, in Water	107.3	76.95	71.71%	5.85	22.55	50.95
Carrots, Diced (3/8"), in Water	108.25	69.1	63.83%	5.05	22.2	51.2
Cheese Sauce for Nachos	104	104	100%*	8.89	35.9	72
Cherries, Dark Sweet, Pitted, Heavy Syrup	112.9	67.9	60.14%	5.9	25.9	54.05
Chili con Carne with Beans	106.75	106.75	100%*	8.9	36	72
Corn, Whole Yellow Kernels, in Water	108.6	73	67.22%	5.55	23.7	48.4
Garbanzo Beans, Whole, in Water	111.3	68.5	61.55%	5.65	23.05	48.4
Green Beans, Blue Lake, Cut 1.5" in Water (Grade A)	107.6	62.5	58.09%	3.75	16.75	33.75
Green Beans, Blue Lake, Cut 1.5" in Water (Grade B)	105.45	62.1	58.89%	3.9	18	37.15
Hominy, White	112.95	79.3	70.21%	5.8	24.1	50
Ketchup	111.95	111.95	100%*	9.6	39.2	76.9
Kidney Beans, Dark Red, in Water	110.7	66.8	60.34%	5.45	23.15	47
Menudo	109.85	109.85	100.00%*	9.154	36.6	73.23
Mushrooms, Sliced, in Water	102.95	65.3	63.43%	5.2	23.15	47.8
Peaches, Halves, 35 Count, in Heavy Syrup	107.8	70.65	65.54%	6.35	27.3	56
Pears, Sliced, in Light Syrup	108.85	63.95	58.75%	7.4	30.5	63.97

(continued)

Canned Foods Weight-to-Volume (Continued)

PRODUCT	TOTAL OUNCES PER #10 CAN	NET OR DRAINED WEIGHT IN OUNCES	DRAINED WEIGHT YIELD PERCENTAGE	OUNCES PER SINGLE CUP	OUNCES PER QUART	OUNCES PER HALF-GALLON
Pears, Diced, in Light Syrup	108.8	61.4	56.43%	6.95	31.9	65
Peas, Whole, Green, Sweet, in Water	108.65	70.65	65.03%	5.65	26.95	54
Pineapple, Crushed, Solid Pack, with Juice	104.55	85.3	81.59%	7.7	32.85	69.25
Pineapple Tidbits, in Juice	109.8	70.5	64.21%	5.9	24.7	51.75
Potatoes, Whole, Peeled, 90 to 110 Count, in Water	109.5	76.65	70%	5.9	22.35	47.1
Sauerkraut, in Water	107.6	72.8	67.66%	4.8	21.6	46
Spinach, Leaf, Stemmed, in Water	102	46.4	45.49%	7.1	30.6	76.35
Tomato Paste	111	111	100%*	9.3	38	76.35
Tomato Purée	107	107	100%*	8.9	36.3	72.9
Tomato Sauce	104.9	104.9	100%*	8.75	36.35	72.8
Tomatoes, Chopped, in Purée	104.25	104.25	100%*	8.45	35.2	79.1
Tomatoes, Crushed, in own Juice	101.55	101.55	100%*	8.58	35.15	70
Tomatoes, Diced, in Juice	108.5	67.05	61.8%	7.15	29.55	67.05
Wax Beans, Yellow, 1.5" Pieces, in Water	104.85	60.95	58.13%	4.35	17.5	36.1

*The last three columns list drained weights unless noted by an asterisk.

Dairy Products

ITEM	OUNCES PER CUP	CUPS PER POUND	OUNCES PER PINT	PINTS PER POUND	POUNDS PER PINT
American Process, Pasteurized, Diced	4.94	3.239	9.880	1.62	0.62
American Process, Pasteurized, Melted	8.60	1.860	17.200	0.93	1.08
American Process, Pasteurized, Shredded	4.00	4.000	8.000	2.00	0.50
Blue, Crumbled	4.75	3.368	9.500	1.68	0.59
Brie or Camembert, Melted	8.47	1.889	16.940	0.94	1.06
Brie or Camembert, Packed	8.67	1.845	17.340	0.92	1.08
Buttermilk, 1%	8.7	1.839	17.400	0.920	1.09
Buttermilk, 2%	8.6	1.86	17.20	0.93	1.08
Cheddar, Diced	4.65	3.441	9.300	1.72	0.58
Cheddar, Melted	8.60	1.860	17.200	0.93	1.08
Cheddar, Shredded	4.00	4.000	8.000	2.00	0.50
Cotijo, crumbled	4.2	3.810	8.400	1.905	0.53
Cottage, Large Curd	7.40	2.162	14.800	1.08	0.93
Cottage, Small Curd	7.90	2.025	15.800	1.01	0.99
Cottage Cheese, Lowfat	8.15	1.963	16.300	0.982	1.02
Cottage Cheese, Nonfat	8.65	1.850	17.300	0.925	1.08
Cream Cheese	8.20	1.951	16.400	0.98	1.03
Crema Mexicana	8.5	1.882	17.000	0.941	1.06
Crème Fraiche	8.15	1.963	16.300	0.982	1.02
Curds, Drained, 1/3" dice	4.75	3.368	9.500	1.684	0.59
Double Devon Cream	8.15	1.963	16.300	0.982	1.02
Dulce de Leche, Canned	10.5	1.524	21.000	0.762	1.31
Edam, Shredded	5.30	3.019	10.600	1.51	0.66
Egg Powder, Sifted	3.00	5.333	6.000	2.67	0.38
Egg Substitute	8.65	1.85	17.3	0.92	1.08
Eggs, Hard-cooked, Chopped	6.00	2.667	12.000	1.33	0.75
Eggs, Whole, Shelled, Pooled	8.57	1.867	17.140	0.93	1.07
Fontina, Shredded	3.80	4.211	7.600	2.11	0.48
Fromage Blanc (Bakers' Cheese)	8.9	1.798	17.800	0.899	1.11
Goat's Milk, Evaporated	8.85	1.808	17.700	0.904	1.11

(continued)

Dairy Products (Continued)

ITEM	OUNCES PER CUP	CUPS PER POUND	OUNCES PER PINT	PINTS PER POUND	POUNDS PER PINT
Gruyere, Shredded	3.80	4.211	7.600	2.11	0.48
Half and Half	8.55	1.87	17.1	0.94	1.07
Half and Half, Fat Free	8.85	1.808	17.700	0.904	1.11
IMO–Sour Cream Substitute	8.25	1.939	16.500	0.970	1.03
Kefir, Low Fat, Fruit Flavored	9.1	1.758	18.200	0.879	1.14
Media Crema	8.4	1.905	16.800	0.952	1.05
Milk, 1%	8.6	1.86	17.20	0.93	1.08
Milk, 2%	8.6	1.86	17.20	0.93	1.08
Milk, Whole, 4%	8.55	1.87	17.1	0.94	1.07
Milk Evaporated	8.90	1.798	17.800	0.90	1.11
Milk, Powdered (from Nonfat Milk)	4.25	3.765	8.500	1.88	0.53
Milk, Powdered (from Whole Milk)	4.50	3.556	9.000	1.78	0.56
Milk, Sweetened Condensed	10.80	1.481	21.600	0.74	1.35
Monterey Jack, Diced	4.65	3.441	9.300	1.72	0.58
Monterey Jack, Shredded	4.00	4.000	8.000	2.00	0.50
Mozzarella, Part Skim, Shredded	4.00	4.000	8.000	2.00	0.50
Muenster, Shredded	4.00	4.000	8.000	2.00	0.50
Non-Dairy Creamer	8.55	1.871	17.100	0.936	1.07
Non-Dairy Creamer, Fat-Free	8.77	1.824	17.540	0.912	1.10
Parmesan, Grated, Dry	3.40	4.706	6.800	2.35	0.43
Parmesan, Grated, Fresh	3.00	5.333	6.000	2.67	0.38
Provolone, Diced	4.65	3.441	9.300	1.72	0.58
Provolone, Shredded	4.00	4.000	8.000	2.00	0.50
Quark–European Bakers' Cheese	8.6	1.860	17.200	0.930	1.08
Queso Fresco, shredded	3.4	4.706	6.800	2.353	0.43
Rice "Milk" Drink	8.6	1.860	17.200	0.930	1.08
Ricotta, Whole or Part Skim	8.68	1.843	17.360	0.92	1.09
Sour Cream	8.54	1.874	17.080	0.94	1.07

Dairy Products (Continued)

ITEM	OUNCES PER CUP	CUPS PER POUND	OUNCES PER PINT	PINTS PER POUND	POUNDS PER PINT
Sour Cream, Fat-Free	8.65	1.850	17.300	0.925	1.08
Sour Cream, Mexican, fresh	8.5	1.882	17.000	0.941	1.06
Soy "Milk"	8.65	1.850	17.300	0.925	1.08
Swiss, Diced	4.94	3.239	9.880	1.62	0.62
Swiss, Melted	8.60	1.860	17.200	0.93	1.08
Swiss, Shredded	4.00	4.000	8.000	2.00	0.50
Whipping Cream	8.4	1.9	16.8	0.95	1.05
Whipping Cream, Heavy	8.4	1.905	16.800	0.952	1.05
Yogurts	8.60	1.860	17.200	0.93	1.08

Notes

1. Eggs:

Size of Eggs	Jumbo	Extra-Large	Large	Medium	Small
Shelled Ounce Weight Each	2.3	2.05	1.777	1.55	1.30

- Whites constitute 66.66%, yolks, 33.33% (or two-thirds white, one-third yolk).

- Yield of 1 large egg, shelled: 1.174 ounces white and .586 ounces yolk (roughly: 1.2 and 0.6).

- 1 dozen large eggs yields about 21 ounces pooled eggs, or 14 ounces whites and 7 ounces yolks.

- 1 quart of pooled eggs equals 19.44 large eggs or 22 medium eggs.

- 1 pound of pooled (shelled) eggs equals 9 large eggs (1.86 cups).

2. Milks and Creams:

- Nonfat, 1%, 2% milks and 2% buttermilk all weigh 34.4 ounces per quart.

- Whole milk weighs 34.2 ounces per quart.

- 1 pound of whole milk equals 15 fluid ounces, or 1.875 cups.

- Heavy whipping cream weighs 33.6 ounces per quart.

- Egg substitutes weigh 34.6 ounces per quart.

Meats

ITEM NAME	NAMP NUMBER	AP WEIGHT IN POUNDS	TRIM LOSS IN POUNDS	PRIMARY-USE YIELD IN POUNDS	YIELD PERCENT	NUMBER OF USEABLE OUNCES PER AP POUND	TRIM: MIS-CELLANEOUS USE IN POUNDS
Beef Ball Tip—Bottom Sirloin Butt	185B	14	3	11	78.6%	12.6	2
Beef Bottom Round	170	21	4.5	16.5	78.6%	12.6	3
Beef Prime Rib (See note 1.)	109	20	10	10	50.0%	8.0	5
Beef Strip Loin (New York) 1″	180	10	3	7	70.0%	11.2	1
Beef Tenderloin, Pismo—Defatted	189A	7	0.9	6.1	87.1%	13.9	0.4
Beef Tenderloin, Whole	189	7.5	3.1	4.4	58.7%	9.4	1.5
Beef Top Round	168	16	4	12	75.0%	12.0	1
Beef Top Sirloin Butt	184	15	4.4	10.6	70.7%	11.3	1
Beef Tri-Tip—Bottom Sirloin Butt	185C	3	1.1	1.9	63.3%	10.1	0.3
Lamb Foreshank	210	1.05	0.00	1.05	100.0%	16.0	0.00
Lamb Leg (Defatted, Leg and Shank Bones Intact)	233A	10.87	3.41	7.46	68.6%	11.0	1.33
Lamb Leg (Trimmed of Fat and Bone), imported	233A	7.5	3.7	3.8	50.7%	8.1	2.5
Lamb Leg, (Fully Defatted, Boned, Shank-Off)	233A	10.87	5.17	5.70	52.4%	8.4	2.77
Lamb Leg, Shank Off, Boneless (Defatted, Detissued)	234A	5.76	0.874	4.89	84.8%	13.6	0.00
Lamb Loin, Short-Cut, Trimmed, 2″ Tail (Trimmed to Loin & Tender)	232A	2.93	1.93	1.00	34.1%	5.4	0.71
Lamb Rack Double (Trimmed French)	204	7	3	4	57.1%	9.1	1.25

Lamb Rack, Split & Chined (Trimmed to French)	204A	3.38	1.84	1.54	45.5%	7.3	0.82
Lamb Ribs, Breast Bones Off	209A	1.02	0.09	0.93	91.4%	14.6	0.14
Lamb Shoulder, Outside (Boned, Detissued, Defatted)	207A	9.37	4.90	4.48	47.8%	7.6	2.40
Pork Leg, Boned, Skinned, Defatted	401A	14	5.25	8.75	62.5%	10.0	3.25
Pork Leg, Skinned, Aitch Removed	401C	14	2.6	11.4	81.4%	13.0	0.75
Pork Loin, Boneless, Defatted	413	7	1.75	5.25	75.0%	12.0	0.75
Pork Loin, Whole, Boned, Defatted	412A	9	4.25	4.75	52.8%	8.4	2.75
Pork Shoulder, Boston Butt	406	8	1.5	6.5	81.3%	13.0	1.5
Pork Tenderloins	415	1	0.125	0.875	87.5%	14.0	0.1
Veal Breast	313	10	2	8	80.0%	12.8	1.5
Veal Double Rack, Frenched	306	16	9.75	6.25	39.1%	6.3	7
Veal Leg (Trimmed of Fat and Bone)	334	35	19.5	15.5	44.3%	7.1	16

Notes

1. Prime Rib (109): 109s in the table had the cap, tail, backstrap, and all bones, plus most exterior connective tissue, removed. Low, slow cooking will reduce the servable weight another 12% at medium-rare. Thus, 1 pound of raw 109 yields 7 ounces, well-trimmed cooked meat—a 44% yield.

2. Convection ovens will increase the shrink (cooking loss) by 20 or 25 percent.

3. All lamb items are U.S. domestic unless noted as Imported.

Seafood

ITEM	FILLET YIELD PERCENT*	EDIBLE OUNCES PER AP POUND
Ahi Tuna, H&G, center cut	60%	9.6
Ahi Tuna, H&G, center cut, bloodline trimmed	45%	7.2
Ahi Tuna H&G	70%	11.2
Albacore Tuna, whole fish	55%	8.8
Bass, H&G	70%	11.2
Bass, whole fish	45%	7.2
Bass, Sea, Drawn	50%	8
Bass, Striped, Drawn	60%	9.6
Catfish, Drawn	45%	7.2
Cod, Atlantic, Drawn	45%	7.2
Cod, Pacific, Drawn	35%	5.6
Flounder, Drawn	45%	7.2
Halibut, Dressed	60%	9.6
Halibut, H&G	70%	11.2
Halibut, whole fish	60%	9.6
John Dory, whole fish	30%	4.8
Ling Cod, H&G	65%	10.4
Mahi Mahi, H&G	70%	11.2
Monkfish Tails	65%	10.4
Ono, Hawaiian Wahoo, H&G	65%	10.4
Opah, Hawaiian Moonfish, H&G	65%	10.4
Petrale Sole, whole fish	35%	5.6
Pompano, Drawn	45%	7.2
Salmon, Dressed	75%	12
Salmon, G&G	70%	11.2
Sand Dabs, whole fish	50%	8
Snappers, whole fish	40%	6.4
Spearfish, H&G	65%	10.4
Swordfish, Dressed	60%	9.6
Swordfish, H&G	65%	10.4
Swordfish, H&G, center cut	60%	9.6
Swordfish, Loin	90%	14.4

Seafood (Continued)

ITEM	FILLET YIELD PERCENT*	EDIBLE OUNCES PER AP POUND
Tilapia, Dressed	45%	7.2
Trout, Drawn	50%	8
Tuna, Loin	95%	15.2

*Fishing/harvesting methods, time, temperature, and distance in shipping all affect yields. The yield percentages listed in this table are, therefore, approximate.

Notes:

1. Finfish: Finfish are seldom sold whole (called "round"). Most are eviscerated (gutted) and are called "drawn" fish. The edible portion yield of drawn finfish ranges from 35% to 75%. Generally, the larger the fish, the higher the yield.

   - *H&G (Headed and Gutted)* This is similar to a dressed fish in that the fish will have had its head removed and been gutted. The scales and fins will not have been removed. Sometimes the tail and collars will have been removed but this is not a given.

   - *G&G (Gilled and Gutted)* This is a fish whose gills and guts have been removed but heads, tails, fins and scales are intact.

   - A dressed fish is scaled, eviscerated, and has had its head and fins removed.

   - A filet may be a whole side or a part of a side. It is boneless and often skinless.

   - A steak is a crosscut of the fish, and includes bones and skin.

2. Crustaceans—Crab, Lobster, Crayfish, Shrimp: These seafood products are commonly sold frozen by weight or a count per pound.

   - Crab yields vary according to species and handling. Fresh Dungeness crab yields 15 to 25% picked, cooked meat. Yield varies with species and handling. Blue crab yields less: 10 to 15%. Lump, claw, and backfin crabmeat are considered the best meat and are priced accordingly. Crabmeat marked "special" is the least expensive.

   - Fresh lobster (Maine or spiny) yields between 20 and 33% picked, cooked meat. Lobsters lose weight very fast once out of water. A 1.5-pound Maine lobster, in good condition, will yield an 8 to 9 ounce tail.

   - Crayfish sized an average of 10 per pound, whole and very fresh, are about as large as is marketed. The count per pound does vary but should not exceed 15 to 18, for restaurant use. Counts up to 30 per pound are not uncommon in very casual "Louisiana style" establishments. Smaller crayfish actually have a 5% larger tail meat yield than larger sizes because of the extra shell weight found with larger sizes. The percentage of tail meat to purchased weight is 15 to 20%, if very fresh.

   - Headless shrimp yield 80 to 85% peeled, deveined meat. Raw, peeled shrimp, when briefly cooked, yield 65 to 70% of their original headless weight.

Poultry

ITEM NAME	PART TYPE	OUNCE WEIGHT	PERCENT OF ORIGINAL WEIGHT	OUNCE WEIGHT OF 1 EACH	1 EACH: PERCENT OF ORIGINAL WEIGHT
Chicken, Large Fryer					
Whole	Whole Bird	58.8	100.0%		
Clean Meat	General	28.2	48.0%		
Bones	General	21.0	35.7%		
Skin	General	7.6	12.9%		
Separable Fat	General	2.0	3.4%		
Gizzard	General			0.60	1.0%
Heart	General			0.25	0.4%
Liver	General			1.10	1.9%
Neck, Skinless	General			1.70	2.9%
Neck, with Skin	General			2.80	4.8%
Water, Blood, and Cutting Loss	General			3.10	5.3%
Back	General			10.90	18.5%
Wings, Whole	Wings	6.3	10.7%	3.15	5.4%
Wings, First Joint (Drumette)	Wings	3.0	5.1%	1.50	2.6%
Wings, Second Joint (Middle)	Wings	2.5	4.3%	1.25	2.1%
Wings, Tips	Wings	0.8	1.4%	0.40	0.7%
Breast, Whole (Both Halves)	Breast	17.4	29.6%		
Breast, Boneless, Skinless	Breast	11.4	19.4%		
Breast, Tenderloins, Both	Breast	2.4	4.1%		
Breast Half, Boneless, Skinless	Breast			5.70	9.7%
Breast Half, Whole, Each	Breast			8.70	14.8%
Breast, Tenderloins, Each	Breast			1.20	2.0%
Legs, Whole (Drum and Thigh)	Legs	16.2	27.6%		
Drumsticks, Whole	Legs	6.7	11.4%		
Drumsticks, Skinless	Legs	6.2	10.5%		
Drumstick Meat, Detissued	Legs	3.2	5.4%		
Legs, Whole, Each	Legs			8.10	13.8%
Drumstick, Skinless, Each	Legs			3.10	5.3%
Cleaned Meat from 1 Drumstick	Legs			1.60	2.7%

Poultry (Continued)

ITEM NAME	PART TYPE	OUNCE WEIGHT	PERCENT OF ORIGINAL WEIGHT	OUNCE WEIGHT OF 1 EACH	1 EACH: PERCENT OF ORIGINAL WEIGHT
Drumstick, Whole, Each	Legs			3.35	5.7%
Thighs, Whole	Thighs	9.6	16.3%		
Thighs, Skinless	Thighs	8.1	13.8%		
Thighs, Skinless, Boneless	Thighs	5.9	10.0%		
Thighs, Whole, Each	Thighs			4.80	8.2%
Thighs, Skinless, Each	Thighs			4.05	6.9%
Thighs, Skinless, Boneless, Each	Thighs			2.95	5.0%
Turkey, Hen (10 Pounds)					
Whole	Whole Bird	160.0	100.0%		
Legs and Thighs	General	47.0	29.4%		
Breast, Whole	General	47.0	29.4%		
Breast, Full, Boneless, Skinless	General	35.0	21.9%		
Back and Ribs	General	32.0	20.0%		
Wings	General	19.0	11.9%		
Gizzard	General	6.1	3.8%		
Neck	General	6.0	3.8%		
Liver	General	3.4	2.1%		
Heart	General	0.6	0.4%		
Turkey, Hen (14 Pounds)					
Hen (14 Pounds)	Whole Bird	224.0	100.0%		
Breast, Whole	General	79.5	35.5%		
Breast, Full, Boneless, Skinless	General	59.6	26.6%		
Legs and Thighs	General	57.0	25.4%		
Back and Ribs	General	46.0	20.5%		
Wings	General	25.3	11.3%		
Gizzard	General	6.3	2.8%		
Neck	General	5.4	2.4%		
Liver	General	3.6	1.6%		
Heart	General	0.7	0.3%		

(continued)

Poultry (Continued)

ITEM NAME	PART TYPE	OUNCE WEIGHT	PERCENT OF ORIGINAL WEIGHT	OUNCE WEIGHT OF 1 EACH	1 EACH: PERCENT OF ORIGINAL WEIGHT
Turkey, Tom (22 pounds)					
Tom	Whole Bird	352.0	100.0%		
Breast, Whole	General	144.0	40.9%		
Breast, Full, Boneless, Skinless	General	106.0	30.1%		
Legs and Thighs	General	84.0	23.9%		
Back and Ribs	General	64.8	18.4%		
Wings	General	33.4	9.5%		
Neck	General	12.0	3.4%		
Gizzard	General	6.3	1.8%		
Liver	General	4.5	1.3%		
Heart	General	1.4	0.4%		

Notes

1. Chicken: 1 cup of cooked chicken liver weighs 5 ounces.

2. Ducklings: Yield 30% bone, 38% skin and fat; meat yield is 32%.

Body Composition	Percent
Breasts	30%
Legs and Thighs	25%
Back and Ribs	25%
Wings	12%
Neck, Gizzard, and Heart	5%
Liver	3%
Whole 7-pound Duckling	100%

Can Sizes

STANDARD U.S. CAN	U.S. FLUID OUNCES	MILLILITERS	U.S. CUPS	IMPERIAL CUPS
Number 10	104.900	3102.263	13.113	10.918
Number 5	56.600	1673.862	7.075	5.891
Number 3 Cylinder	49.560	1465.664	6.195	5.158
Number 2 1/2	28.550	844.324	3.569	2.972
Number 2	19.690	582.303	2.461	2.049
Number 303	16.180	478.500	2.023	1.684
Number 300	14.590	431.478	1.824	1.519
Number 211 Cylinder	12.998	384.397	1.625	1.353
Number 1 Picnic	10.480	309.931	1.310	1.091
8 Ounce	8.320	246.052	1.040	0.866
6 Ounce	5.830	172.414	0.729	0.607

Notes

The fluid ounces and milliliters above specify the capacity of these cans to contain liquids. Food packers sometimes use near-standard custom-size cans that contain less volume. Read the can label for precise weight or volume data regarding its contents.

Scoop or Disher Sizes

DISHER SIZE NUMBER	U.S. FLUID OUNCES	MILLILITERS	CUPS PER SCOOP	SCOOPS PER CUP
4	8.000	236.588	1.000	1.000
5	6.400	189.271	0.800	1.250
6	5.333	157.716	0.667	1.500
8	4.000	118.294	0.500	2.000
10	3.200	94.635	0.400	2.500
12	2.666	78.843	0.333	3.001
16	2.000	59.147	0.250	4.000
20	1.600	47.318	0.200	5.000
24	1.333	39.422	0.167	6.002
30	1.066	31.525	0.133	7.505
40	0.800	23.659	0.100	10.000
60	0.533	15.763	0.067	15.009

Notes

Accurate portioning with dishers requires level scraping across the top of the disher cup. The number imprinted on the blade of the scoop states how many level scoops of that size are in one 32-fluid-ounce quart.

Sizes and Capacities of Hotel Pans

STAINLESS STEAM TABLE PANS GENERAL SIZE	COMMON NAME	DIMENSION (INCHES) WIDTH × LENGTH × DEPTH	BRIMFUL CAPACITY		NON-SPILL CAPACITY	
			QUARTS	LITERS	QUARTS	LITERS
Full	Full Shallow	12 × 20 × 2.5	8.3	7.85	6	5.68
	Full Medium	12 × 20 × 4	14	13.25	12	11.36
	Full Deep	12 × 20 × 6	21	19.87	18	17.03
Standard Half	Half Shallow	12 × 10 × 2.5	4.3	4.07	3	2.84
	Half Medium	12 × 10 × 4	6.7	6.34	6	5.68
	Half Deep	12 × 10 × 6	10	9.46	9	8.52
Long Half	Long Half 1 Inch	6.5 × 20 × 1.25	2.1	1.99	1.5	1.42
	Long Half Shallow	6.5 × 20 × 2.5	3.7	3.50	25	2.37
	Long Half Medium	6.5 × 20 × 4	5.7	5.39	4.5	4.26
	Long Half Deep	6.5 × 20 × 6	8.2	7.76	7	6.62
Two-Thirds	2/3rds—1 Inch	12 × 14 × 1.25	4	3.79	3	2.84
	2/3rds—Shallow	12 × 14 × 2.5	5.6	5.30	4.5	4.26
	2/3rds—Medium	12 × 14 × 4	9.3	8.80	8	7.57
	2/3rds—Deep	12 × 14 × 6	14	13.25	12	11.36
Third	3rd Shallow	12 × 7 × 2.5	2.6	2.46	2	1.89
	3rd Medium	12 × 7 × 4	4.1	3.88	3.5	3.31
	3rd Deep	12 × 7 × 6	6.1	5.77	5.5	5.20
Fourth	4th Shallow	6 × 10 × 2.5	1.8	1.70	1.4	1.32
	4th Medium	6 × 10 × 4	3	2.84	2.5	2.37
	4th Deep	6 × 10 × 6	4.5	4.26	3.75	3.55

Sixth				
1/6th Shallow	6 × 7 × 2.5	1.2	1.14	0.85
1/6th Medium	6 × 7 × 4	1.8	1.70	1.42
1/16th Deep	6 × 7 × 6	2.7	2.56	1.32
Ninth				
1/9th Shallow	4 × 7 × 2.5	0.6	0.57	0.47
1/9th Medium	4 × 7 × 4	1.6	1.51	1.18

Notes

Pan dimensions are rounded. For instance, a full shallow pan is actually 12.75″ × 20.75″ × 2.5″. Hard plastic food pans hold between 5 and 20% less because of thicker wall construction. The larger the pan, the less the plastic material reduces the pan's capacity compared to stainless.

Sheet Trays

Standard sheet trays (also called sheet pans, baking pans, bun pans) are of two primary sizes. A full sheet tray is 18″ × 26″ (46 × 66 cm). A half size measures 13″ × 18″ (33 × 46 cm).

Commercial kitchen ovens, fridges, tables, racks, and so on are designed to fit the full sheet tray. 18″ × 26″ the area of 1 full sheet tray: 468 in² (3,036 cm²).

Full Sheet Tray Cutting Yields

Number of Cuts per Width × Length (Inches)	Piece Size (Square Inches)	Number of Pieces
4.5 × 2	4 × 13	9.0	52
2.5 × 2.6	7 × 10	6.5	70
3 × 2	6 × 13	6.0	78
2.25 × 2	8 × 13	4.5	104
2 × 2	9 × 13	4.0	117

Number of Cuts per Width × Length (Centimeters)	Piece Size (Square Centimeters)	Number of Pieces
11.4 × 5	4 × 13	57.0	52
6.4 × 6.6	7 × 10	42.2	70
7.6 × 5	6 × 13	38.0	778
5.7 × 5	8 × 13	28.5	104
5 × 5	9 × 13	25.0	117

Metric System

THIS UNIT	MEASURES	RELATES TO	ABBREVIATION	EQUIVALENT U.S. MEASURE
Meter	Length	Feet, Yards	m	39.37 Inches (1.09 Yards)
Square Meter	Area	Square Yard	m²	1.1959 Sq. Yards
Liter	Liquid Volume	Quart	L	33.8 Fl. Oz. (1.056 Quarts)
Cubic Meter	Volume	Cubic Yard	m³	35.3 Cu. Ft (1.307 Cu. Yd)
Gram/Kilogram	Mass (Weight)	Ounce/Pound	g/kg	1 G = .035 Oz/1 Kg = 2.2#

Metric System Prefixes (Using Liters as Sample Application)

Prefix	Meaning	Example	Abbreviation	Number Expressions
milli	One-Thousandth	Milliliter	mL	1/1,000th or .001 Liter
centi	One-Hundredth	Centiliter	cL	1/100th or .01 Liter
deci	One-Tenth	Deciliter	dL	1/10th or .1 Liter
The item: Liter	One, Each, Whole	Liter	L	1 Liter
deka	Ten	Dekaliter	daL	10 Liters
hecto	Hundred	Hectoliter	hL	100 Liters
kilo	Thousand	Kiloliter	kL	1,000 Liters

Volume

Items	Equivalent U.S. Fluid Ounces	Equivalent Milliliters	Equivalent Cubic Inches	Notes
U.S. Fluid Ounce	1	29.57353	1.805	Equals 1.041 Imperial Ounce
U.S. Measuring Cup	8	236.58824	14.43	8 Fluid Ounces per Cup
U.S. Tablespoon	0.5	14.786765	0.9023	16 Tablespoons per Cup
U.S. Teaspoon	0.166666	4.92891667	0.3	3 Teaspoons per Tablespoon

U.S. Pint	16	473.17648	28.875	2 Cups per Pint
U.S. Quart	32	946.35296	57.75	2 Pints per Quart
U.S. Gallon	128	3785.41184	231	Equals .833 Imperial Gallon
Liter	33.814	1000	61.02374	Equals 1.056 U.S. quarts

Volume

Items	Imperial Fluid Ounce	Milliliters	Cubic Inches	
Imperial Fluid Ounce	1	28.413	1.734	Equals .961 U.S. Fluid Ounce
Imperial Cup, Liquid	8	227.304	13.87	Equals .9607 U.S. Cup
Imperial Tablespoon	0.5	14.20655	0.86659	16 per Imperial Cup
Imperial Teaspoon	0.1666666	4.73551	0.28886	
U.S. Measuring Cup	8.33	236.58824	14.43	Equals 1.04 Imperial Cup
U.S. Tablespoon	0.52	14.786765	0.9025	
U.S. Teaspoon	0.17	4.92921667	0.3	
Imperial Pint	20	568.26	34.677	2.5 Imperial Cups per Pint
Imperial Quart	40	1136.52	69.354	20 Imperial Cups per Imperial Gallon
Imperial Gallon	160	4545.92	277.42	
Liter	35.19508	1000	61.02374	Equals 0.264 U.S. Gallon

(continued)

Metric System (Continued)

Measuring Spoons . . . contain the following:

Spoon Size (U.S.)	Milliliters (rounded)	Milliliters (exact)	U.S. Fl. Oz.	Imperial Fl. Oz.
Tablespoon	15	14.78680	0.5	0.520421
Teaspoon	5	4.92892	0.166667	0.173474
Half Teaspoon	2.5	2.46446	0.0833335	0.086737
Quarter Teaspoon	1.25	1.23223	0.04166675	0.0433685

Mass (Weight)

Unit of Measurement	Grams	Kilograms	Ounces (avoirdupois)	Pounds (avoirdupois)
Ounce	28.34952313	0.02835	1	0.0625
Pound	453.59237	0.45359	16	1
Gram	1	0.00100	0.03527396	0.002204623
Kilogram	1,000	1.00000	35.27396	2.204623

Temperature

In the Fahrenheit system there are 180 degrees between the freezing and boiling points of water. In Fahrenheit, water freezes at 32 and boils at 212 (at sea level). In the Celsius (Centigrade) system there are 100 degrees between freezing and boiling. In Celsius, water freezes at 0 and boils at 100.

To convert from one system to the other, first reduce the 180 and 100 figures to 9 and 5. The 32 is either added or subtracted within the formulas.

To convert from Celsius to Fahrenheit, multiply the Celsius value by 180/100 (use 9/5), then add 32. For example, to change 22°C to Fahrenheit:

$22 \times 9 = 198$; $198/5 = 39.6$; $39.6 + 32 = 71.6°F$.

To convert Fahrenheit to Celsius, first subtract 32 from the Fahrenheit, then multiply by 5/9. For example, change 72°F to Celsius: $72 - 32 = 40$; $40 \times 5 = 200$; $200/9 = 22.2°C$.

Length

1 mile equals 5,280 feet, or 1,760 yards, or 1,609.3 meters.

1 meter equals 39.37 inches, or 3.28 feet, or 1.0936 yards.

1 yard equals 36 inches, or 3 feet, or .914 meter.

1 foot equals 12 inches, or one-third (.333) yard, or .3048 meter.

1 inch equals 2.54 centimeters.

Note

One 8-U.S.-fluid-ounce (29.573 mL) cup equals 236.6 grams; 8 ounces avoirdupois equals 226.796 grams.

Measurement Conversions

A: TO CONVERT THIS ITEM	B: INTO THIS ITEM	MULTIPLY A BY:	OR, TO CONVERT TO:	MULTIPLY A BY:
LIQUID VOLUME				
U.S. Fluid Ounce	Milliliters	29.57353	Liters	0.02957353
U.S. Cup, Liquid	Milliliters	236.588	Liters	0.236588
U.S. Pint	Milliliters	473.1765	Liters	0.473176
U.S. Quart	Milliliters	946.3529	Liters	0.94635
U.S. Gallon	Milliliters	3785.41	Liters	3.785412
Imperial Fluid Ounce	Milliliters	28.413	Liters	0.028413
Imperial Cup, Liquid	Milliliters	227.304	Liters	0.227304
Imperial Pint	Milliliters	568.26	Liters	0.56826
Imperial Quart	Milliliters	1136.52	Liters	1.13652
Imperial Gallon	Milliliters	4546.08	Liters	4.54608
Milliliters	U.S. Fluid Ounce	0.033814	Imperial Fluid Ounce	0.0351951
Milliliters	Cup of U.S. Fluid Ounce	0.0042265	Cup of Imperial Fluid Ounce	0.00439938
Milliliters	U.S. Pint	0.002113376	Imperial Pint	0.0017952
Milliliters	U.S. Quart	0.001056688	Imperial Quart	0.0008796
Milliliters	U.S. Gallon	0.0002641721	Imperial Gallon	0.00021996124 8
Liters	U.S. Fluid Ounce	33.814	Imperial Fluid Ounce	35.195
Liters	U.S. Cup	4.22675	Imperial Cup	4.33938
Liters	U.S. Pint	2.113	Imperial Pint	1.75975
Liters	U.S. Quart	1.05669	Imperial Quart	0.879877
Liters	U.S. Gallon	0.26417	Imperial Gallon	0.219969
Milliliters	Liters	0.001		
Liters	Milliliters	1000		

Unit	Converts to	Factor	Converts to	Factor
U.S. Fluid Ounce	Imperial Fluid Ounce	1.04084	Imperial Cup, Liquid	0.130105
U.S. Cup, Liquid	Imperial Cup, Liquid	1.04084	Imperial Pint	0.416337
U.S. Pint	Imperial Pint	0.832674	Imperial Quart	0.416337
U.S. Quart	Imperial Quart	0.83267	Imperial Gallon	0.208169
U.S. Gallon	Imperial Gallon	0.832674		
Imperial Fluid Ounce	U.S. Fluid Ounce	0.96076	U.S. Cup, Liquid	0.120095
Imperial Cup, Liquid	U.S. Cup, Liquid	0.96076	U.S. Pint	0.48038
Imperial Pint	U.S. Pint	1.20095	U.S. Quart	0.600475
Imperial Quart	U.S. Quart	1.20095	U.S. Gallon	0.300327
Imperial Gallon	U.S. Gallon	1.20095		
U.S. SYSTEM				
U.S. Fluid Ounce	U.S. Cup	0.125	U.S. Pint	0.0625
U.S. Fluid Ounce	U.S. Quart	0.03125	U.S. Gallon	0.0078125
U.S. Cup	U.S. Fluid Ounce	8	U.S. Pint	0.5
U.S. Cup	U.S. Quart	0.25	U.S. Gallon	0.0625
U.S. Pint	U.S. Fluid Ounce	16	U.S. Cup	2
U.S. Pint	U.S. Quart	0.5	U.S. Gallon	0.125
U.S. Quart	U.S. Fluid Ounce	32	U.S. Cup	4
U.S. Quart	U.S. Pint	2	U.S. Gallon	0.25
U.S. Gallon	U.S. Fluid Ounce	128	U.S. Cup	16
U.S. Gallon	U.S. Pint	8	U.S. Quart	4

(continued)

Measurement Conversions (Continued)

A: TO CONVERT THIS ITEM	B: INTO THIS ITEM	MULTIPLY A BY:	OR, TO CONVERT TO:	MULTIPLY A BY:
IMPERIAL SYSTEM				
Imperial Fluid Ounce	Imperial Cup	0.125	Imperial Pint	0.05
Imperial Fluid Ounce	Imperial Quart	0.025	Imperial Gallon	0.00625
Imperial Cup	Imperial Fluid Ounce	8	Imperial Pint	0.4
Imperial Cup	Imperial Quart	0.2	Imperial Gallon	0.05
Imperial Pint	Imperial Fluid Ounce	20	Imperial Cup	2.5
Imperial Pint	Imperial Quart	0.5	Imperial Gallon	0.125
Imperial Quart	Imperial Fluid Ounce	40	Imperial Cup	5
Imperial Quart	Imperial Pint	2	Imperial Gallon	0.25
Imperial Gallon	Imperial Fluid Ounce	160	Imperial Cup	20
Imperial Gallon	Imperial Pint	8	Imperial Quart	4
CUPS AND SPOONS				
Fluid Ounce	Tablespoons	2	Teaspoons	6
Cups	Tablespoons	16	Teaspoons	48
Teaspoons	Tablespoons	0.33333		
Tablespoons	Teaspoons	3		
Tablespoons	Cup	0.0625		
U.S. Tablespoon	Milliliters	14.78677	U.S. Fluid Ounces	0.5
U.S. Teaspoon	Milliliters	4.92892	U.S. Fluid Ounces	0.16666666
U.S. Tablespoon	Imperial Tablespoons	1.04084	Imperial Fluid Ounces	0.5204
U.S. Teaspoon	Imperial Teaspoons	1.04084	Imperial Fluid Ounces	0.173474
Imperial Tablespoon	Milliliters	14.20655	Imperial Fluid Ounces	0.5
Imperial Teaspoon	Milliliters	4.735516	Imperial Fluid Ounces	0.1666666

Source	Target	Value	Target	Value
Imperial Tablespoon	U.S. Tablespoons	0.96076	U.S. Fluid Ounces	0.4803806
Imperial Teaspoon	U.S. Teaspoons	0.96076	U.S. Fluid Ounces	0.1601268
U.S. Cup (Liquid) 237 mL	Imperial Cup, Liquid	1.04084	Imperial Tablespoons	16.65346195
Imperial Cup (Liquid) 227 mL	U.S. Cup, Liquid	0.96076	U.S. Tablespoons	15.372179107
Milliliters	U.S. Tablespoons	0.067628	U.S. Teaspoons	0.20288
Milliliters	Imperial Tablespoons	0.07039	Imperial Teaspoons	0.21117

WEIGHT (MASS)

Source	Target	Value
Ounce (avoirdupois)	Pound (avoirdupois)	0.0625
Ounce (avoirdupois)	Gram	28.349523125
Ounce (avoirdupois)	Kilogram	0.028349523125
Pound (avoirdupois)	Ounce (avoirdupois)	16
Pound (avoirdupois)	Gram	453.592
Pound (avoirdupois)	Kilogram	0.45359237
Kilogram	Gram	1000
Kilogram	Ounce (avoirdupois)	35.27396
Kilogram	Pound (avoirdupois)	2.204623
Gram	Kilogram	0.001
Gram	Ounce (avoirdupois)	0.035274
Gram	Pound (avoirdupois)	0.002204623

(continued)

Measurement Conversions (Continued)

A: TO CONVERT THIS ITEM	B: INTO THIS ITEM	MULTIPLY A BY:	OR, TO CONVERT TO:	MULTIPLY A BY:
LENGTH				
Inch	Feet	0.0833334	Yard	0.0277778
Inch	Centimeter	2.54	Meter	0.0254
Feet	Inch	12	Yard	0.33333
Feet	Centimeter	30.48	Meter	0.3048
Yard	Inch	36	Feet	3
Yard	Centimeter	91.44	Meter	0.9144
Centimeter	Inch	0.393701	Feet	0.03280841
Centimeter	Yard	0.0109361	Meter	0.01
Meter	Inch	39.3701	Feet	3.28084
Meter	Yard	1.09361	Centimeter	100
AREA				
Square Inch	Square Centimeter	6.4516	Square Meter	0.00064516
Square Inch	Square Feet	0.0069444	Square Yard	0.000771605
Square Feet	Square Centimeter	929.0304	Square Meter	0.0929034
Square Feet	Square Inch	144	Square Yard	0.111111
Square Yard	Square Centimeter	8361.2736	Square Meter	0.836128
Square Yard	Square Inch	1296	Square Feet	9
Square Centimeter	Square Inch	0.155	Square Feet	0.00107639
Square Centimeter	Square Yard	0.000119599	Square Meter	0.0001
Square Meter	Square Inch	1550	Square Feet	10.7639
Square Meter	Square Yard	1.19599	Square Centimeter	10000

LENGTH

Inch	Centimeter	2.54	Meter	0.0254
Centimeter	Inch	0.3937	Foot	0.0328
Foot	Centimeter	30.48	Meter	0.3048
Meter	Inch	39.37008	Foot	3.28

TEMPERATURE CONVERSIONS

Celsius to Fahrenheit:

Multiply the Celsius temperature times 9/5 and add 32 to that answer.

Examples:

Celsius = 60   60 × 9 = 540   540/5 = 108   108 + 32 = 140   So, 60°C = 140°F.

Fahrenheit to Celsius:

Subtract 32 from the Fahrenheit temperature. Multiply the answer times 5/9.

Examples:

Fahrenheit = 140   140 − 32 = 108   108 × 5 = 540   540/9 = 60   So, 140°F = 60°C

Celsius = Fahrenheit Equivalents: (Sample: At 0° Celsius, Fahrenheit equals 32: 0°C = 32°F)

10°C = 50°F   20°C = 68°F   30°C = 86°F   40°C = 104°F   50°C = 122°F

60°C = 140°F   70°C = 158°F   80°C = 176°F   90°C = 194°F   100°C = 212°F

MANAGEMENT BY MENU

PLANNING A MENU∗

Outline

∗ Authored by Lendal H. Kotschevar and Marcel R. Escoffier.

Objectives

After reading this chapter, you should be able to:

■ Identify and characterize various menus used in the foodservice industry.

■ Explain what a meal plan is and how menus may be developed for them.

■ Identify the basic organizational structure of a menu.

■ Identify tools needed to plan menus.

■ Compare and contrast institutional and commercial menus.

INTRODUCTION

For foodservice consumers, a menu is a list, often presented with some fanfare, showing the food and drink offered by a restaurant, cafeteria, club, or hotel. For the manager of a foodservice establishment, however, the menu represents something significantly more: It is a strategic document that defines the purpose of the foodservice establishment and every phase of its operation.

In considering the menu, we may think of it generally in two ways: first, as a working document used by managers to plan, organize, operate, and control back-of-the-house operations, and second, as a published announcement of what is offered to patrons in the front of the house. In its first model it serves a variety of functions: as a guide to purchasing; as a work order to the kitchen, bar, or pantry; and as a service schedule for organizing job duties and charting staff requirements in all departments. In the second case it is a product listing, a price schedule, and the primary means of advertising the food, beverages, and service available to guests.

A good menu should lead patrons to food and beverage selections that satisfy both their dining preferences and the merchandising necessities of the operator. It can serve as public notice of days and hours of operation, inform patrons of special services available, narrate the history of the establishment and significant material concerning its locale; and may even be used to inform patrons of new ways to enjoy the dining experience, including descriptions of unusual or exciting dishes, drinks, or food techniques. The importance of this selling opportunity makes designing a menu that sells critical.

Selling is a goal not only of commercial operations but of noncommercial and semicommercial institutions as well. For instance, a hospital menu must offer items and present them in a manner that pleases patients and leads to favorable impressions. The same applies to other institutional menus. All menus should be an invitation to select something that pleases.

WHAT IS A MENU?

As far as we know, the first coffeehouses and restaurants did not use written menus. Instead, waiters or waitresses simply recited what was available from memory. Some Parisian operations had a sign board posted near the entrance, describing the menu for the day. The maitre d'hôtel stood near the sign and, as guests arrived, described the various offerings to them and took their orders.

Eventually, some restaurants in Paris originated the custom of writing a list of foods on a small sign board. Waiters hung this on their belts to refresh their memories. As menus became more complex, and more items were offered for sale, this method became too confusing—both to the patron and to the servers—and written menus entered into general use. Grimod de La Reyniere, an 18th-century gourmet who outlived both Louis XVI and the French Revolution, noted that many restaurants handed out bills of fare for patrons to take home with them, both as souvenirs and as advertisements to bring in more business. It was at the time of Napoleon that the restaurants in Paris began to reproduce private gourmet dinner menus for use in the public eating rooms. The elaborate menus written traditionally for the great banquets of the nobility thus found their way into the hands of the emerging middle class.

The word *menu* comes from the French and means "a detailed list." The term is derived from the Latin *minutes,* meaning "diminished," from which we get our word *minute.* Based on this, perhaps, we can say that a menu is "a small, detailed list."

Instead of *menu,* some use the term *bill of fare.* A *bill* is an itemized list, while *fare* means food, so we can say the term means "an itemized list of foods." This seems to be just another way of saying the same thing—*menu.*

The Purpose of the Menu

The job of a menu is basically to inform—inform patrons of what is available at what price, and also to inform workers of what is to be produced. But it is much more than that. The menu is the central management document around which the whole foodservice operation revolves.

"Start with the menu" is a familiar byword of the foodservice trade. The menu should be known at the initial stage when planning a foodservice enterprise because it describes the very nature of the undertaking and the scope of the investment.

Management professionals have known for many years that in order for a company to succeed, it needs to have a clear idea of where the company is headed and how it plans to get where it wants to go. This process is known as long-range or *strategic planning*. In addition to developing strategic plans, management must also create short-range or *tactical plans* that define how the various parts of the organization must function in order to achieve strategic plans.

One of the first activities a manager must perform is to create a *mission statement* or statement of purpose for the organization. The mission for a school district's food service might be to provide wholesome, nutritional meals to students from a variety of economic circumstances and faculty during the school year. A commercial operation's mission might be to serve unique Mexican food at moderate prices to customers of all ages. Once the mission statement is developed, the organization must develop objectives or goals the establishment wishes to attain. These may be expressed as *profitability objectives* (written in specific numbers), *growth objectives* (addition of units), market share objectives, or any number of other objectives. Figure 12.1 shows a list of common organizational objectives.

Organizational objectives are turned into specific strategic plans that answer questions concerning exactly how the company intends to fulfill its mission and objectives. Managers often create these plans after studying environmental conditions. The business environment a foodservice establishment finds itself in may be relatively stable and risk-free. More often, though, the environment is one of frequent change and high risk. Patrons' desires change rapidly, while vigorous competition is a constant challenge. Working within these environmental constraints, the manager must develop a strategy to accomplish the objectives found in the mission statement. An example may be the task of converting a productivity objective into a strategy for increasing worker efficiency coupled with a strategy for increasing equipment efficiency.

The next stage in planning is developing the financial plan. This is a complicated activity, whose details are beyond the scope of this book. Successful managers usually consult with a Certified Public Accountant, or other knowledgeable financial advisor when formulating these plans. A numerical "road map" of the future is helpful in many ways, so this important area of planning should not be overlooked. Without long-range financial direction, managers are without a way of measuring operational success.

Related to the financial plan is the establishment of operating budgets. Successful managers perform this task regularly. Budgets are usually created on an annual basis,

Type of Objective	Sample Objective
Diversification	Operate sit down only, takeout only, and mixed units.
Efficiency	Operate using skilled workers and automated processes where possible.
Employee welfare	Provide a workplace free of sexual harassment that encourages maximum growth.
Financial stability	Maintain key financial ratios in accordance with internal standards.
Growth	Increase sales at a rate of 10 percent per year.
Management Development	Train, develop, and provide opportunity for committed emploees to become senior corporate managers within ten years.
Market Share	Enter and compete in markets where the company is likely to have the major market share.
Multinational expansion	Operate throughout North America and Japan.
Product quality and service	Serve wholesome, nutritious foods in a friendly, home-like environment.
Profitability	Operate with an average profitability of 30 percent over total sales.
Social responsibility	Hire and train economically disadvantaged people in the community.

FIGURE 12.1 Common organizational objectives.

and are usually expressed in terms of monthly increments. More detailed information on this topic is found in Chapter 14.

Remember that no manager is so clever or so knowledgeable that he or she can do all of this work alone. A management team, composed of people highly qualified in many skills, will usually be more effective than one manager working alone. While the foodservice industry is one of the few remaining industries where individual achievement is both possible and highly prized, most successful individuals learned to listen to suggestions from others before making the final decisions on their own.

Who Prepares the Menu?

Because the menu is the essential document for successful operation of the establishment, menu planners must be highly skilled in a number of areas. They must know both the operation and the potential market. They should know a great deal about foods; how they are combined in recipes, their origin, preparation, presentation, and description. They must also understand how various recipes can be combined, which menu items go together, and which do not. Planners must also be aware of how operational constraints, such as costs, equipment availability, and the skills of the available labor, affect the final menu selection. They must be able to visualize how the menu will appear graphically, what styles will look good, and which may be inappropriate for the particular operation. Finally, planners must be skilled at communicating successfully with patrons through the menu.

If all of the above requirements seem daunting, perhaps there is a way to compensate. One key skill in managing is the ability to work in groups. Perhaps a management team is the answer (in even a small establishment the cook and host can meet). While no one person in the organization may possess all the skills required to write the perfect menu, it is likely that a group can be formed whose membership combines all of the skills mentioned. Group preparation of the menu has one additional advantage, it gets the various members of the operational staff into a receptive frame of mind, anticipating the changes about to be implemented.

Menu planning is a time-consuming and detailed task, and should not be done quickly or haphazardly. Treat menu planning for what it is—the most critical step in defining the operation.

Tools Needed for Menu Planning

What are the tools needed to prepare a menu? First, a quiet room where one can work without disturbance. A large desk or table is needed so materials can be spread out. These include a file of historical records on the performance of past menus, a menu reminder list, a file of menu ideas, and sales mix data indicating which items may draw patrons away from specials the operation wants to sell. A list of special occasion and holiday menus should be on hand. Costs and the seasonality of possible menu items should also be available.

Market Research

Market research is the link between the consuming public and the seller. This link consists of information concerning the public's buying preferences and how well the seller's business meets those preferences *as viewed by the potential buyer.* Formal market research is a scientific process of collecting data, analyzing the results, and communicating the findings and their implications to the seller. Anyone planning a menu is well advised to use as many forms of market research as possible. The more specific the research (for example, research concerning a specific group of potential patrons in a specific geographic location), the more reliable the results. At the minimum, the menu planner should take advantage of information available from marketing research firms. The federal government, financial newspapers and magazines, stock and investment firms, and many other large companies all provide the results of their market research. Trade associations, such as the National Restaurant Association and state associations, constantly produce market research reports that can be very useful in determining market trends.

Market studies have always been expensive. Traditionally, therefore, only the largest hospitality companies and chains could afford to conduct their own studies. However, it is beginning to be more feasible for smaller organizations to conduct market studies. An operator might, for example, commission a local college or business school to conduct market research. Advertising agencies can sometimes conduct a useful study for just a few thousand dollars. Operators also can conduct studies personally.

A key element of a market study is to determine exactly what kind of information is needed. If a menu planner wants to create a menu that will increase sales from existing customers, then the market data should be gathered only from that group. On the other hand, if the menu planner is attempting to draw in new customers, then that group should be targeted.

The next key element is to decide how the research will be conducted. Methods include conducting interviews, mailing out questionnaires, observing customers, and holding focus groups. Each is useful in eliciting specific kinds of information.

The questions asked of research subjects can be either open-ended or closed. An example of a closed question, which requires only a one-word answer, is "Do you prefer seafood or steak?" An open-ended question, which requires that the respondent supply a longer answer, might be "What are the most important factors that help you decide where to eat out?"

The information obtained through market research must coincide with other menu planning factors, such as purchasing costs and staff capabilities.

MENU PLANNING FACTORS

A new menu should be planned sufficiently in advance of actual production and service to allow time for the delivery of items, to schedule the required labor, and to print the menu. Some menus must be planned six or more months before use. Operational needs will dictate how far in advance of use new menus should be prepared.

Number of Menus

A number of menus may be needed by a single operation. A large hotel might need several different menus for different dining areas and specialty events. Some operations may need even more. Menu needs will differ with differing types of enterprises. Atmosphere, theme, patrons, pricing structure, and type of service must all be taken into account.

TYPES OF MENUS

A La Carte Menu

An *a la carte* menu offers food items separately at a separate price. All entrees, dishes, salads, and desserts are ordered separately; the patron thus "builds" a meal completely to his or her liking. A la carte menus often contain a large selection of food items and, consequently, often lead to increased check averages. Commercial operations often find this menu type highly profitable, provided they can keep food spoilage, which can result from the large number of offerings, under control.

Table d'Hôte Menu

The *table d'hôte* menu groups several food items together at a single price. This often can be a combination, such as a complete meal of several or more courses. Often there is a choice between some items, such as, between a soup or salad or various kinds of desserts. This type of menu often appeals to patrons who are unfamiliar with the cuisine offered by the establishment. It is an excellent way to introduce the new patron to fine dining as perceived by the establishment. Another advantage of this type of menu is the limited number of entrees that must be produced. These menus can combine wine selections with each course, further enhancing the dining experience, especially for those hesitant to order wine. Very often a la carte and table d'hôte menus are combined.

Du Jour Menu

A *du jour* menu is a group of food items served only for that day. (*Du jour* means "of the day.") The term is most often associated with the daily special, the "soup du jour" being one example. Again, these items often are combined on a la carte and table d'hôte menus. One profit-boosting technique is to offer daily specials that use foods purchased at a reduced price or to use surplus goods.

Limited Menu

A *limited* menu is simply one on which selections are limited in some way. Often associated with quick-service operations or cafes, the limited menu allows the manager to concentrate his or her efforts in training, planning, and calculating food cost or other menu analyses. Cost control is one very important benefit available to operators using limited menus. These menus do not look different from those previously mentioned with the exception of offering fewer choices. This is not entirely a new concept; a restaurant in Paris operated from 1729 to the 1800s offering only one menu item, chicken, cooked in several ways.

Cycle Menu

A *cycle* menu refers to several menus that are offered in rotation. A cruise ship, for example, may have seven menus it uses for its seven-day cruise. At the end of the seven-day cycle the menu is repeated. The key idea here is to inject variety into an operation catering to a "captive" patronage.

Some hospitals might use only a three-day cycle. When patron stays are fairly long, such as in nursing homes, prisons, or on long steamship journeys, the cycle must be longer. Some operations of this type may have four cycle menus for a whole year that change with the seasons of the year to allow for seasonal foods as well as menu variety. Figure 12.2 shows a high school's cycle menu for one month. It is important when using a cycle menu over a long period to rotate Sunday and holiday selections since people seem to remember what they had to eat on Sundays and holidays. Holidays require special meals to mark the occasion.

California Menu

The *California* menu, called that because it originated there, offers breakfast, snack, lunch, fountain, and dinner items that are available at any time of the day. (See Figure 12.3.) Thus, if one patron wants hot cakes and sausage at 6:00 PM and another patron wants a steak, french fries, and a salad for breakfast, each is accommodated. It is typically printed on heavy, laminated paper so it does not soil easily. Many hotel roomservice menus are based on this design as well.

CHICAGO PUBLIC SCHOOLS DEPARTMENT OF FOOD SERVICES

High School Lunch Menu
May

WEEK 1

MONDAY (5-3)	Portion	TUESDAY (5-4)	Portion	WEDNESDAY (5-5)	Portion	THURSDAY (5-6)	Portion	FRIDAY (5-7)	Portion
CHICKEN ENCHILADA	2	TURKEY CHOP SUEY W/ VEGETABLES	1 c	SPAGHETTI W/ MEAT SAUCE & CHEESE	1 c	CHICKEN FAJITA W/ TORTILLAS (2)	3 oz	FISH PORTION W/ BUN W/ CHEESE	2 oz
Extra Cheese	1 oz	Fried Rice w/ Eggs	1/4 c	Steamed Broccoli	1/4 c	Carrot Sticks	6	Mixed Vegetables	1 oz
Refried Beans	1/2 c	Fresh Apple	1/2 c	Pear Halves	1/2 c	Blueberries w/ Whipped Topping	1/2 c	Rosy Applesauce	1/2 c
Corn on the Cob	1/4 c	Croissant	1	Garlic Bread	1 oz	Low Fat Milk	1/2 pt	Raisin Spice Bar	1 ser
Fresh Apple		Low Fat Milk	1/2 pt	Low Fat Milk	1/2 pt			Low Fat Milk	1/2 pt
Peanut Butter Brownie	1 ser								
Low Fat Milk	1/2 pt								

WEEK 2

MONDAY (5-10)	Portion	TUESDAY (5-11)	Portion	WEDNESDAY (5-12)	Portion	THURSDAY (5-13)	Portion	FRIDAY (5-14)	Portion
SLICED TURKEY W/ GRAVY	3 oz	BEEF-R-RONI	1 c	CHICKEN NUGGETS	8	BBQ PORK W/ BUN	#8 scp	CHICKEN PATTIES OR	2@1.5 oz
Whipped Potatoes	1/2 c	Peas & Carrots	1/2 c	Tator Rounds	1/2 c	Glazed Carrots	1/2 c	MACARONI & CHEESE	1 c
Seasoned Green Beans	1/2 c	Frozen Juice Cup	1/4 c	Coleslaw	1/4 c	Fresh Apple	1/2 c	Steamed Broccoli	1/2 c
Bread Slice	1	Hot Corn Bread	1	Croissant	1	Sweet Potato Pudding	1	Pear Halves	1/2 c
Peanut Butter Brownie	1 ser	Low Fat Milk	1/2 pt	Low Fat Milk	1/2pt	Low Fat Milk	1/2 pt	Bread Slice	1
Low Fat Milk	1/2 pt							Low Fat Milk	1/2 pt

WEEK 3

MONDAY (5-17)	Portion	TUESDAY (5-18)	Portion	WEDNESDAY (5-19)	Portion	THURSDAY (5-20)	Portion	FRIDAY (5-21)	Portion
CHILI W/ BEANS & CHEESE	1 c	TURKEY EGG ROLL W/ SWEET & SOUR SAUCE	1	SUPER DELUXE BURGER W/ BUN	3.5 oz	MEAT LOAF W/ GRAVY	4 oz	ITALIAN SAUSAGE W/ SAUCE	3 oz
Buttered Corn	1/2 c	Fried Rice w/ Eggs	1/2 c	Cole Slaw	1/2 c	Whipped Potatoes	1/2 c	Mixed Vegetables	1/2 c
Cherry Apple Juice	6 oz	Seasoned Green Beans	1/2 c	Chilled Peaches	1/2 c	Garden Peas	1/2 c	Sweet Potato Pone	1/2 c
Saltine Crackers	8	Fresh Apple	1/2 c	Low Fat Milk	1/2 pt	Croissant	1	Hot Roll	1
Peanut Butter Cookie	1	Low Fat Milk	1/2 pt			Raisin Spice Bar	1 ser	Low Fat Milk	1/2 pt
Low Fat Milk	1/2 pt					Low Fat Milk	1/2 pt		

WEEK 4

MONDAY (5-24)	Portion	TUESDAY (5-25)	Portion	WEDNESDAY (5-26)	Portion	THURSDAY (5-27)	Portion	FRIDAY (5-28)	Portion
CHICKEN NUGGETS W/ BBQ SAUCE	8	SAUSAGE & CHEESE PIZZA	1 sl	HAM & CHEESE ON CROISSANT	3 oz	BAKED CHICKEN	3 pcs	HOT TAMALES	2
Rotini Pasta Salad	1/2 c	Onion Rings	1/2 c	Vegetable Soup	6 oz	Glazed Carrots	1/2 c	W/ CHILI	1/2 c
Mixed Vegetables	1/2 c	Tossed Salad	1/2 c	Fresh Orange	1/2 c	Tossed Salad	6 oz	Buttered Spinach	1/2 c
Frozen Juice Cup	1/4 c	Raisin Spice Bar	1 ser	Peanut Butter Cookie	1 ser	Hot Biscuit	1	Pineapple Orange Juice Drink	6 oz
Hot Roll	1	Low Fat Milk	1/2 pt	Low Fat Milk	1/2 pt	Low Fat Milk	1/2 pt	Low Fat Milk	1/2 pt
Low Fat Milk	1/2 pt								

WEEK 5

MONDAY (5-31)	Portion	TUESDAY (6-1)	Portion	WEDNESDAY (6-2)	Portion	THURSDAY (6-3)	Portion	FRIDAY (6-4)	Portion
MEMORIAL DAY (HOLIDAY)		CHICKEN ENCHILADA	2	SUBMARINE SANDWICH ON CROISSANT	1 oz	TURKEY EGG ROLL	3 oz	CHILI W/ BEANS & CHEESE	1 c
		Extra Cheese	1 oz	Chicken Vegetable Soup	1/2 c	Sweet and Sour Sauce		Corn on Cob	1/4 c
		Tossed Salad	1/2 c	Fresh Apple	1/2 c	Fried Rice w/Eggs	1 c	Mixed Fruit	1/2 c
		Chilled Peaches	1/4 c	Low Fat Milk	1/2 pt	Broccoli	1/2 c	Croissant	1
		Peanut Butter Cookie	1			Chilled Peaches	1/2 c	Low Fat Milk	1/2 pt
		Low Fat Milk	1/2 pt			Low Fat Milk	1/2 pt		

FIGURE 12.2 Cycle menu. Courtesy of Lane Technical High School, Chicago, Illinois.

FIGURE 12.3 California menu. Courtesy of IHOP, Glendale, California.

THE MEAL PLAN AND THE MENU

Menu planners must keep in mind that there are two things a patron decides when selecting a meal from a menu. The first is what the sequence of foods will be, and the second is which items will be selected. The meal planner must do the same: 1) decide on courses and their sequence, and 2) establish the specific foods in each course.

Meal Plan	**Menu**
Fruit or juice	Orange, grape fruit or tomato juice
Cereal	Dry cereal or oatmeal, milk
Entree	Eggs any style or ham timbale with light cheese sauce
Bread	Bran muffins or toast
Beverage	Tea, coffee, or cocoa

FIGURE 12.4 Breakfast meal plan and menu.

These two factors make up the *meal plan* or the manner in which foods are grouped for particular meals throughout a day. The typical American meal plan is three meals a day, with about a fourth of the day's calories eaten at breakfast, a third at lunch, and the balance at dinner, although some may reverse the lunch and dinner percentages. For a table d'hôte menu, some knowledge of the meal plan is required, even for partial meals. On an a la carte menu, a meal plan is not needed since the patron establishes one by selecting the items.

In this country, food patterns have been changing. Some menus may be written for a four or five-meal-a-day plan. This gives a greater division of calories but the same total amount in a day. Most meal plans call for all the food to be consumed within a ten-hour period, but if a snack is served at night, the fasting period may be less than 14 hours.

Staying too strictly with a meal plan can become monotonous, so it is recommended that a plan be varied occasionally. Thus, a supper of vegetable soup with crackers, large fruit plate with assorted cheeses, bran muffins and butter, brownie and a beverage may be a relief from the often consumed meat, potatoes, and hot vegetable dinner. Figure 12.4 shows a breakfast meal plan and a menu based on it.

Menu Organization

Menus usually group foods in some order in which they are intended to be eaten. Typical menus begin with appetizers and end with desserts. Several sample courses are shown in Figure 12.5.

Within the entree category, it is usual to split the various food offerings into the following categories: Seafood and Fish, Meat (beef, lamb and pork), Poultry (chicken, duck, etc.) and Others such as pasta or meatless entrees.

Each menu category should offer choice. It is important to note that the foods within a major category must differ in style of preparation. For example, every entree

Coffee Shop	**French Restaurant**	**Hospital**	**Family Dining**	**Steakhouse**
Appetizers and side dishes	Hors-d'oeuvre	Appetizers and soups	Appetizers	Appetizers
Salads	Potages (soups)	Salads	Soups	Soups and salads
Sandwiches	Salad	Entrees	Salads	Entrees
Hot entrees	Sorbet	Vegetables	Entrees	Side dishes
Fountain items	Entrees	Desserts	Side dishes	Desserts
Desserts	Plateau de fromage (cheese platter)		Desserts	
	Entremets (small desserts)			

FIGURE 12.5 Sample menu organization.

should not be fried. You will notice in the figure that various preparation methods are used such as poaching, roasting, grilling, frying, and baking. While it may not be feasible to include all methods of preparation, a menu heavily weighted toward one method or another will be unbalanced. Only in an operation specializing in one preparation style is repetition recommended.

It is often said that variety is the spice of life. It certainly is the key to creating a menu that sparks patrons' interest and encourages them to come back again and again. The variation of cooking methods is a subtle form of variety. The variations in tastes and textures are less subtle. It is important that foods vary from spicy or hot to bland or mild. Even the most adventurous tongue needs a rest, which is why some dinner restaurants are serving an intermezzo course (often a fruit sorbet) between courses. Both spicy and hot dishes can appear on a menu. (Hot dishes cause burning on the tongue, while spicy items incorporate complex flavors.)

Various cooking methods will affect menu items' texture as much as their ingredients.

Variety extends to the visual as well. There is a saying that people "eat with their eyes." Chefs have been concerned with balancing color as well as texture. Whether an ingredient is cut, ground, minced, cubed, sliced, pared, or kept whole affects its visual appeal as well as its "mouth feel." Even the use and kind of garnish can distinguish items and make an otherwise common dish seem exotic.

How Many Menu Items?

The number of menu items found on menus varies greatly, from the simplest limited menu operations (one operation in London has only one choice each for appetizers, soup, salad, entree, and dessert) to Chinese menus offering many choices. Management usually wants to limit the number of offerings in order to reduce costs and maintain good control. The patron may desire fewer items also. Some studies have shown that people who are confronted with a large number of choices tend to fall back to choices they have made before.

The key word in choosing the items for a menu is balance. There must be a balanced selection that allows for different tastes. It is necessary to balance selections within food groups. Thus, a lunch menu might offer a minimum number of appetizers, say two juices (one fruit and one vegetable), a fruit, and a seafood cocktail. This minimum offering could be expanded on the dinner menu into choices of both hot and cold appetizers. Gourmets may substitute a hot seafood appetizer for the traditional fish course.

The number and kinds of soups also depend on the meal period. Lunch patrons may wish to have a bowl of hearty soup such as vegetable or beef and rice, or a combination cup of soup and half a sandwich. A recent trend has ben toward cold soups. However, the dinner patron may want soup as a beginning for a large meal. This patron may wish to have a light appetite-stimulating soup like a consommé or bouillon. The balance with the amount of food offered with the rest of the meal must be struck between light clear soups, purees, and bisques, and the heavier chowders and stew-like soups.

The typical menu has at least five or six entrees. Theme operations, such as a seafood restaurant, naturally will specialize in certain entree categories. They must still offer some alternative dish; for example, a seafood restaurant may offer a non-seafood item for those patrons who will not (or cannot) eat seafood but who are with a party that does. Non-meat items (like pasta or vegetable dishes) are increasingly being found on menus. If balance is required, the entrees should be divided among beef (more than 50 percent of the meat consumed in this country is beef), pork and ham, poultry, lamb and veal, shellfish, fish, egg dishes, cheese dishes, and non-meat main dishes. If only two food items appear on the menu they may both be beef (one roasted and one a steak dish) or one beef and one of the other meat categories. If two fish or shellfish items appear, they should be quite different, such as a lean, white-fleshed fish like a sole or flounder, and a fatter one such as salmon.

If one is limiting the entrees to five or less, it is best to remember that quantity is no substitute for quality. A few items prepared and served with absolute perfection is as good or better than a dozen mediocre items.

The ideal menu must also provide for balance and variety in the vegetable, sauces, and starch dishes as well. The vegetables must be chosen to complement the entree choices. Thus on a menu in which there is ham, chicken, or turkey an offering of sweet potatoes or yellow winter squash will complement the meal. Potatoes, rice, wild rice, polenta and other starches are today becoming an important part of a menu.

When selecting vegetables, as with all menu items, an eye must be given to the flavors, colors, and texture contrasts. Popular vegetables like asparagus, peas, string beans, and carrots should be offered, but for variety's sake it may be wise to include one or more exotic vegetables as well. Variety can also be achieved by using different cooking, methods, or serving some vegetables raw. Vegetables can be creamed, deep-fried, au gratin, baked, mashed, pureed, diced, julienned, made into fritters, or served with a variety of sauces like honey mustard, hollandaise, tomato, or cheese.

Again, variety must not be carried to extremes; three to five starches, and five or so vegetable dishes are more than sufficient for most menus.

The number of salads offered as meal accompaniments (as opposed to salads as main dishes) has been constantly decreasing. For many years table d'hôte meals included a choice of several salads. Now the typical menu includes only one or two salads, usually a tossed green salad, or a special salad like a Caesar salad. However, as more people strive for a healthy lifestyle, many menus are offering several generous-portioned salads as main courses; seafood, egg, fruit, and chicken salads are common. Variety is achieved, not only in ingredients—arugula, endive, spinach, artichokes, hearts of palm, nuts, seeds, patés, and edible flowers are just some of the creative ingredients being used—but also by offering a variety of salad dressings, featuring flavored vinegars and oils, honey, cheese, and mustards. The self-serve salad bar trend has crested as more operators become concerned with the labor and food costs associated with salad bars as well as heightened fears concerning the potential health risks associated with self-serve environments.

Another trend in menus has been to offer fewer dessert choices but to make those items offered more elaborate. Some operators see dessert as a low-profit area and confine the choices to an ice cream or sherbet dish. Others see dessert as a selling opportunity, where desserts can be coupled with unusual after-dinner drinks to generate sales. The dessert course must be treated as seriously by the manager as any of the other food courses. One way to free up tables for other diners is to offer dessert in a separate room. Several operations have turned this into a very profitable way to keep diners on site and spending additional dollars on after-dinner drinks.

We look at breakfast menus last. This important meal is characterized by high volumes and low profits. The public seems to have a clear idea of what it is willing to pay for breakfast, and operators emphasizing this meal period are often caught in the price trap. While there is no study to support the idea, it may be conjectured that the reason for this price problem is that breakfast, more than any of the other meals, consists of food items that patrons regularly prepare at home. This knowledge of the costs and skills required may cause a patron to develop firm attitudes concerning what is a fair price to pay for breakfast.

One way to boost check averages is to offer breakfast items that the public cannot relate to their at-home cost. Elaborate breakfasts or unusual breakfast combinations may be the answer to this dilemma. An analysis of 50 breakfast menus from hotels around the country showed that the average breakfast menu offers about 100 choices. There are usually five to ten juices, five or more fruit dishes, a hot cereal, and a dozen or more cold ones offered with and without a variety of toppings. One may find eggs cooked in perhaps four to six different ways (there are about 100 ways to cook an egg in formal French cooking). Meat dishes may include hams, steaks, bacon, sausages, and other meats. French toast, pancakes, toasts, and rolls may add another five to ten items to the menu, with side dishes increasing the menu size even further. Finally there are the requisite beverages from coffee and tea to milk shakes smoothees, herbal teas, and vegetable shakes.

It is possible to limit the breakfast menu to perhaps 30 or 40 items, but many menu planners feel the need to offer more.

MENUS FOR VARIOUS MEALS AND OCCASIONS

Many operations have one menu that covers choices for the three meals of the day with the times of availability included. Other operations have separate menus for each meal, and still others have separate menus for special occasions and parties. The types and number of menus needed will vary from one operation to another.

Each operation must rely on patron preference, costs, and operational goals to dictate what items should be included on each menu. The following section should serve as a guide to meeting general menu requirements.

Breakfast Menus

Some breakfast menus are printed on the regular menu, while others appear as separate menus. An example of a breakfast menu is shown in Figure 12.6. A California menu that offers breakfast during all hours of operation may list the breakfast on a side panel or in a special space. Some have breakfast items on the back of a placement. A children's breakfast menu may be offered in some units.

Both a la carte items and table d'hôte breakfasts should be on a breakfast menu. Table d'hôte offerings should list a *continental breakfast* that includes a juice (usually orange), a bread item (usually a sweet roll or toast, but a croissant is common), and a beverage (usually coffee). It should also list heavier breakfasts. A juice or fruit may or may not be included with these. Meat, eggs, or other main dishes will be accompanied by toast, hot breads or rolls, and perhaps hashed brown potatoes or grits (in the South).

BREAKFAST

Eggs

Start the day right with your favorite breakfast. Served with your choice of buttermilk biscuits, wheat or white toast, or one blueberry or Country Morning muffin.

Early Riser - Two eggs any style, with choice of bread $1.99

Daybreak - Two eggs any style, with sausage patties or links or thick-sliced bacon and your choice of bread $3.69

Rise and Shine - Two eggs any style, with sausage patties or links or thick-sliced bacon, home fries and your choice of bread $4.69

NEW! Country Skillet - Hot from the skillet. A layer of eggs, with home fries, Bob Evans Farms' sausage, country gravy and shredded cheese. Served with your choice of bread $4.79

NEW! Border Skillet - Hot from the skillet. A layer of eggs, Bob Evans Farms' sausage, diced potatoes, green and red peppers, onions, mild picante sauce and shredded cheese. Served with your choice of bread $4.79

NEW! Country Fried Steak and Eggs - Our famous country fried steak topped with country gravy served with two eggs any style, home fries and your choice of bread $5.99

Homestead Breakfast - Two large eggs, choice of Bob Evans Farms' sausage patties or links or thick-sliced bacon, sausage gravy, home fries and your choice of bread $6.29

You may substitute egg beaters, on any egg combination for an additional .... $.40

Omelettes

Served with our home fries and your choice of buttermilk biscuits, wheat or white toast, or one blueberry or Country Morning muffin.

Sausage & Cheese - Featuring our own Bob Evans Farms' sausage $4.49

Western Omelette - Diced ham, onions, green peppers and sharp American cheese $4.49

Ham & Cheese Omelette - An all-American favorite $4.49

You may substitute egg beaters, on any omelette for an additional .... $.40

© 1995 BEF & Co'Pb, Inc. c. 11/95

Hotcakes with Strawberry Topping

It's the tops

Border Skillet

Get that great wake-up taste

Hotcakes & Such

These complete breakfasts are served with warm syrup and margarine. Low calorie syrup and real butter available upon request.

Hotcakes - Three buttermilk hotcakes served with your choice of thick-sliced bacon or sausage patties or links $3.69

French Toast - Thick-sliced, dipped in cinnamon batter, served with your choice of thick-sliced bacon or sausage patties or links $3.69

Fried Mush - A Bob Evans' Special Recipe. Three slices, fried golden brown and served with your choice of thick-sliced bacon or sausage patties or links $3.69

Blueberry Hotcakes - Full of blueberries. Served with your choice of thick-sliced bacon or sausage patties or links $4.20

Hotcakes & Eggs - Three buttermilk hotcakes and two eggs any style, served with your choice of thick-sliced bacon or sausage patties or links $4.79

The Tops - Juicy strawberry topping crowned with whipped topping or delicious cinnamon apples on your choice of hotcakes or our classically prepared French Toast. Served with your choice of thick-sliced bacon or sausage patties or links $4.90

Bob Evans' Own

Sausage Gravy & Biscuits - With home fries $3.79

Border Scramble - A large flour tortilla filled with sausage, fresh vegetables and potatoes. Topped with Texas-style chili and melted cheese $4.39

The Lighter Side

Try one of our lighter alternatives for breakfast. They'll satisfy you without slowing you down.

Fruit Bowl - Half a melon filled with fresh mixed fruit and two Country Morning muffins. Fruit selection will change seasonally to give you the best available $4.99

Oatmeal Breakfast - **QUAKER** Oatmeal, orange juice and blueberry muffin. Until 11 A.M. Brown sugar available upon request $3.19

egg beaters Breakfast - With fresh sliced tomatoes and wheat toast $2.40

A La Carte

One Fresh Egg - Any style	$.69
Grits - Southern style Until 11 A.M.	$.69
Home Fries	$1.19
QUAKER Oatmeal - Served with milk. Until 11 A.M. Brown sugar available upon request	$1.59
Toast & SMUCKERS Jelly - Choose white or wheat bread	$.99
Two Hot Buttermilk Biscuits	$.99
Two Blueberry or Country Morning Muffins	$1.69
Breakfast Fruit (seasonal fresh fruit)	$1.80
Breakfast Fruit Bowl - Half a melon filled with seasonal fresh fruit	$3.30
Hotcakes - Three hotcakes with warm syrup	$1.99
Golden Brown Waffle - With warm syrup	$1.99
French Toast - Dipped in cinnamon batter. With warm syrup	$1.99
Fried Mush - Three slices with warm syrup	$1.99
Blueberry Hotcakes - Three hotcakes with warm syrup	$2.50
Sausage Gravy	$1.89
Bob Evans Farms' Sausage - Patties or links	$1.80
NEW! Bob Evans Farms' Smoked Sausage - Our larger 5 oz. link	$2.10
Honey-Cured Ham	$1.80
Bacon - Three thick-sliced strips	$1.80
Sausage Gravy & Biscuits - With two buttermilk biscuits	$2.79

FIGURE 12.6 Breakfast menu. Courtesy of Bob Evans Farms, Columbus, Ohio.

Hot cakes or waffles are served sometimes with bacon, ham, sausage, or even an egg. A beverage is usually included with these breakfasts.

The offerings of table d'hôte breakfasts should be balanced. A familiar one includes eggs fixed any style and priced for one or two eggs. Omelets on eggs with bacon, ham, or sausage are other offerings. Egg-white omelets are becoming common. Meats alone may also be offered. A pancake and a waffle breakfast offering is usual. Sometimes a corned beef hash main dish with or without poached eggs, steak plain or with eggs, creamed chipped beef, or other main dishes may be on table d'hôte listings. Items such as shirred eggs with chicken livers, Huevos Rancheros, Eggs Florentine, or other "occasion" foods can be included, depending on the type of operation. Specials, such as a low-calorie breakfast, a steak-and-egg breakfast, a high-protein breakfast, or a child's breakfast may be offered. Remember to also have something low-calorie, low-fat, low-cholesterol, or low-salt that guests might want.

Table d'hôte and similar breakfasts that bring in a higher check average should be in the most prominent place on the menu. These higher-income items should also be given as effective a presentation as possible, with large type, bracketing, and effective description.

Numbering breakfasts on a large menu makes them easier to order, both for the patron and the server. Specialties, such as a variety of syrups, jellies, and jams, may serve to encourage choices of desirable menu items.

The menu order for breakfast items is usually as follows.

1. Fruits and juices
2. Cereals
3. Eggs alone or combined with something else
4. Omelets
5. Meat and other main dishes
6. Pancakes, waffles, and french toast
7. Toast, rolls, and hot breads
8. Beverages

Side orders must be placed in available space. On the a la carte menu, items should be grouped together in the same order.

Breakfast menus should cover less space than the other meal menus and usually should have larger type, because people are not yet awake. Do not list items only as "juices" or "cereals," but list each offering separately. Also list essential information, such as how long breakfast is served and special breakfast facilities.

Special breakfasts may have to be catered. A wedding breakfast may start with champagne, silver gin fizzes, a fruit punch, or juices. A fresh fruit cup is often served as the first course. If not, then a fruit salad is appropriate. A typical main dish is served with a high-quality sweet roll. Eggs Benedict would be suitable for the early party-type breakfast, or eggs with sausage, ham, or bacon. An omelet of some type would be suitable. If the affair is held late in the morning, the main dish can be creamed chicken, sautéed ham and mushrooms, or creamed sweetbreads and ham in a patty shell with a few green peas. A beverage choice is offered. If wine is served at the table, it is usually a chilled white or a semi-dry rose.

Buffet breakfasts are popular. Some may be offered to allow guests to quickly obtain what they want and leave. Many people at conventions or meetings are in a hurry and may patronize the foodservice operation if they feel they can get what they want quickly. Some who usually skip breakfast may still be enticed to come in and get a quickly served buffet continental breakfast.

A buffet breakfast for more leisurely dining may be much more elaborate and feature a wide choice of juices and fruits; cold items, such as cheese, sliced baked ham, lox and bagels, chopped chicken livers, even salads; and hot dishes, such as scrambled eggs, assorted breakfast meats, chicken livers and mushrooms, pepper steaks, corned beef hash, creamed chipped beef, and different kinds of omelettes. Side dishes, such as hashed brown potatoes or grits, can be included. Assorted hot breads and sweet rolls are offered, along with a beverage choice. The drink is often poured at the table by servers, the guests selecting the other foods they wish at the buffet. On the most elaborate buffets, a dessert may be offered. Depending upon the occasion, champagne or alcoholic beverages may be provided. Wine service may be offered with a white wine (not completely dry) or a rose.

A *hunt breakfast* is an elaborate buffet that may include broiled lamb chops, steaks, roast beef, grilled pork chops, pheasant, hare, venison, or other items. A hunt breakfast originated as a meal before or after a hunt and was intended for hearty eaters leading a vigorous life—they had to eat that way.

A *chuck wagon breakfast* should feature sourdough hot cakes, steaks, eggs, hashed brown potatoes, and perhaps freshly cooked doughnuts. Grits or biscuits can be substituted for the breads. (A chuck wagon is the meal wagon that was used to feed cowboys when they were away from the ranch.)

A *family-style breakfast* is one in which the food is brought to the table and guests serve themselves.

Group breakfasts should be planned carefully. Eggs and other breakfast items cool rapidly. Do not attempt difficult egg preparations, such as omelets or Eggs Benedict, unless they can be prepared and served properly. Toast is difficult to serve because it gets cold and chewy quickly. Hot breads, such as biscuits, muffins, and cinnamon rolls, are easier to handle. American service is usually used, but Russian service is sometimes also used.

Brunch Menus

A brunch menu should combine items usually found on breakfast and lunch menus and provide for substantial meals. (See Figure 12.7.) A fruit juice or fruit should be offered. The main dish should be substantial—omelets; creamed chicken on toast points; a soufflé; a small steak with hashed brown potatoes; chicken livers and bacon; or a mixed grill of lamb chop; sausage, bacon, grilled tomato slice, and potato. Hot breads and a beverage choice should be offered, and vegetables can be served. A fruit salad or molded gelatin salad filled with fruit is a favorite.

Luncheon Menus

Luncheon menus may contain a wide assortment of foods, from complete table d'hôte meals to snacks. Offer a wide number of a la carte items with combinations, such as a sandwich and a beverage, or a cup of soup, salad, and dessert with beverage. A few casserole dishes can be offered. Items such as sandwiches, salads, soups, and fountain products can be stressed. A lunch menu can more easily feature economical purchases than the dinner menu.

Many units have modestly priced items and attempt to cover costs with volume and fast turnover. Occasion foods—specialty items created to enhance a special occasion—may be profitable, if especially slanted to the trade. Executives, and expense account patrons, may wish more elaborate menus. Thus, a club, better hotel, or fine restaurant that they patronize may feature a higher priced list for lunch. Alcoholic beverages may or may not be offered.

Lunch menus should have permanent a la carte offerings on the cover, but may also present daily offerings. The permanent menu will offer sandwiches, salads, fountain items, and desserts. Flexible menus are more typical of lunch than any other meal. Inserts or table displays may be used to call attention to specials.

Lunches for groups usually are complete meals. Clubs or organizations may meet at lunchtime, and a main dish with vegetables, salad, dessert, and beverage will be included. A first course will be included for a more elaborate luncheon. Party or occasion

THE RITZ-CARLTON
CHICAGO
A Four Seasons Hotel

The Dining Room
Sunday Brunch Buffet

Array of Breakfast Pastries
Large Assortment of Different Salads
Smoked Salmon Station
Smoked Trout
Shellfish on Ice
Pates and Terrines
Omelette Station
Waffle Station
Eggs Benedict
Poached Fish in Wine Sauce
Baked Salmon in Puff Pastry
Chicken Stir Fry
Roasted Prime Rib
Assorted Fresh Vegetables and Potatoes
International Cheese Display
Fresh Fruit
Extensive Pastry Buffet
Coffee or Tea

Children's Buffet available for kids 12 and under

Adults - $31.00   *Children - $15.50*
Seating times are at 10:30 a.m. and 1:00 p.m.
Reservations are recommended.

160 East Pearson Street • Chicago, Illinois 60611-2124 • USA • Telephone (312) 266-1000
Hotel FAX (312) 266-1194 • Executive Office FAX (312) 266-9501

FIGURE 12.7 Brunch menu. Courtesy of The Ritz-Carlton, Chicago, Illinois.

foods may be offered, but since most diners have little time, the menu must allow them to eat quickly, have time for scheduled events, and get back to work. However, if the luncheon is to last for a longer period and features an important speaker, the group may want to have more luxury-type food. The foods should fit the occasion and the particular group.

Afternoon Menus

Many operations have little or no business in the afternoon following lunch. To bring in business, a different menu can be designed with specials to catch the afternoon and shopper trade. These menus should appeal to people most likely to eat out in the late afternoon—retirees, shoppers, artists, students, and homemakers. The foods should be snack-type with considerable occasion appeal, different from the usual menu. Thus, after the lunch-hour rush, an operation might put on a special menu with small sandwiches, desserts, fountain items, fruit plates, cookies, and pastries.

Many people can be induced to eat dinner early if a lower-priced menu is offered. A variety of items should be included; along with one or two light desserts.

Dinner Menus

The dinner menu usually has more specialty items than others and attempts to feature more occasion foods. The menu must be carefully directed to patrons.

The typical American meal plan of a main dish, potato, vegetable, salad, dessert, and beverage is probably most appropriate for the family market, though some variation of it is probably appropriate in all operations. (See Figures 12.8 and 12.9.)

This traditional family dinner is as well liked as it once was; however, families often cannot eat together, so meals like pizza or hamburgers might be in order. However, for the usual family dinner meal, a soup, fruit cocktail, juice, or other appetizer may be served as a first course. Regular meats such as steaks or roasts are popular but other kinds of meat, chicken and other poultry, fish and shellfish should also appear. Interesting casserole dishes, Italian pastas, and some specialty foods may also be offered.

Families dining out very often know exactly what they want from any menu. Familiar items, such as mashed potatoes, macaroni and cheese, and chocolate cake are often appealing.

Menus are also needed that feature moderately priced items for those who want convenience foods rather than the luxury type. Some individuals who live in modest housing or patrons eating in a downtown cafeteria will not have much to spend but will

THESE RECIPES MEET THE FAT AND CHOLESTEROL GUIDELINES OF THE AMERICAN HEART ASSOCIATION OF METROPOLITAN CHICAGO • QUALITY FIRST •

Soft Shell Crabs —From Virginia— May 15—Sept. 15	Royster With the Oyster —Oyster Hour: 4:30-6:30, Mon-Fri— Blue Crab Lounge	Private Party —Facilities Available— See Manager

WELCOME TO SHAW'S

Our commitment to quality is a source of pride at Shaw's. We have exact standards of freshness and we work with oyster growers, distributors and fishermen to insure that we receive the highest quality in seafood.

PROPRIETORS

Kevin Brown   Chef Yves Roubaud   Stephen LaHaie

REGIONAL OYSTERS

–See Daily Specials for Today's Selections–

Eastern (Crassostrea virginica)

Long Island Sound	Connecticut
Malpeque	Prince Edward Island, Canada
Caraquet	New Brunswick, Canada
Chincoteague Salts	Virginia
Cotuit	Massachusetts
Wellfleet	Massachusetts
Pemaquid	Maine

European Flats (Ostrea edulis)

Wescott Flats	Washington
Casco Bay Flats	Maine

Olympia (Ostrea lurida)

Grown in the Bays of Southern Puget Sound

Kumamoto (Crassostrea gigas var. kumomoto)

Grown in Northern California & Washington

Pacific (Crassostrea gigas)

Totten Inlet	Washington
Malaspina	British Columbia
Fanny Bay	Vancouver Island, Canada
Hamma Hamma	Washington
Hog Island	California
Pearl Point	Oregon
Rock Point	Washington
Shoalwater Bay	Washington
Coromandel	New Zealand
Tomales Bay	California
Wescott Bay Selects	Washington
Crescent Beach	Washington

Chilean (Ostrea chilensis)

Chiloé	Chile

OYSTER ACCOMPANIMENTS

Oyster-Friendly Wines

A selection of dry, crisp, clean-finishing wines are available by the glass or bottle to complement Shaw's oysters.

Or try a malty beer from our list below.

BEER

–DRAUGHT–

Anchor Steam	Guiness Stout
Bass Ale	Harp
Samuel Adams	

–BOTTLED–

Amstel Light	Miller Lite
Heineken	Pike's Place Ale
Michelob Dry	Point Special
Miller Genuine Draft	Budweiser
Samuel Smith's Nut Brown Ale	Beck's
Clausthaler—non-alcohol	Rolling Rock

BOTTLED WATER

La Croix	Evian
Canadian Spring Berry	

FRESH OYSTERS

On The Half Shell	6.95 ♥
Daily Regional Varieties, Shucked to Order	
Oysters Rockefeller	6.95
Pan Fried Pacific Oysters	6.95

SEAFOOD APPETIZER PLATTERS

–serves two–

Cold Combo—Shrimp, Oysters, Clams, & Blue Crab Fingers	14.90
Hot Combo—Mini Crab Cakes, Popcorn Shrimp, & Calamari	12.90
(additional servings available)	

COLD APPETIZERS

Blue Crab Fingers	4.95
Topneck Clams, Half-Dozen	5.95
Charred Sashimi Tuna	6.95
Shrimp Cocktail (in shell)	7.95

HOT APPETIZERS

Escargot, Garlic Butter	5.95
Steamed Blue Mussels	4.95 ♥
French Fried Calamari	4.95
Popcorn Shrimp	6.95
Baked Clams Casino	6.50
Shaw's Crab Cake	7.50

SOUPS

	CUP/BOWL
Seafood Gumbo	2.50/3.95
New England Clam Chowder	2.50/3.95
Soup of the Day	2.50/3.95

SALADS

Shaw's Caesar Salad	3.95
Mixed Greens	2.95
Cole Slaw	1.95
Iceberg Wedge	2.50
Sliced Tomato & Onion	2.50
Boursin Cheese with Mixed Greens	3.95

Dressings: 1000 Island, Mustard Vinaigrette, Italian, Maytag Blue Cheese, & Ranch

VEGETABLES

–serves two–

Creamed Spinach	2.50
Baked Ratatouille en Casserole	2.95
Steamed Fresh Broccoli	2.50 ♥
Steamed Green Beans	2.25 ♥
Green Beans and Mushrooms	2.50
Steamed Carrots	2.25 ♥
Carrots and Mushrooms	2.50

POTATOES & RICE

–serves two–

Hashed Browns	2.50
Hashed Browns w/Onions	2.95
Au Gratin Potatoes	2.95
Boiled Red Potatoes	1.95
Cajun Rice	1.95
Four Grain Wild Rice	1.95 ♥

–single–

Charred Baked Potato	2.50

Cigar and Pipe —Smoking— In Bar Only	All Major —Credit Cards— Accepted	Margarine —And Salt Substitute— On Request

FIGURE 12.8 Dinner menu. Courtesy of Shaw's Crab House, Chicago, Illinois.

*** * * ***

GRAND MENU DEGUSTATION 75.00

Smoked Maine Salmon with Petite Lobster-Tomato Salad & Chilled Smoked Salmon Broth
or
New York State Foie Gras & Oxtail Terrine with Arkansas Short Grain Rice Salad,
Sherry Wine Vinaigrette & Yellow Bell Pepper Juice

Grilled Hamachi with Peeky Toe Crab & Cardamom Infused Carrot Juice
or
Potato Wrapped Veal Sweetbreads with Soy-Bacon Vinaigrette & Shiitake Mushroom Essence

Hawaiian Spot Prawns & Hand-Harvested Sea Scallop with Spring Pea Shoots, Cashews
& Spicy Coconut Milk Broth
or
Belgian Endive, Frisee, Roasted Hazelnuts, Goat Cheese, Japanese Pears & Dried Mission Figs

Gulf of Maine Swordfish with Carmelized Walla Walla Shallots, Artichokes,
Olive Oil Poached Tomato, Basil & Black Olives
or
Duck Confit with Lamb's Tongue, Mushroom & Pig's Feet Ragout
& Szechwan Peppercorn Infused Reduction

Spicy Seared Yellow Fin Tuna with Morel Mushrooms, Black-Eyed Peas, Sweet Peas,
Celery Root Coulis & Veal Stock Reduction
or
Texas Baby Antelope Saddle with Rosemary Polenta, Ratatouille
& Thyme Reduction

* * *

This Evening's Progression of Desserts

FIGURE 12.9 Dinner menu. Courtesy of Charlie Trotter's, Chicago, Illinois.

want enough to eat. Some operations may find this type of trade a satisfactory source of revenue.

In attempting to develop memorable dinner menus, it is important to remember that service and decor are as essential as food. Complete follow-through on *all* details of an idea is a requirement. Too many menus attempt to create a food *atmosphere* on the menu only to have the rest of the performance a dismal failure, or vice versa. The operation must be constantly watched to see that there is complete follow-through. If a menu features Mexican, Greek, or other ethnic foods, the cuisine should be absolutely authentic. Research is necessary to verify authenticity, yet modification may also be needed to suit the nonauthentic palate. A very hot Indian curry served with Bombay duck and all the side dishes may be delicious to one who knows this food, but it could be a disappointment to one who isn't familiar with it.

Foreign foods have become popular and help give menu interest and variety. If offered, make them correctly with high quality ingredients. An important consideration in featuring many foreign foods is that they usually do not require the most expensive ingredients. Thus, they may be more profitable to serve than some American foods. The fact that these foods are not normally a part of our diet makes them good occasion foods. Novelty can also be achieved by unique service.

It is becoming more and more common on table d'hôte meals to omit appetizers and desserts and have these selected from a special menu. Figure 12.10 shows an attractive dessert menu. Furthermore, many operations today serve the salad as a first course, making it fairly substantial, and then omit the salad with the main course.

Formal Dinner Menus

Very few formal dinners are served today that follow the traditional French style of three settings with a progression of courses for each setting. People now find it difficult to eat this much food. Even a more simple formal meal can be exhausting unless properly planned. Portions for a formal dinner should be adequate but restrained. The food should be selected to give a progression of flavor sensations, avoiding any heavy sweetness until the very end of the meal to "silence the appetite."

A formal meal should give time for the guests to appreciate the food and service and to converse. The most formal meal today usually does not have more than eight courses.

The first course of a formal meal can be oysters or clams on the half shell, a seafood cocktail of some type, a canapé, or fruit, such as melon or mango. Some may omit this course if cocktails and appetizers are served before the meal. Soup is the next course, and this should be quite light, such as a consommé, bouillon, or a light cream soup.

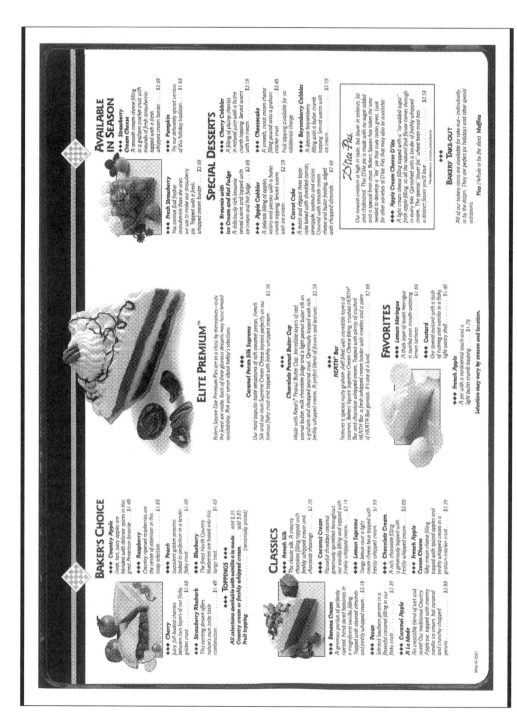

FIGURE 12.10 Dessert menu. Courtesy of Bakers Square Restaurants and Pie Shops, Matteson, Illinois.

On all courses, garnishes are important and may be the only item accompanying the food.

The fish course comes after the soup, followed by the poultry course. The roast or joint, with perhaps potato and vegetable, is the next course. The salad course is next, followed in turn by cheese and then the dessert course. In some formal meals an ice cream or sherbet (sorbet) may come right after the roast course. The meal may end with demitasse coffee, nuts, mints, and bonbons.

Wines and alcoholic beverages for this type of meal should be selected carefully. Cocktails made from spirits, such as martinis, manhattans, or scotch and sodas, blunt the taste buds. Apertif wines, such as Dubonnet or cocktail sherry, are usually better before meals. Nevertheless, many Americans prefer the former. The trend, however, is toward moderation. Sweet wines and drinks should be avoided, except at the end of the meal, where a sweet dessert wine may be served with the dessert or sweet liqueurs after the dessert. A wine such as dry white may be served with the first course and with the soup. A light, white dinner wine would typically follow with the fish and poultry. A full-bodied dry red wine is served with the main course. A very good dry red wine may be served with the cheese, followed by a light-bodied but refreshing sweet dessert wine. The tendency in formal meals is to limit the number of wines. As few as one or two wines may suffice.

A properly planned formal dinner should compare with a symphony or concerto. It should have different movements and themes in the foods and wines. The progression of themes should lead to a taste climax and then gradually recede in intensity as other foods follow. There are places in the meal where the foods should be modest, quiet, and subtle in flavor; and in other places the food and flavor should stand out in a loud crescendo. Each course should lead into the next, and the meal should be a continuity of related taste themes just as a good symphony has continuity of musical themes.

Lightness and delicacy should be the motif of the appetizer and soup. Both should be refreshing and somewhat zestful to whet the appetite. They should be a good introduction to the foods to follow. The fish course should be bland but sufficiently pronounced to give contrast with the courses that preceded it and with the more pronounced flavors in the poultry course that follows. The poultry course should be light and delicate. The roast can be a rod meat, and should be the peak of the meal. At this point, flavor values are more pronounced, but excessive richness, sweetness, or sharpness in flavor should be avoided.

The salad should be a relieving course. Its cool crispness and slightly tart flavor should be delicate and possess a distinctive quality, giving a respite from the heavier foods that have gone before. The Italians say the salad should "clean the palate and the teeth." Any oiliness or sharpness in flavor should be avoided. The tangy cheese course that may follow the salad should renew the jaded palate and set it properly for the concluding course—a sweet dessert. This should be light and delicate, and not too sweet. It should end the meal with some finality, but not on a heavy note.

The most formal meals are served without bread or butter. Salt and pepper are not on the table. In fine commercial dining rooms, bread and butter, salt and pepper, and ashtrays are often found on the table.

A less formal dinner may have only three to five courses. The first course can be a fruit cup, juice or seafood cocktail, melon, canapé, oysters, or clams, followed by a soup and then a fish course. Or there can be either an appetizer or a soup and no fish course. The main course is next, accompanied by a potato and vegetable and, often, a salad. A dessert follows. Rolls and butter are usually also served.

Evening Menus

Some places operate after dinner hours. This may be to catch those who are out at the theater, an athletic event, or some other attraction, or it may be that there is a good trade potential for night workers. Some operations may attempt to attract business with entertainment that may either be quite elaborate, such as a night club, or very basic such, as a juke box for dancing. Still others may be simple places that function more as social gathering places with limited menus. It is possible to combine night customers, with separate areas for those who want entertainment and those who only want food.

Many operations remove specials and table d'hôte dinners from the a la carte dinner menu and offer customers a selection of a la carte items on their evening menu. Some may "pull" the whole dinner menu and have another completely different evening menu emphasizing snack foods, such as hamburgers, milk shakes, pies, cakes, and fountain goods. Such a menu may be desirable to reduce the need for a large kitchen staff after dinner hours. In this case, most menu items are available up front where servers can get them. If an operation is near a theater or similar entertainment area, the menu should emphasize good desserts, specialty coffees, and perhaps light supper items.

A good analysis of the desires of the market and its potential should be made if one hopes to draw a trade that seeks entertainment as well as food and drink. The menu should then be designed to meet the demand for the appropriate kind of entertainment rather than attempting to create a demand. The menu should be supported by proper emphasis on props, such as servers' uniforms, dishes and glassware, table decorations (candles, vases, flowers), lighting, and entertainment.

Special Occasion Menus

Quite a few hotels, clubs, and restaurants are able to develop a substantial party and catering business. This can be profitable and considerably supplement overall income.

The special occasion menu must be planned in detail to suit the occasion. While there is a general pattern to events such as wedding receptions, buffet dinners, and cocktail parties, each should be characterized by its own special arrangements. Thus, some menus may have to be planned for a special occasion and not used again. Some events will be quite formal, while others will be very informal. The menu should carry out the spirit of the theme. Creativity and ingenuity are required. Color, decor, props, foods, and service should be combined to give correctness and novelty to the occasion.

A menu for a party should be carefully analyzed to see that it does not make excessive demands on servers, dishware, or equipment. The foods selected should be easy and fast to serve and be kinds that can stand delays in service and stay sanitary and of good quality. They should be suitable for the occasion.

Party Menus

A wide variety of parties may be served by a catering department. These may be as simple as punch for a dance or coffee for a meeting. Others may provide a simple dessert before a bridge game or a fashion show. Others may cater formal luncheons or dinners.

Party menus will vary according to the event, and thus there may be constant change. However, some operations pre-plan menus and have these available for clients when they call to discuss an event. This helps to standardize production and allows an accurate calculation of costs. If a special menu has to be developed, the individual planning the menu and establishing the price should know the complete costs so that an adequate price can be charged.

Most operations doing party business use standard printed forms for details. (See Figure 12.11.) The number of copies distributed may vary, and each operation will have its own special procedure. It is important that all departments affected have a coordinated plan and that the system work smoothly. There are many details in planning parties. Party business can be profitable, but it is very demanding of personnel. Good use of forms and procedures and a smooth, efficient system can do much to add to profits and make party business less demanding.

Some food services have catering departments that do considerable amounts of business and set up a sales office to handle booking arrangements. They often have pre-planned menus with prices for patrons. It may be necessary to quote a tentative price at first and then later, when menu and service costs can be more accurately determined, set

BANQUET PROSPECTUS

Name of Organization

Address   Phone

Nature of Function

Day   Date   Time

Room   Rent

Name of Engager

Address   Phone

Responsibility of Party

Price per Person $ Gratuity: Minimum Number Guaranteed: Maximum Attendance:

FOOD MENU

BEVERAGE MENU

Cash: Charge:

Corkage:

Room for Bar:

Open: Close:

Bartender Charge:

Minimum Charges (4 hrs):

Types of Beverages:

Staff:

Table Arrangements: Set Up By: Head:

Flowers: Centerpiece: Checking:

Ticket Table: Blackboard: P.A.:   Stage:

Lectern: Screen: Piano: Music:

Cigars: Cigarettes: Platform: Dance Floor:

REMARKS: _____   **COMMENTS BY PARTY:** _____

Copies to:

☐ Customer   ☐ Chef   **Date Booked:** _____

☐ Catering Office   ☐ Houseman

☐ Maitre D'   ☐ Bar

☐ Accounting   BY _____

FIGURE 12.11 Typical catering form.

Dinner Menu

Passed Hors d'oeuvres

Fillet of Beef Bruschetta
Slice tenderloin of beef on a garlic crouton
with a remoulade sauce and a mixed pepper garnish

Endive Spear with Gorgonzola and Walnuts

Artichoke Strudel

*

First Course

Fresh Buffalo Mozzarella
with
Red and Yellow Beefsteak Tomatoes
Served on a bed of raddichio and boston lettuces
with a pesto vinaigrette

Entree

Broiled Whitefish with Lemon Chive Butter
Garnished with fresh chive and lemon zest

Fresh Steamed Asparagus
With a balsamic cherry tomato relish

Three Grain Pilaf
Wild rice, brown rice and bulgar cooked in a
rich chicken stock and seasoned with fresh herbs

Dinner Rolls
Thyme and onion muffins
Au Pain French dinner rolls
Served with butter

*

Dessert

Feuillite of Fresh Mixed Berries
with
Champagne Sabayon

Coffee Service
Regular and decaffeinated Caravelli coffees,
tea, cream, sugar and sweet & low

27.00 per person

FIGURE 12.12 Catering menu. Courtesy of Blue Plate Catering, Chicago, Illinois.

a firmer price. Catering menus can be developed for breakfasts, continental breakfasts, buffet breakfasts, luncheons, luncheon buffets, dinners, buffet dinners, receptions, holiday celebrations, cocktail parties, birthdays, weddings, liquor (spirits), and wine lists; prices can be established for shows and for entertainment. Figure 12.12 is an example of a catering menu selected for a dinner party held in someone's honor.

Tea Menus

There are both high and low teas. A low tea might be simply tea, served perhaps with lemon, milk, and sugar. A more elaborate tea would add a simple dessert item or cookies, mints, bonbons, and nuts; a still more elaborate one might add a frozen dessert and fancy sandwiches. Often a tea is served from a table with the beverages served by an honored host or hostess at one end of the table and coffee served at the opposite end by another honored host. These hosts are usually changed at set times. The placement of the tea table should be considered carefully because traffic can be a problem.

A high tea is somewhat like a meal. It is served late in the afternoon and is more substantial than a low tea. It is often used as a light meal between lunch and a late evening dinner. It is most frequently served in homes, and in some luxury hotels.

A tea may be used as a reception event for a speaker, followed by a talk in a room nearby. It can also be used to honor individuals or an event.

Reception Menus

Receptions resemble teas but may feature alcoholic beverages. A wedding reception usually has a punch made of fruit juice, brandy, and champagne; although almost any other drink combination would be satisfactory. At some receptions, cocktails with canapés, hors d'oeuvres, and other tangy foods may be served. These may be picked up by guests at a buffet or they may be passed among guests. When passed, the service is called "flying service," from the Russians who originated the service and called the trays of foods carried around "flying platters."

It is usual to estimate that each guest will eat two to eight pieces of food at such gatherings. The type of function, its length of time, and the variety of food offered will affect the amount of food required. Bowls of dips and platters of crisp foods, such as pickles, olives, or crudites, that are easy to replenish may assist in giving flexibility. Some operations plan a runout time and toward the end of service leave only a few foods remaining. The number of drinks served may also vary. Men usually consume, on the average, two to four alcoholic drinks per hour of service, and women one to three. Again, the occasion decides but on the average the number consumed is falling. Light wines, beer and non-alcoholic drinks are to be included.

Since the preparation of a large quantity of fancy canapés, sandwiches, and hors d'oeuvres can require much labor, ways should be found to reduce preparation time by purchasing items such as tiny pre-baled puff shells or other items that can be filled and made to look as if they were prepared on the premises. A tray of canapés or fancy sandwiches need not have every item on it highly decorated. If a few fancy ones are properly spaced among plain ones that take little labor to produce, the effect is still appealing.

Buffet Menus

Originality in buffet menu items should be sought; too many foodservice operations today provide little in the way of novelty. They may use kidney bean and onion salad, picked corn, whole or sliced pickled beets, or fruit gelatin salads made with canned fruit cocktail. This may be fine if other, more unique foods are served with them. Otherwise, patrons may find the buffet resembles a home-style potluck buffet.

The effectiveness of a buffet depends almost wholly on the originality and presentation of the food. If these are not emphasized, the purpose is lost. Foods that look tired or that are excessively garnished, off-color, or messy will not satisfy guests.

The number of foods to put on a buffet can differ with the occasion and the meal. For a simple meal, only a few items may be used along with rolls and butter, beverage, and dessert. A dessert buffet may be only a dessert and beverage. A slightly more elaborate luncheon buffet may have four to six cold foods, including appetizers and salads, several hot entrees with a vegetable and perhaps a potato dish, a selection of bread or rolls and butter, and perhaps several desserts. There may be 20 or more cold foods on an elaborate buffet, eight or more hot entrees, a number of different hot vegetable dishes, potatoes, a variety of hot breads and rolls, and various cakes, puddings, pies, and other desserts.

The number of items and their presentation will dictate the degree of elaborateness. Certainly an overabundance of items is a mistake and can lead to waste. The purpose of a buffet is to give guests the feeling that they can help themselves as much as they desire, but it should not lead to waste.

Some buffets require particular foods. A *smorgasbord,* the Swedish buffet, must have pickled herring or other fish, rye bread, and *mysost* or *gjetost* cheese, in addition to many cold foods; hot foods, such as *lefse* and Swedish meat balls; and a dessert, such as a pancake in which lingonberries are rolled. Cold foods are eaten before the hot ones. Dessert is served last with coffee. A true smorgasbord usually includes *aquavit* or *schnapps* (Swedish liquers) served in tiny glasses.

A Russian buffet should have caviar served from the table center, either from a beautiful glass bowl or from a bowl made of ice. Dark rye bread and butter must accompany it. The other foods should be typical of a buffet.

A Russian buffet includes small glasses of vodka. The use of buffets for breakfast has been mentioned previously.

MENUS FOR PATRONS' SPECIAL NEEDS

The menu planner must focus closely on the patrons to be served. Special menus may have to be developed to meet the special tastes of specific groups. Restaurants and institutions that serve the elderly or the very young may want to consider how to meet the desires and needs of these groups, either with the regular menu or by developing a new one.

Children's and Teenagers' Menus

Menus for young people are generally limited because youngsters usually lack food experience and tend to stick to a small number of food selections that they know and like.

Small children tend to have smaller appetites. While the food selections for children should vary, the portions should be small. Teenagers often have hearty appetites and usually are hard to fill up. Therefore, the teenage menu should be filling as well as appealing.

A highly visible, entertaining menu appeals to children and makes them feel they are getting special attention. Colorful menus, perhaps with reproductions of animals or cartoon characters (you may need permission for the latter) will appeal to young children. Figure 12.13 shows an appealing children's menu. If the menu is planned to entertain and keep the child busy working at some puzzle, drawing, or reading a cartoon, the child will be quiet and not annoy other guests.

Offering children's food on the regular menu is often not as successful as having a special menu. A special menu is advantageous to the operation as well as to the child. That is because with a regular menu parents may be tempted to share their own food with children and not order anything special for them. If a moderately priced child's menu is available, it leads to the child selecting his or her own food, and the total size of the guest check is increased.

The child's menu should limit selections to favorites, because confusion results if too many items are offered. The wording should be simple and straightforward. Following through on a theme is a good way to introduce interest. A menu that gives the child something to do that can be taken home—or an operation that gives a child a gift—is a big help in winning young customers. Some operations give a child something that is one of a group of items, such as a cartoon glass. Then the child wants to return to the operation to complete the set.

The selections offered on a children's or teenagers' menu should also be set for the geographic location and/or ethnic base. Some Mexican foods that are successful on a Texas menu may not be successful on a Wisconsin menu. A menu offering Cuban and South American items that is popular in Miami, Florida, might not be popular in Portland, Oregon. It is a case of suiting a menu to patrons, something covered more thoroughly in Chapter 13.

Senior Citizens' Menus

Many operations can sell successfully to senior citizens, especially by offering discounts during slow periods and using the menu as a marketing tool. Senior citizens generally eat less than younger adults, so food portions for menu items should be smaller. In some areas seniors may be on fixed incomes so, for the most part, prices should be moderate.

FIGURE 12.13 Children's menu. Courtesy of Red Lobster, Orlando, Florida.

369

A few seniors will not chew well and soft foods should be included on the menu. In a recent Gallup survey of a group over 65, 20 percent were on some sort of diet. Thus, a menu that offers items that are low-fat, low-calories, and low-sodium will likely satisfy most senior dietary needs.

It is helpful to know something about the background of seniors who patronize an operation. For example, if they are of Germanic stock from a rural area in Wisconsin, they will probably enjoy things like chicken, pork, liver, sausage, sauerkraut, and other similar items. Knowing customers is the first and most important step to satisfying them.

MENUS FOR NONCOMMERCIAL AND SEMICOMMERCIAL ESTABLISHMENTS

Noncommercial and semicommercial operations will have menu requirements that differ from those prepared for full commercial operations. If the operation is institutional, and the patrons are a "captive market," the menu must be evaluated, based on the fact that most patrons will eat two or more meals a day at the facility. This means that nutritional needs of patrons must be considered along with their preferences.

Institutional Menus

Commercial operations are in business to make a profit. Institutions usually are not. They often operate from a budget that indicates the limits of expenditures, and they try to keep within these limits. *Breaking even*—bringing in enough sales to cover costs—is often the goal of an institutional operation. Most institutional food services work with a cost per meal allowance or a per-day allowance. A hospital will have a cost per bed per day. Hospital budgets for operating the entire department usually take about 12 percent of the hospital's expenditures.

Many institutional menus do not have to merchandise or sell to the extent that commercial operations do, since they have a built-in clientele. Many patrons in institutions do not see the menu, since it is written for the foodservice staff only. Some menus will not have a selection. Others will offer slight variations in the basic menu, such as a choice of dessert or beverage. Institutional operations that have to compete with commercial establishments may have to do some merchandising. In any event, dull menus are not a necessary characteristic of institutional operations.

Nutritional considerations about the food served are important; however, these will depend on whether patrons must get all their meals from the operation or eat there only occasionally. If they dine there less often and have other options, the nutritional consideration becomes less important, although still a factor. Certain patrons will be on special diets, and this will have to be considered.

Whether patrons eat in the institutional food service all the time or just sporadically, it is important to have good variety in the foods and avoid too much repetition. Long-term cycle menus are best for institutional operations. To break monotony there should be occasional meals that feature such themes as "Night in Venice" "Circus Days," "Old-fashioned Picnic," and so forth. Such menu breaks build interest, and patrons look ahead to such events. It is also important to occasionally break the meal plan and offer items different from the typical groups of foods served.

Most institutional menus are built around the three-meal-a-day plan. Breakfast is a fruit or juice, cereal and milk, a main dish such as eggs and/or meat, hot cakes, Danish pastry, toast or bread with margarine or butter, jelly or jam, and beverage. Lunch may include an entree, vegetable, salad, bread and butter, beverage, and dessert, or be smaller, including only a soup, beverage, and dessert. The institutional dinner may have a first course, such as an appetizer or soup, followed by a main dish with vegetable and potato or other starchy food, a salad, bread and butter or margarine, a dessert, and beverage. There is usually a 10-hour span between breakfast and dinner and a 14-hour fast between dinner and breakfast, unless a snack is served in the evening.

Some institutions find that a four-meal or five-meal plan is more suitable for patron needs. In the four-meal plan, a light continental breakfast is served at 7:00 AM. A substantial brunch follows it at 10:30 AM. At 3:30 PM, the main meal is served, followed by a light supper in the evening.

The five-meal plan is similar, with a continental breakfast at 7:00 AM, a brunch at 10:00 AM, a light snack about 12:30 or 1:00 PM with the main meal following between 3:00 and 4:00 PM. A light snack is served between 6:30 and 8:00 PM.

The four-meal or five-meal plans may reduce the labor required in the kitchen since the two big meals of the day are close enough together for one shift to prepare them. The other meals are light enough to be prepared by skeleton crews.

Some hospitals also find that the four-meal or five-meal plans permit patients to be gone for early morning examinations without losing out on a full meal. Even though the patient has had to miss breakfast, he or she can return to the room when one of the heavier meals of the day is being served.

Not all attempts to change to a four-meal or five-meal plan have been successful. The failures have usually been ones of planning. It is essential that the plan be thoroughly discussed with staff members who may be affected. The plan must have the complete support of the staff and solid backing from the administration. Communication of the plan and a complete discussion of the advantages and disadvantages should occur well in advance of implementation.

The change must consider every factor. For instance, a large state mental hospital changed to a five-meal plan and found that the patients felt they were not getting a breakfast because cereal was not a part of the first meal served. Adding either cold or hot cereals to the continental breakfast resulted in eliminating most complaints. In implementing these four-meal or five-meal plans in health institutions, the amount of sugar, flour, and fat should be watched. Sweet rolls or other breakfast pastries and evening snacks such as cookies, cakes, or other rich products cause undesirable weight gain and result in a failure on the part of patients to eat adequate amounts of other essential nutrients.

Planning Institutional Menus It is best in planning the institutional menu to work with a sheet large enough to hold the menu for an entire period, usually a month. Obviously, if the menu is to run for a long period, such as three months or more, this cannot be done. The proper headings should be set up with days, dates, and meals designated by columns and rows.

Most planners start with the main dishes for a meal, beginning first with dinners, then lunches, and then breakfasts. Balance and variety must be sought between days and also between the meals in a day. The frequency of the types of meats to use should be established, and a table can be set up for this. For instance, in a week, beef, pork or cured pork, poultry, fish or shellfish, and a casserole dish should each be served only once. These may be varied with an occasional selection of variety meat, veal, lamb, or sausage, and eggs, cheese, or other nonmeat dishes. Similarly, a table may be set up to indicate the frequency desired for various vegetables and other foods. A frequency table can also be set up for breakfast and lunch items.

After the main dishes are selected, the vegetables and potato, or other starch items, are added, followed by salads and dressings. Again, these must be balanced against the various foods used in a day and from day to day. Breads for each meal, desserts, and beverages can follow in that order. After adding these major items, the planner may select the breakfast fruits and cereals.

After this, the menu should be checked to see that balance has been maintained, nutritional needs met, cost restraints not exceeded, and other factors, such as balanced use of equipment and skill of labor, considered. If modified diets based on this general menu are required, they should be planned by someone trained in nutrition.

Different institutions will need different menus to suit varying operational requirements and meet patron demands. Regulatory agencies and other outside influences may also affect menu planning.

Health Facilities

A great number of menus will be required to meet the needs of various kinds of health facilities, such as hospitals, convalescent centers, nursing homes, and retirement homes, in which there may be available nursing and medical care.

Hospitals must provide food for staff, nurses, doctors, visitors, and catered events, as well as patient meals. Different menus will be required. Some of these menus will be typical of those in other food services. Generally, only patient menus will differ.

A hospital's dietary staff will first prepare a general or *house* menu. This includes what foods will be served at a particular meal on a specific day. The planner must keep in mind that from this menu will be prepared the various diets the hospital must serve. The dietitian goes through this general menu selecting items, changing them, or substituting others to meet dietary requirements. It takes an experienced, skilled person to modify a general diet. For instance, if a hospital put the items in Figure 12.14 on the general menu, the dietitian would modify the menu for the various patient diets.

In the diabetic diet the cranberry jelly, potatoes and dressing, and coconut cream pie might not be allowed. The sherbet and baked apple sweetened with an artificial sweetener probably would be allowed. Some carbohydrates would be allowed on this menu because even the most severe diabetic must still have some carbohydrates. The amount of calories allowed in the other choices on the diabetic menu would also be evaluated.

Roast Turkey, Cranberry Jelly

Vegetable Plate with Lamb Chop

Poached Salmon Fillet, Lemon Butter

Spaghetti Tettrazini

Mashed Potato   Baked Potato   Cornbread Dressing

Buttered Broccoli   Glazed Onions   Pureed Carrots

Sliced Tomato Salad, French Dressing

Soft Diced Vegetables in Aspic, Mayonnaise

Bran Muffin   Parkerhouse Roll

Butter

Sherbet   Coconut Cream Pie   Baked or Fresh Apple

FIGURE 12.14 Typical hospital menu.

For a low-fiber diet, the dietitian would eliminate the bran muffin and would scrutinize the fiber content in the vegetables and salad, substituting something else if it is considered too high. The baked potato would be served without the skin.

A diet of soft foods would probably require the cook to process the turkey or lamb chop by chopping them up. The salmon might still be served whole because it flakes easily. The broccoli might be finely chopped, and the tomatoes might not appear in a salad but as a cooked vegetable.

A low-fat diet would substitute plain lemon juice for the lemon butter. The butter on the broccoli and that added to the carrots would be omitted, and the glazed onions might not be permitted because they are usually glazed in butter or margarine. The butter would be omitted from the menu and a low-fat French or other dressing would be substituted for the regular French dressing. The sherbet or the apple would probably be allowed.

Menu planning for other health facilities may follow that of a typical hospital, with a general menu being written and modified diets taken from it. However, if only a few in the institution need special diets, the general menu would be written and special foods provided for those few. These foods could come largely from the general diet along with some specialized items.

Retirement homes may provide three meals a day, but some do not. Retirees may live in small apartments, which have facilities for preparing meals. Usually two meals a day are prepared there by the retirees. Thus, the foodservice facility may serve only one meal; this is often a heavy meal at noon or dinner. Patrons often need assistance during the meal. When full nursing care is given and the patron is confined to his or her rooms, a meal service might resemble that of a hospital or even hotel room service. In many retirement homes there will be a small coffee shop that serves breakfast, lunch, dinner, and snacks. Some operations may cater special events for retirees and their guests. Theme meals such as those mentioned earlier are also very popular with retirees and help to give a feeling of change and variety.

Business and Industrial Feeding Operations

A factory or other in-plant food service or one in an office building may have to have several kinds of dining services. There may be an executive dining room where top executives can meet at mealtime, bring guests, and have meetings over a meal. A staff dining service and sometimes a coffee shop are also used. Meals are usually paid for by the employees, but at a marked-down rate because of employer subsidization. Some serve the meal free. Metropolitan Life Insurance was one of the leaders in providing meals under dietary supervision to employees. If a cycle menu is used, it must cover a fairly long period because patrons probably eat there regularly. Worker dining units are often cafeterias. Snacks, sandwiches, salads, beverages, entrees, vegetables, potatoes, breads, and desserts are served. Some foodservice operations may offer breakfast and lunch, and a few executive dining rooms may even serve dinner for late evenings.

Normally, a menu board in the cafeteria announces offerings and prices. A set meal only may be served, and no menu is prepared for patrons. The full-service dining rooms usually have printed or written menus.

College and University Food Services

Many college residence halls have cafeterias. Self-bussing is usual. Some residence halls have foods priced individually or together in a table d'hôte menu, and the students pay for the foods chosen. Other students may be on a board plan where the meals for a period are paid for at a set rate. There is usually no credit for missed meals. Some have systems that combine the two features, with the student purchasing a block of tickets and using these occasionally. The menu must reflect such conditions.

Students are usually young and active and often want substantial meals. They will ignore foods that are not popular, and there may be a limited number of foods that they will eat. Some students still try to live on hamburgers, pizza, sodas, and french fries. Most residence hall units bow to student demands and serve what they want, hoping that in some way the student will get the nutrition needed. In other cases, students will demand natural or healthful foods, and these will have to be on the menu. Student wants vary and it is important to try to meet nutritional needs and student desires for quality, while providing variety as well.

Many colleges and universities operate student unions. A variety of foodservice operations can exist here, including snack shops, quick-service operations, seated-service dining rooms, and even banquet and catering facilities. The menus must satisfy the students' culinary desires and, usually, meet a limited budget. However, these now may even be a part of a fixed residence hall menu.

Many campuses also have a faculty club or dining center. Usually the club's big meal is lunch, but some offer dinners. A few may have bars. Prices must be modest, and often the college or university subsidizes the operation by providing space and equipment and even some occupancy costs. These clubs are usually operated like other clubs, and promotional programs will be a part of the merchandising system to bring in members. A faculty dining center differs from a club in that it does not require membership but is open to all faculty.

Elementary and Secondary School Food Services

Elementary schools usually are on the federal school lunch program. If so, a meal following the general meal pattern will be served. (Some schools may be only on the federal milk program.) Many school systems are now computerized and feed a large number of children. Others serve children from diverse racial, religious, and ethnic backgrounds and their menus should reflect this patronage.

In secondary schools, students either pay for foods as they get them or purchase tickets to exchange for meals. Because of the preferences of teenagers for special foods, secondary schools may often serve popular as well as nutritious foods. For a variety of reasons, many secondary schools try to prevent students from leaving the grounds to seek meals elsewhere, such as nearby quick-service outlets. Some have "closed campuses," and the students are forbidden to leave during lunch.

At some schools, such as military academies or religion supported schools, students live on campus. These are usually for older children and teenagers. The menus for these children should be planned with strong consideration given to their nutritional needs, since they will eat most of their meals in the school food service. The foods should also be those young people like. The service may be cafeteria- or even family-style, where foods are brought to the table in dishes and passed around. Some provide table service.

Many elementary, secondary, and private schools will have faculty dining rooms. For the most part, the foods will be somewhat similar to those served to students but modified for adult tastes. These foods may also be supplemented by other offerings, or there may be a completely different menu from that for the students.

Miscellaneous Institutional Food Services

There are many kinds of institutions, such as orphanages, prisons, associations, and religious and charitable groups, that provide food for people. Often these units function on a very limited budget. There may be no payment by the patrons. Either the government or the operation may provide the funds. In general, the provisions presented previously for institutional menu planning will apply.

THE FINAL STEPS IN MENU PLANNING

Once the various marketing, operational, and strategic decisions have been made, the pricing of the menu may begin. This is covered in more detail in Chapter 14. In addition, operators must continually research the popularity of menu items. Customer feedback is essential. The menu that was ideal for the establishment in the spring may be a real turn-off in the fall. Menu analysis is the key to heading off such marketing disasters. Financial considerations must also be considered.

SUMMARY

The task of planning and writing a menu is a daunting one. The most successful operations seem to have a menu that is "right"; that is, a menu that fits the operation. That does not happen by accident.

The menu should be thought of as the single most important document that defines the purpose, strategy, market, service, and theme of the operation. It should help "sell" items to patrons, and offer choices that will please a variety of tastes.

Before a menu can be planned, managers must develop a *mission statement* for the operation. This statement sums up the ultimate purpose of the operation, such as to provide high-quality, mid-priced food to a family market, or to provide three daily wholesome and nutritious meals to hospital patients. Next, managers set the objectives the operation is to attain. These can be in the form of profitability objectives, growth objectives, and market share objectives, among others. These objectives are used to develop *strategic plans* for the ongoing growth of the business.

A long-range financial plan is necessary before the menu can be developed. These long-range financial objectives are used to create budgets that must be followed by owners, managers, and supervisors.

One of the first decisions the menu planner (or menu planning team) must make is the number of menus an operation will use. This decision is based on the mission and scope of the operation.

Next the type or types of menus to be used is decided. Menus fall in the following categories:

- *A la carte* menus, which offer foods separately at separate prices
- *Table d'hôte* menus, which group several items together at a single price
- *Du jour* menus, which change daily
- *Limited* menus, which offer only a few selected items
- *Cycle* menus, which rotate after a set amount of time
- *California* menus, which offer items from all meal periods at all times of the day.

Menu planners must take into account the meal plan to be followed—one that will fit the needs of most patrons. Each meal—breakfast, brunch, lunch, dinner, and evening meals—and each type of operation has a typical meal plan that leads to a fitting menu organization for the operation. Each course of the meal plan, such as Entrees, is normally divided on the written menu into subcategories, such as Fish and Seafood, Meat, Poultry, Pasta, and Meatless Entrees.

It is essential that patrons be given a variety of choices within each category and subcategory on a menu. Menu planners should vary flavors, textures, and cooking preparations as much as possible to please guests. This variety will also dictate the number of items included on the menu.

Breakfast menus are often associated with low prices and low profits unless care is taken to offer unique items to guests. Planners should strike a balance between items that patrons expect to find and those that will spark interest. *Brunch menus* traditionally combine items from both the breakfast and lunch menus. *A lunch menu* should offer guests items that are light enough, priced moderately enough, and prepared quickly enough for people who have to get back to work. (Leisurely items should also be available for people out on special occasions.) *Afternoon menus* are intended to attract people with free afternoons—retired persons, shoppers, artists, students, and homemakers—and typically offer sandwiches, fountain items, fruit plates, and pastries and desserts.

Dinner menus normally require the most elaborate dishes, organization, and variety. Of all menus, dinner menus should reflect the atmosphere, decor, theme, and patronage of the operation. *Formal dinner menus* follow the traditional courses included in classic French menus. They are appropriate for very formal occasions and may follow some—not necessarily all—classic French customs. *Evening menus* cater primarily to theater-goers and others who are out for the evening. They typically offer such items as hamburgers, fountain items, and desserts. *Special occasion, party,* and *reception menus* must fit the occasion and specific clientele of the event. Every effort should be made to make the meal a very unique experience for guests. *Tea menus* can range from simply tea and a light snack to an elaborate combination of teas, sandwiches, and pastries.

Buffet menus rely as much on presentation and service as they do on the food served, and the food itself should be unique. A traditional Swedish *smorgasbord* must have pickled herring, rye bread, Swedish cheeses, meat balls, a traditional dessert, such as lingonberry pancakes, and small glasses of aquavit or schnapps. A traditional Russian buffet includes small glasses of vodka, caviar, and dark rye bread.

Special considerations must be taken into account when planning menus for specific groups of people. Children's menus should include a manageable number of simple items that will appeal to children. The menu itself should be fun for young diners. Menus for teenagers should include filling items that will appeal to teenage tastes. Many older guests will appreciate menu items that reflect dietary concerns, such as those low in fat, cholesterol, salt, and sugar. Softer textures and mild flavors might also be appropriate.

Menus for *institutional operations*—hospitals, nursing homes, schools, universities—require some special menu planning considerations. While commercial operations are in business to earn a profit, institutional food services often have as their goal to *break even,* or earn enough in sales to cover all costs. (If a not-for-profit organization does earn more than it spends, that money must go back in the business rather than being paid out to stockholders.) Since many institutional operations have a "captive" clientele, they are more driven to offer nutritious and varied menu choices.

Business and industrial food services must plan menus that will induce workers to stay on the premises rather than taking their business outside the office or factory. The types of food served should reflect the tastes of the patrons. College and university food services must please young people and their tastes. Elementary and secondary school food services, whether or not they follow the federal government's school lunch program guidelines, are obligated to offer children nutritious meals at costs their families can afford.

QUESTIONS

1. Collect five menus from different types of operations. Which type of menu is each one? Are items on each menu organized well? Are there too many items? Too few items?

2. For each menu type below, discuss the types of operations in which they are appropriate, the tone they set, and the markets to which they normally appeal.

    A la carte   Limited
    Table d'hôte   Cycle
    Du jour   California

3. What typical items do consumers expect to find on a breakfast menu? On a brunch menu? On a lunch menu? What specialty items have you seen on these menus that balanced expected items?

4. What are some of the important factors to consider when developing a children's menu? A menu for teenagers? A menu for senior citizens?

5. How is planning and developing an institutional menu different from developing a commercial menu? How is it similar? What are the *specific* considerations involved in developing menus for business and industry operations? For colleges and universities? For elementary and secondary schools?

CONSIDERATIONS AND LIMITS
IN MENU PLANNING*

Outline

Objectives

After reading this chapter, you should be able to:

- Identify cost constraints in menu planning and explain what considerations relate to cost.

- List labor constraints in menu planning and explain what considerations relate to labor.

- Identify food purchasing constraints in menu planning and explain what considerations relate to availability.

---

* Authored by Lendal H. Kotschevar and Marcel R. Escoffier.

- Identify patron expectations and preferences and explain what considerations relate to variety, psychology, and health concerns in menu planning.

- Explain how Truth in Menu Standards relates to menu planning and patron expectations

INTRODUCTION

The creator of a new menu almost never has a free hand in developing the menu. Constraints are placed on the development from various constituencies, including customers, owners, investors, lenders, suppliers, employees, and regulators. Like a politician, the menu planner must be able to address the concerns of these constituencies while not seeming to pander to one group at the expense of another. Thus, the menu developer must take into account the financial constraints placed on the foodservice facility (like limits to equipment purchasing, or required minimum net return on investment) from the lenders and investors while attempting to maximize menu choice and variety within the limits of current employee abilities and training. This balancing act is a fine art, and may be what distinguishes the truly great foodservice operator from the merely competent.

Let us look at the various forces that tug at the creator of a new menu. These forces, when in balance, pull the foodservice operation down the road to success. When one or more factors are neglected or slighted, the disharmony created is as annoying as one singer in a chorus who sings flat. A menu must be in tune with all of the constraining factors.

What is the perfect balance? Figure 13.1 shows the forces that effect the menu and that, when in harmony, create exceptional menus.

PHYSICAL FACTORS

The menu must be written to fit the physical operation; that is, the facilities must be capable of supporting production of menu items of the right quality and quantity, and the menu must not be written so as to allow some of the facility to be overburdened while other parts are underused. A menu featuring a dinner of baked ham and escalloped potatoes, candied squash, baked tomatoes, corn bread, and chocolate cake may sound appealing, but it places too much demand on the ovens and underutilizes other

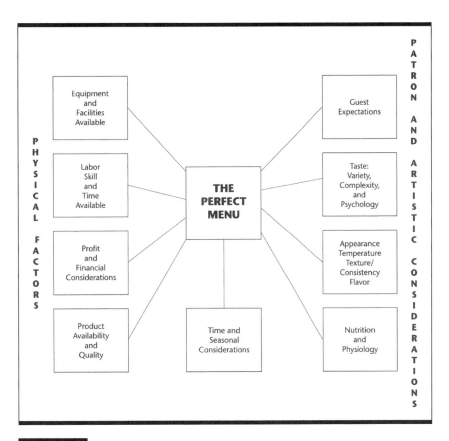

FIGURE 13.1 Considerations in designing a menu.

equipment. This sample menu also places too much demand on too few personnel; perhaps the fry cook may wish to take this night off. Such failure to take into account physical factors may result in service breakdowns (such as making patrons wait while the over-busy oven has room enough to cook the meal).

Equipment and Facilities Available

Clearly, some thought must be given to what equipment is on hand (or could be reasonably purchased). Menus must be planned to suit equipment capacity. A 20-gallon steam kettle is capable of producing only 16 gallons of soup. Similarly, a 24-by-36-inch griddle can produce a maximum of 80 orders of pancakes (at three cakes per order) in an hour. Planning for production of 100 orders per hour plus bacon, ham, hash browns, or eggs is another example of a failure to consider equipment limitations.

The time required to process foods through the equipment must also be considered. Using the soup example from before, to plan for more soup than the capacity of the steam kettle (say, having a menu with only one soup item on it and more than 16 gallons of soup served per meal period) means the soup must be cooked in batches. If you are cooking two batches of soup, why not have two different soups on the menu? Similarly, one must consider whether there is even enough time to cook more than one batch of soup before the meal is to begin. Workbenches, mixers, ovens, sinks, cooking ranges, and other equipment can handle a limited amount of work in a given time. (Not to mention the cooks who can process food only so fast.) If a menu planner asks for more than the kitchen is able to produce, the whole system might collapse because employees cannot meet quotas if they have to wait for equipment to become available.

Storage capacity also must be considered when planning a menu. Refrigerated, frozen, and dry storage areas must accommodate foods both before and after preparation. A menu that overburdens storage abilities is disastrous for sanitation and cost reasons.

The menu planner can often alleviate equipment overload by getting input from production personnel. Planning or procedural changes regarding equipment use can often overcome perceived limitations. For instance, a seeming lack of griddle space may be overcome by cooking hashed brown potatoes in an oven and finishing smaller portions quickly on the griddle. A stew could be cooked in a roasting pan in the oven just as easily as on top of the stove. A change in the menu can avoid overload problems by using alternative cooking methods (a double bonus, as this also increases menu variety). This is a perfect example of why the menu planner must be familiar with cooking methods and preparations.

Naturally, equipment overload can often be reduced by buying preprepared, or *convenience* foods. Many planners find that they can considerably extend the production capacity of the kitchen by purchasing this kind of food. Convenience foods have the advantage of overcoming deficiencies in labor skills as well. But nothing is free, and convenience foods may add considerably to the operation's food cost and might not offer the appropriate quality.

Of course, the menu planner's equipment considerations do not end in the kitchen. Service equipment is also a consideration. A dining area can support service for only a given number of people; and many believe that the quality of service begins to deteriorate whenever the room nears capacity. The kind of service offered matters as well. Full continental service, complete with flambé dishes prepared tableside, requires more square feet per guest than would a cafeteria-style operation. A service with numerous courses or one that requires long periods of time between ordering and serving can greatly affect the capabilities of a given space to accommodate a given number of people. Too many menu selections can also slow down service, both because of the time it takes patrons to make up their minds, and the time it takes to serve them. Even the type of staffing—experienced servers versus part-timers, how they are organized (teams versus individual servers), and what levels of expertise they bring to the job—all have an effect. These considerations must be taken into account by the menu planner. Most importantly, the menu planner must maintain a certain continuity. The service area's decor must match the menu's theme.

LABOR CONSIDERATIONS

Labor constitutes one of the single largest expenses in most foodservice operations. Just as the equipment, furnishings and fixtures are assets, the smart modern manager will think of labor as a valuable asset. Just as we don't abuse the equipment by demanding more than the capacity of the griddle, so, too, we must be careful not to exceed the capabilities of our people. Unlike equipment, however, people have the capacity to learn. A good training program can teach new skills (and, hence, new capacities) to an employee. The menu planner must know what skills people possess in order to: 1) create a menu that requires labor skills currently possessed by the people; 2) not to create a menu which does not underuse employees' skills.

Training of labor can overcome the first problem, at some cost in terms of money and time. It takes time to train a person in new skills, and that time must be allocated from an already busy schedule. Also, training costs money. Even with on-the-job training the operator must realize that the person being trained and the person who is doing the training will work at a much slower pace during the training period than they may be expected to do once the trainee has been trained. By conducting a *skills inventory*—an assessment of the skills employees currently possess—the menu planner can create a menu that minimizes the amount of training and retraining required, and hence reduce the cost of implementing a new menu and the time it will take before the staff is operating as efficiently as possible under the new menu.

Underusing employees' conversely, will most likely lead to boredom. Studies have consistently shown that bored employees are less productive and more prone to absenteeism and turnover than are employees whose jobs provide challenges and are interesting. Similarly, many operations note that on slow nights they often have more problems and employees seem less coordinated than on busy evenings. The essential question is why pay a person who has more skills than you need? In simple economic terms, it makes sense for the menu planner to write a menu that uses as many skills as possible.

The menu planner must also consider the most effective use of employees' time. If equipment is unavailable or inefficient, valuable labor is wasted. The menu planner must seek to avoid bottle-necks in production when writing the menu.

Before a food item can be placed on a menu, its availability must be known. It makes little sense to have a dish listed on the menu that is unavailable. This only causes aggravation on the part of the patron and the server. Menus planned to last a long time must be careful to avoid too many seasonal items. One way of getting around this is to list a product group with the words "in season." For instance, breakfast menus usually offer some kind of fresh fruit, so the listing could be "melon, in season" or "seasonal melon." In the United States, many food items are available year-round. But even these items often have very different prices and quality at different times of year. Cost and, often, profitability are important considerations.

Often seasonal availability or price fluctuations can be overcome by volume purchasing. For instance, beef is often cheaper in early winter as the ranchers thin their herds. Buying large quantities and freezing them for later use will save money and ensure a constant supply of these items throughout the year. However, the costs of freezer storage, interest on the investment, and other elements must also be covered.

One subject whose importance cannot be overstated is that of food quality. Nearly every food item has a season in which its flavor, color, and texture are at their peak. Menu planners working in operations whose patrons are expecting a high level of quality must be especially sensitive to seasonal and quality constraints.

A menu should maximize revenue and minimize costs. Even in a nonprofit operation like a school or hospital there are cost and budgetary restraints that a menu planner must weigh when creating the perfect menu.

PATRON AND ARTISTIC CONSIDERATIONS

The menu must fulfill the patron's physiological, social, and psychological needs—especially when it comes to food. To plan a menu that adequately does this requires a considerable amount of knowledge and ability.

Guest Expectations

Often what a guest expects from a menu is unknown, even to the guest. Marketing experts use a variety of techniques to measure guest expectations. They may survey guests (or potential guests), either by telephone or through some written survey form. They may interview selected guests, either individually or in focus groups. However it is done, the goal is to encourage patrons to reveal their innermost thoughts and desires. This process is inexact, though it can help foodservice managers spot trends and have at least an idea of what patrons expect.

There is a very important difference between customers' *needs* and their *wants*. A need can be either a *physiological need*, such as the need for liquid when one is dehydrated, or a *perceived need*, such as a "need" to own a new television set. Perceived needs are quite strong, and many industries successfully cater to these needs. The wine industry, for instance, has experienced poor overall sales in recent years, yet premium wine sales remain strong. Perhaps some of this strength is due to a perceived need on the part of some wine drinkers to project a certain image. Wants are often expressed through impulse sales. A patron may be quite full after a large meal, but a dessert cart display may spark a desire within the patron to order something which surely cannot fulfill a physiological (and probably no perceived) need. The menu should try to satisfy both customer needs and customer wants.

In an attempt to satisfy their guests, foodservice operations have invoked ad campaigns saying everything from "You deserve a break today" to "Sometimes you have to break the rules." These slogans appeal to people's vague needs and expectations for satisfaction. When these needs and expectations cannot be satisfied or when they are in conflict—such as when a guest wants a hearty breakfast of eggs, bacon, toast smothered in butter, and milk, but also wants to reduce cholesterol, the menu planner must make the best effort possible toward meeting patrons needs, but of course, one can't satisfy all of them.

Variety and Psychological Factors

One theory of human behavior, called Hedonism, states that people try to maximize pleasure and minimize pain. Certainly maximizing pleasure as it relates to food means maximizing sensory input. This can be done by varying taste. Just as film makers have found that with each new thriller they must create more and more elaborate special effects, so too do menu planners find that they must have greater variety in their menus if they are to spark patrons' interest. There is a physiological reason for this: the brain reacts to new situations with much greater interest than to similar or routine situations. Just as we spend much time thinking about how to dance the latest step and almost no time thinking about how we place our feet when we walk down the street, we spend more time savoring and thinking about new taste combinations than we do when we are eating something common.

At the same time, too much variety, like too much ice cream, can cause dissatisfaction. Patrons must feel that there are items on the menu with which they are familiar, especially those who are slow to experiment. However, even these common items can be prepared exceptionally well or in an unusual way. The sizzle platter (a metal plate that is heated in a broiler and causes food to sizzle when served) offers little added taste to fairly common menu items, but the unusual presentation has been used for years to add variety to menus.

There are other ways that food can meet people's psychological needs. Hunger creates anxiety and restlessness that eating quiets. This property of food is often called its *satiety value*. Some people overeat when they are frustrated or disturbed. The satiety or peaceful feeling they get from food tends to quiet and soothe them.

College and university students often show distinct ties between frustration and food. Arriving at school in the fall, most students' spirits are high and they enjoy their new experiences. At this time, the food is considered good. (This is the time for the foodservice department to save money on its budget.) However, as the newness wears off, students miss home and loved ones more. The rigors of classes, assignments, term papers, and examinations begin to create problems and frustrations, and students begin to take out their frustrations on the food they are getting. Food riots can even result. It is no coincidence that troubles on campus rise with such pressures. Students don't realize that the food they once liked now is cause for great dissatifaction. (This is now the time to put the money previously saved on the budget into better food.)

The ties between food are even more apparent in prisons or reform institutions. For example, one prison riot was traced to the fact that the foodservice operation ran out of fried chicken for its weekend special and had to substitute another entree.

Many otherwise well-balanced adults have the same reactions. A business executive at a club may find the food totally unacceptable while others at the table enjoy it. The problem may be that frustrations at the office did more to ruin the person's appreciation of the food than the chef, the waiter, or anything else the club did.

Many times institutional food of much higher quality than individuals get at home is criticized for its quality because diners miss the home atmosphere, family, or friends. People learn to eat amid familiar surroundings. There is comfort and security in eating at home. This security and comfort is deeply ingrained and is tied to food. These feelings have been built up over a long period of time. Then, when an individual must eat elsewhere, away from his or her loved ones, there is insecurity, loneliness, and a desire for familiar things associated with food. So it is not always the food that is lacking in quality but environmental factors associated with it. An individual who is hospitalized and ill has little appetite anyway; under the stress of illness and the loss of familiar surroundings, it is not surprising that the hospital's food is often not liked.

Fact and fancy are frequently interwoven in our notions about food. Thus, some individuals eat certain foods because they feel they are beneficial. For instance, many feel they must eat the foods they associate with home in order to stay healthy. Others believe that raw oysters stimulate the sex drive, while saltpeter (potassium nitrate) depresses it.

Food can also have deep religious significance. Unleavened Passover bread (matzo) and kosher foods have deep religious meaning for Jews. Many Seventh Day Adventists and Roman Catholics will not eat meat at certain times. The fact that food and drink may have deep religious significance for some people should be respected.

Appearance, Temperature, Texture, and Consistency

People use senses other than their taste buds to evaluate food. It is often said that people "eat with their eyes," so visual factors may take precedence in creating appealing food. Certainly we often see a dish well before we can smell it. Smelling the aroma actually stimulates our appetite. Once we taste a food, its temperature, texture, and consistency help determine whether we like it. Since the menu planner's goal is to provide the patron with maximum pleasure, making the food appealing is a key part of creating a menu.

Appearance includes things like color, form, and texture (in its visual sense). Colors like red, red-orange, butter-yellow, pink, tan, light or clear green, white, or light brown are said to enhance the appeal of food while purple, violet, yellow-green, mustard yellow, gray, olive, and orange-yellow do not. Colors naturally associated with foods are appealing while colors associated with spoilage or unnaturalness are not. But the chef as artist does not stop here. Contrasts in acceptable colors should be sought. A myriad of bright color in a fruit cup or salad can heighten interest. Freshness in color (what an artist calls "vibrancy") is important, while unnatural combinations or too vivid colors should be avoided.

The form taken by the food is also important. There should be an interesting assortment of forms or shapes on a plate. Variations on a plate, like a slice of baked ham dressed with a small quantity of light brown raisin and shredded pineapple sauce, diced beets with a light gloss or sheen of butter and mashed sweet potatos graced with a small spray of parsley will achieve a difference in form and color that is agreeable. But too many mounds, cubes, balls, or similar shapes can cause loss of appeal, just as a picture that is "too busy" may distract from its central theme.

It is also important to have foods of different heights. While this may not be possible to do on one plate (or may cause one plate to become "busy"), it can be achieved by having different heights on different foods. Thus, a salad can be in a deep bowl, or a dessert can be in a parfait glass in order to give contrast to the relatively flat entree plate. Tall garnishes (provided they do not overpower the dish) can also be used to give height to a plate.

It is important that at least some of the foods served retain their natural shapes. Too much pureed or chopped food can resemble baby food. This does not mean that planners should avoid interesting or novel presentations, but they should be sure not to get carried away.

Texture, such as that of kiwi fruit or fork-split English muffins, provides pleasant visual experience. A glistening or shimmering food is often more appealing than a dull one. Butter or other oil coatings, aspics and jellies, and egg liaisons all provide a shimmer.

The creative use of garnishing provides important visual activity, and having different garnishes for different food items is one sign that the menu planner has really put thought into planning the perfect menu.

The *flavor* of food is distinguished primarily by its aroma or odor. While our taste buds can distinguish only four tastes—sweet, sour, bitter, and salt—our noses can distinguish among different aromas with incredible accuracy. The great chef Escoffier was said to have made omelets by stirring them with a fork on which a clove of garlic was impaled. The taste of garlic was so minor, yet so compelling that he single-handedly introduced this "forbidden vegetable" into haute cuisine.

Which flavors are pleasing or disagreeable differ with individuals. Naturally, we are more pleased with flavors with which we are familiar. The intensity of the flavor is also subject to individual preference. Intensity or "sharpness" may be an acquired taste. One may dislike sharp cheeses until one becomes used to the flavor, then one may cherish the flavor. There is a saying about olives that one must eat seven of them before one can appreciate their distinct flavor.

Our senses are sharpest around the ages of 20 to 25 and after this they slowly decline. Actually, there is a spot devoid of taste in the middle of the tongue which grows in size as we age, perhaps from the ingestion of hot or cold foods which kill off the nerve endings. Thus as people age, they may seek out more intensely flavored foods. Because the sweet- and salt-sensing nerves deteriorate first, older people lose their ability to taste these first.

An interesting phenomenon associated with flavor and taste is that some people cannot identify different food items by taste alone; they must see the food while tasting it. This may be why food displays engender such a favorable response from patrons. Certainly the display of foods at a buffet or in a salad bar is apt to stimulate the appetites of many patrons.

Another aspect of taste is the texture and consistency of food. Cooking food breaks down cell walls within the food, softening the texture. Unless carefully monitored, this can result in mushy, overcooked food. *Texture* is the resistance food gives to the crushing action of the jaw. Texture contrasts bring variety to the menu. A crisp salty cracker contrasts nicely with soups; chewy turkey meat is complemented by soft mashed potatoes.

Consistency refers to the surface texture of food. Okra can often have a slimy feel or consistency. Mashed potatoes made without enough liquid have a grainy, pasty consistency that is unappealing. Years ago we drastically overcooked vegetables; for instance, at the turn of the century, most recipes called for cooking asparagus for 40 minutes in water and baking soda. Most people today would find this most unappealing.

Finally, taste is a function of temperature. Serving cold foods cold and hot foods hot not only helps ensure sanitation and food quality, but also can give a pleasing temperature contrast to a meal. Mixing hot and cold foods in a meal provides further contrast and variety. A small cup of cold cucumber salad is an excellent contrast to a hot Indian curry. By serving hot foods hot and cold foods cold, we are not simply providing variety, we are showing the patron that we really care.

Time and Seasonal Considerations

A menu planner is very much constrained by the time of day that the menu will be for. A menu with traditional breakfast items on it will find little appeal at dinner time, except as a novelty. Others menu items are identified with certain times of the year. Others are associated with certain holidays, for example turkey and its fixings with Thanksgiving and ham with Easter. These items can be put on menus long before and after the specific date of the holiday. Some items are associated with the four seasons. A winter menu may have a very different selection of salads, for example, than might a summer menu. Also, we would expect fewer salad offerings in winter but more soup offerings, especially up north. In sunny Florida and California, we tend not to have this constraint. In fact, tourists tend to enjoy the surprise of seeing summer salads available in the dead of winter.

RATING FOOD PREFERENCES

Patron's food preferences must be considered when planning menus. People are influenced in their preferences by food habits acquired over a long period. Studies have been made of what foods individuals most prefer. These may be helpful to menu planners, but what people say they prefer is not always what they select from menus. The military asked what its personnel said they preferred and, then, in a further study, found that there were significant differences between what people said and what they actually selected. (See Figure 13.2.)

FOOD PREFERENCE QUESTIONNAIRE

Now I am going to ask you to rate the food you just ate. For each food, will you tell me if you liked it extremely, liked it very much, liked it moderately, liked it slightly, neither liked nor disliked it, disliked it slightly, disliked it moderately, disliked it very much, or disliked it extremely. This card has a list of these ratings. (Interviewer circle number.)

a. What main dish? _____ 1 2 3 4 5 6 7 8 9
   _____

b. Any other main dish? _____ 1 2 3 4 5 6 7 8 9
   _____

c. Vegetables? _____ 1 2 3 4 5 6 7 8 9
   _____

d. Drinks? _____ 1 2 3 4 5 6 7 8 9
   _____

e. Breads or cereals? _____ 1 2 3 4 5 6 7 8 9
   _____

f. Potatoes or starches? _____ 1 2 3 4 5 6 7 8 9
   _____

g. Salads? _____ 1 2 3 4 5 6 7 8 9
   _____

h. Soup? _____ 1 2 3 4 5 6 7 8 9
   _____

i. Desserts? _____ 1 2 3 4 5 6 7 8 9
   _____

For breakfast, ask only for main dishes, beverages, breads and cereals, and fruits.

Overall, how would you rate the meal you just ate, using the same scale? (Circle)

1   2   3   4   5   6   7   8   9

How did this meal compare with other Army meals you have had?

❑ Much better?   ❑ About the same?   ❑ Much worse?
❑ A little better?   ❑ A little worse?

Respondent's name _____   Number _____

Interviewer _____

FIGURE 13.2 Food preference questionnaire.

Source: United States Army.

The Taste Panel

Many operations use taste panels made up of their own personnel to taste various dishes—which might be placed on menus—to see if the item has good enough appeal to warrant featuring it. Often score sheets are used with the tasters rating factors such as flavor, consistency, temperature, color and others with a scale of 0 to 9. The individual scores are totalled to see how well each dish did. If there seems to be a problem, individual scores can be checked as well as the scores for specific factors to see why a dish might not score well. Often scores are interpreted as follows:

9: Like extremely

8: Like very much

7: Like moderately

6: Like slightly

5: Neither like nor dislike

4: Dislike slightly

3: Dislike moderately

2: Dislike very much

1: Dislike extremely

0: Discard

Not all individuals have a good sense of taste, so panel members should be selected carefully. Some lack good taste perception, lacking in the ability to distinguish subtle differences of one or more of the four basic tastes: sweet, sour, bitter, and salt. They might have poor flavor memories so they cannot carry over flavors to evaluate differences between two or more samples. Smokers often have taste problems. Alcohol dulls the taste, as can a cold, fatigue, and stress.

A young person may not be as good a judge as an older one, because flavor identification requires experience. Older people often have more experience, but may lose good taste perception.

Taste panels are best conducted at 11:30 AM and 4:30 PM, when people are apt to be most hungry. The room in which the taste panel works should be quiet and comfortable. Each judge should have a separate place in which to taste and score, unless a panel discussion is desired as the tasting goes on. A score sheet is shown in Figure 13.3.

Where only flavor, consistency, texture, and temperature are being evaluated, judges may be blindfolded and asked to judge a food without being able to see it. Often the elimination of appearance can make quite a difference in what a judge thinks of a food. The saying, "We eat with our eyes" is often proven in such a test, since a food

| ITEM: OLD FASHIONED RICE PUDDING | | Menu No. 512 |
| | | Scorer LHK |

Date _____

Characteristic to Score	Score (0 to 9)	Comments
Flavor	6	Little eggy; may have been baked at too high a temperature; either lower temperature or bake in pan of water.
Color	8	Good; also good sheen.
Texture	5	Slightly watery and some openness showing some syneresis. Rice soft and visible in grains.
Form	8	Good; soft, yet solid. Holds shape.
Temperature	7	Served at room temperature; try chilled or try slightly warm.
Consistency	6	Slightly tough perhaps because of too high a baking temperature or too much heat.
Total Score	40	

General Comments:

Cut down on baking temp. and bake in another pan of water; try reducing eggs & replacing with yolks to tighten up more and retard breakdown.

FIGURE 13.3 Scoring rice pudding.

found not too appealing when tasted and seen, may be quite acceptable when one does not see it. However, the opposite result can also be true.

Sometimes guests are given complimentary samples to see how well they like new items. Sampling is usually not as formal as a taste panel, and guests may simply be asked if they like it or not.

PATRON EXPECTATIONS

When most patrons look at a menu they conjure up specific images of what they expect. They should not be disappointed. If the menu item is accompanied with a short description, this can help in indicating what it is. Service personnel can also do much to indicate this.

The purpose of a menu is to communicate to the patron what is offered and, in commercial operations, the price. Menu terms should be simple, clear, and graphic, presenting an exact description of what the patron is to get. Patron expectations and what is served should coincide. The patron may misinterpret unclear terms. Or, the menu writer may mean one thing, but the patron may interpret it differently, especially if there are regional meanings applied to cooking terms or certain food flavors. There should be no ambiguity or confusion. Nor should descriptions be overstated.

If foreign words are used, the writer must be sure they are explained and understood by patrons. An explanation may be given in English below a term. *Keep the language simple* is a cardinal rule in menu planning.

Restaurant critic, John Rosson, has said that readers should be able to read menus quickly and understand what the terms mean. He discourages the use of ambiguous terms, such as "meats dressed with the chef's special sauce" or, for a pie, "(made from) a secret family recipe, handed down from generation to generation" saying that they seem phony and may even be insulting. Rosson advises the use of descriptions that clarify without being overdone. Terms that mean little or nothing to the average diner should not be used. Examples are items called "The Navajo Trail" and "The Ocean Blue." The first might involve beef and the other fish, but they could just as easily be a Mexican platter and a plate of tempura seaweed. The item's name should leave little doubt as to what will be served.

When a menu says the steaks come from prime beef, this means the beef grade is prime and not a lower grade. A bisque is a cream soup flavored with shellfish, yet one company producing canned soup named a product "tomato bisque," which is essentially false. Every menu planner should be sure of the terms used and what they mean when they are put on the menu. Too much license is frequently taken in the use of terms when writing menus. Selecting words to describe a food and give it glamor must be

---

SHRIMP GUMBO WITH RICE

A thick soup made of shrimp, tomatoes, okra, and other vegetables delicately seasoned with filé, a seasoning containing ground sassafras leaves, served over a mound of fluffy rice.

FIGURE 13.4 Menu description.

done carefully. To try to make a curry sound more exotic by calling it "Curry of Chicken, Bombay," without using the typical ingredients that a curry from Bombay, India, would have, is not encouraged. The right curry seasoning, accompaniments (especially the small, dried fish called Bombay duck), and other factors should be a part of the menu item.

Often menus present the name of the item in large letters and then give a description in smaller letters, as in Figure 13.4. Further clarification of the dish might also come from the server, who might explain that the rice is served as a mound in the middle of a large bowl containing the gumbo. (The method of item presentation will be discussed further in Chapter 15.)

Truth-in-Menu Standards

In some cities and states, *truth-in-menu*, or *accuracy-in-menu* laws govern menu descriptions. These laws are the results of governmental interest in menu irregularities, such as indicating one item on a menu and actually serving another. These irregularities are perceived as *misrepresentations*, which may benefit the operator at the expense of the consumer. Some of these laws are state interpretations of Food and Drug Administration standards. They act as guides to menu planners when it is time to write the menu.

The federal government's Pure Food, Drug, and Cosmetic Act of 1938 forbids the use of any pictorial or language description on a container that misrepresents the contents. No grade or other term can be used that is not representative of the product. For example, if a label says, "Georgia's choicest peaches" and the peaches are not graded at least Choice, the product is considered mislabeled because of the similiarity of the terms *choicest* and *Choice*. If any nutritional claim or nutrient value is given, the claimed value must be a part of the food. (For more on truth-in-menu guidelines, see Chapter 16.)

The federal government also has developed *standards of identity* that define what a food must be if a specific name is used for it. For example, food labeled as egg noodles *must* contain at least 3.25 percent dry egg solids. If the word *cheese* is used, the product must contain 51 percent or more of milk fat, based on the dry weight of the product. The term *cheese food* must be used for food that cannot meet this regulation. Standards of identity exist for all foods except those considered common, such as sugar.

Many states have adopted laws that clarify federal standards.

Menu Pricing

Menus must often be priced to meet patrons' expectations. Some patrons are very price conscious and a menu must meet their expectations of perceived value. Most patrons want to get the most for their money. The price paid should represent an adequate value in patrons' minds, and they will not be happy if they do not feel the value is there. However, menus must also be priced to cover costs in institutional operations and ensure profits for commercial ones. Since pricing is such an important subject, it will be discussed in depth in Chapter 14.

SUMMARY

The perfect menu is one that weighs the various constraining factors to menu design and ideally balances them. Many of the factors involved in menu planning seem contradictory. Like a good politician, the menu planner must produce a solution that keeps everyone happy. Patron needs and desires must be balanced against the needs of the owners to make money. Equipment constraints must be balanced against the need for a variety of choices.

QUESTIONS

1. What are some personnel considerations that will affect the success of a menu?

2. Why might a patron who orders a meal of broiled chicken, mashed potatoes and gravy, and a fruit cup in a restaurant not like the same meal when in a hospital?

3. Why is it important to know whether a menu can help satisfy the physiological, psychological, and social needs of patrons?

4. Look at an actual menu's items and descriptions. Are explanations adequate in predicting what will be served? How do items served together complement each other?

5. Name five important considerations discussed in this chapter that anyone planning a menu should give special emphasis. Why did you select these and not others?

14

MENU PRICING *

Outline

Objectives

After reading this chapter, you should be able to:

- Discuss several theories of menu pricing.

- Characterize the most common pricing techniques used in the foodservice industry.

---

* Authored by Lendal H. Kotschevar and Marcel R. Escoffier.

INTRODUCTION

After a menu is planned, each item on it has to be priced. There are a number of items to be considered in pricing, including the market, type of operation, and costs. The market is a major factor in pricing. Some patrons want only low prices; others seek moderate ones; some will be willing to pay higher prices. *Perception of value* will also vary from patron to patron; it is sometimes difficult to meet the desires of each group. Many operators use a "what the market will bear" approach to pricing. However, such a simplistic strategy may not meet operational needs and may drive patrons to competitors.

Prices must cover costs. This requirement is essential for both commercial operations and institutions that charge for meals but do not operate for profit. However, noncommercial operations may seek supplemental help from government or charitable contributions to balance losses. Commercial units do not have this insurance and will pay the consequences of faulty pricing.

Different kinds of operations use different markups. A *markup* is the difference between the cost of a product and its selling price. Some operations can use a low markup and depend on high volume to give an adequate income with which to operate and make a profit. Others will use a higher markup and require a lower volume to attain a profit. Others will use a higher markup and require a lower volume to attain a profit. Menu prices must not only cover the costs of food and labor but must often include other significant cost factors such as atmosphere, rent—especially in a prime location—and advertising.

VALUE PERCEPTION

Value perception, or what patrons think of the desirability of a product compared with its menu price, is an important factor in menu pricing, since prices largely influence patrons' thinking about the value of menu items. A high price may be associated with high quality and a low price with lower quality. Some people want to go to a place where prices are high. A sales person may take potential customers to a luxury restaurant to impress them with a lavish display and hope this will transfer into a sale. The menu must tie in with any attempt to present high prices and luxury by using proper menu wording and item presentation.

The way patrons perceive value is often manipulated favorably by getting buyers to think there is something special about a product that competing products do not have. We call this development of special, unique characteristics in a product *differentiation*. If a product can be favorably differentiated, buyers will want it rather than a competing product, and the seller has better control of the market and pricing. Through good advertising, buyers can be persuaded that it is better in some way than competitive products.

Food services can differentiate products and services in various ways. Often several special items on the menu can do the trick, such as a special, thick cut of roast beef, a marvelous dessert, or a unique cocktail. Location, atmosphere, and decor are also used to differentiate one operation from another. A smiling host or hostess to greet guests can differentiate the service from that of a competing operation. When a food service achieves differentiation, patrons may pass up competitors, going a long way just to eat at a particular place.

PRICING PSYCHOLOGY

Studies have been made on how menu planners set, and how patrons react to, menu prices. Menu planners usually try to avoid whole numbers and try to shade numbers just below them. A price of $6.95 or even $3.99 is perceived as less than $7 and $4. Number "5" is the most-often used ending digit. Some authorities say that a price ending in .99 is more suited to quick-service menus, and 0 and 5 as ending digits suit full-service menus better.

Price length also has importance; many menu pricers hesitate to use four-digit prices if they can avoid it. They feel that a price of $9.95 is better than $10.95, that three digits appear as a much lower price to patrons than four.

Patrons tend to group prices by range and think of them as single-figure amounts. For example, prices from $0.86 to $1.39 are considered to be about $1.00; from $1.80 to $2.49, about $2.00; and from $2.50 to $3.99, about $3.00. Prices from $4.00 to $7.95 are thought of being about $5.00. Instead of raising a price into the next full price range, the menu planner may try to hit the upper limit of the range the present price is in. That is, instead of raising a menu price from $2.25 to $2.55, the planner may raise it to $2.45. A price of $7.75 is preferable to $8.25 because the former is in the "about $5" range and the $8.25 is in the "about $10" range.

Patrons do not seem to like wide ranges in menu prices. They want prices grouped together within the price range they want to pay. If too wide a price range occurs, patrons will tend to select the lower-priced items.

Price increases are not always viewed in the same way by patrons. A price increase from $5.95 to $6.25 is seen as a bigger jump than from $6.25 to $6.75.

Some operations find that patrons resist buying after a certain price is reached. A Montana restaurant built as a stockade and located on a high mountain pass found there was resistance to any dinner price above $10. Items priced below $10 were popular.

Items over that were not. To resolve this, the table d'hôte meals were dropped and all items were listed a la carte at seemingly lower prices. The highest priced steak with a salad, potato, and roll and butter was $8; a popular Trappers Stew was priced at $6.50; and so on. If patrons wanted appetizers, soup, or desserts, they paid extra. The strategy worked. Most checks now came to more than $10 per person. The $10 wall was broken.

MARKET RESEARCH

Market research should point out what kind of market exists and what consumers will pay. If a gourmet restaurant opens, it must be located where it will attract customers with money. Adequate market information can lead to greater precision in setting prices. A *base price* and *top price* can be defined and the menu planned to work within this range.

ECONOMIC INFLUENCES

A variety of methods is used to price menus. Many managers calculate their costs as a percentage of sales and then calculate the selling price. Others base selling prices on factors other than costs.

Foodservice operators should be aware that laws of economics and commerce work both to the benefit and harm of businesses. The two most basic economic laws affect every operation: 1) when supply is limited, prices tend to rise, and when supply is plentiful, prices tend to drop; and 2) when demand is high, prices rise, and when demand is low, prices drop. Thus, every menu planner should try to plan menus that create a high demand. Operations that can restrict supply are few, but when they can, they have a chance to charge enough to make a good profit and still hold their business. Dropping prices may create more demand, while raising them may reduce it. Some operations reduce prices and hope to increase demand and, while making a smaller profit, make it up with increased volume. Some others raise prices, reducing demand but earning a greater profit. The south Florida area has a large demand for food and housing during the winter months and many operations do well as a result. Many restaurants have waiting lines for lunch and dinner. In the summer, both tourists and residents leave, and many operations drop their prices, cut their staffs, and retrench in every way because of the lack of demand. Some even close.

A smart menu planner knows that when the supply of an item is plentiful, costs will be low, and a better profit made. Thus, in the late spring and early summer, lamb is plentiful, relatively inexpensive, and of good quality and, therefore, a good menu item. Turkey is plentiful in the late fall and early winter. When the smelt run is on in the Great

Lakes, local foodservices offer them as "all you can eat" menu items. This promotion works well.

If possible, menu planners should be sure that there will be an adequate supply of all menu ingredients. Sometimes, an item is on the menu and the price of the items used for it become so costly that the item is served at a loss. Often such an item is removed from the menu. Some prices, such as those for fresh fish and seafood, vary so much that a menu planner lists only "market price," indicating the price will depend on the operation's cost.

Competition can reduce the number of patrons coming to an operation. Some may try to increase demand by dropping prices. However, this can be dangerous, because if competitors also drop their prices, both fight for the same demand, earning less from it.

Advertising and special promotions create demand. When a big-name star appears in a Las Vegas casino, food, beverage, and room prices may rise because the demand for them is high. Many operations put on special promotions to create a higher demand during certain periods. A quick-service chicken operation may advertise a special price on a bucket of fried chicken during a slow period. Such a promotion not only increases demand but also introduces the product to people who might not otherwise become patrons.

When forces of supply and demand increase and decrease in relation to increases and decreases in prices, the market is called *elastic*. If it does not, it is *inelastic*. If patrons pay no attention to prices and purchase items regardless of price, or refuse to purchase others even when prices drop, the market is said to be *inflexible*. However, if patrons increase demand when prices drop, and decrease demand when prices rise, the market is called *flexible*. A market where neither the price, nor supply and demand are changing is called *steady*.

A menu pricer should know whether a market is flexible, inflexible, or steady, and how patrons are likely to respond to price changes. Often a test can be made by changing prices briefly to test patron response. The test should be repeated several times over a period.

Thus, menus must reflect and take advantage of the influence of economic laws. A menu is one of the best contacts a food service can have with its patrons. It can help create demand and take advantage of the flow of supplies.

Pricing Based on Costs

One of the most common methods of pricing is for the planner to list costs of the food for each item on the menu and then mark up the final figure to obtain a selling price. For instance, if an a la carte item has a food cost of $2 and the operation wants a

40 percent food cost percentage, the selling price would be $5 ($2 ÷ 0.40 = $5). If the operator wants to maintain a 30 percent food cost in relation to sales, all foods would be marked up by dividing food cost percentage into food cost.

The disadvantage of this method is that it assumes that other costs associated with preparing menu items remain the same with every menu item. On the contrary, menu items vary considerably in the cost of labor, energy, and other factors needed to produce and serve them.

For instance, a steak is on the menu for $16.95 while a pasta dish sells for $11.95. The labor cost to make a portion of pasta is $1.10, and the labor cost to prepare the steak is $0.97. The pasta's labor cost is 9.2 percent of the selling price and the steak's is 5.7.

Assuming other costs have the same ratio to food cost for each menu item is a mistake. This leads to undesirable pricing that fails to cover costs on some items and can work against the sale of others. Costs must still be a paramount factor in pricing, and all costs—not just one or two—should be considered.

Pricing experts also claim that food cost pricing does not work because many foodservice operators do not know *all* their costs. These critics point out that many hidden food costs, such as spoilage, are not determined.

Derived Food Cost Percentage   The most common method is to use a *derived food cost percentage*. Simply divide the dollar cost of the food by the desired percentage (food cost ÷ food cost percentage). If a menu item has a total food cost of $2.73, and the operator wants a food cost percentage of 35 percent, the calculation is $2.73 ÷ 0.35 = $7.80.

Pricing Factor or Multiplier   Sometimes managers may convert a desired food cost percentage into a *pricing factor* or *multiplier*. For this, one divides the desired food cost percentage into 100 percent. This formula gives a factor by which a food cost is multiplied to get a selling price. For instance, suppose a food's cost is $2.73 and the desired food cost percentage is 35 percent; 100% ÷ 35% = 2.86, and 2.86 × $2.73 = $7.81, the selling price. (See Figure 14.1.) If a manager uses a food cost percentage plus other cost percentages, a multiplier can be calculated and used. Thus, if an operator wants a multiplier based on a combined food and labor cost of 65 percent, the calculation would be 100% ÷ 65% = 1.54. If the combined food and labor costs were $5.20, the selling price would be 1.54 × $5.20 = $8.01, probably set at $8.00 or $7.95.

Variable Cost Pricing   A *variable food cost pricing* method is sometimes used for a la carte menu items. For instance, an operator may assign food cost markups for menu items, as shown in Figure 14.2. This variable pricing can also be used to arrive at table d'hôte

Food Cost %	Factor	Food Cost %	Factor	Food Cost %	Factor
20	5.00	30	3.33	40	2.50
21	4.76	31	3.23	41	2.43
22	4.55	32	3.13	42	2.38
23	4.35	33	3.00	43	2.32
24	4.17	34	2.94	44	2.27
25	4.00	35	2.85	45	2.22
26	3.85	36	2.78	46	2.17
27	3.70	37	2.70	47	2.12
28	3.57	38	2.63	48	2.08
29	3.45	39	2.56	49	2.04
				50	2.00

FIGURE 14.1 Pricing factors (multipliers).

Appetizers	25%	Beverages	40%
Salads	40	Desserts	35
Entrees	35	Breads and butter	30
Vegetables	40	Miscellaneous	35

FIGURE 14.2 Variable cost markups.

	Food Cost	Allocated Menu Price
Tomato juice cocktail	$0.14	$ 0.56
Salad	0.48	1.20
Entree	2.84	8.11
Vegetables	0.94	2.35
Roll and Butter	0.27	0.90
Dessert	1.24	3.54
Beverage	0.46	1.15
Total Food Cost	$6.37	$17.81

FIGURE 14.3 Table d'hôte meal costed out.

menu prices. Using the food costs in Figure 14.3 and the percentages in Figure 14.2, a table d'hôte meal can be priced out.

Labor costs also can vary with different menu items. In these cases, menu items are divided into high (H), medium (M), and low (L) labor costs. Different percentages

Item	Labor Requirements	% Food Cost Assigned	$ Food Cost	Selling Price Based on Labor and Food Costs	33.3 % Selling Price
Stew	H	25	$1.10	$4.40	$ 3.67
Steak	M	35	3.20	9.14	10.67
Milk	L	40	0.17	0.43	0.57

FIGURE 14.4 Selling costs based on various labor costs.

of food costs are then assigned to these. For instance, for high-labor menu items, a food cost of 25 percent is assigned; for medium, 35 percent; and for low, 40 percent. Figure 14.4 indicates how the pricing of three items turns out compared with an overall price based on a 33.3 percent food cost only calculation. This method weighs food cost and allows the price to reflect the influence of labor as a cost.

Prime Cost Pricing A selling price can be based on a dollar *prime cost value* divided by a cost percentage or multiplied by a factor. Prime cost is raw food cost plus *direct labor,* or labor spent in preparing an item. Thus, if food cost is $1.00 and direct labor $0.25, prime cost is $1.25. An operation that wants a 35 percent food cost and a 10 percent direct labor cost has a prime cost percentage of 45 percent.

Direct labor time is usually obtained by timing work. A selling price based on prime cost is obtained by establishing a desired combined food and direct labor cost percentage. Suppose 45 percent is the desired prime cost percentage and a cook takes 1.5 hours to make a recipe that gives 30 portions. The food cost of the recipe is $126.56. The cook is paid $5.75 per hour; 1.5 × $5.75 gives a direct labor cost of $8.62 which, added to $126.56, gives a prime cost of $135.18. The prime cost per portion is then $135.18 ÷ 60 portions = $2.25. The selling price would be $2.25 ÷ 0.45 = $5.00.

Combined Food and Labor Costs Prices may sometimes be based on a combined food and labor cost percentage. For instance, if one wants a 35 percent food cost and a 30 percent labor cost, the combined cost would be 65 percent. If food cost for an item is $2.11 and all labor costs are $1.60, then using these figures the selling price would be ($2.11 + $1.60) ÷ 0.65 = $5.71. For this method, it is necessary to use a dollar value for labor cost that is based neither on a percentage of the selling price nor on a percentage of sales. It has to be a labor cost that is specific for the item.

All or Actual Cost Pricing Sometimes a method called by several different names—*all cost, actual cost,* or *pay-yourself-first*—is used in operations that keep detailed and accurate

Food cost	35%	$3.12
Labor cost	30	2.44
Operating cost	25	1.81
Total	90%	$7.37
Profit	10	
Total	100%	

FIGURE 14.5 All cost pricing.

cost records. Cost are divided into food, labor, and operating cost units. A dollar value and the desired food and labor costs are obtained for each item. A desired profit percentage is also established. Figure 14.5 illustrates how this method is used to price a dinner choice on a menu. Selling price equals 100 percent; desired profit equals 10 percent; food cost equals 35 percent, or $3.12; labor cost equals 30 percent, or $2.44; operating cost equals 25 percent, or $1.81. The formula is food cost plus labor cost plus operating cost plus 10% profit equals 100% (or selling price). Using these figures, $3.12 + $2.44 + $1.81 = 100% − 10% = $7.37 ÷ 0.90, a selling price of $8.20 is obtained. Again, *actual cost figures must be used,* not percentages based on sales or selling price. This method is useful only with a good cost accounting system. Many smaller food services do not have the accounting system to do this.

Gross Profit Pricing

In this method, a gross profit dollar figure is taken, usually from the profit and loss statement for a certain period. This is divided by the number of guests served during that time to get an average dollar gross profit per guest as the following example shows.

Sales	$851,322.14
Cost of food	261,110.36
Gross profit	$590,211.78

Suppose the number of guests served during this period was 108,113. The average gross profit per guest is $590,211.78 ÷ 108,113 = $5.46. The dollar cost for an item is then added to this average dollar gross profit to get a selling price. For instance, suppose the food cost for four items was $2.10, $3.13, $2.85, and $4.00. The selling price would then be calculated as in Figure 14.6.

When using a gross profit average, be sure it includes an adequate profit. If it does not, the desired profit should be added in a dollar value. Thus, if the profit were $0.27 of

Item	Food Cost	Average Gross Profit	Selling Price
A	$2.10	$5.46	$7.56
B	3.13	5.46	8.59
C	2.85	5.46	8.31
D	4.00	5.46	9.46

FIGURE 14.6 Gross profit pricing.

the $5.46 figure but management wants a per patron profit of $0.85, about 10 percent of sales, $0.85 – $0.27, or $0.58, is added to the $5.46 figure, bringing the gross profit average up.

Gross profit pricing is a useful method because in many operations the cost of serving each patron is much the same after food costs are considered. It tends to even out prices in a group.

The One-Price Method

In some food services, the overall cost of menu items is the same, such as a doughnut shop where all doughnuts and beverages cost about the same. This is the one-price method in action. The operation can charge just one or a few prices to simplify things. The small differences will usually even out. A nightclub with a cover charge can also use the one-price method because the cost of what is served is nominal when compared to other costs, such as entertainment, music, and decor.

Another kind of operation that might charge one price regardless of the item selected is one in which selling food is not a primary purpose. This operation could be a tavern that makes all sandwiches one price. Or it could be a casino, where the objective is to get people in to gamble, and if food at one price helps do that, the operation benefits.

Marginal Analysis Pricing

Retail operations, including food services, may use the *marginal analysis* pricing method. This is an objective method in which the maximum profit point is calculated. The selling price chosen will be the one that establishes this maximum.

Say a quick-service operation wants to set the most favorable price for its milk-shakes in order to maximize profit. It estimates that at various prices it will sell a certain

Selling Price	Number Sold*	Total Sales	Total Costs	Marginal Profit
$1.50	100	$150.00	$120.00†	$ 30.00
1.40	190	266.00	156.00	100.00
1.30	275	357.50	190.00	167.50
1.20	340	408.00	216.00	192.00
1.10	410	451.00	244.00	207.00
1.00	450	450.00	260.00	190.00
0.90	525	472.50	290.00	182.50

*Estimated from marketing studies or by other means.
†$80 fixed cost + 100 sold × $0.40 each = $120.

FIGURE 14.7 Marginal analysis projection.

number of milkshakes, as shown in Figure 14.7. Fixed costs are $80.00, and variable cost per milkshake is $0.40. Thus, 100 shakes cost 100 × $0.40 food cost plus $80.00 (100 × 0.40 + $80 = 120).

From the marginal profit column we can see that the best selling price is $1.10 with 410 sold; the next best is $1.20. Figure 14.8 shows how this marginal analysis problem appears in graph form.

The greatest distance between the costs and sales lines is at points a and b, where 410 milk shakes are sold to bring $451 in sales at a cost of $244, giving a profit of $207.

Cost-Plus-Profit Pricing

In the *cost-plus-profit* pricing method, a food service may decide it needs a standard profit from every patron who enters. The rationale behind this is that every customer who comes through the door costs the operation money, no matter what is ordered. The operation may reason that it wants a set amount of profit from every patron. Therefore, total costs may be calculated and then a set amount added to this. For instance, an operation that wants to make $600 a day in profit may find it has an average of 400 patrons per day. Then, when pricing, the food, labor, and operating costs are added together, plus a $1.50 profit. This covers all costs and should result in the desired profit of $600. Next, an average labor cost value and operating cost value are determined from the profit and loss statement, and both are divided by the number of patrons served in that period. If labor costs, including all benefits, are $80,511, operating costs are $50,336, and 83,001 patrons are served, the average labor cost per patron is $0.97 and the operating cost per patron is $0.61. The selling price calculations for items A, B,

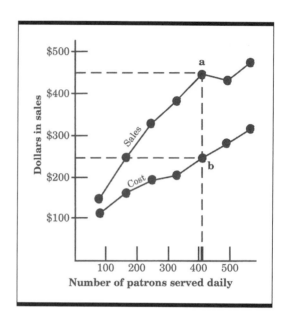

Marginal analysis graph.

	Food Cost		Labor Cost Per Patron		Operating Cost Per Patron		Profit	Selling Price
A	$2.10	+	$0.97	+	$0.61	+	$0.50	$5.18 (or $5.25)
B	3.13	+	0.97	+	0.61	+	0.50	6.21 (or $6.25)
C	2.85	+	0.97	+	0.61	+	0.50	5.93 (or $5.95)
D	4.00	+	0.97	+	0.61	+	0.50	7.08 (or $7.15)

FIGURE 14.9 Cost-plus-profit pricing.

C, and D are shown in Figure 14.9. In this case, the low operating and labor costs give low selling prices. This method tends to even out selling prices.

Minimum Charge Pricing

Pricing based on a minimum price to cover costs and give a desired profit in a commercial operation is much the same as calculating the price based on costs plus profit. The rationale for this method is also much the same. That is, every customer costs a certain amount to serve, and by having a minimum charge these costs will be covered. A hospital or nursing home might use such a method. A private club might have a minimum for certain rooms where food and beverage service is provided. Such a policy is also common in some commercial dining rooms. Pricing may be "arranged" on a menu card to make it impossible to obtain service below a certain price. For instance, the Four Seasons Restaurant in New York City may not want to see a price of less than $15 for lunch. Such operations will test out various luncheon combinations. Whatever menu items produce the desired result will be priced accordingly.

Some operations may state on the menu card that there is a check minimum. If a customer does not order enough food to cover this minimum, the check will still include the minimum payment. Clubs may require that members spend a specific amount during a certain period on foods and beverages so that the foodservice department receives enough income to operate. If the member does not spend this amount during the period, the balance is added to the bill.

Cover Charge A cover charge is a set price that is added to customers' bills, regardless of what menu items are purchased. This cover charge establishes a base from which costs for space, atmosphere, entertainment, and other costs will be paid. The cover charge is used by nightclubs and other operations where entertainment or dancing may be an attraction.

Pricing Based on Sales Potential

Some menu planners believe that pricing should reflect factors in addition to food and labor costs, including how an item is expected to sell. They divide items into High Cost (HC) or Low Cost (LC), High Risk (HR) or Low Risk (LR), and High Volume (HV) or Low Volume (LV). Thus, one menu item may be labeled HC-LR-HV and another LC-HR-HV. Ratings of LC, HV, and LR are considered favorable, while HR, HC, and LV are considered unfavorable. Eight combinations are possible.

HR-LC-HV	HR-HC-LV	
HR-LC-LV	HC-LR-LV	HV-LR-LC
HR-HC-HV	HC-HV-LR	LR-LC-LV

If a plus (+) is assigned for a favorable factor and a minus (−) for an unfavorable one, the following matrix is obtained, based on the previous combinations.

− + +	− − −	
− + −	− + −	+ + +
− − +	− + +	+ + −

Menu Item	Risk High	Risk Low	Cost High	Cost Low	Volume High	Volume Low	% Markup
N.Y. steak	—		—		+		30
Apple pie		+	—			—	30
Chicken noodle soup		+		+	+		40
Vanilla ice cream		+		+	+		40
Lobster	—		—			—	25
Vegetables		+		+		—	35
Fish	—			+	+		35
Swiss steak		+		+		—	35

These figures can be used to record risk, cost, and volume levels so the menu pricer can properly allocate markup.

FIGURE 14.10 Menu item markups.

If this formula is used when pricing, the highest markup would be assigned to any item having three minuses or two minuses and a plus, a lower markup to one minus and two pluses, and the lowest markup to the one with three pluses. Perhaps the highest markup would be based on a 25 percent food cost, the next on a 30 percent food cost, and the last on a 40 percent cost. This type of pricing would ensure a profit margin for low-volume, high-cost items. Figure 14.10 shows how an operation might classify items, and then indicates the desired food cost markup.

Non-Cost Pricing

A number of methods that are not based on cost are used in pricing menus. Some, such as pricing based on tradition, competition, or what the market will bear, are in this category. Surprisingly, they are used often and, even more surprisingly, are often successful. Some are difficult to use. Finding a price based on what the market will bear is an involved process and takes a lot of testing and observation. Others, such as traditional pricing or pricing against competition, are relatively simple.

Pricing Based on Tradition Many operations look at traditional prices, often those set by the-leaders in a particular market niche. Tradition in pricing may relate not only to a specific price but also to the pricing structure and market. Many operations find they cannot get as high a markup for California wines as they can for European ones, although the California product may be of equal or better quality. Tradition has it that the domestic product is usually lower in price, and, therefore, the pricing structure must be varied to suit this tradition.

Different types of operations will also find that the pricing structure they must follow is one in which a lower markup must be taken than that of another type of food service. Thus, while two operations may sell exactly the same thing, one will be able to set prices on a different basis from the other.

Some operators may find that they have prices established for such a long time that they become traditional with their customers. When they attempt to change these prices, they may meet with very strong sales resistance and customer dissatisfaction. Thus, they may decide not to change the price on a particular item but get a higher markup on other items that are not so restricted.

It is possible for a price to become traditional because a leader in the market charges this price. Thus, an industry leader like McDonald's sees its prices often become traditional among similar operations.

Competitive Pricing One of the most common pricing methods is to base the price on what the competition charges. While it is wise to pay attention to competitors' prices, it is unwise to base your prices *completely* on them. What the competition charges may bear no relation to your costs. Pricing in this manner wrongly assumes that prices satisfactory for the competition's customers are also satisfactory for your customers. In addition, copying competitors' prices does little to differentiate an operation.

If competitive pricing is indicated by research, a unit should study its own costs to analyze how it can price menu items to produce a more favorable response than that produced by the competition's prices. A competitor's price may be studied to reveal information about the food, labor, and other operating costs. Food and labor are standard commodities that will have a known price. A close scrutiny of these may indicate what the competition might be doing to achieve a favorable price structure. A study of competitive pricing and its effect on one's own business is also warranted. In some cases, prices may have to reflect the influence of competition.

What the Market Will Bear A method of pricing used by some companies is to design a product and then test market it at a given selling price. The product may be put on several markets at different prices and the reaction of customers studied to ascertain which is the best price.

Marketing specialists state that one of the best routes to success is to develop a product the market wants and then price it so that a healthy profit is made. With proper pricing and strong demand, these specialists say there is a high chance of developing a successful market. They recommended studying the value of the product *in the minds of patrons* and then charging accordingly. Some products may have to be priced only slightly above cost to be accepted by the market; others can have a much higher margin.

Establishing a selling price based on what the market will bear is not just a trial-and-error method; it should be based on sound research. To some extent, prices can be tested to see how well patrons accept them; however, not too much experimentation per item can be done. Perhaps three or four prices can be tested on an item, and then the testing must be stopped. Otherwise, patrons might reject the item because of the instability of the price.

Pricing according to what the market will bear has become a popular method with a number of industries. It is a perfectly legitimate system. If a customer attaches a certain value to a product and is willing to pay a higher price for it, there is no reason why it should not be marketed at that price. Patrons are not interested in what it costs to produce and market an item. They are much more interested in getting something they feel represents a good value for their money. If the price is within the value they have in mind, they are happy. If they feel that a meal priced at $4 is worth $4, even though its cost to the operation is only $2.50, there is nothing wrong with charging $4 for it. However, if a meal costs $4 to produce and patrons consider it worth only $2.50, they will not buy it, even if it represents a lower price than cost.

Success in pricing according to what the market will bear is enhanced if some attempt is made to show patrons the true value of the menu item. It is often difficult for patrons to see the value in atmosphere, service, fine tableware, linens, and foods that are somewhat out of season or brought in at an additional cost. If a way is found to get patrons to understand that these increased costs increase value, they may be willing to pay a higher price. Having a differentiated product is one way of getting patrons to see and appreciate value. The food service with exceptional service, decor, atmosphere, and dining experience, including fine presentation that patrons would not get at home, can make patrons feel they are getting good value for their money.

Pricing Aids

Pricing can be made easier if tables and computer printouts are available that give food costs of various menu items, with prices based on the operation's food cost percentage already tabulated.

EVALUATING PRICING METHODS

Few operations use only one pricing method. Most use a combination to best meet their needs and those of their patrons. Pricing based only on competition is generally not a good method, but considering the prices of competition when establishing one's own menu prices is advisable. Using a standard markup over an accurate cost determination can give a fairly precise price basis and usually assures an adequate profit or budget performance. However, varying margins over cost based on what the market will bear should also be considered, as should pricing based on volume. Perhaps markups should be varied to promote merchandising and entice patrons with *loss leaders* (items that have a low markup but can bring in extra business). Additionally, various items may be priced to cover the high food cost of items with prices based on local competition. Thus, a quick-service operation may not make as much on its hamburgers and hot dogs as it does on milkshakes, carbonated beverages, and fries, but in the end the achieved sales mix can give a very desirable markup level.

And, finally, *pricing is not something that is done and then is over.* There is a need to evaluate prices constantly, to study how customers react to prices, and to gather data on costs. Too often no follow-up occurs, and when a revision of menus and prices is necessary, there is a lack of adequate information to do a good job. Gathering, compiling, collating, and filing pricing data are as much a part of the pricing function as the establishment of prices.

PRICING FOR NONPROFIT OPERATIONS

The previous discussions on menu pricing for commercial operations are applicable to noncommercial operations, except that most do not need to allow for profit in their prices. They are required to meet costs and perhaps make a slight margin above costs as a safety factor that can be accumulated to bridge times when costs are not covered. Costs may be paid for in part by the patron, with the balance paid by federal and state agencies or charitable donations.

When nonprofit operations price items to break even, accurate cost information must be available from operating records. The information must be timely, so that action can be taken promptly to bring costs into line when they vary from desired levels. Many institutional operations estimate costs for a period and are given an allowance from the total budget to cover this estimate. When costs for a future period must be estimated, information from federal agencies, economic indexes, and economic price predictions of price changes can be used.

Nonprofit operations usually establish prices on the basis of a budgeted amount per meal, per day, per week, per month, or per period. Some non-profit budgets allow a cash allowance for food, labor, and operating expenses. Other operations may first get a total food allowance for a period, and then a cash allowance is worked out from this. This food and cash allowance system is called a *ration allowance.*

PRICING EMPLOYEE MEALS

In in-plant and company food services, whether employees pay for meals or the company subsidizes them, costs must be calculated so they are covered.

Employee meals are often considered a benefit and, thus, an operating expense. Therefore, when the cost of food for employee meals is included in the cost of food used, it must be deducted. The value for deducting an employee meal can be based on: 1) the actual cost of the meal, 2) experience, 3) a standard charge made in the area for meals, or 4) an arbitrary amount. The actual cost may be standard menu prices with a discount given to employees. In some areas it may be a practice to deduct a predetermined amount for employee meals, a plan followed by a number of food services. The arbitrary amount may be only a nominal charge but one that the food service thinks is adequate to cover its costs, or at least a major portion of them.

Most operations take their average food cost percentage and use this to arrive at a value for the food used for employees' meals. Thus, if the food cost is 35 percent and a total value of $800.19 is assigned to the employees' meals for a certain period, the value of the food in these meals for that period will be 0.35 × $800.19, or $280.07. When employees eat the same meals as patrons, the cost of employees' meals may be calculated using the following steps.

1. Ascertain the number of employee meals.
2. Calculate the individual meal cost.
3. Consider food for employees' meals as 50 percent of the cost of a meal.
4. Multiply the number of employees' meals by the food cost per employee meal.

For example, say an operation has total costs of $11,400 for 15,571 meals served during the period, with 1,240 of these meals being eaten by employees.

$11,400 ÷ 15,571 = $0.732 per meal
73.2 × 0.50 = $0.366 food cost per employee meal
1,240 × $0.36 = $453.84 total food cost for employee meals

Based on this information, it would be simple to arrive at a nominal menu price. This information will also provide the data for deducting meals as a benefit.

CHANGING MENU PRICES

At times a change in a menu price is necessary. If it is an item that appears frequently on the menu and has good acceptance, repricing may present difficulties. Customers may resent the change and may stop ordering the item to show their displeasure.

Sometimes this does not last long, and the item gradually assumes its former importance as a seller.

During periods in which food prices increase rapidly, most customers recognize the need for an operation to increase prices and will accept them. In periods when prices are stable but some other factor makes a change necessary, customers may not be so willing to accept a price change.

Some operators attempt to change a price by removing an item from the menu for a time and then bringing it back at a new price. If it also comes back in a somewhat new form, or in a new manner of serving and with a slightly different menu name, the price change may be less noticeable. Changes also tend to be noticed less if they are made when volume is down.

Prices are frequently changed when a new menu is printed. In fact, the need to change menu prices may sometimes stimulate a menu change more than the need to change items. Some authorities advise against changing format and prices at the same time, saying that customers will notice price changes *because* of the new format.

Using menu clip-ons removes the need to print new menus just to incorporate new prices, and avoid the needs to cross out old prices to put new ones in. The latter, especially, can give patrons a negative impression.

In some instances, if a general rise in menu prices must occur, an operation may change several items $0.05 or $0.10 each and then let these remain at this price while changing a few others. In this manner prices are gradually worked up to the desired level. This has the disadvantage of giving customers the impression that the menu and its prices are unstable.

Some announcement, on a clip-on or table tent, can be helpful in indicating to patrons why a change in menu prices in necessary. However, this can also have the undesirable effect of calling attention to changes that some customers might not otherwise notice.

Patrons are likely to be especially aware of price changes for popular items. Also, as noted, they are more aware of a change when the dollar price changes than when the cents price does. Rather than changing a dollar amount, it might be wiser to make a price change in cents, or make it gradually.

Only a few items' prices at a time should be changed if prices are changed frequently. More items can be repriced if the change is less frequent. Some say that only two price changes should be made in a year; others feel that only one is advisable. With some items for which prices may have to change frequently, such as lobster or stone crab, it is advisable to list the price as "Seasonable Price" or "Market Price."

Printed menus that rarely change their items offered or format are more difficult to change in price than are blackboards, panels, or handwritten menus. If a menu has a daily insert in which items change daily or frequently, it is easier to make price changes on this than on the permanent hardcover menu. It is not a good practice to cross out an old price and write in a new one. Even whiting out a price and writing in a new one is not recommended.

Specials and highly promoted items are difficult to change in price because buyer attention is often centered on them. They should be dropped from promotion for a time and then reinserted with the prices changed. In all cases, price changes on menu items should end up within the range expected by patrons.

PRICING PITFALLS

All foodservice managers should beware of the following pitfalls committed by inefficient menu planners.

1. Pricing should not be based entirely on just one cost, such as food cost, giving a price that may not reflect actual costs. Other costs vary considerably from a direct relationship with food cost. Thus, pricing only on the basis of food cost can lead to pricing errors.

2. Foodservice pricing should not ignore the economic laws of supply and demand.

3. Value perception on the part of patrons in equating price to value of a menu item should have greater emphasis in pricing.

4. More attention needs to be given to market information in establishing prices.

SUMMARY

Pricing a menu is a complex process, and a number of factors need to be considered when doing it. Anyone pricing a menu should know about and experiment with some of the latest pricing methods, such as marginal analysis pricing and market testing of prices.

Studies have shown that consumers view prices somewhat differently than operators might expect. They want prices grouped within a price range they expect to pay, and resist purchasing items outside this range.

A number of pricing methods are used by the foodservice industry. Probably no one method is used alone, but a combination of them can be used in establishing menu prices.

Patrons will pay only so much for certain kinds of foods in certain kinds of operations, and the menu planner must be sure to meet this restriction. Menu prices are often not easy to change, and planners should be aware of the techniques that make price changes less noticeable by patrons.

QUESTIONS

1. What is a *differentiated* product? What is its purpose? How is it used?

2. A food service wants a 25 percent food cost. What multiplier or pricing factor should be used?

3. The food cost of a table d'hôte group of foods is $8.20, and a 30 percent food cost is desired. What is the selling price?

4. If raw food cost is $1 and direct labor cost is $0.30, what is the prime cost? If a 40 percent price based on this prime cost is wanted, what is the selling price?

MENU MECHANICS*

Outline

Objectives

After reading this chapter, you should be able to:

- Identify the basic requirements to make a menu an effective communication and merchandising medium.

- Discuss aspects of using type: typeface, type size, line length, spacing between lines and letters, blank space, weight, and type style.

- Indicate how to give menu items prominence by using displays in columns, boxes, or clip-ons.

- Indicate how to best use color in menus.

- Discuss paper use, construction of covers, and other physical factors.

- Indicate how menus are commonly printed, how to work with professional menu printers, and how to self-print.

---

* Authored by Lendal H. Kotschevar and Marcel R. Escoffier.

INTRODUCTION

Certain mechanical factors must be considered in menu planning. No matter how well the menu is planned and priced, it must also be properly presented so that it is understood quickly and leads to satisfactory sales. *Communicating* and *selling* are the main functions of a successful menu. Good use of mechanical factors will enhance a menu's appearance, make a favorable impression on patrons, and advance the overall aims of the operation.

Professional menu printing companies can be of considerable help in developing a menu that is attractive and achieves its purpose as a merchandising medium. For this reason, the material in this chapter is designed to teach readers how to work with professionals as well as how to do the job without assistance.

MENU PRESENTATION

How a menu is presented is important to most operations. For commercial establishments the menu does much to convey the type of operation and its food and service. If the menu communicates accurately through design and layout, as well as through the copy, it can "sell" the items on it.

While most menus are printed on paper and given to patrons to look at, some might not be presented in this way. A cafeteria menu board may show items for sale and list prices. A quick-service operation might have menu signs or cards on tables. Some operations have hand-written menus to give a homey and personal touch. A menu may be made to resemble a small newspaper and list the latest news along with menu items.

The manner in which menu items are presented should be selected to best meet the needs of the operation. A hospital may have selected menus printed on colored paper, each color indicating a different diet. On some, special instructions concerning selections by patients may be used. The sales department of a hotel or catering department may need a special menu to give to people interested in arranging special functions at the hotel.

Some operations need a number of different menus, such as breakfast, brunch, lunch, matinee, dinner, or evening. A country club may need a menu for its bar where steaks, sandwiches, and snack foods are served; another for a coffee shop or game room; a small snack and beverage service near the swimming pool; and another for the main dining room. A hotel or motel might use a special room service menu. As these menus vary in their purpose and requirements, so must they vary in the manner in which menu items are presented.

The most common menu is the one presented on firm paper, the front being used for some logo, design, or motif. Inside, on the left and right sides of the fold, a la carte

	A la carte	Table d'hôte
Southern Fried Chicken with Country Gravy and Corn Fritter	$ 6.50	$ 8.95
Crab Cakes Mornay en Coquille with Steamed Rice	9.50	11.95
Fillet of Cod en Papillote, French Fried Zucchini	7.95	10.50
Hawaiian Ham Steak, with Mashed Sweet Potatoes	5.95	8.50
New York Strip Steak, French Fried Potatoes	11.50	14.95
Breaded Veal Cutlet, Sauteed Mushrooms, and Baked Potato	5.95	7.95

With the table d'hôte dinner you have a choice of salad or vegetable, beverage, and desert. Roll and butter are served with a la carte and table d'hôte orders.

FIGURE 15.1 A la carte and table d'hôte entrees.

offerings (items selected and paid for individually) are listed. The back may also contain a la carte items and alcoholic beverages, or give information about hours of operation and short notes of interest about the operation, locale, or some of the special items served. The items on this heavy paper are permanent.

Often, menu items that change, including table d'hôte meals (foods or meals sold together at one price), are printed on lighter paper and attached to the more rigid menu.

Sometimes menus list two prices for an item, one including the entree as the main dish in a table d'hôte meal, the other offering it a la carte. Even as an a la carte item, it may be served with such foods as bread and butter. (See Figure 15.1.) While this listing shows about a $2.50 difference between table d'hôte and a la carte items, it is not unusual to see prices between table d'hôte and a la carte vary widely.

Some menus may offer specials. These can be attached as clip-ons or inserts. If they are used, the basic menu should provide space for them. They should not cover other menu items.

Clip-ons or inserts are used to give greater emphasis to items management wants to push. They should not repeat what is on the menu but offer variety. The use of these clip-ons or inserts make it possible to change a permanent menu for weekly specials or holidays. Clip-ons and inserts may be of the same color as the menu, but if special effects and heightened patron attention are desired, the use of a different color can help focus attention on them.

Menu Format

Regardless of how a menu is presented to patrons, certain rules in format should be observed. Wording and its arrangement should be such that the reader quickly understands what is offered. If foods are offered in groups, it should be clear what foods are included. Clarity is promoted by making menu items stand out. Simplicity helps avoid clutter. Foods usually should be on the menu in the order in which they are eaten. An exception might be a cafeteria menu board, listing items as they appear in the line. Some offer cold foods first and hot foods last. This avoids the hot foods cooling off while a customer selects cold foods. Some menus also indicate the location of foods, such as in a takeout operation, where different counters offer different foods, or a cafeteria where patrons move from one section to another to get different foods. In this case, the menu board can be helpful by indicating a counter number or using a diagram to show where foods are found.

Production Menus

Some menus may never be seen by patrons. They are written principally for back-of-the-house workers to inform them about what must be produced. This requires a different form and terminology, selling words and fancy descriptions are not required. The term "carrot pennies," which sounds good on a menu read by patrons, instead will be "sliced carrots." Production information is included, such as the amount to prepare, the recipe number to use, preparation time, distribution to service units, designation of the worker to prepare items, and portion sizes. Service instructions, such as the portioning instructions and the dishes and utensils to use, may also be added.

MENU DESIGN

The design of a menu contributes greatly to its legibility and patron reaction. Therefore, the design should be well planned. Menus, like individuals, should have personalities. They should reflect the atmosphere and "feel" of the operation. The eye should be pleased with what it sees on the menu and patrons should quickly grasp what is offered and the price. The printing and coloring should blend in with the logo or trademark of the operation as well as with the type of establishment. The menu should be to a foodservice operation what a program is to a play or opera—an indication of what is to come. It should be an invitation to a pleasing experience and should not promise too much. Patrons should clearly understand what they are to get and the price they are to pay for it.

Using Type

Typefaces One of the most important factors in accomplishing a menu's purpose is the style of type, or *typeface*, used. There are many different kinds of type, and some are more legible and more easily and quickly read than others. The type most often used for

---

Serif

Times Roman
Braised in butter and then simmered in red wine with shallots and other herbs, this dish has been for centuries one of the most typical of the Bretony area. Braised in butter and then simmered in red wine with shallots and other herbs, this dish has been for centuries one of the most typical of the Bretony area.

Palatino
Braised in butter and then simmered in red wine with shallots and other herbs, this dish has been for centuries one of the most typical of the Bretony area. Braised in butter and then simmered in red wine with shallots and other herbs, this dish has been for centuries one of the most typical of the Bretony area.

Bookman
Braised in butter and then simmered in red wine with shallots and other herbs, this dish has been for centuries one of the most typical of the Bretony area. Braised in butter and then simmered in red wine with shallots and other herbs, this dish has been for centuries one of the most typical of the Bretony area.

Sans Serif

Helvetica
Braised in butter and then simmered in red wine with shallots and other herbs, this dish has been for centuries one of the most typical of the Bretony area. Braised in butter and then simmered in red wine with shallots and other herbs, this dish has been for centuries one of the most typical of the Bretony area.

Helvetica Black
Braised in butter and then simmered in red wine with shallots and other herbs, this dish has been for centuries one of the most typical of the Bretony area. Braised in butter and then simmered in red wine with shallots and other herbs, this dish has been for centuries one of the most typical of the Bretony area.

Futura
Braised in butter and then simmered in red wine with shallots and other herbs, this dish has been for centuries one of the most typical of the Bretony area. Braised in butter and then simmered in red wine with shallots and other herbs, this dish has been for centuries one of the most typical of the Bretony area.

FIGURE 15.2 Samples of type.

---

menus is a *serif* type, or one in which letters are slightly curved. These are some of the easiest types to read. Letters set in *sans serif* type may look blocky. (See Figure 15.2.)

Studies have been made to ascertain legibility, reading speed, and comprehension using different typefaces. Unfortunately, not all of the types that have been studied are used in menus, and some that have considerable popularity in menu use were not included in the studies. Nevertheless, from such studies we can get some idea of the best typefaces to use for menus.

Print comes in plain (regular), bold (heavy print), italics, and script. Any type of italic or script print is more difficult to read than plain type. However, in some cases,

Point Size	Name
3 1/2 point	Brilliant
4 1/2 point	Diamond
5 point	Pearl
5 1/2 point	Agate
6 point	Nonpareil
7 point	Minion
8 point	Brevier
9 point	Bourgeois
10 point	Long Primer
11 point	Small Pica
12 point	Pica
14 point	English
16 point	Columbian
18 point	Great Primer

FIGURE 15.3 Type sizes.

these might be preferable to others because of special effects desired. A fine-dining restaurant may want to use these typefaces because they imply elegance in dining. Italic, script, or specialty types can bring on fatigue in reading, but this may not be a factor in short menus.

Type Size The size of type is important to both understanding and speed of reading. Type that is too small makes reading difficult, but type that is too large takes up too much space; it might actually inhibit comprehension because it spreads the words out too much.

Type size is measured in *points.* There are 72 points to an inch. Thus, 18-point type is nearly a fourth of an inch high. Most menu designers use 10- or 12-point type for listing menu items and 18-point type for headings. This can be varied for descriptions.

Readers have ranked their preferences for type size. Three sizes—10-point, 11-point, and 12-point—ranked together as first choices, followed by 9-point, 8-point, and 6-point. Various type sizes are shown in Figure 15.3.

Menus printed in a single type size can be monotonous. The sizes are usually varied on a page to give relief. (See Figure 15.4.) Thus, menu items may be listed

```
                        Appetizers

Shrimp Cocktail .......................................5.50
Six ice-cold jumbo shrimp served with tangy cocktail sauce.

Cold Seafood Platter..............................6.50
Oysters, shrimp, lobster, and clams served with Brooklyn
Navy Yard sauce or drawn butter.

Steak Tidbits ........................................5.50
Strips of sirloin breaded, served with marinara sauce.

Chicken Fingers...................................4.75
Fried chicken pieces with blue cheese dipping sauce.

Mozzarella Marinara..............................4.95
Fried cheese smothered in marinara sauce.

Arugula Salad .......................................6.95
Crunchy arugula with mushrooms, red onions,
hearts of palm, and raspberry vinaigrette dressing.

Mozzarella Salad ..................................5.95
Fresh Buffalo mozzarella, fresh basil, and sliced tomatoes
dribbled with extra virgin olive oil.
```

FIGURE 15.4 Menu showing type size contrast.

in 12-point type, with 9- or 10-point type used for the description just below the item.

Menu headings may be in capital letters in bolder and larger type. Different type from that for the items sometimes may be used, but some mixtures may give an undesirable effect.

At times, to draw attention to an item, larger type is used than that used for regular items. For instance, the menu may use normal-weight 12-point type rather than heavy or light printing for regular items, and then change to a 14-point boldface (heavy) type to give emphasis to a special item. Additional emphasis can be given to an item by placing it in a box and putting an ornamental border around it. (See Figure 15.5.) If the box is in a different color from the regular menu, the emphasis may be greater.

Spacing of Type Another factor affecting ease of reading and comprehension of menus is the spacing between letters in a word and between words. If letters and words are set too close together or too far apart, reading is difficult. Associated with this horizontal dimension in typography is the width of the individual characters in a particular style of type. They may be condensed (narrow), regular, or extended (wide), and this quality

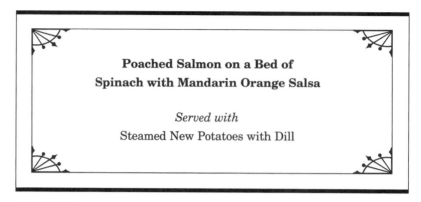

FIGURE 15.5 Menu item emphasis.

has its effect on readability and the scanning rate. Thus, a condensed word would be printed "**condensed**," a regular would be printed "regular," and an extended would be printed **"extended."**

Vertical spacing between lines of type, called *leading* (pronounced "ledding", is also important to ease of reading. If no leading is used, the type is said to be set *solid*. The thickness of these leads is also measured in points. Figure 15.6 shows lines set solid and lines with 1-, 2-, 3-, and 5-point leading.

A general rule for easy reading is that leading should be two to four points larger than the typeface being used.

Weight *Weight* is a term used to indicate the heaviness or lightness of print. Light print may appear gray, rather than black, and does not stand out well; normal or *medium* is what is normally used on menus; *bold* or heavy print is quite dense; sometimes even extra bold is used. Bold or extra bold weight helps give emphasis, but too much can lead to a cluttered look.

Emphasis can be given and items made to stand out by the wise use of light and bold type. Very light type will not give good emphasis and is sometimes difficult to read. Extra bold type is extremely dark and black. Bold or heavy may be desirable to draw attention but should be used sparingly. Too bold a typeface is not suitable for the menu of a refined, quiet dining room. (Sometimes a printer may refer to the weight of the type as its color, but this is a term used mostly by professionals.) The light level of an operation must be considered. A restaurant with a low level of light should use a slightly larger, bolder type to keep the menus readable. Light type is best used in an operation with a normal or higher than normal level of light.

Use of Uppercase and Lowercase Letters Another factor that influences ease of reading and comprehension is effectively combining the use of *uppercase* (capitals) and

---

Solid

A treat to the palate sends taste buds soaring. A treat to the palate sends taste buds soaring. A treat to the palate sends taste buds soaring. A treat to the palate sends taste buds soaring.

1-Point Leading

A treat to the palate sends taste buds soaring. A treat to the palate sends taste buds soaring. A treat to the palate sends taste buds soaring. A treat to the palate sends taste buds soaring.

2-Point Leading

A treat to the palate sends taste buds soaring. A treat to the palate sends taste buds soaring. A treat to the palate sends taste buds soaring. A treat to the palate sends taste buds soaring.

3-Point Leading

A treat to the palate sends taste buds soaring. A treat to the palate sends taste buds soaring. A treat to the palate sends taste buds soaring. A treat to the palate sends taste buds soaring.

5-Point Leading

A treat to the palate sends taste buds soaring. A treat to the palate sends taste buds soaring. A treat to the palate sends taste buds soaring. A treat to the palate sends taste buds soaring.

FIGURE 15.6 Samples of leading.

lowercase (small) letters. Uppercase gives emphasis and can set words out more clearly. Lowercase is easier to read. Uppercase is used with lowercase to begin sentences, or for proper nouns. It is usually desirable to capitalize all first letters of proper names and main words in item titles on the menu. For instance, the following would be normal: "Top Sirloin Steak Sandwich." Words such as *or, the, a la, in, and,* or *with* are usually not capitalized. Descriptive information, such as "A combination of shrimp, scallops, halibut and oysters in Newburg sauce," will not have capitalized letters, except for the name of the sauce, because it is a proper noun. Uppercase may be used to emphasize words, as in "Includes French Fries, Tossed Green Salad with your Favorite Dressing, and Choice of Beverage and Dessert."

Often menu items are put in large, bold caps to stand out, and lowercase type is used for descriptive material below the name of the item, capitalizing only proper nouns and first letters of sentences in the descriptive material. Special effects can be obtained at times by setting all descriptive words in small caps.

Descriptions should be in keeping with the menu theme and set in a typeface compatible with other type on the page. All elements should blend together if a maximum effect is to be achieved.

Special Effects Using Type   The use of type may give special effects. For instance, a nation's script—Javanese, Russian, Greek, Hebrew, Arabic, Japanese, Chinese, Thai—can be used to reflect a restaurant's cuisine.

Page Design

Page layout and design is an essential element of menu development. A good menu will "grab" patrons and attract them to items the operation wants to sell. A poorly designed menu will do the opposite—lead patrons into a maze with more than enough items to confuse them.

The amount of copy on a page affects how quickly a menu can be read and understood. In one study, readers indicated they wanted fairly wide margins and disliked copy that ran too close to the edge of the paper, a warning to menu planners who tend to overcrowd areas. Normally, just slightly more than 52 percent of a printed page has print on it, and slightly over 47 percent is margins. Figure 15.7 shows a page with a black area in the center indicating print and margins around it. One would not suspect that

FIGURE 15.7 Spacing for margins.

the white margin area accounts for nearly half the page. Perhaps, in menu copy, slightly more than this could be covered but not much more. Providing as much margin space as possible without squeezing menu copy should be the objective.

If space is a problem, extra pages should be added rather than crowding a single menu page. A good margin should remain on the left and right sides and on the top and bottom. Figures 15.8 and 15.9 are an example of good use of space in the menu.

Making a menu so large that a patron has difficulty holding it should be avoided. Some operations make menus very large to give the feeling of luxury, but many guests find them difficult to handle. An operation should give strong consideration to the menu size before deciding on a large one. If a large menu must be used, extra panels that fold open should be considered. Separate menus may also be used. For instance, if desserts are not given with the table d'hôte dinner, or if there is an additional group of a la carte desserts, a special dessert menu may be set up, thus saving space on the main menu. Likewise, alcoholic beverages and wines can be on separate menus. A special fountain menu may be placed on tables and counters where customers can find them, leaving the main menu free to offer a more standard list of items.

Line Width The width of a line affects reading comprehension. One study found that about two-thirds of menu readers preferred a double-column page to a single-column page. It has also been found that students learn better when reading two column pages. Most readers like a line length of about 22 *picas* (about 3 $^2/_3$ inches—there are approximately 6 picas to an inch). Long lines may cause readers to lose their place. Figure 15.10 shows a line length of 24 picas. Most books are printed in columns of line length around 25 picas (about 4 $^3/_{16}$ inches), which gives the page only one column.

The eye does not normally flow evenly across a page. It grasps a certain group of words, comprehends these, and then jumps to the next group. If the line is too long, there is a chance of losing the reading place when the eye jumps. Also, too long a line may cause a reader to lose his or her place since the eyes must refocus as the gaze passes from the end of one long line to the beginning of the next line. One menu authority recommends that, rather than putting prices a distance from menu items in a column to the right, prices should be put right next to or directly under the item.

Emphasizing Menu Items There are a number of ways to emphasize menu items that management wants to sell. An item is set apart by separating it by a bit of space and special type, and then giving some description of the item, as shown in Figure 15.11. This description can help give emphasis, especially if it is italicized or set in different type. Items can also be shifted slightly for emphasis. Note that when this is done in a column, items will stand out well, as in Figure 15.12.

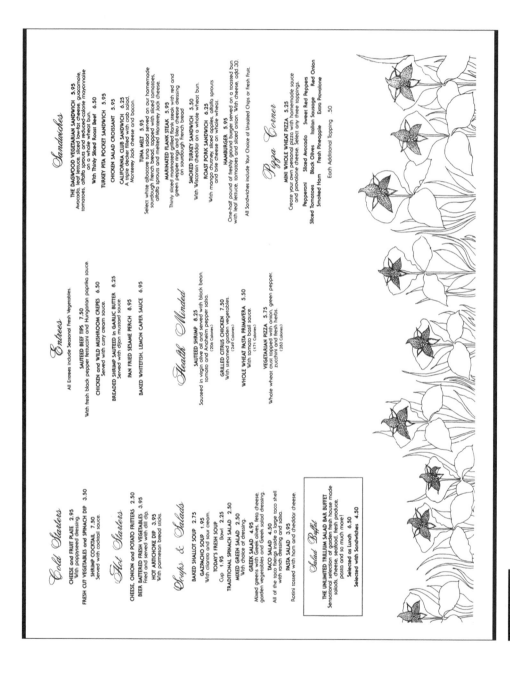

FIGURE 15.8 Menu using space effectively. Courtesy of Trillium Restaurant, Grand Traverse Resort, Grand Traverse Village, Michigan.

EXECUTIVE COCKTAILS

A Drake Hotel specialty which has earned the respect of members worldwide for many years. Our creative cocktail feature four ounces of your preferred brand of gin, vodka, scotch or bourbon, expertly mixed and served in a logoed snifter to create Chicago's best Martini, Manhattan or Rob Roy.

HOUSE 6.50   CALL 7.95   PREMIUM 7.25

MOCKTAILS
Non-alcoholic variations of your favorite refresher

STRAWBERRY DAQUIRI   PINA COLADA
SEA BREEZE   ORANGE FIZZ
4.50

FRESHLY BREWED 100% COLOMBIAN
REGULAR OR DECAFFEINATED COFFEE
1.75

SELECTION OF TEAS
1.75

PERRIER, SAN PELLEGRINO AND EVIAN WATERS
2.75

SAN PELLEGRINO NATURAL SOFT DRINKS
Aranciata: A lightly Carbonated Orange Soft Drink with Real Fruit Juice or
Limonata: A Lightly Carbonated Lemon Soft Drink with Real Fruit Juice
1.85

SOFT DRINKS
2.25

APPETIZERS

FRESH OYSTERS OR CLAMS
On the Half-Shell with
Traditional Accompaniments
7.50

DRAKE CHICKEN WINGS
In a Tangy Sauce with Celery Sticks
and Blue Cheese Dip
5.95

CHICKEN SATAY
With Peanut Sauce, Sliced Cucumbers and Onions
8.25

CAPE COD MINI CRAB CAKES
With Mustard Sauce
8.75

COQ FOR QUESADILLAS
With Guacamole
7.75

SOUPS
These Drake Hotel soup traditions are created
fresh daily from our original recipes

	Cup	Bowl
BOOKBINDER SOUP	3.50	6.50
NEW ENGLAND CLAM CHOWDER	3.50	6.95
TODAY'S FRESH SOUP	3.25	5.50

SANDWICHES
Served with potato chips and pickles

TUNA OR CHICKEN SALAD
7.95

CALIFORNIA SANDWICH
Fresh-Sliced Breast of Turkey,
Tomato, Avocado and Alfalfa Sprouts
Served on Whole Wheat Bread
7.95

BROILED CHICKEN BREAST
Skinless Breast of Chicken Served
in a Sesame Roll with Mayonnaise,
Iceberg Lettuce and Tomato
7.75

THE COQ FOR CLUB
Undoubtedly Chicago's Best!
7.95

CHICAGO REUBEN
Wafer-Thin Slices of Corned Beef
Grilled on Rye Bread with Swiss Cheese,
Sauerkraut and Thousand Island Dressing
7.25

BLACKENED SWORDFISH
On a Sesame Roll with Iceberg
Lettuce, Tomatoes
and Onion
8.75

SOUP AND SANDWICH COMBINATION
A Cup of Bookbinder, Clam Chowder or Soup of The Day Served with Your
Choice of a Half of Chicken Salad, Tuna Salad or Sliced Turkey Sandwich
7.95

ENTREES

CHOPPED SIRLOIN STEAK
9.75

GRILLED STRIP STEAK
With French Fries and a Grilled Tomato
14.95

FRENCH-FRIED SHRIMP
Rémoulade Sauce and French Fries
11.50

▼ BREAST OF TURKEY
Thinly Sliced with a Salad of Wild Rice, Apples and Walnuts
8.75

SPAGHETTI
With Meat or Marinara Sauce
and Garlic Bread
7.75

HOT TURKEY
A Coq d'Or Favorite,
with Mashed Potatoes and Gravy
8.75

MOSTACCIOLI PRIMAVERA
Fresh Vegetables and a Three Cheese Sauce, or
Fresh Vegetables, Olive Oil, Garlic and Herbs
8.25

▼ BROILED WHITEFISH
A Local Favorite
11.50

CAPE COD CRAB CAKES
Mustard Sauce
16.75

COQ FOR HAMBURGER
Seasoned with Green Pepper and Onion
on a Rye Bun, with Sliced Tomatoes,
Pickle and French Fries
7.50

AVOCADO DELIGHT
Half Avocado with Your Choice of
Shrimp, Tuna or Chicken Salad
on a Bed of Lettuce
8.50

CAESAR SALAD
With Grilled Breast of Chicken and Caesar Dressing or
Low Calorie Caesar Dressing
8.95

SEARED SALMON SALAD
Oriental Style Salmon on Baby
Greens, Sesame Ginger Dressing
8.95

▼ SEASONAL FRUITS AND BERRIES
With Cottage Cheese
or Yogurt
7.50

SIDE DISHES
1.75

CAPE COD COLESLAW
FRENCH FRIES   MIXED GREENS SALAD

BAKED POTATO
With Traditional Accompaniments

The items marked with a ▼ are prepared with a reduction of
calories, cholesterol and sodium

DESSERTS

All of our desserts are made fresh daily in the
renowned Drake bake shop

DRAKE BREAD PUDDING
2.95

CARAMEL CUSTARD
2.95

CLUB RICE PUDDING
2.95

ICE CREAM PIE
Chocolate or Strawberry
3.50

APPLE PIE
2.95

With Melted Cheese or à la mode   3.75

CHEESECAKE
Chocolate Marbled or Strawberry
3.50

SELECTION OF SHERBET OR ICE CREAM
2.95

AFTER DINNER DRINKS

MEXICAN COFFEE
Espresso or Cappucino with Kahlúa
6.75

CAPPUCCINO
2.75

ESPRESSO
2.50

FRENCH COFFEE
Espresso or Cappucino with Grand Marnier
6.75

COGNAC

MARTELL CORDON BLEU
10.00

COURVOISIER V.S.O.P.
7.50

HENNESSY X.O.
55.00

REMY MARTIN V.S.O.P.
8.00

REMY MARTIN NAPOLEON
12.00

REMY MARTIN X.O.
19.95

REMY MARTIN EXTRA PERFECTION
50.00

REMY MARTIN LOUIS XIII
80.00

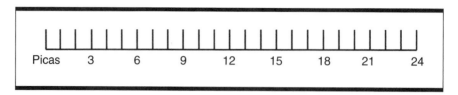

FIGURE 15.10 Line length of 24 picas.

Special Texas Sirloin Steak, Charbroiled $14.50
*Served with Baked or French Fried Potatoes and a Tossed Green Salad
with your choice of French, Thousand Island or Blue Cheese Dressing*

FIGURE 15.11 Item emphasis.

FRUITS OF THE SEA

Sauteed English Sole, Almandine with Spinach Soufflé $14.00
   with Cole Slaw and French Fried Potatoes

New England Crab Cakes with Braised French Garden Vegetables $13.50
   with Sun-dried Tomatoes and Creamy Basil Dressing

Curry of Shrimp on a Bed of Rice, Chutney Sauce $16.95
   with Fresh Fruit Salad and Creamy French Dressing

Cold Boiled Salmon with Mayonnaise and Potato Salad $12.95
   with Sliced Cucumbers in Balsamic Vinaigrette Sauce

FIGURE 15.12 Emphasis on columns.

In spite of the fact that menu space is valuable, some blank space is needed to set items apart and to avoid having the menu so crowded that the reader is confused. From one-fourth to one-third of the printed area should be blank, in addition to margin space. Headings and lines can be used to indicate separations and help draw attention to items.

The first and last items in a column are seen first and best. This is the place to put menu items one wishes to sell. It is possible to lose items in the middle of a column, so items that management may be less interested in selling, but that have to be on the menu, might be placed here.

Readers tend to skip items. Indenting items presented in a column can help make readers look at all the items in the column. Items too deeply indented, however, are often lost; this is where items are put when one does not wish to give them emphasis. Arranging items in a column from highest in price to lowest, or vice versa, is usually not desirable. People looking for price tend to go to the least expensive one and never look at the others. Mixing up prices makes people look through all the items to see what is there and what the prices are. This is more likely to make even price-conscious buyers see something they find very desirable, and that they will purchase not on the basis of price but on the desirability of the product.

Where the eyes focus is also important in menu design. When patrons open up a two-page menu, their eyes usually go to the right, often to the center of the page or, if not there, to the upper right hand corner of the page and then counterclockwise to the right bottom. If a menu is a single page, readers will tend to go to the middle of the page and then to the upper right, left and down, then across to the lower right, and then up again. Figure 15.13 indicates this eye movement. It is important that items management most wants to sell be put into those positions where readers look first, known as *emphasis areas*. These items need not necessarily be specials, since many people will come in for specials and will hunt for them on the menu.

Color

Besides making an artistic contribution, color can affect legibility and speed of reading. The use of white print on a black background may get more attention than black print on a white background, but it is harder to read. Black on white is read 42 percent more rapidly than white on dark gray, and black type on light color is read less easily than

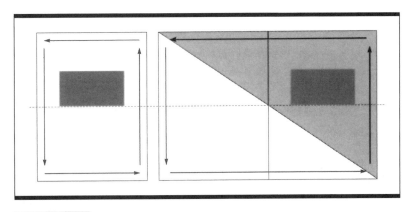

FIGURE 15.13 Menu view.

Quite Easy to Read	Fairly Easy to Read	Poorly Read
Black on light cream	Black on light yellowish green	Black on fairly saturated
Black on light sepia cream	Black on light blue green	yellowish red
Black on deep cream	Black on yellowish red	Black on reddish orange
Black on very light buff		
Black on fairly saturated		
yellow		

FIGURE 15.14 Color test results.

Quite Easy to Read	Fairly Easy to Read	Poorly Read
Black print on yellow	Blue print on white	White print on black
Green print on white	Yellow print on blue	Red print on yellow
Red print on white	White print on red	Green print on red
Black print on white	White print on green	Red print on green
White print on blue		

FIGURE 15.15 Colored ink test.

black on white. However, black on some light colors is about equal to black on white. The results of a test with various colors of paper stock are summarized in Figure 15.14. The order is from easiest to most difficult in each of the three classifications.

In another test, to ascertain how well different colored inks stood out against various tints, the results shown in Figure 15.15 were obtained. Again the order is from easiest to most difficult to read.

Color and design can enhance a menu and make it a better merchandising tool. Color and design are as important to menus as they are to dinnerware. A large amount of color can run into a sizable cost. Yet, as with china, considerable decorative effect can be obtained with only a small amount of color and design. Similarly, plain-colored paper can achieve good effect and is not too expensive. Some paper has color on one side and white or another color on the reverse. As special effects are added, costs increase. Adding silver or gold to a menu can be expensive.

Too much color and design can distract a customer's attention from the menu. Colors that are intense should be avoided unless some special effect is desired. Perhaps a bold color could be used in a club wishing to project a bold atmosphere, but it might not be appropriate in other food services.

Clip-ons can be used to give a different design and color. Some menus may be given additional color by using ribbons, silken cords, or tassels, but these are usually restricted to menus for special dining areas.

The variety of colors in paper suitable for menus is wide, including metallic papers. Since the basic color must serve as a background for the print that describes menu items and lists prices, the print stands out very clearly. Dark browns, blacks, reds, and other deep colors should not be used with dark colored type, nor should light-colored print be used with light-colored paper.

Color can be used for special events and holidays. Mandarin red with black print is appropriate to commemorate a Chinese New Year; a Halloween menu would traditionally use orange and black; black type on emerald green is suitable for St. Patrick's Day.

All colors and designs on the menu should blend with and complement the operation's decor. However, they also can be contrasted with good effect. For example, an outstanding color may be used to give a vivid color splash, just as an interior decorator might use a vividly colored vase to give a highlight to a more subtle color scheme in a room.

Complementary colors are usually those that come from the same primary colors (red, yellow, and blue), but some contrasts between colors coming from different primaries also can be pleasing. For instance, some shades of red and green go well together. While some greens and blues clash, others do not. Knowing how to use colors in design is inherent in some people, but not all. Some menus may seek to achieve a desirable color effect by using colors in stripes, squares, triangles, or circles in some unique pattern. Obtaining a balance takes an ability to blend colors, lightness, and darkness. The services of a menu designer can do much to produce a menu with striking color effects that are in good taste.

Color in pictures results from colored inks used in different combinations. The basic four-color-process inks—black, yellow, magenta, and cyan (bright blue)—will give almost any combination of color desired in printing. A fifth color is sometimes added for special effect.

To reproduce a picture, at least four pieces of film, called *separations,* must be made. This is done by taking all yellow tones in one separation. All cyan (blue) tones are taken from the picture in another separation. All shades of magenta and all blacks are taken in other separations. If a special color effect is wanted, other color separations will be made. When these are printed, the human eye will combine and reproduce all colors in the picture. Sophisticated machines now do separations by laser beam, which gives almost perfect placement of one color over another.

A four-color separation can cost more than $500. However, high quality might be worth the expense. Also, buying a large quantity of printed menus can reduce color costs considerably because the cost is spread or *prorated* over time to bring down the unit cost.

There frequently are other costs in using colored pictures. A photograph must be of good quality—a corned beef sandwich must look like a *good* corned beef sandwich. It is usually worthwhile to pay a professional photographer and perhaps even a food stylist to take photographs.

There are companies that maintain libraries of stock photos. It may be possible to find an appropriate photograph and save the cost of taking a picture. Clip-out sheets can also be used.

Color separations can be "doctored" to enhance colors and make them appear as close as possible to the color of the actual food. Color *proofs* must be checked carefully before any printing is done. Poor color reproductions of food items will ruin an otherwise well-developed menu.

Sketches and line drawings, which may or may not have a background, can also be effective. It is possible to add a color within a line picture to help give color contrast, rather than have the line drawing appear filled with the basic color of the paper. Further emphasis can be obtained by putting the picture in a box. Additional color and design can be obtained with decorative borders.

Paper

The Menu Cover The paper used for a menu cover should be chosen carefully. First, consider how often the menu will be used. If it is designed to be disposable, such as a menu on a place mat, then the paper can be a lightweight stock. If it is going to be used regularly, a coated grease-resistant stock is important. Texture is also a factor to consider, since customers hold the menu in their hands, which can soil menus.

A heavy paper, *cover stock*, is used for most menu covers. This is usually a paper stiff enough to be held in the hands without bending. It may be laminated with a soil-resistant material, such as plastic. The surface is frequently shiny and smooth, but cover materials may have different textures. Laminated covers last for a longer time than untreated ones, and they are easily wiped clean. The weight of cover stock should be in heavy cover, bristol, or tag stock at least .006 inch thick.

Some operations may use a hard-cover folder in which a menu is placed. Often these are highly decorative and represent a considerable investment. Some may be padded, to give a softness in the hand, and covered with strong plastic or other materials that can give the feeling of silk, linen, or leather. These are often laminated onto base materials. Sometimes a foil inlay is stamped on the front. This is done using heat and pressure to set the inlay in the cover material. Gold, silver, and other colors may be stamped onto covers. Using inlays can be expensive.

Bond paper	Used for letterheads, forms, and business uses
Book paper	Has characteristics suitable for books, magazines, brochures, etc.
Bristol	Cardboard of .006 of an inch or more in thickness (*index, mill,* and *wedding* are types of bristol)
Coated	Paper and paperboard whose surface has been treated with clay or some other pigment
Cover stock	Variety of papers used for menu covers, catalogs, booklets, magazines, etc.
Deckle edge	Paper with a feathered, uneven edge
Dull-coat	Coated paper with a low-gloss surface
Enamel	Coated paper with a high-gloss surface
English finish	Book paper with a machine finish and uniform surface
Grain	Weakness along one dimension of the paper—paper should be folded with the grain
Machine finish	Book paper with a medium finish—rougher than English finish but smoother than eggshell
Matte-coat	Coated paper with a little- or no-gloss surface
Offset paper	Coated or uncoated paper suitable for offset lithography printing
Vellum finish	Similar to eggshell but from harder stock, with a finer grained surface

FIGURE 15.16 Basic paper terms.

When these heavy covers are used, it is possible to use a printed menu inside that will not be too expensive. With such a cover, the menu is usually printed on lighter-weight paper.

Characteristics of Paper The weight of paper for interior pages can be of a lighter weight than that for covers. Usually, strong, heavy book paper is used. Its finish should be such that it resists dirt. Paper can be given different finishes to make it more suitable for menu use. If novelty or striking effects are desired, some specialty papers can be used. (See Figure 15.16) The type of ink and printing process used will also help determine the type of paper used. The operator should also investigate the texture and the opacity of the paper.

Paper Textures Paper textures can range from slight rises, such as is seen in a wood grain, to a rough, coarse surface. It is possible today to give paper almost any texture desired, even that of velvet or suede. *Opacity* of paper (inability to see through it) may depend on the strength of the ink or the use of color. Heavy ink should be used on highly opaque paper. Some transparency may be desirable for some artistic effects. Most often, a maximum opacity is desirable.

Menu Shape and Form

The shape and form of the menu can help create interest and sales appeal. A wine menu may be in the shape of a bottle, while one featuring seafood can be in the shape of a crab, lobster, or fish. When using food shapes, however, the shape chosen must be one that is distinctly recognizable to the average diner. A steakhouse bill of fare may feature a menu in the shape and coloring of a Black Angus steer. A child's menu can be in the shape of a clown whose picture is on the front cover. A pancake house can have its menu shaped like a pancake or waffle.

The fold given the menu may also create an effect. Instead of having a right and left side to a cover, a fold may be used that gives a half page on the right and a half page on the left, so that the menu opens up like a gate. If folds are used, the "foldability" of the paper should be investigated; most coated papers crack easily at a fold.

If a special fold or shape is required, a special *die* (a form used to cut out shapes, such as in jigsaw puzzles) may have to be made. This costs money and should be the property of the one who pays for making the die. Also, the film and/or plates used for printing the menu should become the property of the individual paying for them. This must be clearly understood in making the original agreement. In some instances, printers may not be required to give them up unless such an agreement is made beforehand.

Printing the Menu

Development of Typesetting and Printing Around 700 A.D., the Chinese began the process of setting type. They carved each block of type from blocks of wood, rubbing ink over the top surfaces and blotting them on cloth or paper. In Europe, it was not until the early 1400s when Gutenberg invented movable type on the printing press. Letters were cast in metals and put together to make words, then blotted much like their wood counterparts. This method of hand typing was long and laborious.

The *linotype*, a machine that enabled a person to stroke a keyboard that dropped carved letters into place, made the process faster and easier.

The Modern Printing Process Today, handset and linotype is seldom done except for special purposes. Typesetting has become highly computerized, and in many instances copy is typed into a computer system. The typesetting program reproduces the type either on transparent film or on photographic paper. A copy of the type called a galley proof or reader proof is made. It is not in pages and is usually in long sheets. This is used to proofread and note corrections. The corrections are made on the computer and either a new galley proof or a finish proof is made. An artist will either take the finish proof and make up the menu pages or they may be made up in the computer with special programs. Page proofs with the illustrations and everything else in place, just as it would be in the final form, must be checked. The page proof is usually a print (similar to a photograph) made from a film. A photographic process is used to make a printing plate from the film once the proof has been okayed. The plate is put onto a press that produces the printed material. Most printing nowadays is produced by the offset method. Ink is applied to the plate, the image is then transferred to another roller, and this image is then "offset" onto a piece of paper. The paper never touches the plate.

Silk-screening is another method of printing. A stencil is made, either photographically or by hand, and applied to a silk screen. Ink is forced through the silk onto the paper. This method produces a very intense, dense image and is often used for metallic inks (gold, silver, and copper). It is usually a hand process and consequently expensive, but the effect may be worth the cost. It is most often used on covers.

Hot foil stamping, *embossing* (creating a raised image with a stamp), and *diecutting* (by which shapes are cut into paper) are other methods of printing and enhancing a menu. As with silk-screening, they are all more expensive to do and the cost must be weighed against the effect produced.

Permanent menus are usually printed in sizable lots. Some large chains using the same menu can have printings in the hundreds of thousands. A single, small restaurant may print only 500, having only 100 with prices and leaving the remainder with the printer, who can add new prices later when more menus are needed. This takes care of price changes. Most printers want a minimum order of 500.

Self-printing Many operations today print their own menus. This can be done almost as well as a professional printing company and often at a huge savings. Having personal control of menu production has some advantages also. Some large operations set up sophisticated systems and produce remarkable results. However, even with much less equipment and facilities, it is possible to print one's own menus which are adequate for use in the facility. The minimum equipment is usually a computer and a desktop-publishing program with a number of different typefaces and design features. A good laser, color printer is also desirable on which to print camera-ready pages to send to a printer. Figure 15.17 shows an attractive self printed menu produced on a personal computer.

Luncheon Orchestrations

Sauteed Lake Perch
Sauce Remoulade.

$12.95

Broiled Great Lakes Pickerel
Served with a leek and horseradish coulis and sour cream.

$12.95

Pan Fried Salmon Cakes
Garnished with asparagus tips.
Sauce Bearnaise.

$11.95

Broiled Halibut
Accompanied with veal bits and haricot vert.
Served with lemon champagne sauce.

$12.95

Spicy Broiled Sea Scallops Shish Kabob
Served over lobster rice and ratatouille.

$13.95

Sauteed Rock Shrimp
Served with oregano-garlic angel hair pasta,
julienne sundried tomatos and poppy seeds.

$12.95

Broiled Breast of Chicken
Served with a honey mustard sauce.

$9.95

Sauteed Breast of Chicken
Topped with a roasted corn and bacon compote and served with
a chicken reduction that is accented with chervil and shallots.

$10.95

Includes the Vegetable or Potato of the Day

FIGURE 15.17 Self-printed menu. Courtesy of Opus One, Detroit, Michigan.

Working with Professional Printers Certain companies specialize in assisting food services in planning, developing ideas for, and printing menus. They usually can be extremely helpful in producing an effective menu. Their experience in setting up menus to do the best merchandising job will be greater than that of the individual menu planner, and this experience can be helpful in avoiding mistakes.

Professional menu printing companies exist to do special art and design work. They may blend printing techniques with special effects to make the best possible menu. If special artwork is needed, either the company will have a staff artist to do it or they will know where to find freelance artists. Often, using a professional menu printer is just one more facet of producing a successful menu.

SUMMARY

Mechanical factors can be extremely important in making a menu an effective communication and merchandising medium. This, of course, can translate into a profit for the operation.

Dark, simple typefaces are easiest to read, but italic and adorned prints can be used for special effects. The best size print for regular menu items is 10 to 12 points, and headings are usually 18 points.

Usually menu items are typeset in 10- to 12-point type with smaller type right below giving the description. The price should be close to the item, either immediately to the right or underneath, so there is no confusion.

Menus should have about two-thirds to three-fourths of their space covered with print. Some blank space is desirable on a menu. The size of a menu should allow it to be handled easily by patrons. If more space is needed, additional pages should be used, or separate menus might be used for such items as wines, alcoholic beverages, or desserts.

There are many ways to give emphasis to menu items; a change from one size of type to a larger size, for example. Using bold type can make a menu item stand out. Using all uppercase letters will give emphasis. Setting items in a box with an ornamental border and giving enough space around the box to make it stand out can attract patrons to items, as can clip-ons.

Placement of menu items is also important. The most prominent place on a two-page menu is the middle of the right-hand side. The most prominent place on a single sheet is the middle. The top and bottom are the most prominent places in a column. Items can be lost in the middle of a column. It is not desirable to offer menu items in order of price, because price-conscious patrons will quickly go to the lowest-priced items and not look at other offerings. Mixing prices requires price-conscious patrons to hunt and may lead to their selection of higher-priced items.

The printing of menus has become quite sophisticated. The offset method is mostly used to print menus. The use of full-color pictures requires making four-color separations. This can be expensive, but color is a good selling medium.

The contrast between the color of the type used and the background should be considered carefully. If either the type or background does not give a good contrast, the menu is not read as easily and may be confusing to patrons. Color is also important in giving good decorative effect.

Type, color, and other factors designed to achieve harmony must not be mixed together haphazardly.

The papers used for covers should be either bristol, cover, or tag stock and should be surfaced with some kind of soil-resisting substance. Covers should be made of heavy paper covered with a durable coating. Covers can be given very effective decoration by stamping or high-pressure lamination.

The shape and form of a menu should be carefully considered. Often a desirable effect can be obtained by working in the logo of the operation, a trademark, or a major food item sold.

Many operations today have computers and laser printers to print menus. This makes the process much faster as well as giving operators more control over design and content.

QUESTIONS

1. Look at any menu and note its typeface, size of printing, legibility, spacing, use of color, and leading. How effectively are these elements used?

2. Look at the same menu. Where is your eye drawn first? Which items does management want to sell?

3. On the same menu, what do its colors tell you? Does its design match the operation's theme and clientele?

4. What is the difference between manual and cast typesetting/printing, offset printing, block printing, and silk screen printing?

5. What are the advantages of desktop publishing over traditional printing in regard to menus? What are the advantages of traditional printing?

16

ACCURACY IN MENUS AND MENU EVALUATION*

REPRESENTATION OF QUANTITY

If standard recipes and portion control are strictly adhered to, no quantities of menu items should ever by misrepresented. For instance, it is perfectly acceptable to list precooked weight of a steak on a menu; double martinis must be twice the size of a regular drink; jumbo eggs must be labeled as such; petite and supercolossal are among the official size descriptions for olives.

While there is no question about the meaning of a "three-egg omelet" or "all you can eat," terms such as "extra large salad" or "extra tall drink" may invite problems if they are not qualified. Also remember the implied meaning of words: a bowl of soup should contain more than a cup of soup.

REPRESENTATION OF QUALITY

Federal and state quality grades exist for many foods including meat, poultry, eggs, dairy products, fruits, and vegetables. Terminology used to describe grades include Prime, Grade A, Good, No. 1, Choice, Fancy, Grade AA, and Extra Standard.

Care must be exercised in preparing menu descriptions when these terms are used. In some uses, they imply a definite quality. An item appearing as "choice sirloin beef" should be USDA Choice Grade Sirloin Beef. One recognized exception is the term *prime rib*. Prime rib is a long-established, well-understood, and accepted description for a cut of beef (the "primal" ribs, the 6th to 12th) and does not represent the grade quality, unless USDA is used also.

---

* Authored by Lendal H. Kotschevar and Marcel R. Escoffier.

Ground beef must contain no extra fat (no more than 30 percent), water, extenders, or binders. Seasonings may be added as long as they are identified. Federally approved meat must be ground and packaged in government-inspected plants.

REPRESENTATION OF PRICE

If your pricing structure includes a cover charge, service charge, or gratuity, these must be brought to your customers' attention. If extra charges are made for special requests, guests should be told when they order.

Any restriction when using a coupon or premium promotion must be clearly defined. If a price promotion involves a multi-unit company, clearly indicate which units are or are not participating.

REPRESENTATION OF BRAND NAMES

Any product brand that is advertised must be the one served. A registered or copy-written trademark or brand name must not be used generically to refer to a product.

A house brand may be so labeled even when prepared by an outside source, if its manufacturing was to your specifications. Contents of brand-name containers must be the labeled product.

REPRESENTATION OF PRODUCT IDENTIFICATION

Because of the similarity of many food products, substitutions are often made. These substitutions may be due to stockouts, the substitutions' sudden availability, merchandising considerations, or price. When substitutions are made, be certain these changes are reflected on your menu. Substitutions that *must* be spelled out as such include the following.

Maple-flavored syrup for maple syrup

Nondairy creamer for cream

Boiled ham for baked ham

Ground beef or chopped beef for ground sirloin

Capon for chicken

Veal pattie for veal cutlet

Ice milk for ice cream

Cod for haddock (or vice versa)

Sole for flounder

Whipped topping for whipped cream

Processed cheese or cheese food for cheese

Chicken for turkey (or vice versa)

Nondairy cream sauce for cream sauce

Hereford beef for Black Angus beef

Peanut oil for corn oil (or vice versa)

Bonita for tuna fish

Powdered eggs for fresh eggs

Picnic-style pork for pork shoulder or ham

Light-meat tuna for white-meat tuna

Skim milk for milk

Pollack for haddock

Pectin jam for pure jam

Blue cheese for Roquefort cheese

Beef liver for calf's liver (or vice versa)

Diced beef for tenderloin tips

Half & half for cream

Salad dressing for mayonnaise

Margarine for butter

REPRESENTATION OF POINTS OF ORIGIN

A potential area of error is in describing the point of origin of a menu offering. Claims may be substantiated by the product, by packaging labels, invoices or other documentation provided by your supplier. Mistakes are possible as sources of supply change and availability of product shifts. The following are common assertions of points of origin.

Lake Superior whitefish

Bay scallops

Gulf shrimp

Idaho potatoes

Florida orange juice

Maine lobster

Imported Swiss cheese

Smithfield ham

Wisconsin cheese

Puget Sound sockeye salmon

Danish blue cheese

Louisiana frog legs

Alaskan king crab

Colorado brook trout

Imported ham

Colorado beef

Florida stone crabs

Long Island duckling

Chesapeake Bay oysters

There is widespread use of geographic names used in a generic sense to describe a method of preparation of service. Such terminology is readily understood and accepted by the customer and their use should in no way be restricted. Examples of acceptable terms follow.

Russian dressing

French toast

New England clam chowder

Country fried steak

Denver sandwich

Irish stew

French dip

Country ham

Swiss steak

French fries

German potato salad

Danish pastries

Russian service

French service

English muffins

Manhattan clam chowder

Swiss cheese

REPRESENTATION OF MERCHANDISING TERMS

A difficult area to clearly define as right or wrong is the use of merchandising terms. "We serve the best gumbo in town" is understood by the dining-out public for what it is—boasting for advertising sake. However, to use the term "we use only the finest beef" implies that USDA prime beef is used, as a standard exists for this product.

Advertising exaggerations are tolerated if they do not mislead. When ordering a "mile high pie" a customer would expect a pie heaped tall with meringue or similar fluffy topping, but to advertise a "foot long hot dog" and to serve something less would be in error.

Mistakes are possible in properly identifying steak cuts. Use industry standards such as provided in the National Association of Meat Purveyors *Meat Buyer's Guide*.

"Homestyle" or "our own" are suggested terminology rather than "homemade" in describing menu offerings prepared according to a home recipe. Most foodservice sanitation ordinances prohibit the preparation of foods in home facilities.

The following terms should be used carefully.

Fresh daily	Aged steaks
Corn fed	Milk fed
Fresh roasted	Low-calorie
Flown in daily	Low-fad
Kosher	Low-sodium
Black Angus beef	Low-cholesterol
Center cut ham	

REPRESENTATION OF MEANS OF PRESERVATION

The accepted means of preserving foods are numerous, including canned, chilled, bottled, frozen, and dehydrated. If you choose to describe your menu selections with these terms, they must be accurate. Frozen orange juice is not fresh, canned peas are not frozen, and bottled applesauce is not canned.

REPRESENTATION OF FOOD PREPARATION

The means of food preparation is often the determining factor in the customer's selection of a menu entrée. Absolute accuracy is a must. Readily understood terms include the following.

Charcoal broiled	Broiled
Deep fried	Prepared from scratch
Sauteed	Roasted
Barbecued	Fried
Baked	Poached
Smoked	

REPRESENTATION OF VERBAL AND VISUAL PRESENTATION

When your menu, wall placards, or other advertising contains a pictorial representation of a meal or platter, it should portray the actual contents with accuracy. Following are several examples of *visual misrepresentation*.

- Using mushroom pieces in a sauce when the picture shows mushroom caps

- Using sliced strawberries on a shortcake when the picture shows whole strawberries

- Using numerous thin sliced meat pieces when the picture shows a single thick slice

- Using four or five shrimp when the picture shows six

- Omitting vegetables or other entree extras when the picture shows them

- Using a plain bun when the picture shows a sesame topped bun

  Examples of *verbal misrepresentation* include the following.

- A server asking whether a guest would like sour cream or butter with a potato, but serving an imitation sour cream and margarine

- A server telling guests that menu items are prepared on the premises when in fact they are purchased preprepared

REPRESENTATION OF DIETARY OR NUTRITIONAL CLAIMS

Potential public health concerns are real if misrepresentation is made of the dietary or nutritional content of food. For example "salt-free" or "sugar-free" foods must be exactly that to assure the protection of your customers who may be under particular dietary restraints. "Low-calorie" or nutritional claims, if made, must be supportable by specific data.

Sections adapted form *Accuracy in Menus*, Copyright © 1984 by the National Restaurant Association, Washington, DC.

MENU EVALUATION

Menu Profitability

	4 Excellent	3 Good	2 Fair	1 Poor	Comments
1. Does the menu have an adequate number of high gross profit items?					
2. Has there been a good selection of popular items?					
3. Is there a good balance between high- and low-priced items and no concentration of either?					
4. Does pricing meet competition?					
5. Are menu prices changed frequently enough to reflect costs?					
6. Are portion costs based on reliable cost information?					
7. Is portion size adequate?					
8. Are menu items selected with a view toward reducing waste and other risks?					
9. Are menu items selected to reflect labor requirements?					
10. Are menu items selected to reflect energy needs?					
11. Does the menu encourage a higher check average?					
12. Can items be controlled in cost?					

Presentation of Wording

1. Are menu items described accurately and truthfully?				
2. Does the menu avoid indicating weight, size, using pictures and other factors that can cause problems in patron interpretation?				

	4 Excellent	3 Good	2 Fair	1 Poor	Comments
3. Are effective descriptive words used to indicate menu item qualities?					
4. Is the choice of words adequate to describe the item? Is overkill avoided? Is the wording simple and easy to understand?					
5. Are truth-in-menu requirements met?					
6. If pictures are used, are they exact replicas of what is delivered?					
7. Does the wording do a good job of merchandising?					
8. Do menu items appear at consistent quality, size, etc.					

Menu Comprehension

	Excellent	Good	Fair	Poor
1. Do menu items stand out?				
2. Is print legible and easy to read quickly?				
3. Are words and reading matter not crowded?				
4. Is the menu free from clutter?				
5. Are menu items presented in a logical order, usually the order of eating?				
6. Are prices presented clearly?				
7. Are items easy to find?				
8. Will patrons know what to expect from items from the menu listing?				
9. Are any unexplained words used?				
10. Is there adequate space between menu items so there is no confusion?				
11. Does the print stand out from the background?				
12. Are headings prominent and of sufficient size?				

Physical Support

		Excellent 4	Good 3	Fair 2	Poor 1	Comments
1.	Is the menu appropriate for kitchen equipment production capability?					
2.	Is the menu appropriate for servers to do a proper job?					
3.	Is the layout of the operation adequate to meet menu needs?					
4.	Is there adequate storage to support the menu?					
5.	Are menu items matched to the ability of kitchen employees to produce them?					
6.	Is the distance between the kitchen and dining area reasonable?					
7.	Is lighting adequate to read the menu?					
8.	Is a comfortable environment provided for guest comfort?					
9.	Do patrons have adequate space?					

Menu Mechanics

		Excellent 4	Good 3	Fair 2	Poor 1	Comments
1.	Is the cover durable, easily cleaned, and attractive?					
2.	Is the menu on strong, sturdy paper?					
3.	Is the menu easily read and understood?					
4.	Can one find things easily?					
5.	Does the menu appear neat and clean?					
6.	Is the menu free of cross-outs and handwritten changes?					
7.	Are color combinations effective?					
8.	Is the shape appropriate?					
9.	Is it of proper size to be easily handled?					
10.	Are decorative features appropriate?					

	4 Excellent	3 Good	2 Fair	1 Poor	Comments
11. Are the style of the menu and wording in keeping with the decor, atmosphere, and logo of the operation?					
12. Does the menu have good symmetry and form?					
13. Does the menu look professional?					
14. Is accurate information given on operation time, address and telephone number, special catering, and so on?					
15. Is the logo prominent on the menu?					

Item Selection

	4 Excellent	3 Good	2 Fair	1 Poor
1. Is the selection of items balanced and varied?				
2. Do items offer variety of form, color, taste, and temperature?				
3. Are seasonal foods offered?				
4. Is preparation of items varied among broiling, frying, steaming, boiling, etc.?				
5. Are items suited to the preparation skill of employees and servers?				
6. Has simplicity been preserved and not too much lavish ornateness attempted?				
7. Is there a wide enough selection to appeal to most patrons?				
8. Is the sales mix effective?				
9. Are specials prominent but not over-emphasized?				
10. Are items balanced in their popularity?				

Menu and Item Presentation

	4 Excellent	3 Good	2 Fair	1 Poor
1. Are items served in the proper dish or the right packaging?				

	4 Excellent	3 Good	2 Fair	1 Poor	Comments
2. Are they well garnished?					
3. Are items attractive?					
4. Are items offered in the order of eating or in an otherwise logical manner?					
5. Are items not offered by order of price?					
6. Are long columns broken up in some way?					
7. Are items that management most wants to sell put in the most prominent places?					
8. Are children's menus offered, and do they suit children's need?					
9. Is the use of clip-ons or inserts well done (not repeating menu items, covering other material, etc.)?					
10. Do menu items reflect a manageable inventory?					
11. Are high-risk items limited?					
12. Do most items have a high volume potential?					
13. Is there good presentation of high gross profit items?					
14. Is attention given to nutritional concerns?					
15. Is there a proper balance between a la carte and table d'hôte items?					
16. Are ingredients available on the market without great price fluctuations?					
17. If it is a cycle menu, is the cycle repeated at a reasonable length of time?					
18. Is there good and adequate presentation of alcoholic beverages?					

To score: Analyze the menu, evaluating each factor in the menu evaluation form and scoring 4 for excellent, 3 for good, 2 for fair, and 1 for poor. Do not score factors that are not applicable, but keep track of those omitted.

When scoring is completed, total the numbers. Next, count the factors not considered applicable and deduct the number from 85 (the total factors in this menu evaluation form). Multiply the result by 4, which will be the total of a perfect score. Divide the actual score given the menu by the perfect score. This will give a menu evaluation percentage.

Anything from 90 percent to 100 percent is excellent; anything from 60 percent to 69 percent is poor. Most likely, any menu scoring below 80 percent should be redone.

The comment column should be used to indicate how a factor might be improved if it is not satisfactory.

PLANNING AND COST CONTROL

17

MANAGING REVENUE AND EXPENSE*

Overview

This chapter presents the relationship among foodservice revenue, expense, and profit. As a professional foodservice manager, you must understand the relationship that exists between controlling these three areas and the resulting overall success of your operation. In addition, the chapter presents the mathematical foundation you must know to express your operating results as a percentage of your revenue or budget, a method that is the standard within the hospitality industry.

Chapter Outline

Professional Foodservice Manager

Profit: The Reward for Service

Fun on the Web!

Getting Started

Understanding the Profit and Loss Statement

Fun on the Web!

Understanding the Budget

Apply What You Have Learned

Key Terms and Concepts

Highlights

At the conclusion of this chapter, you will be able to:

- Apply the basic formula used to determine profit.

- Express both expenses and profit as a percentage of revenue.

- Compare actual operating results with budgeted operating results.

---

* Authored by Jack E. Miller, Lea R. Dopson, and David K. Hayes.

PROFESSIONAL FOODSERVICE MANAGER

There is no doubt that to be a successful foodservice manager you must be a talented individual. Consider, for a moment, your role in the operation of an ongoing profitable facility. As a foodservice manager, you are both a manufacturer and a retailer. A professional foodservice manager is unique because all the functions of product sales, from item conceptualization to product delivery, are in the hands of the same individual. As a manager, you are in charge of securing raw materials, producing a product, and selling it—all under the same roof. Few other managers are required to have the breadth of skills that effective foodservice operators must have. Because foodservice operators are in the service sector of business, many aspects of management are more difficult for them than for their manufacturing or retailing management counterparts.

A foodservice manager is one of the few types of managers who actually has contact with the ultimate customer. This is not true of the manager of a tire factory or automobile production line. These individuals produce a product, but do not sell it to the person who will actually use their product. In a like manner, grocery store or computer store managers will sell their product lines, but they have had no role in actually producing their goods. The face-to-face guest contact in the hospitality industry requires that you assume the responsibility of standing behind your own work and that of your staff, in a one-on-one situation with the ultimate consumer, or end-user, of your products and services.

The management task checklist in Figure 17.1 shows just some of the areas in which foodservice, manufacturing, and retailing managers vary in responsibilities.

Task	Foodservice Manager	Manufacturing Manager	Retail Manager
1. Secure raw materials	Yes	Yes	No
2. Manufacture product	Yes	Yes	No
3. Distribute to end-user	Yes	No	Yes
4. Market to end-user	Yes	No	Yes
5. Reconcile problems with end-user	Yes	No	Yes

FIGURE 17.1 Management task checklist.

In addition to your role as a food factory supervisor, you must also serve as a cost control manager. Unless you can perform this vital role successfully, your business might well cease to exist. Foodservice management provides the opportunity for creativity in a variety of settings. The control of revenue and expense is just one more area in which the effective foodservice operator can excel. In most areas of foodservice, excellence in operation is measured in terms of producing and delivering quality products in a way that assures an appropriate operating profit for the owners of the business.

PROFIT: THE REWARD FOR SERVICE

There is an inherent problem in the study of cost control or, more accurately, cost management. The simple fact is that management's primary responsibility is to deliver a quality product or service to the guest, at a price mutually agreeable to both parties. In addition, the quality must be such that the consumer, or end-user of the product or service, feels that excellent value was received for the money spent on the transaction. When this level of service is achieved, the business will prosper. If management focuses on controlling costs more than on servicing guests, problems will certainly surface.

It is important to remember that guests cause businesses to incur costs. You do not want to get yourself in the mind-set of reducing costs to the point where it is thought that "low" costs are good and "high" costs are bad. A restaurant with $5,000,000 in revenue per year will undoubtedly have higher costs than the same size restaurant with $200,000 in revenue per year. The reason is quite clear. The food products, labor, and equipment needed to serve $5,000,000 worth of food are likely to be greater than that required to produce a smaller amount of revenue. Remember, if there are fewer guests, there are likely to be fewer costs, but fewer profits as well! Because that is true, when management attempts to reduce costs, with no regard for the impact on the balance between managing costs and guest satisfaction, the business will surely suffer.

In addition, efforts to reduce costs that result in unsafe conditions for guests or employees are never wise. While some short-term savings may result, the expense of a lawsuit resulting from a guest or employee injury can be very high. Managers who, for example, neglect to spend the money to salt and shovel a snowy restaurant entrance area may find that they spend thousands more dollars defending themselves in a lawsuit brought by an individual who slipped and fell on the ice.

As an effective manager, the question to be considered is *not* whether costs are high or low. The question is whether costs are too high or too low, given management's view of the value it hopes to deliver to the guest and the goals of the foodservice operation's owners. Managers can eliminate nearly all costs by closing the operation's doors. Obviously, however, when you close the doors to expense, you close the doors to profits. Expenses, then, must be incurred, and they must be managed in a way that allows the operation to achieve its desired profit levels.

Some people assume that if a business purchases a product for $1.00 and sells it for $3.00, the profit generated equals $2.00. In fact, this is not true. As a business operator, you must realize that the difference between what you have paid for the goods you sell and the price at which you sell them does not represent your actual profit. Instead, all expenses, including advertising, the building housing your operation, management salaries, and the labor required to generate the sale, to name but a few, are expenses that must be subtracted before you can determine your profits accurately.

Every foodservice operator is faced with the following profit-oriented formula:

Revenue − Expenses = Profit

Thus, when you manage your facility, you will receive revenue, the term used to indicate the money you take in, and you will incur expenses, the cost of the items required to operate the business. The dollars that remain after all expenses have been paid represent your profit. For the purposes of this book, the authors will use the following terms interchangeably: revenues and sales; expenses and costs.

This formula holds even in the "nonprofit" sector of foodservice management. For example, consider the situation of Hector Bentevina. Hector is the foodservice manager at the headquarters of a large corporation. Hector supplies the foodservice to a large group of office workers, each of whom is employed by the corporation that owns the facility Hector manages. In this situation, Hector's employer clearly does not have "profit" as its primary motive. In most business dining situations, food is provided as a service to the company's employees either as a no-cost (to the employee) benefit or at a greatly reduced price. In some cases, executive dining rooms may be operated for the convenience of management. In all cases, however, some provision for profit must be made. Figure 17.2 shows the flow of business for the typical foodservice operation. Note that profit must be taken out at some point in the process, or management is in a position of simply trading cash for cash.

If you find that, in your own operation, revenue is less than or equal to real expense, with no reserve for the future, you will likely also find that there is no money for new equipment, needed equipment maintenance may not be performed, employee raises (as well as your own) may be few and far between, and, in general, the foodservice facility will become outdated due to a lack of funds needed to remodel and upgrade. The truth is, all foodservice operations need revenue in excess of expenses if they are to

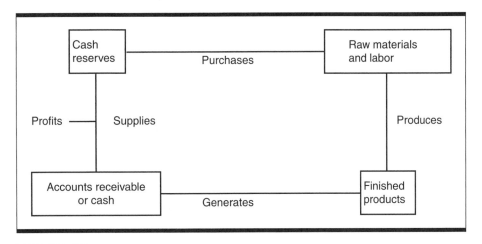

FIGURE 17.2 Foodservice business flowchart.

thrive. Whether you manage a foodservice operation in a profit or a nonprofit setting, it will be your responsibility to communicate this message to your own staff.

Profit is the result of solid planning, sound management, and careful decision making. The purpose of this text is to give you the information and tools you need to make informed decisions with regard to managing your operation's revenue and expense. If these tools are utilized properly, the potential for achieving the profits you desire is greatly enhanced.

Profit should not be viewed as what is left over after the bills are paid. In fact, careful planning is necessary to earn a profit. Obviously, investors will not invest in businesses that do not generate enough profit to make their investment worthwhile. Because that is true, a more appropriate formula, which recognizes and rewards the business owner for the risk associated with business ownership or investment, is as follows:

Revenue − Desired Profit = Ideal Expense

Ideal expense, in this case, is defined as management's view of the correct or appropriate amount of expense necessary to generate a given quantity of revenue. Desired profit is defined as the profit that the owner wants to achieve on that predicted quantity of revenue.

This formula clearly places profit as a reward for providing service, not a leftover. When foodservice managers deliver quality and value to their guests, anticipated revenue levels can be achieved and desired profit is attainable. Desired profit and ideal expense levels are not, however, easily achieved. It takes an astute foodservice operator to consistently make decisions that will maximize revenue while holding expenses to the ideal or appropriate amount. This book will help you to do just that.

Revenue

To some degree, you can manage your revenue levels. Revenue dollars are the result of units sold. These units may consist of individual menu items, lunches, dinners, drinks, or any other item produced by your operation. Revenue varies with both the number of guests frequenting your business and the amount of money spent by each guest. You can increase revenue by increasing the number of guests you serve, by increasing the amount each guest spends, or by a combination of both approaches. Adding seating or drive-in windows, extending operating hours, and building additional foodservice units are all examples of management's efforts to increase the number of guests choosing to come to the restaurant or foodservice operation. Suggestive selling by service staff, creative menu-pricing techniques, as well as discounts for very large purchases are all examples of efforts to increase the amount of money each guest spends.

It is the opinion of the authors that management's primary task is to take the steps necessary to deliver guests to the foodservice operation. This is true because the profit formula begins with revenue. Experienced foodservice operators know that increasing revenue through adding guests, suggestive selling, or possibly raising menu prices is an extremely effective way of increasing overall profitability, but only *if* effective cost management systems are in place.

The focus of this text is on managing and controlling expense, not generating additional revenue. While the two are clearly related, they are different. Marketing efforts, restaurant design and site selection, employee training, and food preparation methods are all critical links in the revenue-producing chain. No amount of effective expense control can solve the profit problems caused by inadequate revenue result-ing from inferior food quality or service levels. Effective cost control, however, when coupled with management's aggressive attitude toward meeting and exceeding guests' expectations, can result in outstanding revenue and profit performance.

FUN ON THE WEB!

www.restaurant.org   Link to "Industry Research," then "Industry at a Glance" to see the National Restaurant Association's revenue projections for the industry.

Expenses

There are four major foodservice expense categories that you must learn to control. They are:

1. Food costs
2. Beverage costs
3. Labor costs
4. Other expenses

Food Costs Food costs are the costs associated with actually producing the menu items a guest selects. They include the expense of meats, dairy, fruits, vegetables, and other categories of food items produced by the foodservice operation. When computing food costs, many operators include the cost of minor paper and plastic items, such as the paper wrappers used to wrap sandwiches. In most cases, food costs will make up the largest or second largest expense category you must learn to manage.

Beverage Costs Beverage costs are those related to the sale of alcoholic beverages. It is interesting to note that it is common practice in the hospitality industry to consider beverage costs of a nonalcoholic nature as an expense in the food cost category. Thus, milk, tea, coffee, carbonated beverages, and other nonalcoholic beverage items are not generally considered a beverage cost.

Alcoholic beverages accounted for in the beverage cost category include beer, wine, and liquor. This category may also include the costs of ingredients necessary to produce these drinks, such as cherries, lemons, olives, limes, mixers like carbonated beverages and juices, and other items commonly used in the production and service of alcoholic beverages.

Labor Costs Labor costs include the cost of all employees necessary to run the business. This expense category would also include the amount of any taxes you are required to pay when you have employees on your payroll. Some operators find it helpful to include the cost of management in this category. Others prefer to place the cost of managers in the other expense category. In most operations, however, labor costs are second only to food costs in total dollars spent. If management is included as a labor cost rather than an other expense, then this category can well be even larger than the food cost category.

Other Expenses Other expenses include all expenses that are neither food, nor beverage, nor labor. Examples include franchise fees, utilities, rent, linen, and such items as china, glassware, kitchen knives, and pots and pans. While this expense category is sometimes incorrectly referred to as "minor expenses," your ability to successfully control this expense area is especially critical to the overall profitability of your foodservice unit.

GETTING STARTED

Good managers learn to understand, control, and manage their expenses. Consider the case of Tabreshia Larson, the Food and Beverage director of the 200-room Renaud Hotel, located in a college town and built near an interstate highway. Tabreshia has just received her end-of-the-year operating reports for the current year. She is interested in comparing these results to those of the prior year. The numbers she received are shown in Figure 17.3.

Tabreshia is concerned, but she is not sure if she should be. Revenue is higher than last year, so she feels her guests must like the products and services they receive. In fact, repeat business from corporate meetings and special-events meals is really beginning to develop. Profits are greater than last year also, but Tabreshia has the uneasy feeling that things are not going as well as they could. The kitchen appears to run smoothly. The staff, however, often runs out of needed items, and there seems to be a large amount of leftover food thrown away on a regular basis. Sometimes, there seem to be too many staff members on the property; at other times, guests have to wait too long to get served. Tabreshia also feels that employee theft may be occurring, but she certainly doesn't have the time to watch every storage area within her operation. Tabreshia also senses that the hotel general manager, who is Tabreshia's boss, may be less than pleased with her department's performance. She would really like to get a handle on the problem (if there is one), but how and where should she start?

	This Year	Last Year
Revenue	$1,106,040	$850,100
Expense	1,017,557	773,591
Profits	**88,483**	**76,509**

FIGURE 17.3 Renaud Hotel operating results.

The answer for Tabreshia, and for you, if you want to develop a serious expense control system, is very simple. You start with the basic mathematics skills that you must have to properly analyze your expenses. The mathematics required, and used in this text, consist only of addition, subtraction, multiplication, and division. These tools will be sufficient to build a cost control system that will help you professionally manage the expenses you incur.

What would it mean if a fellow foodservice manager told you that he spent $500 on food yesterday? Obviously, it means little unless you know more about his operation. Should he have spent $500 yesterday? Was that too much? Too little? Was it a "good" day? These questions raise a difficult problem. How can you equitably compare your expenses today with those of yesterday, or your foodservice unit with another, so that you can see how well you are doing?

We know that the value of dollars has changed over time. A restaurant with revenue of $1,000 per day in 1954 is very different from the same restaurant with daily revenue of $1,000 today. The value of the dollar today is quite different from what it was in 1954. Generally, inflation causes the purchasing power of a dollar today to be less than that of a dollar from a previous time period. While this concept of changing value is useful in the area of finance, it is vexing when one wants to answer the simple question, "Am I doing as well today as I was doing five years ago?"

Alternatively, consider the problem of a multiunit manager. Two units sell tacos on either side of a large city. One uses $500 worth of food products each day; the other unit uses $600 worth of food products each day. Does the second unit use an additional $100 worth of food each day because it has more guests or because it is less efficient in utilizing the food?

The answers to all of the preceding questions, and many more, can be determined if we use percentages to relate expenses incurred to revenue generated. Percentage calculations are important for at least two major reasons. First and foremost, percentages are the most common standard used for evaluating costs in the foodservice industry. Therefore, knowledge of what a percentage is and how it is calculated is vital. Second, as a manager in the foodservice industry, you will be evaluated primarily on your ability to compute, analyze, and control these percentage figures. Percentage calculations are used extensively in this text and are a cornerstone of any effective cost control system.

Review of Percentages

Understanding percentages and how they are mathematically computed is important. The following review may be helpful for some readers. If you thoroughly understand the concept of a percentage, you may skip both this section and the following one on Computing Percentages and proceed directly to the section on Using Percentages.

Form	Percent		
	1%	10%	100%
Common	1%	10%	100%
Fraction	1/100	10/100	100/100
Decimal	0.01	0.10	1.00

FIGURE 17.4 Forms of expressing percent.

The word "percent" (%) means "out of each hundred." Thus, 10 percent would mean 10 out of each 100. If we ask how many guests will buy blueberry pie on a given day, and the answer is 10 percent, then 10 people out of each 100 we serve will select blueberry pie. If 52 percent of your employees are female, then 52 out of each 100 employees are female. If 15 percent of your employees will receive a raise this month, then 15 out of 100 employees will get their raise.

Figure 17.4 shows three ways to write a percent.

Common Form In its common form, the "%" sign is used to express the percentage. If we say 10%, then we mean "10 out of each 100," and no further explanation is necessary. The common form, the "%," is equivalent to the same amount expressed in either the fraction or the decimal form.

Fraction Form In fraction form, the percent is expressed as the part, or a portion, of 100. Thus, 10 percent is written as 10 "over" 100 (10/100). This is simply another way of expressing the relationship between the part (10) and the whole (100).

Decimal Form A decimal is a number developed from the counting system we use. It is based on the fact that we count to 10, then start over again. In other words, each of our major units—10s, 100s, 1,000s, and so on—are based on the use of 10s, and each number can easily be divided by 10. Instead of using the % sign, the decimal form uses the decimal point (.) to express the percent relationship. Thus, 10% is expressed as 0.10 in decimal form. The numbers to the right of the decimal point express the percentage.

Each of these three methods of expressing percentages is used in the foodservice industry, and to be successful you must develop a clear understanding of how a percentage is computed. Once that is known, you can express the percentage in any form that is required or that is useful to you.

Computing Percentages

To determine what percent one number is of another number, divide the number that is the part by the number that is the whole. Usually, but not always, this means dividing the smaller number by the larger number. For example, assume that 840 guests were served during a banquet at your hotel; 420 of them asked for coffee with their meal. To find what percentage of your guests ordered coffee, divide the part (420) by the whole (840).

The process looks as follows:

$$\frac{\text{Part}}{\text{Whole}} = \text{Percent} \quad \text{or} \quad \frac{420}{840} = 0.50$$

Thus, 50% (common form), 50/100 (fraction form), or 0.50 (decimal form) represents the proportion of people at the banquet who ordered coffee. A large number of new foodservice managers have difficulty computing percentage figures. It is easy to forget which number goes "on the top" and which number goes "on the bottom." In general, if you attempt to compute a percentage and get a whole number (a number larger than 1), either a mistake has been made or costs are extremely high!

Many people also become confused when converting from one form of a percentage to another. If you are one of those, remember the following conversion rules:

1. To convert from common form to decimal form, move the decimal two places to the left, that is, 50.00% = 0.50.

2. To convert from decimal form to common form, move the decimal two places to the right, that is, 0.40 = 40.00%.

In a restaurant, the "whole" is usually a revenue figure. Expenses and profits are the "parts," which are usually expressed in terms of a percentage. It is interesting to note that, in the United States, the same system in use for our numbers is in use for our money. Each dime contains 10 pennies, each dollar contains 10 dimes, and so on. Thus, it is true that a percentage, when discussing money, refers to "cents out of each dollar" as well as "out of each 100 dollars." When we say 10% of a dollar, we mean 10 cents, or "10 cents out of each dollar." Likewise, 25% of a dollar represents 25 cents, 50% of a dollar represents 50 cents, and 100% of a dollar represents $1.00.

Sometimes, when using a percentage to express the relationship between portions of a dollar and the whole, we find that the part is, indeed, larger than the whole. Figure 17.5 demonstrates the three possibilities that exist when computing a percentage.

Possibilities	Examples	Results
Part is smaller than the whole	$\frac{61}{100} = 61\%$	Always less than 100%
Part is equal to the whole	$\frac{35}{35} = 100\%$	Always equals 100%
Part is larger than the whole	$\frac{125}{50} = 250\%$	Always greater than 100%

FIGURE 17.5 Percent computation.

Great care must always be taken when computing percentages, so that the percentage arrived at is of help to you in your work and does not represent an error in mathematics.

Using Percentages

Consider a restaurant that you are operating. Imagine that your revenues for a week are in the amount of $1,600. Expenses for the same week are $1,200. Given these facts and the information presented earlier in this chapter, your profit formula for the week would look as follows:

> **Revenue − Expense = Profit**
>
> *or*
>
> $1,600 − $1,200 = $400

If you had planned for a $500 profit for the week, you would have been "short." Using the alternative profit formula presented earlier, you would find

> **Revenue − Desired Profit = Ideal Expense**
>
> *or*
>
> $1,600 − $500 = $1,100

Note that expense in this example ($1,200) exceeds ideal expense ($1,100) and, thus, too little profit was achieved.

These numbers can also be expressed in terms of percentages. If we want to know what percentage of our revenue went to pay for our expenses, we would compute it as follows:

$$\frac{\text{Expense}}{\text{Revenue}} = \text{Expense \%}$$

$$or$$

$$\frac{\$1,200}{\$1,600} = 0.75, \quad or \quad 75\%$$

Another way to state this relationship is to say that each dollar of revenue costs 75 cents to produce. Also, each revenue dollar taken in results in 25 cents profit.

$$\$1.00 \text{ Revenue} - \$0.75 \text{ Expense} = \$0.25 \text{ Profit}$$

As long as expense is smaller than revenue, some profit will be generated, even if it is not as much as you had planned. You can compute profit % using the following formula:

$$\frac{\text{Profit}}{\text{Revenue}} = \text{Profit \%}$$

In our example:

$$\frac{\$400 \text{ Profit}}{\$1,600 \text{ Revenue}} = 25\% \text{ Profit}$$

We can compute what we had planned our profit % to be by dividing desired profit ($500) by revenue ($1,600):

$$\frac{\$500 \text{ Desired Profit}}{\$1,600 \text{ Revenue}} = 31.25\% \text{ Desired Profit}$$

In simple terms, we had hoped to make 31.25% profit, but instead made only 25% profit. Excess costs could account for the difference. If these costs could be identified and corrected, we could perhaps achieve the desired profit percentage. Most foodservice operators compute many cost percentages, not just one. The major cost divisions used in foodservice are as follows:

1. Food and beverage cost

2. Labor cost

3. Other expenses

A modified profit formula, therefore, looks as follows:

Revenue − (Food and Beverage Cost + Labor Cost
+ Other Expenses) = Profit

Put in another format, the equation looks as follows:

Revenue (100%)
  − Food and Beverage Cost %
  − Labor Cost %
  − Other Expenses %
  = Profit %

Regardless of the approach used, foodservice managers must evaluate their expenses, and they use percentages to do so.

UNDERSTANDING THE PROFIT AND LOSS STATEMENT

Consider Figure 17.6, an example from Pat's Steakhouse. All of Pat's expenses and profits can be computed as percentages by using the revenue figure, $400,000, as the whole, with expenses and profit representing the parts, as follows:

$$\frac{\text{Food and Beverage Cost}}{\text{Revenue}} = \text{Food and Beverage Cost \%}$$

or

$$\frac{\$150,000}{\$400,000} = \$37.50\%$$

Revenue	$400,000
Expenses	
Food and Beverage Cost $150,000	
Labor Cost 175,000	
Other Expense 25,000	
Total Expense	$350,000
Profit	**$ 50,000**

FIGURE 17.6 Pat's Steakhouse.

Similarly,

$$\frac{\text{Labor Cost}}{\text{Revenue}} = \text{Labor Cost \%}$$

or

$$\frac{\$175,000}{\$400,000} = \$43.75\%$$

And of course,

$$\frac{\text{Other Expenses}}{\text{Revenue}} = \text{Other Expenses \%}$$

or

$$\frac{\$25,000}{\$400,000} = \$6.25\%$$

Thus,

$$\frac{\text{Total Expenses}}{\text{Revenue}} = \text{Total Expenses \%}$$

or

$$\frac{\$350,000}{\$400,000} = \$87.50\%$$

Finally, we can compute:

$$\frac{\text{Profit}}{\text{Revenue}} = \text{Profit \%}$$

or

$$\frac{\$50,000}{\$400,000} = \$12.50\%$$

An accounting tool that details revenue, expenses, and profit, for a given period of time, is called the income statement, commonly referred to as profit and loss (P&L) statement. It lists revenue, food and beverage cost, labor cost, and other expense. The P&L also identifies profits since, as you recall, profits are generated by the formula:

Revenue − Expense = Profit

Figure 17.7 is a simplified P&L statement for Pat's Steakhouse. Note the similarity to Figure 17.6. This time, however, expenses and profit are expressed both in terms of dollar amounts and percent of revenue.

Another way of looking at Pat's P&L is shown in Figure 17.8. The pieces of the pie represent Pat's cost and profit categories. Costs and profit total 100%, which is equal

Revenue		$400,000	100%
Food and Beverage Cost	$150,000		37.50%
Labor Cost	175,000		43.75%
Other Expense	25,000		6.25%
Total Expense		**$350,000**	**87.50%**
Profit		**$ 50,000**	**12.50%**

FIGURE 17.7 Pat's Steakhouse P&L.

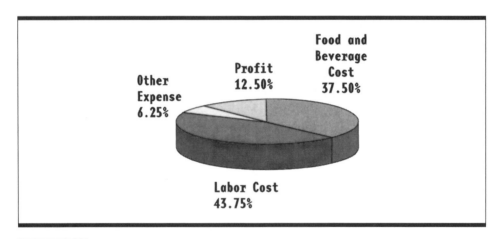

FIGURE 17.8 Pat's Steakhouse costs and profit as a percentage of revenues.

to Pat's total revenues. To put it in another way, out of every revenue dollar that Pat generates, 100% is designated as either costs or profit.

Pat knows from the P&L that revenues represent 100% of the total dollars available to cover expenses and provide for a profit. Food and beverage cost is 37.50%, and labor cost percentage in the steakhouse equals 43.75%. Other expense percentage equals 6.25%, and thus Pat's total expense percentage is 87.50% (37.50 + 43.75 + 6.25 = 87.50%). The steakhouse profit equals 12.50%. Thus, for each dollar in revenue, Pat earns a profit of 12.50 cents. Pat's revenue, expense, and profit information is contained in the steakhouse's P&L.

In restaurants that serve alcohol, food costs and beverage costs are most often separated into two categories in the P&L. Likewise, food revenues and beverage revenues are reported separately. This is done so that the food cost can be compared to food revenues, and the beverage cost can be compared to beverage revenues. Suppose, for example, that one manager is responsible for controlling food cost % in the restaurant, and another manager is responsible for controlling beverage cost % in the bar. Separation of these two "departments," then, is especially helpful when evaluating the performance of these two managers. It also helps these managers to quickly identify and anticipate problems associated with their costs and identify ways to correct these problems.

The P&L is important because it indicates the efficiency and profitability of an operation. Because so many individuals and groups are interested in a food facility's performance, it is important that the P&L and other financial statements are prepared in a manner that is consistent with other facilities. If, for example, you owned two Italian restaurants, it would be very confusing if one of your managers used a particular method for preparing his or her unit's P&L, while the other manager used an entirely different method. You, your investors, your accountant, governmental taxing entities, and your creditors may all be interested in your operational results, and unless you report and account for these in a manner they can easily understand, confusion may result.

To avoid such a set of circumstances, the Uniform System of Accounts is used to report financial results in most foodservice units. This system was created to ensure uniform reporting of financial results. A Uniform System of Accounts exists for restaurants, and there are also Uniform Systems of Accounts for hotels and for clubs. The Uniform System of Accounts is discussed in greater detail in Chapter 22 of this text.

The primary purpose of preparing a P&L is to identify revenue, expenses, and profits for a given time period. As a manager, your efforts, more than any other factor, will influence your operation's profitability. Good managers provide excellent value to their guests, which causes guests to return, and thus increases revenue. In addition, good managers know how to analyze, manage, and control their costs. For these managers, expenses are held to the amount that was preplanned. The result is the desired profit level. Good managers influence the success of their units and their own employees. The results for them personally are promotions, added responsibilities, and salary increases. If you wish to succeed in the hospitality industry, it is important to remember that your performance will be evaluated primarily on your ability to achieve the profit levels your operation has planned for.

In addition to your own efforts, many factors influence profit dollars and profit percentage, and you must be aware, and in control, of all of them. All of the factors that impact profit percentage are discussed in later chapters of this text.

FUN ON THE WEB!

www.restaurant.org   Link to "Industry Research," then "Reports" to see how you can get industry averages for P&Ls.

UNDERSTANDING THE BUDGET

Some foodservice managers do not generate revenue on a daily basis. Consider, for a moment, the foodservice manager at a summer camp run for children. In this case, parents pay a fixed fee to cover housing, activities, and meals for a set period of time. The foodservice director, in this situation, is just one of several managers who must share this revenue. If too many dollars are spent on providing housing or play activities, too few dollars may be available to provide an adequate quantity or quality of meals. On the other hand, if too many dollars are spent on providing foodservice, there may not be enough left to cover other needed expense areas. In a case like this, foodservice operators should prepare a budget. A budget is simply an estimate of projected revenue, expense, and profit. In some hospitality companies, the budget is known as the plan, referring to the fact that the budget details the operation's estimated, or "planned for," revenue and expense for a given time period.

Both commercial and noncommercial foodservice operators may use budgets. They are most frequently used, however, by effective managers, whether in the commercial (for profit) or nonprofit sector. Budgeting is simply planning for revenue, expense, and profit. If these items are planned for, you can determine how close your actual performance is to your plan or budget. In the summer camp example, the following information is known:

1.   Number of campers: 180
2.   Number of meals served to each camper per day: 3
3.   Length of campers' stay: 7 days

With 180 campers eating 3 meals each day for 7 days, 3,780 meals will be served (180 campers × 3 meals per day × 7 days = 3,780 meals).

Generally, in a case such as the summer camp, the foodservice director is given a dollar amount that represents the allowed expense for each meal to be served. For example, if $1.85 per meal is the amount budgeted for this director, the total revenue budget would equal $6,993 ($1.85 per meal × 3,780 meals = $6,993).

From this figure, an expense budget can begin to be developed. In this case, we are interested in the amount of expenses budgeted and the amount actually spent on

Weekday	Budgeted Amount	% of Total
Monday	$1.00	14.28%
Tuesday	$1.00	14.28%
Wednesday	$1.00	14.28%
Thursday	$1.00	14.28%
Friday	$1.00	14.28%
Saturday	$1.00	14.28%
Sunday	$1.00	14.28%
Total	$7.00	100.00%

FIGURE 17.9 Candy purchases.

expenses. Equally important, we would be interested in the percentage of the budget actually used, a concept known as performance to budget.

A simple example may help to firmly establish the idea of budget and performance to budget. Assume that a child has $1.00 per day to spend on candy. On Monday morning, the child's parents give the child $1.00 for each day of the week, or $7.00 total ($1.00 × 7 days = $7.00). If the child spends only $1.00 per day, he or she will be able to buy candy all week. If, however, too much is spent in any one day, there may not be any money left at the end of the week. A good spending pattern could be tabulated in Figure 17.9. The % of Total column is computed by dividing $1.00 (the part) by $7.00 (the whole). Notice that we can determine the percentage of total that should have been spent by any given day; that is, each day equals 14.28%, or $\frac{1}{7}$ of the total.

This same logic applies to the foodservice operation. Figure 17.10 represents commonly used budget periods and their accompanying proportion amount.

Many foodservice operations are changing from "one month" budget periods to periods of 28 days. The 28-day-period approach divides a year into 13 equal periods of 28 days each. Therefore, each period has four Mondays, four Tuesdays, four Wednesdays, and so on. This helps the manager compare performance from one period to the next without having to compensate for "extra days" in any one period. The downside of this approach is that you can no longer talk about the month of March, for example,

Budget Period	Portion	% of Total
One week	One day	1/7 or 14.3%
Two-week period	One day One week	1/14 or 7.1% 1/2 or 50.0%

One month
  28 days
  30 days
  31 days
	One week
	One day
	One day
	One day
		1/4 or 25.0%
		1/28 or 3.6%
		1/30 or 3.3%
		1/31 or 3.2%

Six months	One month	1/6 or 16.7%
One year	One day One week One month	1/365 or 0.3% 1/52 or 1.9% 1/12 or 8.3%

FIGURE 17.10 Common foodservice budget periods.

because "period 3" would occur during part of February and part of March. Although using the 28-day-period approach takes a while to get used to, it is an effective way to measure performance and plan from period to period.

In the summer camp, after one week's camping was completed, we found the results shown in Figure 17.11. Assume that we used the expense records from last summer, as well as our solid industry knowledge and experience, to develop the expense budget figures for this summer.

In this case, we are interested in both our plan (budget) and our actual performance. Figure 17.12 shows a performance to budget summary with revenue and expenses presented in terms of both the budget amount and the actual amount. In all cases, percentages are used to compare actual expense with the budgeted amount, using the formula:

$$\frac{\text{Actual}}{\text{Budget}} = \text{\% of Budget}$$

In this example, revenue remained the same although some campers skipped (or slept through!) some of their meals. This is often the case when one fee or price buys a

Item	Budget	Actual
Meals Served	3,780	3,700
Revenue	$6,993	$6,993
Food Expense	$2,600	$2,400
Labor Expense	$2,800	$2,900
Other Expense	$ 700	$ 965
Profit	$ 893	$ 728

FIGURE 17.11 Camp Eureka one-week budget.

Item	Budget	Actual	% of Budget
Meals Served	3,780	3,700	97.9%
Revenue	$6,993	$6,993	100.0%
Food Expense	$2,600	$2,400	92.3%
Labor Expense	$2,800	$2,900	103.6%
Other Expense	$ 700	$ 965	137.9%
Total Expense	$6,100	$6,265	102.7%
Profit	$ 893	$ 728	81.5%

FIGURE 17.12 Camp Eureka performance to budget summary.

number of meals, whether they are eaten or not. In some other cases, managers will only receive revenue for meals actually served. This, of course, is true in a traditional restaurant setting. In either case, budgeted amount, actual expense, and the concept of % of budget, or performance to budget, are important management tools. In looking at the Camp Eureka performance to budget summary, we can see that the manager served fewer meals than planned and, thus, spent less on food than estimated, but spent more on labor than originally thought necessary. In addition, much more was spent than estimated for other expenses (137.9% of the budgeted amount). As a result, the profit dollars were lower than planned. This manager has some problems, but they are not everywhere in the operation.

How do we know that? If our budget was accurate and we are within reasonable limits of our budget, we are said to be "in line," or "in compliance," with our budget, because it is difficult to budget exact revenue and expenses. If, as management, we decided that plus (more than) or minus (less than) 10% of budget in each category would be considered in line, or acceptable, then an examination of Figure 17.12 shows we are in line with regard to meals served, food expense, labor expense, and total expense. We are not in line with other expenses because they were 137.9% of the amount originally planned. Thus, they far exceed the 10% variation that was reasonably allowed. Profit also was outside the acceptable boundary we established because it was only 81.5% of the amount budgeted. Note that figures over 100% mean too much (other expense), while figures below 100% mean too little (profit).

Many operators use the concept of "significant" variation to determine whether a cost control problem exists. In this case, a significant variation is any variation in expected costs that management feels is an area of concern. This variation can be caused by costs that were either higher or lower than the amount originally budgeted or planned for.

When you manage a foodservice operation and you find that significant variations with your planned results occur, you must:

1.   Identify the problem.
2.   Determine the cause.
3.   Take corrective action.

It is crucial to know the kind of problem you have if you are to be an effective problem solver. Management's attention must be focused on the proper place. In this case, the proper areas for concern are other expense and profit. If, in the future, food expense became too low, it, too, would be an area of concern. Why? Remember that expenses create revenue; thus, it is not your goal to eliminate expense. In fact, those managers who focus too much on eliminating expense, instead of building revenue, often find that their expenses are completely eliminated when they are forced to close their operation's doors permanently because guests did not feel they received good value for the money spent at that restaurant! Control and management of revenue and expense are important. Elimination of either is not desired.

As you have seen, revenue and expense directly impact profit. Your important role as a hospitality manager is to analyze, manage, and control your costs so that you achieve planned results. It can be done, and it can be fun.

The remainder of this text discusses how you can best manage and account for foodservice revenue and expense. With a good understanding of the relationship among revenue, expense, and profit, and your ability to analyze using percentages, you are ready to begin the cost control and cost management process.

APPLY WHAT YOU HAVE LEARNED

Jennifer Caratini has recently accepted the job as the Foodservice Director for Techmar Industries, a corporation with 1,000 employees. As Foodservice Director, Jennifer's role is to operate a company cafeteria, serving 800 to 900 meals per day, as well as an executive dining room, serving 100 to 200 meals per day. All of the meals are provided "free of charge" to the employees of Techmar. One of Jennifer's first jobs is to prepare a budget for next year's operations.

1.   In addition to food products and foodservice employees, what are other expenses Techmar will incur by providing free meals to its employees?

2.   Since employees do not pay for their food directly, what will Jennifer use as the "revenue" portion of her budget? How do you think this number should be determined?

3.   In addition to her know-how as a foodservice manager, what skills will Jennifer need as she interacts with the executives at Techmar who must approve her budget?

KEY TERMS AND CONCEPTS

The following are terms and concepts discussed in the chapter that are important for you as a manager. To help you review, please define the terms below.

Revenue	Food costs	Income statement
Expense	Beverage costs	Profit and loss statement (P&L)
Profit	Labor costs	Budget/plan
Ideal expense	Other expenses	Performance to budget
Desired profit	Percent	28-day-period approach

DETERMINING SALES FORECASTS*

Overview

This chapter presents the methods and procedures you must learn to create accurate histories of what you have sold in the past as well as projections of how much you will sell in the future. This includes the total revenue you will generate, the number of guests to be served, and the number of dollars each guest will spend. Knowledge of these techniques is critical if you are to analyze sales trends in the facility you manage and be prepared to serve your future guests well.

Chapter Outline

---

* Authored by Jack E. Miller, Lea R. Dopson, and David K. Hayes.

Highlights

At the conclusion of this chapter, you will be able to:

- Develop a procedure to record current sales.
- Compute percentage increases or decreases in sales over time.
- Develop a procedure to estimate future sales.

IMPORTANCE OF FORECASTING SALES

The first question operating managers must ask themselves is very simple: "How many guests will I serve today? This week? This year?" The answers to questions such as these are critical, since these guests will provide the revenue from which the operator will pay basic operating expenses. Simply put, if too few guests are served, total revenue may be insufficient to cover costs, even if these costs are well managed. In addition, purchasing decisions regarding the kind and quantity of food or beverage to buy are dependent on knowing the number of guests who will be coming to consume those products. Labor required to serve the guests is also determined based on the manager's "best guess" of the projected number of guests to be served and what these guests will buy. Forecasts of future sales are normally based on your sales history since what has happened in the past in your operation is usually the best predictor of what will happen in the future.

In the hospitality industry, we have many ways of counting or defining sales. In its simplest case, sales are the dollar amount of revenue collected during some predetermined time period. The time period may be an hour, shift, day, week, month, or year. When used in this manner, sales and revenue are interchangeable terms. When you predict the number of guests you will serve and the revenues they will generate in a given future time period, you have created a sales forecast.

You can determine your actual sales for a current time period by using a computerized system called a point of sales (POS) system that has been designed to provide specific sales information. Alternatively, a standard cash register, or even manually produced guest checks or head counts, will help you establish how many sales were completed. Today, however, even the smallest of foodservice operations should take advantage of the speed and accuracy provided by computerized programs for recording sales.

It is important to remember that a distinction is made in the hospitality industry between sales (revenue) and sales volume, which is the number of units sold. Consider Manuel, a bagel shop manager, whose Monday business consists of $2,000 in sales (revenue) because he actually sold 3,000 bagels (sales volume). Obviously, it is important for Manuel to know how much revenue is taken in, so he can evaluate the expenses required to generate his revenue and the number of units that have been sold. With this information, he can properly prepare to serve additional guests the next day.

In many areas of the hospitality industry, for example, in college and university dormitory foodservice, it is customary that no cash actually changes hands during a particular meal period. Of course, the manager of such a facility still created sales and would be interested in sales volume, that is, how much food was actually consumed by the students on that day. This is critical information because, as we have seen, a manager must be prepared to answer the question, "How many individuals did I serve today, and how many should I expect tomorrow?"

In some cases, your food and beverage operation may be a blend of cash and noncash sales. Consider Tonya Brown, a hospital foodservice director. It is very likely that Tonya will be involved in serving both cash guests (public cafeteria) and noncash patients (tray line). In addition, employee meals may be cash sales but at a reduced or subsidized rate. Clearly, Tonya's operation will create sales each day, and it will be important to her and her staff to know, as accurately as possible, how many of each type guest she will serve.

An understanding of anticipated sales, in terms of revenue dollars or guest counts, will help you have the right number of workers, with the right amounts of product available, at the right time. In this way, you can begin to effectively manage your costs. In addition to the importance of accurate sales records for purchasing and staffing, sales records are valuable to the operator when developing labor standards. Consider, for example, a large restaurant with 400 seats. If an individual server can serve 25 guests at lunch, you would need 400/25, or 8, servers per lunch shift. If management keeps no accurate sales histories or forecasts, too few or too many servers could possibly be scheduled on any given lunch period. With accurate sales records, a sales history can be developed for each foodservice outlet you operate, and better decisions will be reached with regard to planning for each unit's operation. Figure 18.1 lists some of the advantages that you gain when you can accurately predict the number of people you will serve in a future time period.

SALES HISTORY

A sales history is the systematic recording of all sales achieved during a predetermined time period. It is no less than an accurate record of what your operation has sold. Before you can develop a sales history, however, it is necessary for you to think about the definition of sales that is most helpful to you and your understanding of how the

1. Accurate revenue estimates
2. Improved ability to predict expenses
3. Greater efficiency in scheduling needed workers
4. Greater efficiency in scheduling menu item production schedules
5. Better accuracy in purchasing the correct amount of food for immediate use
6. Improved ability to maintain proper levels of nonperishable food inventories
7. Improved budgeting ability
8. Lower selling prices for guests because of increased operational efficiencies
9. Increased dollars available for current facility maintenance and future growth
10. Increased profit levels and stockholder value

FIGURE 18.1 Advantages of precise sales forecasts.

facility you manage functions. The simplest type of sales history records revenue only. The sales history format shown in Figure 18.2 is typical for recording sales revenue on a weekly basis.

Notice that, in this most basic of cases, you would determine daily sales either from your POS system, from sales revenue recorded on your cash register, or from adding the information recorded on your guest checks. You would then transfer that number, on a daily basis, to the sales history by entering the amount of your daily sales in the column titled "Daily Sales." "Sales to Date" is the cumulative total of sales reported in the unit. Thus, Sales to date is the number we get when we add today's sales to the sales of all prior days in the reporting period.

Sales to date on Tuesday, January 2, is computed by adding Tuesday's sales to those of the prior day ($851.90 + $974.37 = $1,826.27). The "Sales to Date" column is a running total of the sales achieved by Rae's Restaurant for the week. Should Rae's manager prefer it, the sales period could, of course, be defined in blocks other than one week. Common alternatives are meal periods, days, weeks, two-week periods, four-week (28-day) periods, months, quarters (three-month periods), or any other unit of time that is helpful to the manager.

Sometimes, you will not have the ability to consider sales in terms of revenue generated. Figure 18.3 is the type of sales history you can use when no cash sales are typically reported. Notice that, in this case, the manager is interested in recording sales

Rae's Restaurant			
Sales Period	Date	Daily Sales	Sales to Date
Monday	1/1	$ 851.90	$ 851.90
Tuesday	1/2	974.37	1,826.27
Wednesday	1/3	1,004.22	2,830.49
Thursday	1/4	976.01	3,806.50
Friday	1/5	856.54	4,663.04
Saturday	1/6	1,428.22	6,091.26
Sunday	1/7	1,241.70	7,332.96
Week's Total			7,332.96

FIGURE 18.2 Sales history.

based on meal periods rather than some alternative time frame such as a 24-hour (one-day) period. This approach is often used in such settings as extended care facilities, nursing homes, college dormitories, correctional facilities, hospitals, camps, or any other situation where knowledge of the number of actual guests served during a given period is critical for planning purposes.

Given the data in Figure 18.3, the implications for staffing service personnel at the camp are very clear. Fewer service personnel are needed from 9:00 to 11:00 A.M. than from 7:00 to 9:00 A.M. The reason is obvious. Fewer campers eat between 9:00 and 11:00 A.M. (40) than between 7:00 and 9:00 A.M. (121). Notice that, as a knowledgeable manager, if you were operating this camp, you could either reduce staff during the slower service period or shift those workers to some other necessary task. Notice also that you might decide not to produce as many menu items for consumption during the 9:00 to 11:00 A.M. period. In that way, you could make more efficient use of both labor and food products. It is simply easier to manage well when you know the answer to the question, "How many guests will I serve?"

Sales histories can be created to record revenue, guests served, or both. It is important, however, that you keep good records of how much you have sold because doing so is the key to accurately predicting the amount of sales you will achieve in the future.

Eureka Summer Camp								
	Guests Served							
Serving Period	Mon	Tues	Wed	Thurs	Fri	Sat	Sun	Total
7:00–9:00 A.M.	121							
9:00–11:00 A.M.	40							
11:00–1:00 P.M.	131							
1:00–3:00 P.M.	11							
3:00–5:00 P.M.	42							
5:00–7:00 P.M.	161							
Total Served	506							

FIGURE 18.3 Sales history.

Computing Averages for Sales Histories

In some cases, knowing the average number of revenue dollars generated in a time period, or the average number of guests served in that period, may be a real benefit to you. This is because it may be helpful to know, for example, the number of dollar sales achieved on a typical day last week or the number of guests served on that same typical day. Since future guest activity can often be expected to be very similar to the activities of guests in the past, using historical averages from your operation can be quite useful in helping you project future guest sales and counts.

An average is defined as the value arrived at by adding the quantities in a series and dividing the sum of the quantities by the number of items in the series. Thus, 11 is the average of $6 + 9 + 18$. The sum of the quantities in this case equals 33 ($6 + 9 + 18 = 33$). The number of items in the series is three, that is, 6, 9, and 18. Thus, $33/3 = 11$, the average of the numbers.

Sometimes, an average is referred to as the mean in a series of quantities. The two major types of averages you are likely to encounter as a foodservice manager are as follows:

1. Fixed average
2. Rolling average

Dan's Take-Out Coffee	
Day	**Daily Sales**
1	$ 350.00
2	320.00
3	390.00
4	440.00
5	420.00
6	458.00
7	450.00
8	460.00
9	410.00
10	440.00
11	470.00
12	460.00
13	418.00
14	494.00
14-day total	$5,980.00

$$\frac{\$5,980}{14} = \$427.14 \text{ per day}$$

FIGURE 18.4 14-day fixed average.

Fixed Average A fixed average is an average in which you determine a specific time period, for example, the first 14 days of a given month, and then you compute the average amount of sales or guest activity for that period. Note that this average is called fixed because the first 14 days of the month will always consist of the same days and, thus, the same revenue numbers, as shown in Figure 18.4, the sales activity of Dan's Take-Out Coffee for the first 14 days of this month. This average (total revenue / number of days) is fixed or constant because Dan's management has identified 14 specific days that are used to make up the average. The number $427.14 may be very useful because it might, if management wants it to, be used as a good predictor of the revenue volume that should be expected for the first 14 days of next month.

Rolling Average The rolling average is the average amount of sales or volume over a changing time period. Essentially, where a fixed average is computed using a specific or constant set of data, the rolling average is computed using data that will change regularly. To illustrate, consider the case of Ubalda Salas, who operates a sports bar in

Ubalda's Sports Bar			
Date	Sales	Date	Sales
1	$350	8	$460
2	320	9	410
3	390	10	440
4	440	11	470
5	420	12	460
6	458	13	418
7	450	14	494

FIGURE 18.5 14-day sales levels.

a university town in the Midwest. Ubalda is interested in knowing what the average revenue dollars were in her operation for *each prior seven-day period*. Obviously, in this case, the prior seven-day period changes or rolls forward by one day, each day. It is important to note that Ubalda could have been interested in her average daily revenue last week (fixed average), but she prefers to know her average sales for the last seven days. This means that she will, at times, be using data from both last week and this week to compute the last seven-day average. Using the sales data recorded in Figure 18.5, the seven-day rolling average for Ubalda's Sports Bar would be computed as shown in Figure 18.6.

Note that each seven-day period is made up of a group of daily revenue numbers that changes over time. The first seven-day rolling average is computed by summing the first seven days' revenue (revenue on days 1–7 = $2,828) and dividing that number by seven to arrive at a seven-day rolling average of $404.00 ($2,828/7 = $404.00). Each day, Ubalda would add her daily revenue to that of the prior seven-day total and drop the day that is now eight days past. This gives her the effect of continually rolling the most current seven days forward. The use of the rolling average, while more complex and time consuming than that of a fixed average, can be extremely useful in recording data to help you make effective predictions about the sales levels you can expect in the

Ubalda's Sports Bar								
	Seven-Day Period							
Date	1–7	2–8	3–9	4–10	5–11	6–12	7–13	8–14
1	$350	–						
2	320	$320	–					
3	390	390	$390	–				
4	440	440	440	$440	–			
5	420	420	420	420	$420	–		
6	458	458	458	458	458	$458	–	
7	450	450	450	450	450	450	$450	–
8		460	460	460	460	460	460	$460
9			410	410	410	410	410	410
10				440	440	440	440	440
11					470	470	470	470
12						460	460	460
13							418	418
14								494
Total	2,828	2,938	3,028	3,078	3,108	3,148	3,108	3,152
7-Day Rolling Average	404.00	419.71	432.57	439.71	444.00	449.71	444.00	450.29

FIGURE 18.6 Seven-day rolling average.

future. This is true because, in many cases, rolling data are more current and, thus, more relevant, than some fixed historical averages. You may choose to compute fixed averages for some time periods and rolling averages for others. For example, it may be helpful to know your average daily sales for the first 14 days of last month and your average sales for the last 14 days. If, for example, these two numbers were very different, you would know if the number of sales you can expect in the future is increasing or declining. Regardless of the type of average you feel is best for your operation, you should record your sales history because it is from your sales history that you will be better able to predict future sales levels.

Recording Revenue, Guest Counts, or Both?

As previously mentioned, some foodservice operations do not regularly record revenue as a measure of their sales activity. For them, developing sales histories by recording the number of individuals they serve each day makes the most sense. Thus, guest count, the term used in the hospitality industry to indicate the number of people served, is recorded on a regular basis. For many other foodservice operations, sales are recorded in terms of sales revenue generated. Not surprisingly, you may decide that your operation is best managed by tracking both revenue and guest counts. In fact, if you do decide to record both revenue and guest counts, you have the information you need to compute average sales per guest, a term commonly known as check average.

Average sales per guest is determined by the following formula:

$$\frac{\text{Total Sales}}{\text{Number of Guests Served}} = \text{Average Sales per Guest}$$

Consider the information in Figure 18.7 in which the manager of Brothers' Family Restaurant has decided to monitor and record the following:

1.   Sales

2.   Guests served

3.   Average sales per guest

Most POS systems are programmed to tell you the amount of revenue you have generated in a selected time period, the number of guests you have served, and the average sales per guest. If the operation you manage does not have a POS system in place, the revenue generated may be determined by a cash register, and the number of guests served may be determined by an actual guest head count, by a count of the number of utensils or trays issued, or by adding the number of individuals listed on the guest checks used. In the case of Brothers' Family Restaurant, Monday's revenue was \$1,365, the number of guests served was 190, and the average sales per guest for Monday, January 1, was determined to be \$7.18 (\$1,365/190 = \$7.18). On Tuesday, the average sales per guest was \$8.76 (\$2,750/314 = \$8.76).

Brothers' Family Restaurant				
Sales Period	Date	Sales	Guests Served	Average Sales per Guest
Monday	1/1	$1,365.00	190	$7.18
Tuesday	1/2	2,750.00	314	8.76
Two-Day Average		2,057.50	252	8.16

FIGURE 18.7 Sales history.

To compute the two-day revenue average, the Brothers' manager would add Monday's revenue and Tuesday's revenue and divide by 2, yielding a two-day revenue average of $2,057.50 [($1,365 + $2,750)/2 = $2,057.50]. In a like manner, the two-day guests-served average is computed by adding the number of guests served on Monday to the number served on Tuesday and dividing by 2, yielding a two-day average guests-served number of 252 [(190 + 314)/2 = 252].

It might be logical to think that the manager of Brothers' could compute the Monday and Tuesday combined average sales per guest by adding the averages from each day and dividing by 2. It is important to understand that this *would not* be correct. A formula consisting of Monday's average sales per guest plus Tuesday's average sales per guest divided by 2 [($7.18 + $8.76)/2] yields $7.97. In fact, the two-day average sales per guest is $8.16 [($1,365 + $2,750)/(190 + 314) = ($4,115/504) = $8.16].

While the difference of $0.19 might, at first glance, appear to be an inconsequential amount, assume that you are the chief executive officer (CEO) of a restaurant chain with 4,000 units worldwide. If each unit served 1,000 guests per day and you miscalculated average sales per guest by $0.19, your daily revenue assumption would be "off" by $760,000 per day [(4,000 × 1,000 × $0.19) = $760,000]!

Returning to the Brothers' Family Restaurant example, the correct procedure for computing the two-day sales per guest average is as follows:

$$\frac{\text{Monday Sales} + \text{Tuesday Sales}}{\text{Monday Guests} + \text{Tuesday Guests}} = \text{Two-Day Average Sales per Guests}$$

or

$$\frac{\$1,365 + \$2,750}{190 + 314} = \$8.16$$

	Sales	Guests Served	Average Sales per Guest
Day 1	$ 100	20	$ 5.00
Day 2	4,000	400	10.00
Two-Day Average	2,050	210	???

FIGURE 18.8 Weighted average.

The correct computation is a weighted average, that is, an average that weights the number of guests with how much they spend in a given time period.

To demonstrate further, assume that you were to answer the question, "What is the combined average sales per guest?" using the data in Figure 18.8. From the data in Figure 18.8, it is easy to "see" that the two-day average would not be $7.50 ($5.00 + $10.00/2) because many more guests were served at a $10.00 average than were served at the $5.00 average. Obviously, with so many guests spending an average of $10.00, and so few spending an average of $5.00, the overall average should be quite close to $10.00. In fact, the correct weighted average sales per guest would be $9.76, as follows:

$$\frac{\text{Day 1 Sales} + \text{Day 2 Sales}}{\text{Day 1 Guests} + \text{Day 2 Guests}} = \text{Two-Day Average Sales per Guests}$$

or

$$\frac{\$100 + \$4,000}{20 + 400} = \$9.76$$

MAINTAINING SALES HISTORIES

While a sales history may consist of revenue, number of guests served, and average sales per guest, you may want to use even more detailed information, such as the number of a particular menu item served, the number of guests served in a specific meal or time period, or the method of meal delivery (e.g., drive-through vs. counter sales). The important concept to remember is that you have the power to determine the information that best suits your operation. That information should be updated at least daily, and a cumulative total for the appropriate time period should also be maintained.

In most cases, your sales histories should be kept for a period of at least two years. This allows you to have a good sense of what has happened to your business in the recent past. Of course, if you are the manager of a new operation, or one that has recently undergone a major concept change, you may not have the advantage of reviewing meaningful sales histories. If you find yourself in such a situation, it is imperative that you begin to build and maintain your sales histories as soon as possible so you will have good sales information on which to base your future managerial decisions.

SALES VARIANCES

Once an accurate sales history system has been established, you may begin to see that your operation, if it is like most, will experience some level of sales variation. These sales variances, or changes from previously experienced sales levels, will give you an indication of whether your sales are improving, declining, or staying the same. Because that information is so important to predicting future sales levels, many foodservice managers improve their sales history information by including sales variance as an additional component of the history.

Figure 18.9 details a portion of a sales history that has been modified to include a "Variance" column, which allows the manager to see how sales are different from a prior period. In this case, the manager of Quick Wok wants to compare sales for the first three months of this year to sales for the first three months of last year. Note, of course, that the manager would find this revenue information in the sales histories regularly maintained by the operation.

Quick Wok			
Month	Sales This Year	Sales Last Year	Variance
January	$ 54,000	$ 51,200	$ 2,800
February	57,500	50,750	6,750
March	61,200	57,500	3,700
First-Quarter Total	172,700	159,450	13,250

FIGURE 18.9 Sales history and variance.

The variance in Figure 18.9 is determined by subtracting sales last year from sales this year. In January, the variance figure is obtained as follows:

Sales This Year — Sales Last Year = Variance

or

$54,000 — $51,200 = $2,800

Thus, the manager of Quick Wok can see that sales are improving. All three months in the first quarter of the year showed revenue increases over the prior year. The total improvement for the first quarter was $13,250 ($172,700 − $159,450 = $13,250).

While the format used in Figure 18.9 does let a manager know the dollar value of revenue variance, good managers want to know even more. Simply knowing the dollar value of a variance has limitations. Consider two restaurant managers. One manager's restaurant had revenue of $1,000,000 last year. The second manager's restaurant generated one half as much revenue, or $500,000. This year both had sales increases of $50,000. It is clear that a $50,000 sales increase represents a much greater improvement in the second restaurant than in the first. Because that is true, effective managers are interested in the percentage variance, or percentage change, in their sales from one time period to the next.

Figure 18.10 shows how the sales history at Quick Wok can be expanded to include percentage variance as part of that operation's complete sales history. Percentage variance is obtained by subtracting sales last year from sales this year and dividing the

	Quick Wok			
Month	Sales This Year	Sales Last Year	Variance	Percentage Variance
January	$ 54,000	$ 51,200	$ 2,800	5.5%
February	57,500	50,750	6,750	13.3
March	61,200	57,500	3,700	6.4
First-Quarter Total	172,700	159,450	13,250	8.3

FIGURE 18.10 Sales history, variance, and percentage variance.

resulting number by sales last year. Thus, in the month of January, the percentage variance is determined as follows:

$$\frac{\text{Sales This Year} + \text{Sales Last Year}}{\text{Sales Last Year}} = \text{Percentage Variance}$$

or

$$\frac{\$54,000 - \$51,200}{\$51,200} = 0.055$$

(in common form, 5.5%)

Note that the resulting decimal-form percentage can be converted to the more frequently used common form discussed in Chapter 17 by moving the decimal point two places to the right (or multiplying by 100).

An alternative, and shorter, formula for computing the percentage variance is as follows:

$$\frac{\text{Variance}}{\text{Sales Last Year}} = \text{Percentage Variance}$$

or

$$\frac{\$2,800}{\$51,200} = 0.055$$

(in common form, 5.5%)

Another way to compute the percentage variance is to use a math shortcut, as follows:

$$\frac{\text{Sales This Year}}{\text{Sales Last Year}} - 1 = \text{Percentage Variance}$$

or

$$\frac{\$54,000}{\$51,200} - 1 = 0.055$$

(in common form, 5.5%)

The computation of percentage variance is invaluable when you want to compare the operating results of two foodservice operations of different sizes. Return to our previous example of the two restaurant managers, each of whom achieved $50,000 revenue increases. Using the first percentage variance formula presented before, you will see that the restaurant with higher sales increased its revenue by 5.0%, [($1,050,000 − $1,000,000)/$1,000,000 = 5.0%], while the restaurant with lower revenue achieved a 10% increase [($550,000 − $500,000)/$500,000 = 10.0%]. As your level of expertise increases, you will find additional areas in which knowing the percentage variance in revenue, and later in this text, in expense areas, will be instrumental to your management decision making.

PREDICTING FUTURE SALES

It has been pointed out that truly outstanding managers have an ability to see the future in regard to the revenue figures they can achieve and the number of guests they expect to serve. You, too, can learn to do this when you apply the knowledge of percentage variances you now have to estimating your own operation's future sales. Depending on the type of facility you manage, you may be interested in predicting, or forecasting, future revenues, guest counts, or average sales per guest levels. We examine the procedures for all three of these in detail.

Future Revenues

Erica Tullstein is the manager of Rock's Pizza Pub on the campus of State College. Her guests consist of college students, most of whom come to the Rock to talk, listen to music, eat, and study. Erica has done a good job in maintaining sales histories in the two years she has managed the Rock. She records the revenue dollars she achieves on a daily basis, as well as the number of students frequenting the Rock. Revenue data for the last three months of the year are recorded in Figure 18.11.

	Rock's Pizza Pub			
Month	Sales This Year	Sales Last Year	Variance	Percentage Variance
October	$ 75,000	$ 72,500	$ 2,500	3.4%
November	64,250	60,000	4,250	7.1
December	57,500	50,500	7,000	13.9
Fourth-Quarter Total	196,750	183,000	13,750	7.5

FIGURE 18.11 Revenue history.

As can easily be seen, fourth-quarter revenues for Erica's operation have increased from the previous year. Of course, there could be a variety of reasons for this. Erica may have extended her hours of operation to attract more students. She may have increased the size of pizzas and held her prices constant, thus creating more value for her guests. She may have raised the price of pizza, but kept the pizzas the same size. Or perhaps a competing pizza parlor was closed for renovation during this time period. Using all of her knowledge of her own operation and her market, Erica would like to predict the sales level she will experience in the first three months of next year. This sales forecast will be most helpful as she plans for next year's anticipated expenses, staffing levels, and profits.

The first question Erica must address is the amount her sales have actually increased. Revenue increases range from a low in October of 3.4%, to a high in December of 13.9%. The overall quarter average of 7.5% is the figure that Erica elects to use as she predicts her sales revenue for the first quarter of the coming year. She feels it is neither too conservative, as would be the case if she used the October percentage increase, nor too aggressive, as would be the case if she used the December figure.

If Erica were to use the 7.5% average increase from the fourth quarter of last year to predict her revenues for the first quarter of this year, a planning sheet for the first quarter of this year could be developed as presented in Figure 18.12.

Revenue forecast for this time period is determined by multiplying sales last year by the % increase estimate and then adding sales last year. In the month of January, revenue forecast is calculated using the following formula:

Rock's Pizza Pub				
Month	Sales Last Year	% Increase Estimate	Increase Amount	Revenue Forecast
January	$ 68,500	7.5%	$ 5,137.50	$ 73,637.50
February	72,000	7.5	5,400.00	77,400.00
March	77,000	7.5	5,775.00	82,775.00
First-Quarter Total	217,500	7.5	16,312.50	233,812.50

FIGURE 18.12 First-quarter revenue forecast.

Sales Last Year + (Sales Last Year × % Increase Estimate) = Revenue Forecast

or

$68,500 + ($68,500 × 0.075) = $73,637.50

An alternative way to compute the revenue forecast is to use a math shortcut, as follows:

Sales Last Year × (1 + % Increase Estimate) = Revenue Forecast

or

$68,500 × (1 + 0.075) = $73,637.50

Erica is using the increases she has experienced in the past to predict increases she may experience in the future. Monthly revenue figures from last year's sales history plus percent increase estimates based on those histories give Erica a good idea of the revenue levels she may achieve in January, February, and March of the coming year. Clearly, Erica will have a better idea of the sales dollars she may achieve than managers who did not have the advantage of sales histories to help guide their planning.

Future Guest Counts

Using the same techniques employed in estimating increases in sales, the noncash operator or any manager interested in guest counts can estimate increases or decreases in the number of guests served. Figure 18.13 shows how Erica, the manager of Rock's Pizza Pub, used a sales history to determine the percentage of guest count increases achieved in her facility in the fourth quarter of last year.

If Erica were to use the 6.1% average increase from the fourth quarter of last year to predict her guest count for the first quarter of the coming year, a planning sheet could be developed as presented in Figure 18.14. It is important to note that Erica is not required to use the same percentage increase estimate for each month. Indeed, any forecasted increase management feels is appropriate can be used to predict future sales.

Notice that in January, for example, the guest increase estimate is rounded from 769.82 to 770. This is because you cannot serve "0.82" people!

The guest count forecast is determined by multiplying guest count last year by the % increase estimate and then adding the guest count last year. In the month of January, guest count forecast is calculated using the following formula:

Guest Count Last Year
+ (Guest Count Last Year × % Increase Estimate)
= Guest Count Forecast

or

12,620 + (12,620 × 0.061) = 13,390

Rock's Pizza Pub				
Month	Guests This Year	Guests Last Year	Variance	Percentage Variance
October	14,200	13,700	+ 500	3.6%
November	15,250	14,500	+ 750	5.2
December	16,900	15,500	+1,400	9.0
Fourth-Quarter Total	46,350	43,700	+2,650	6.1

FIGURE 18.13 Guest count history.

Rock's Pizza Pub				
Month	Guests Last Year	% Increase Estimate	Guest Increase Estimate	Guest Count Forecast
January	12,620	6.1%	770	13,390
February	13,120	6.1	800	13,920
March	13,241	6.1	808	14,049
First-Quarter Total	38,981	6.1	2,378	41,359

FIGURE 18.14 First-quarter guest count forecast.

This process can be simplified if desired by using a math shortcut, as follows:

Guest Count Last Year × (1.00 + % Increase Estimate)
= **Guest Count Forecast**

or

12,620 × (1.00 + 0.061) = 13,390

You should choose the formula with which you feel most comfortable.

Future Average Sales per Guest

Recall that average sales per guest (check average) is simply the amount of money an average guest spends during a visit. The same formula is used to forecast average sales per guest as was used in forecasting total revenue and guest counts. Therefore, using data taken from the sales history, the following formula is employed:

Last Year's Average Sales per Guest
+ **Estimated Increase in Sales per Guest**
= **Sales per Guest Forecast**

Alternatively, you can compute average sales per guest using the data collected from revenue forecasts (Figure 18.12) and combining that data with the guest count forecast (Figure 18.14). If that is done, the data presented in Figure 18.15 result.

Average sales per guest forecast is obtained by dividing the revenue forecast by the guest count forecast. Thus, in the month of January, the average sales per guest forecast is determined by the following formula:

$$\frac{\text{Revenue Forecast}}{\text{Guest Count Forecast}} = \text{Average Sales per Guest Forecast}$$

or

$$\frac{\$75,657.50}{13,390} = \$5.50$$

It is important to note that sales histories, regardless of how well they have been developed and maintained, are not sufficient, used alone, to accurately predict future sales. Your knowledge of potential price changes, new competitors, facility renovations, and improved selling programs are just a few of the many factors that you must consider when predicting future sales. There is no question, however, that you must develop and

Rock's Pizza Pub			
Month	Revenue Forecast	Guest Count Forecast	Average Sales per Guest Forecast
January	$ 73,637.50	13,390	$5.50
February	77,400.00	13,920	5.56
March	82,775.00	14,049	5.89
First-Quarter Total	233,812.50	41,359	5.65

FIGURE 18.15 First-quarter average sales per guest forecast.

monitor, daily, a sales history report appropriate for your operation. They are easily developed and will serve as the cornerstone of other management systems you will design. Without accurate sales data, control systems, regardless of their sophistication, are very likely to fail.

When added to your knowledge of the unique factors that impact your unit, properly maintained histories will help you answer two important control questions, namely, "How many people are coming tomorrow?" and "How much is each person likely to spend?" The judgment of management is critical in forecasting answers to these questions. Since you can now answer those questions, you are ready to develop systems that will allow you to prepare an efficient and cost-effective way of serving those guests, be they guests in a hotel lounge, tourists visiting your restaurant as part of their overall travel experience, or any other foodservice guest. You want to be ready to provide them with quality food and beverage products and enough staff to serve them properly. You have done your homework with regard to the number of individuals who may be coming and how much they are likely to spend; now you must prepare for their arrival!

FUN ON THE WEB!

Look up the following sites to see examples of point of sales (POS) systems used to record historical sales information.

www.micros.com   Click on "Products" and continue linking from here to see product descriptions.

www.squirrelsystems.com   Click on "The Solutions" to find product information.

www.datatrakpos.com   Click on "Screens" to see how sales are displayed on the system.

APPLY WHAT YOU HAVE LEARNED

Pauline Cooper is a Registered Dietitian (RD) and the Foodservice Director at Memorial Hospital. Increasingly, the hospital's marketing efforts have emphasized its skill in treating diabetic patients. As a result, the number of diabetic meals served by Pauline's staff has been increasing. As a professional member of the hospital's management team, Pauline has been asked to report on how the hospital's diabetic treatment marketing efforts have affected her area.

1. How important is it that Pauline have historical records of the "type" of meals served by her staff, and not merely the number of meals served? Why?

2. Assume that Pauline's "per meal" cost has been increasing because diabetic meals are more expensive to produce than regular meals. Could Pauline use sales histories to estimate the financial impact of serving additional diabetic meals in the future? How?

3. What are other reasons managers in a foodservice operation might keep detailed records of meal "types" (i.e., vegetarian or low-sodium) served, as well as total number of meals served?

KEY TERMS AND CONCEPTS

The following are terms and concepts discussed in the chapter that are important for you as a manager. To help you review, please define the terms below.

Sales forecast	Average (mean)	Average sales per guest (check average)
Point of sales (POS) system	Fixed average	Weighted average
Sales volume	Rolling average	Sales variance
Sales history	Guest count	Percentage variance
Sales to date		

MANAGING THE COST
OF FOOD*

Overview

This chapter presents the professional techniques and methods used to effectively purchase, receive, and store food products. It teaches the formulas used to compute the true cost of the food you provide your guests, as well as a process for estimating the value of food you have used on a daily or weekly basis, by applying the food cost percentage method, which is the standard in the hospitality industry.

Chapter Outline

---

* Authored by Jack E. Miller, Lea R. Dopson, and David K. Hayes.

Highlights

At the conclusion of this chapter, you will be able to:

■ Use sales histories and standardized recipes to determine the amount of food products to buy in anticipation of forecasted sales.

■ Purchase, receive, and store food products in a cost-effective manner.

■ Compute the cost of food sold and food cost percentage.

MENU ITEM FORECASTING

When they get hungry, many potential guests ask the question, "What do you feel like eating?" For many, the answer is "a meal prepared away from home!" The U.S. Bureau of Labor Statistics consumer expenditure surveys reported that sales of food consumed away from home grew an average of 5% per year in the 1990s, with annual increases expected to be even higher in the 2000s. According to the National Restaurant Association's current forecasts, continued economic growth, gains in consumers' real disposable income, and changes in the lifestyles of today's busy American families are all spurring the sustained rise in the number of meals served away from home. This is good news for your career as a hospitality manager.

All this growth, activity, and consumer demand, however, will also create challenges for you. Consider the situation you would encounter if you used sales histories (Chapter 18) to project 300 guests for lunch today at the restaurant you manage. Your restaurant serves only three entrée items: roast chicken, roast pork, and roast beef. The question you would face is this, "How many servings of each item should we produce so that we do not run out of any one item?"

If you were to run out of one of your three menu items, guests who wanted that item would undoubtedly become upset. Producing too much of any one item would, on the other hand, cause costs to rise to unacceptable levels unless these items could be sold for their full price at a later time.

Clearly, in this situation, it would be unwise to produce 300 portions of each item. While you would never run out of any one item, that is, each of your 300 estimated guests could order the same item and you would still have enough, you would also have 600 leftovers at the end of your lunch period. What you would like to do, of course, is instruct your staff to make the "right" amount of each menu item. The right

Date: 1/1–1/5		Menu Items Sold					
Menu Item	Mon	Tues	Wed	Thurs	Fri	Total	Week's Average
Roast Chicken	70	72	61	85	77	365	73
Roast Pork	110	108	144	109	102	573	115
Roast Beef	100	140	95	121	106	562	112
Total	280	320	300	315	285	1,500	

FIGURE 19.1 Menu item sales history.

amount would be the number of servings that minimize your chances of running out of an item before lunch is over, while also minimizing your chance of having excessive leftovers.

The answer to the question of how many servings of roast chicken, pork, and beef you should prepare lies in accurate menu forecasting.

Let us return to the example cited previously. This time, however, assume that you were wise enough to have recorded last week's menu item sales on a form similar to the one presented in Figure 19.1. An estimate of 300 guests for next Monday makes sense because the weekly sales total last week of 1,500 guests served averages 300 guests per day (1,500/5 days = 300/day). You also know that, on an average day, you sold 73 roast chicken (365 sold/5 days = 73/day), 115 roast pork (573 sold/5 days = 115/day), and 112 roast beef (562 sold/5 days = 112/day).

Once you know the average number of people selecting a given menu item, and you know the total number of guests who made the selections, you can compute the popularity index, which is defined as the percentage of total guests choosing a given menu item from a list of alternatives. In this example, you can improve your "guess" about the quantity of each item to prepare if you use the sales history to help guide your decision. If you assume that future guests will select menu items much as past guests have done, given that the list of menu items remains the same, that information can be used to improve your predictions with the following formula:

$$\text{Popularity Index} = \frac{\text{Total Number of a Specific Menu Item Sold}}{\text{Total Number of All Menu Items Sold}}$$

Menu Item	Guest Forecast	Popularity Index	Predicted Number to Be Sold
Roast Chicken	300	0.243	73
Roast Pork	300	0.382	115
Roast Beef	300	0.375	112
Total			300

FIGURE 19.2 Forecasting item sales.

In this example, the popularity index for roast chicken last week was 24.3% (365 roast chicken sold/1,500 total guests = 0.243, or 24.3%). Similarly, 38.2% (573 roast pork sold/1,500 total guests = 38.20%) preferred roast pork, while 37.50% (562 roast beef sold/1,500 total guests = 37.50%) selected roast beef.

If we know, even in a general way, what we can expect our guests to select, we are better prepared to make good decisions about the quantity of each item that should be produced. In this example, Figure 19.2 illustrates your best guess of what your 300 guests are likely to order when they arrive.

The basic formula for individual menu item forecasting, based on an item's individual sales history, is as follows:

> **Number of Guests Expected × Item Popularity Index**
> **= Predicted Number of That Item to Be Sold**

The predicted number to be sold is simply the quantity of a specific menu item likely to be sold given an estimate of the total number of guests expected.

Once you know what your guests are likely to select, you can determine how many of each menu item your production staff should be instructed to prepare. It is important to note that foodservice managers face a great deal of uncertainty when attempting to estimate the number of guests who will arrive on a given day because a variety of factors influence that number. Among these are the following:

1. Competition
2. Weather
3. Special events in your area

4.   Facility occupancy (hospitals, dormitories, hotels, etc.)

5.   Your own promotions

6.   Your competitor's promotions

7.   Quality of service

8.   Operational consistency

These, as well as other factors that affect sales volume, make accurate guest count prediction very difficult.

In addition, remember that sales histories track only the general trends of an operation. They are not able to estimate *precisely* the number of guests who may arrive on any given day. Sales histories, then, are a guide to what can be expected. In our example, last week's guest counts range from a low of 280 (Monday) to a high of 320 (Tuesday). In addition, the percentage of people selecting each menu item changes somewhat on a daily basis. As a professional foodservice manager, you must take into account possible increases or decreases in guest count and possible fluctuations in your predicted number to be sold computations when planning how many of each menu item you should prepare.

In Chapter 18, we began to discuss the concept of sales forecasting. Forecasting can involve estimating the number of guests you expect, the dollar amount of sales you expect, or even what those guests may want to purchase. This forecasting is crucial if you are to effectively manage your food expenses. In addition, consistency in food production and guest service will greatly influence your overall success.

STANDARDIZED RECIPES

While it is the menu that determines what is to be sold and at what price, the standardized recipe controls both the quantity and the quality of what your kitchen will produce. A standardized recipe consists of the procedures to be used in preparing and serving each of your menu items. The standardized recipe ensures that each time a guest orders an item from your menu, he or she receives exactly what you intended the guest to receive.

Critical factors in a standardized recipe, such as cooking times and serving size, must remain constant so the menu items produced are always consistent. Guests expect to get what they pay for. The standardized recipe helps you make sure that they do. Inconsistency is the enemy of any quality foodservice operation. It will make little difference to the unhappy guest, for instance, if you tell him or her that while the menu item he or she purchased today is not up to your normal standard, it will be tomorrow, or that it was the last time the guest visited your operation.

Good standardized recipes contain the following information:

1.   Menu item name
2.   Total yield (number of servings)
3.   Portion size
4.   Ingredient list
5.   Preparation/method section
6.   Cooking time and temperature
7.   Special instructions, if necessary
8.   Recipe cost (optional)*

Figure 19.3 contains a standardized recipe for roast chicken. If this standardized recipe represents the quality and quantity management wishes its guests to have and if it is followed carefully each time, then guests will indeed receive the value management intended.

Interestingly, despite their tremendous advantages, many managers refuse to take the time to develop standardized recipes. The excuses used are many, but the following list contains arguments often used against standardized recipes:

1.   They take too long to use.
2.   My people don't need recipes; they know how we do things here.
3.   My chef refuses to reveal his or her secrets.
4.   They take too long to write up.
5.   We tried them but lost some, so we stopped using them.
6.   They are too hard to read, or many of my employees cannot read English.

Of the preceding arguments, only the last one, staff's inability to read English, has any validity. The effective operator should have the standardized recipes printed in the language of his or her production employees. Standardized recipes have far more advantages than disadvantages. Reasons for incorporating a system of standardized recipes include:

1.   Accurate purchasing is impossible without the existence and use of standardized recipes.
2.   Dietary concerns require some foodservice operators to know exactly the kinds of ingredients and the correct amount of nutrients in each serving of a menu item.

---

* This information is optional. If the recipe cost is *not* included in the standardized recipe, a standardized cost sheet must be developed for each recipe item.

Roast Chicken

Special Instructions: <u>Serve with</u>
<u>Crabapple Garnish (see Crabapple Garnish</u>
<u>Standardized Recipe).</u>
<u>Serve on 10-in. plate.</u>

Recipe Yield: <u>48</u>
Portion Size: <u>¼ chicken</u>
Portion Cost: <u>See cost sheet</u>

Ingredients	Amount	Method
Chicken Quarters (twelve 3–3½-lb. chickens)	48 ea.	Step 1. Wash chicken; check for pinfeathers; tray on 24 in. × 20 in. baking pans.
Butter (melted)	1 lb. 4 oz.	Step 2. Clarify butter; brush liberally on chicken quarters; combine all seasonings; mix well; sprinkle all over chicken quarters.
Salt	¼ C	
Pepper	2 T	
Paprika	3 T	
Poultry Seasoning	2 t	
Ginger	1½ t	
Garlic Powder	1 T	Step 3. Roast at 325°F in oven for 2½ hours, to an internal temperature of at least 165°F.

FIGURE 19.3 Standardized recipe.

3. Accuracy in menu laws require that foodservice operators be able to tell guests about the type and amount of ingredients in their recipes.

4. Accurate recipe costing and menu pricing is impossible without standardized recipes.

5. Matching food used to cash sales is impossible to do without standardized recipes.

6. New employees can be better trained with standardized recipes.

7. The computerization of a foodservice operation is impossible unless the elements of standardized recipes are in place; thus, the advantages of advanced technological tools available to the operation are restricted or even eliminated.

In fact, standardized recipes are so important that they are the cornerstones of any serious effort to produce consistent, high-quality food products at an established cost. Without them, cost control efforts become nothing more than raising selling prices, reducing portion sizes, or lessening quality. This is not effective cost management. It is hardly management at all. Without established standardized recipes, however, this happens all too frequently.

Any recipe can be standardized. The process can sometimes be complicated, however, especially in the areas of baking and sauce production. It is always best to begin with a recipe of proven quality. Frequently, you may have a recipe designed to serve 10 guests, but you want to expand it to serve 100 people. In cases like this, it may not be possible to simply multiply each ingredient used by 10. A great deal has been written regarding various techniques used to expand recipes. Computer software designed for that purpose is now on the market. As a general rule, however, any item that can be produced in quantity can be standardized in recipe form and can be adjusted, up or down, in quantity.

When adjusting recipes, it is important that measurement standards be consistent. Weighing with a pound or an ounce scale is the most accurate method of measuring any ingredients. The food item to be measured must be recipe ready. That is, it must be cleaned, trimmed, cooked, and generally completed, save for its addition to the recipe. For liquid items, measurement of volume (i.e., cup, quart, or gallon, etc.) may be preferred. Some operators like to weigh all ingredients, even liquids, for improved accuracy.

When adjusting recipes for quantity (total yield), two general methods may be employed. They are:

1. Factor method
2. Percentage technique

Factor Method

When using the factor method, you must use the following formula to arrive at a recipe conversion factor:

$$\frac{\text{Yield Desired}}{\text{Current Yield}} = \text{Conversion Factor}$$

If, for example, our current recipe makes 50 portions, and the number of portions we wish to make is 125, the formula would be as follows:

$$\frac{125}{50} = 2.5$$

Ingredient	Original Amount	Conversion Factor	New Amount
A	4 lb.	2.5	10 lb.
B	1 qt.	2.5	2½ qt.
C	1½ T	2.5	3¾ T

FIGURE 19.4 Factor method.

Thus, 2.5 would be the conversion factor. To produce 125 portions, we would multiply each ingredient in the recipe by 2.5 to arrive at the required amount of that ingredient.

Figure 19.4 illustrates the use of this method for a three-ingredient recipe.

Percentage Method

The percentage method deals with recipe weight, rather than with a conversion factor. It is sometimes more accurate than using a conversion factor alone. Essentially, the percentage method involves weighing all ingredients and then computing the percentage weight of each recipe ingredient in relation to the total weight of all ingredients.

To facilitate the computation, many operators convert pounds to ounces prior to making their percentage calculations. These are converted back to standard pounds and ounces when the conversion is completed. To illustrate the use of the percentage method, let us assume that you have a recipe with a total weight of 10 pounds and 8 ounces, or 168 ounces. If the portion size is 4 ounces, the total recipe yield would be 168/4, or 42, servings. If you want your kitchen to prepare 75 servings, you would need to supply it with a standardized recipe consisting of the following total recipe weight:

75 Servings × 4 oz. per Serving = 500 oz.

You now have all the information necessary to use the percentage method of recipe conversion.

Figure 19.5 details how the process would be accomplished. Note that % of total is computed as ingredient weight/total recipe weight. Thus, for example, ingredient

Ingredient	Original Amount	Ounces	% of Total	Total Amount Required	% of Total	New Recipe Amount
A	6 lb. 8 oz.	104 oz.	61.9%	300 oz.	61.9%	185.7 oz.
B	12 oz.	12	7.1	300 oz.	7.1	21.3 oz.
C	1 lb.	16	9.5	300 oz.	9.5	28.5 oz.
D	2 lb. 4 oz.	36	21.5	300 oz.	21.5	64.5 oz.
Total	10 lb. 8 oz.	168	100.0	300 oz.	100.0	300.0 oz.

FIGURE 19.5 Percentage method.

A's % of total is computed as follows:

$$\frac{\text{Item A Ingredient Weight}}{\text{Total Recipe Weight}} = \%\text{ of Total}$$

or

$$\frac{104\text{ oz.}}{168\text{ oz.}} = 0.619\,(61.9\%)$$

To compute the new recipe amount, we multiply the % of total figure times the total amount required. For example, with ingredient A, the process is:

$$\text{Item A \% of Total} \times \text{Total Amount Required}$$
$$= \text{New Recipe Amount}$$

or

$$61.9\% \times 300\text{ oz.} = 185.7\text{ oz.}$$

The proper conversion of weights and measures is important in recipe expansion or reduction. The judgment of the recipe writer is critical, however, since such factors as cooking time, temperature, and utensil selection may vary as recipe sizes are increased or decreased. In addition, some recipe ingredients, such as spices or flavorings, may not respond well to mathematical conversions. In the final analysis, it is your assessment of product taste that should ultimately determine ingredient ratios in standardized recipes. All recipes should be standardized and used as written. It is your responsibility to see that this is done.

INVENTORY CONTROL

With a knowledge of what is likely to be purchased by your guests (sales forecast) and a firm idea of the ingredients necessary to produce these items (standardized recipes), you must make decisions about desired inventory levels. A desired inventory level is simply the answer to the question, "How much of each needed ingredient should I have on hand at any one time?"

It is clear that this question can only be properly answered if your sales forecast is of good quality and your standardized recipes are in place so you do not "forget" to stock a necessary recipe ingredient. Inventory management seeks to provide appropriate working stock, which is the amount of an ingredient you anticipate using before purchasing that item again, and a minimal safety stock, the extra amount of that ingredient you decide to keep on hand to meet higher than anticipated demand. Demand for a given menu item can fluctuate greatly between delivery periods, even when the delivery occurs daily. With too little inventory, you may run out of products and therefore reduce guest satisfaction. With too much inventory, waste, theft, and spoilage can become excessive. The ability to effectively manage the inventory process is one of the best skills a foodservice manager can acquire.

Determining Inventory Levels

Inventory levels are determined by a variety of factors. Some of the most important ones are as follows:

1. Storage capacity
2. Item perishability
3. Vendor delivery schedule
4. Potential savings from increased purchase size
5. Operating calendar
6. Relative importance of stock outages
7. Value of inventory dollars to the operator

Storage Capacity Inventory items must be purchased in quantities that can be adequately stored and secured. Many times, kitchens lack adequate storage facilities. This may mean more frequent deliveries and holding less of each product on hand than would otherwise be desired. When storage space is too great, however, the tendency by some managers is to fill the space. It is important that this not be done, as increased inventory of items generally leads to greater spoilage and loss due to theft. Moreover, large quantities of goods on the shelf tend to send a message to employees that there is "plenty" of everything. This may result in the careless use of valuable and expensive products. It is also unwise to overload refrigerators or freezers. This not only can result in difficulty in finding items quickly, but also may cause carryovers (those items produced for a meal period but not sold) to be "lost" in the storage process.

Item Perishability If all food products had the same shelf life, that is, the amount of time a food item retains its maximum freshness, flavor, and quality while in storage, you would have less difficulty in determining the quantity of each item you should keep on hand. Unfortunately, the shelf life of food products varies greatly.

Figure 19.6 demonstrates the difference in shelf life of some common foodservice items when properly stored in a dry storeroom or refrigerator. Figure 19.7 demonstrates the difference in shelf life of some common foodservice items when properly stored in a freezer.

Because food items have varying shelf lives, you must balance the need for a particular product with the optimal shelf life of that product. Serving items that are "too old" is a sure way to develop guest complaints. In fact, one of the quickest ways to determine the overall effectiveness of a foodservice manager is to "walk the boxes." This means to take a tour of a facility's storage area. If many products, particularly in the refrigerated area, are moldy, soft, overripe, or rotten, it is a good indication of a foodservice operation that does not have a feel for proper inventory levels based on the shelf lives of the items kept in inventory. It is also a sign that sales forecasting methods either are not in place or are not working well.

Vendor Delivery Schedule It is the fortunate foodservice operator who lives in a large city with many vendors, some of whom may offer the same service and all of whom would like to have the operator's business. In many cases, however, you will not have the luxury of daily delivery. Your operation may be too small to warrant such frequent stops by a vendor, or the operation may be in such a remote location that daily delivery is simply not possible. Consider, for a moment, the difficulty you would face if you were the manager of a foodservice operation located on an offshore oil rig. Clearly, in a case like that, a vendor willing to provide daily doughnut delivery is going to be hard to find! In all cases, it is important to remember that the cost to the vendor for frequent deliveries will be reflected in the cost of the goods to you.

Vendors will readily let you know what their delivery schedule to a certain area or location can be. It is up to you to use this information to make good decisions regarding

Item	Storage	Shelf Life
Milk	Refrigerator	5–7 days
Butter	Refrigerator	14 days
Ground Beef	Refrigerator	2–3 days
Steaks (fresh)	Refrigerator	14 days
Bacon	Refrigerator	30 days
Canned Vegetables	Dry Storeroom	12 months
Flour	Dry Storeroom	3 months
Sugar	Dry Storeroom	3 months
Lettuce	Refrigerator	3–5 days
Tomatoes	Refrigerator	5–7 days
Potatoes	Dry Storeroom	14–21 days

FIGURE 19.6 Shelf life.

the quantity of that vendor's product you must buy to have both working stock and safety stock.

Potential Savings from Increased Purchase Size Sometimes, you will find that you can realize substantial savings by purchasing needed items in large quantities. This certainly makes sense if the total savings actually outweigh the added costs of receiving and storing the larger quantity. For the large foodservice operator, who once a year buys canned green beans by the railroad car, the savings are real. For the smaller operator, who hopes to reduce costs by ordering two cases of green beans rather than one, the savings may be negligible. Generally, however, reduced packaging and shipping costs result in lower per unit costs when larger bags, boxes, or cartons of ingredients are purchased.

Remember, too, that there are costs associated with extraordinarily large purchases. These may include storage costs, spoilage, deterioration, insect or rodent

Food	Maximum Storage Period at −10 to 0°F (−2.23 to −17.7°C)
Meat	
Beef, ground and stewing	3–4 months
Beef, roasts and steaks	6 months
Lamb, ground	3–5 months
Lamb, roasts and chops	6–8 months
Pork, ground	1–3 months
Pork, roasts and chops	4–8 months
Veal	8–12 months
Variety meats (liver, tongue)	3–4 months
Ham, frankfurters, bacon, luncheon meats	2 weeks (freezing not recommended)
Leftover cooked meats	2–3 months
Gravy, broth	2–3 months
Sandwiches with meat filling	1–2 months
Poultry	
Whole chicken, turkey, duck, goose	12 months
Giblets	3 months
Cut-up cooked poultry	4 months
Fish	
Fatty fish (mackerel, salmon)	3 months
Other fish	6 months
Shellfish	3–4 months
Baked Goods	
Cakes, prebaked	4–9 months
Cake batters	3–4 months
Cookies	6–12 months
Fruit pies, baked or unbaked	3–4 months
Pie shells, baked or unbaked	1 to 1½–2 months
Yeast breads and rolls, prebaked	3–9 months
Yeast breads and rolls, dough	½ month
Other	
French-fried potatoes	2–6 months
Fruit	8–12 months
Fruit juice	8–12 months
Precooked combination dishes	2–6 months
Vegetables	8 months

Source: HACCP Reference Book, SERVSAFE program of the Educational Foundation of the National Restaurant Association.

FIGURE 19.7 Recommended freezer storage period maximums.

infestation, or theft. As a general rule, you should determine your ideal product inventory levels and then maintain your stock within that need range. Only when the advantages of placing an extraordinarily large order are very clear should such a purchase be undertaken.

Operating Calendar When an operation is involved in serving meals seven days a week to a relatively stable number of guests, the operating calendar makes little difference to inventory level decision making. If, however, the operation opens on Monday and closes on Friday for two days, as is the case in many school foodservice accounts, the operating calendar plays a large part in determining desired inventory levels. In general, it can be said that an operator who is closing down either for a weekend (as in school foodservice or a corporate dining situation) or for a season (as in the operation of a summer camp or seasonal hotel) should attempt to reduce overall inventory levels as the closing period approaches. This is expecially true when it comes to perishable items. Many operators actually plan menus to steer clear of highly perishable items near their closing periods. They prefer to work highly perishable items, such as fresh seafood and some meat items, into the early or middle part of their operating calendar. This allows them to minimize the amount of perishable product that must be carried through the close-down period.

Relative Importance of Stock Outages In many foodservice operations, not having enough of a single food ingredient or menu item is simply not that important. In other operations, the shortage of even one menu item might spell disaster. While it may be all right for the local French restaurant to run out of one of the specials on Saturday night, it is not difficult to imagine the problem of the McDonald's restaurant manager who runs out of French fried potatoes on that same Saturday night!

For the small operator, a mistake in the inventory level of a minor ingredient that results in an outage can often be corrected by a quick run to the grocery store. For the larger facility, such an outage may well represent a substantial loss of sales or guest goodwill. Whether the operator is large or small, being out of a key ingredient or menu item is to be avoided, and planning inventory levels properly helps prevent it. In the restaurant industry, when an item is no longer available on the menu, you "86" the item, a reference to restaurant slang originating in the early 1920s (86 rhymed with "nix," a Cockney term meaning "to eliminate"). If you, as a manager, "86" too many items on any given night, the reputation of your restaurant as well as your ability to manage it will suffer.

A strong awareness and knowledge of how critical this outage factor is help determine the appropriate inventory level. A word of caution is, however, necessary. The foodservice operator who is determined never to run out of anything must be careful not to set inventory levels so high as to actually end up costing the operation more than if realistic levels were maintained.

Value of Inventory Dollars to the Operator In some cases, operators elect to remove dollars from their bank accounts and convert them to product inventory. When this is done, the operator is making the decision to value product more than dollars. When it is expected that the value of the inventory will rise faster than the value of the banked dollar, this is a good strategy. All too often, however, operators overbuy or "stockpile" inventory, causing too many dollars to be tied up in non-interest-bearing food products. When this is done, managers incur opportunity costs. An opportunity cost is the cost of foregoing the next best alternative when making a decision. For example, suppose you have two choices, A and B, both having potential benefits or returns for you. If you choose A, then you lose the potential benefits from choosing B (opportunity cost). In other words, you could choose to use your money to buy food inventory that will sit in your storeroom until it is sold, or you could choose not to stockpile food inventory and invest the money. If you stockpile the inventory, then the opportunity cost is the amount of money you would have made if you had invested rather than holding the excess inventory.

If the dollars used to purchase inventory must be borrowed from the bank, rather than being available from operating revenue, an even greater cost to carry the inventory is incurred since interest must be paid on the borrowed funds. In addition, a foodservice company of many units that invests too much of its money in inventory may find that funds for acquisition, renovation, or marketing are not readily available. In contrast, a state institution that is given its entire annual budget at the start of its fiscal year (a year that is 12 months long but may not follow the calendar year) may find it advantageous to use its purchasing power to acquire large amounts of inventory at the beginning of the year and at very low prices.

Setting the Purchase Point

A purchase point, as it relates to inventory levels, is that point in time when an item should be reordered. This point is typically designated by one of two methods:

1. As needed (just in time)
2. Par level

As Needed When you elect to use the as-needed, or just-in-time, method of determining inventory level, you are basically purchasing food based on your prediction of unit sales and the sum of the ingredients (from standardized recipes) necessary to produce those sales. Then, no more than the absolute minimum of needed inventory level is secured from the vendor. When this system is used, the buyer compiles a list of needed ingredients and submits it to management for approval to purchase. For example, in a hotel foodservice operation, the demand for 500 servings of a raspberries and cream torte dessert, to be served to a group in the hotel next week, would cause the responsible person to check the standardized recipe for this item and, thus, determine the amount of raspberries that should be ordered. Then that amount, and no more, would be ordered from the vendor.

Par Level Foodservice operators may set predetermined purchase points, called par levels, for some items. In the case of the raspberries and cream torte dessert referred to previously, it is likely that the torte will require vanilla extract. It does not make sense, however, to expect your food production manager to order vanilla extract by the tablespoon! In fact, you are likely to find that you are restricted in the quantity you could buy due to the vendor's delivery minimum, namely, bottle or case, or the manufacturer's packaging methods. In cases such as this, or when demand for a product is relatively constant, you may decide to set needed inventory levels for some items by determining purchase points based on appropriate par levels.

When determining par levels, you must establish both minimum and maximum amounts required. Many foodservice managers establish a minimum par level by computing working stock, then adding 25 to 50% more for safety stock. Then, an appropriate purchase point, or point at which additional stock is purchased, is determined. If, for example, you have decided that the inventory level for coffee should be based on a par system, the decision may be made that the minimum (given your usage) amount that should be on hand at all times is four cases. This would be the minimum par level. Assume that you set the maximum par level at ten cases. While the inventory level in this situation would vary from a low of four cases to a high of ten cases, you would be assured that you would never have too little or too much of this particular menu item.

If cases of coffee were to be ordered under this system, you would always attempt to keep the number of cases between the minimum par level (four cases) and the maximum par level (ten cases). The purchase point in this example might be six cases; that is, when your operation had six cases of coffee on hand, an order would be placed with the coffee vendor. The intention would be to get the total stock up to ten cases before your supply got below four cases. Since delivery might take one or two days, six cases might be an appropriate purchase point.

Whether we use the as-needed or the par level method, or, as in the case of most operators, a combination of the two, each ingredient or menu item should have a management-designated inventory level. As a rule, highly perishable items should be ordered on an as-needed basis, while items with a longer shelf life can often have their inventory levels set using a par level system. The answer to the question "How much of each ingredient should I have on hand at any point in time?" must come from you. Many factors will impact this decision. The decision, however, must be made and compliance monitored on a regular basis.

PURCHASING

Placing Products in Storage

Food products are highly perishable items. As such, they must be moved quickly from your receiving area to the area selected for storage. This is especially true for refrigerated and frozen items. An item such as ice cream, for example, can deteriorate substantially if it is at room temperature for only a few minutes. Most often, in foodservice, this high perishability dictates that the same individual responsible for receiving the items is the individual responsible for their storage.

Consider the situation of Kathryn, the receiving clerk at the Fairview Estates, an extended care facility with 400 residents. She has just taken delivery of seven loaves of bread. They were delivered in accordance with the purchase order and now must be put away. While Kathryn stores these items, she must know whether management requires her to use the LIFO (last in, first out) or FIFO (first in, first out) method of product rotation.

LIFO System When using the LIFO storage system, the storeroom operator intends to use the most recently delivered product (last in) before he or she uses any part of that same product previously on hand. If Kathryn decides, for example, to use the new bread first, she would be using the LIFO system. In all cases, you must strive to maintain a consistent product standard. In the case of *some* bread, pastry, dairy, fruit, and vegetable items, the storeroom clerk could practice the LIFO system, if these items have been designed as LIFO items by management. With LIFO, you will need to take great care to order only the quantity of product needed between deliveries. If too much product is ordered, loss rates will be very high. Costs can rise dramatically when LIFO items must eventually be discarded or used in a way that reduces their revenue-producing ability. For most items you will buy, the best storage system to use is the FIFO system.

DETERMINING ACTUAL FOOD EXPENSE

Assume that you own and manage a small ice cream store that makes its own products. You have reviewed your records for the past month and found the following:

JANUARY REVENUE AND EXPENSE

Ice Cream Sales = $98,000

Food Expense = $39,000

You have determined your revenue figure from the sales history you maintained (see Chapter 18). You have determined your food expense by adding the dollar value of all the properly corrected delivery invoices that you accumulated for the month. That is, you totaled the value of all food purchased and delivered between the first day of the month and the last. Would it be correct to say that your food expense for the month of January is $39,000? The answer is No. Why not? Because you may have more, or less, of the food products required to make your ice cream in inventory on the last day of January than you had on the first day. If you have more food products in inventory on January 31 than you had on January 1, your food expense is less than $39,000. If you have fewer products in inventory on January 31 than you had on January 1, your food expense is higher than $39,000. To understand why this is so, you must understand the formula for computing your actual food expense. The correct formula is shown in Figure 19.8.

Cost of food sold is the dollar amount of all food actually sold, thrown away, wasted, or stolen.

Beginning Inventory

Beginning inventory is the dollar value of all food on hand at the beginning of the accounting period. It is determined by completing an actual count and valuation of the products on hand.

Purchases

Purchases are the sum cost of all food purchased during the accounting period. It is determined by adding all properly tabulated invoices for the accounting period.

```
Beginning Inventory

PLUS

Purchases
= Goods Available for Sale

LESS

Ending Inventory
= Cost of Food Consumed

LESS

Employee Meals

= Cost of Food Sold
```

FIGURE 19.8 Formula for cost of food sold.

Goods Available for Sale

Goods available for sale is the sum of the beginning inventory and purchases. It represents the value of all food that was available for sale during the accounting period.

Ending Inventory

Ending inventory refers to the dollar value of all food on hand at the end of the accounting period. It also is determined by completing a physical inventory.

Cost of Food Consumed

The cost of food consumed is the actual dollar value of all food used, or consumed, by the operation. Again, it is important to note that this is not merely the value of all food sold, but rather the value of all food no longer in the establishment and includes the value of any meals eaten by employees.

Cost of Food Sold

Accounting Period: _____ to _____

Unit Name: _____

Beginning Inventory $ _____

PLUS

Purchases $ _____

Goods Available for Sale $ _____

LESS

Ending Inventory $ _____

Cost of Food Consumed $ _____

LESS

Employee Meals $ _____

Cost of Food Sold $ _____

FIGURE 19.9 Recap sheet.

Employee Meals

Employee meal cost is a labor-related, not food-related cost. Free or reduced-cost employee meals are a benefit much in the same manner as medical insurance or paid vacation. Therefore, the value of this benefit, if provided, should be transferred and charged not as a cost of food but as a cost of employee benefits. The dollar value of food eaten by employees is subtracted from the cost of food consumed to yield the cost of food sold.

Cost of Food Sold

As stated earlier, the cost of food sold is the actual dollar value of all food expense incurred by the operation except for those related to employee meals. It is not possible to determine this number unless a beginning inventory has been taken at the start of the month, followed by another inventory taken at the end of the month. Without these two numbers, it is impossible to accurately determine the cost of food sold. Figure 19.9 illustrates a recap sheet used to determine the cost of food sold.

In the ice cream store example, had you completed such a form, you would have known your actual cost of food sold. Every manager should, on a regular basis, compute the actual cost of food sold because it is not possible to improve your cost picture unless you first know what your costs are.

Variations on the Basic Cost of Food Sold Formula

While Figure 19.9 demonstrates the format most commonly used to determine cost of food sold, some operators prefer slightly different formulas, depending on the unique aspects of their units. The important point to remember, however, is that all of these formulas should seek to accurately reflect actual cost of food sold by the operation for a given time period. Two variations of the formula follow.

Food or Beverage Products Are Transferred from One Foodservice Unit to Another   This is the case when, for example, an operator seeks to compute one cost of food sold figure for a bar and another for the bar's companion restaurant. In this situation, it is likely that fruits, juices, vegetables, and similar items are taken from the kitchen for use in the bar, while wine, sherry, and similar items may be taken from the bar for use in the kitchen. The formula for cost of food sold in this situation would be as follows:

Beginning inventory	$ _____	
PLUS		
Purchase	$ _____	
= Goods Available for Sale		$ _____
LESS		
Ending Inventory		$ _____
LESS		
Value of Transfers Out		$ _____
PLUS		
Value of Transfers In		$ _____
= Cost of Food Consumed		$ _____
LESS		
Employee Meals		$ _____
= Cost of Food Sold		$ _____

No Employee Meals Are Provided   When an operation has no employee meals at all, the computation of cost of food sold is as follows:

Beginning inventory	$ _____	
PLUS		
Purchase	$ _____	
= Goods Available for Sale		$ _____
LESS		
Ending Inventory		$ _____
= Cost of Food Sold		$ _____

It is important for you to know exactly which formula or variation is in use when analyzing cost of food sold. The variations, while slight, can make big differences in the interpretation of your cost information. In all cases, it is critical that accurate beginning and ending inventory figures be maintained if accurate cost data are to be computed.

In the following example, both beginning inventory and ending inventory figures are known, thus enabling you to determine your actual cost of food sold.

Beginning Inventory	$22,500
Purchases	$39,000
Ending Inventory	$27,500
Employee Meals	$725

You are now able to complete your recap sheet, as illustrated in Figure 19.10.

In this example, employee meals were determined by assigning a food value of $1.00 per employee ice cream treat consumed. It is important to note that food products do not have to be consumed as a meal in order to be valued as an employee benefit. Soft drinks, snacks, and other food items consumed by employees are all considered employee meals for the purpose of computing cost of food sold.

If records are kept on the number of employees eating per day, monthly employee meal costs are easily determined. Some operators prefer to estimate the dollar value of employee meals each month rather than record actual employee meals. This is not a good practice, both from a control point of view and in terms of developing accurate cost data.

It is important to note that ending inventory for one accounting period becomes the beginning inventory figure for the next period. For example, in the case of your ice cream store, the January 31 ending inventory figure of $27,500 will become the February 1 beginning inventory figure. In this manner, it is clear that physical inventory need only be taken one time per accounting period, not twice. Again, while the physical inventory process can be time consuming, it must be performed in order to determine actual food expense.

Cost of Food Sold		
Accounting Period:	1/1 to 1/31	
Unit Name:	Your Ice Cream Store	
Beginning Inventory	$22,500	
PLUS		
Purchases	$39,000	
Goods Available for Sale		$61,500
LESS		
Ending Inventory		$27,500
Cost of Food Consumed		$34,000
LESS		
Employee Meals		$ 725
Cost of Food Sold		$33,275

FIGURE 19.10 Recap sheet.

While there is no reliable method for replacing the actual counting of inventory items on hand, there are many computer programs on the market programmed to allow an individual to scan the bar codes on products using a handheld scanning device and, thus, perform both the counting and the price extension process necessary to develop actual inventory valuations. Using technology in this manner can make a time-consuming task much less tedious and more efficient.

Food Cost Percentage

You know from Chapter 17 that food expense is often expressed as a percentage of total revenue or sales. Since you can now determine your actual cost of food sold, you can also learn to compute and evaluate your operation's food cost percentage. Again, this is both the traditional way of looking at food expense and generally the method used by most operators when preparing the profit and loss statement.

The formula used to compute actual food cost percentage is as follows:

$$\frac{\text{Cost of Food Sold}}{\text{Food Sales}} = \text{Food Cost \%}$$

Food cost % represents that portion of food sales that was spent on food expenses.

In the case of the ice cream store discussed previously, you know that the cost of food sold equals $33,275 (Figure 19.10). If the store experienced $98,000 in food sales for the period of January 1 to January 31, the food cost percentage for the period would be

$$\frac{\text{Cost of Food Sold}}{\text{Food Sales}} = \frac{\$33,275}{\$98,000} = 34\% \text{ Food Cost}$$

Thus, 34% of the dollars in sales revenue taken in was needed to buy the food used to generate that revenue. Put another way, 34 cents of each dollar in sales were used to buy the products needed to make the ice cream.

Estimating Daily Cost of Food Sold

Many operators would like to know their food usage on a much more regular basis than once per month. When this is the case, the physical inventory may be taken as often as desired. Again, however, physical inventories are time consuming.

It would be convenient if you could have a close estimate of your food usage on a weekly or daily basis without the effort of a daily inventory count. Fortunately, such an approximation exists. Figure 19.11 illustrates a seven-column form that you can use for a variety of purposes. One of them is to estimate food cost % on a daily or weekly basis.

As an example, assume that you own an Italian restaurant that serves no liquor and caters to a family-oriented clientele. You would like to monitor your food cost percentage on a more regular basis than once a month, which is your regular accounting period. You have decided to use a six-column form to estimate your food cost percentage. Since you keep track of both daily sales and purchases, you can easily do so. In the space above the first two columns, write the word Sales. Above the middle two columns, write Purchases, and above the last two columns, enter the term Cost %.

You then proceed each day to enter your daily sales revenue in the column labeled Sales Today. Your invoices for food deliveries are totaled daily and entered in the column

Date: _____

Weekday	Today	To Date	Today	To Date	Today	To Date

FIGURE 19.11 Six-column form.

titled Purchases Today. Dividing the Purchases Today column by the Sales Today column yields the figure that is placed in the Cost % Today column. Purchases to Date (the cumulative purchases amount) is divided by Sales to Date (the cumulative sales amount) to yield the Cost % to Date figure. A quick summary is as follows:

SIX-COLUMN FOOD COST % ESTIMATE

1. $\dfrac{\text{Purchases Today}}{\text{Sales Today}} = \text{Cost \% Today}$

2. $\dfrac{\text{Purchase to Date}}{\text{Sales to Date}} = \text{Cost \% to Date}$

Figure 19.12 shows this information for your operation for the time period January 1 to January 7.

As can be seen, you buy most of your food at the beginning of the week, while sales are strongest in the later part of the week. This is a common occurrence at many foodservice establishments. As can also be seen, your daily cost percentage ranges from a high of 130% (Monday) to a low of 0% (Sunday), when no deliveries are made. In the Cost % to Date column, however, the range is only from a high of 130% (Monday) to a low of 39.20% (Sunday).

What is your best estimate about what your food cost % actually is as of Sunday? The answer is that it will be slightly less than 39.20%. Why? Let us go back to the formula for cost of food sold. Before we do, we must make one important assumption: "For any time period we are evaluating, the beginning inventory and ending inventory amounts are the same." In other words, over any given time period, you will have approximately the same amount of food on hand at all times. If this assumption is correct, the six-column food cost estimate is, in fact, a good indicator of your food usage. The reason is very simple. The formula for cost of food sold asks you to add beginning inventory and then later subtract ending inventory (see Figure 19.8). If these two numbers are assumed to be the same, they can be ignored, since adding and subtracting the same number to any other number will result in no effect at all. For example, if we start with $10, add $50, and subtract $10, we are left with $50. In terms of the cost of food sold formula, when beginning inventory and ending inventory are assumed to be the same figure, it is the mathematical equivalent of adding a zero and subtracting a zero. To continue our example, if we start with $0, add $50, and subtract $0, we are left, again, with $50. Thus, when beginning inventory and ending inventory are equal, or assumed to be equal, the cost of food sold formula would look as follows:

	Beginning Inventory	$0
+	Purchases	
=	Purchases	
−	Ending Inventory	$0

	=	Purchases
	−	Employee Meals
	=	Cost of Food Sold

Date: 1/1–1/7

	Sales		Purchases		Cost %	
Weekday	**Today**	**To Date**	**Today**	**To Date**	**Today**	**To Date**
Monday	$ 850.40	$ 850.40	$1,106.20	$1,106.20	130.00%	130.00%
Tuesday	920.63	1,771.03	841.40	1,947.60	91.40	110.00
Wednesday	1,185.00	2,956.03	519.60	2,467.20	43.80	83.50
Thursday	971.20	3,927.23	488.50	2,955.70	50.30	75.30
Friday	1,947.58	5,874.81	792.31	3,748.01	40.70	63.80
Saturday	2,006.41	7,881.22	286.20	4,034.21	14.30	51.20
Sunday	2,404.20	10,285.42	0	4,034.21	0	39.20
Total	10,285.42		4,034.21		39.20	

FIGURE 19.12 Six-column food cost estimate.

Therefore, as stated earlier, if you assume that your inventory is constant, your cost of food sold for the one-week period is a little less than $4,034.21, or 39.20% of sales. Why a little less? Because we must still subtract the value of employee meals, if any are provided, since they are an employee benefit and not a food expense. How accurate is the six-column form? For most operators, it is quite accurate and has the following advantages:

1. It is very simple to compute, requiring 10 minutes or less per day for most operations.

2. It records both sales history and purchasing patterns.

3. It identifies problems before the end of the monthly accounting period.

4. By the ninth or tenth day, the degree of accuracy in the To Date column is very high.

5. It is a daily reminder to both management and employees that there is a very definite relationship between sales and expenses.

The use of a six-column food cost estimator is highly recommended for the operator who elects to conduct a physical inventory less often than every two weeks.

The control of food expense is critical to all foodservice operations. From the purchase of the raw ingredient to its receiving and storage, the effective foodservice operator strives to have the proper quality and quantity of product on hand at all times. Food represents a large part of your overall expense budget. Protecting this product and accounting for its usage are extremely important in helping to manage overall costs.

APPLY WHAT YOU HAVE LEARNED

Tonya Johnson is the Regional Manager of Old Town Buffets. Each of the five units she supervises is in a different town. Produce for each unit is purchased locally by each buffet manager. One day, Tonya gets a call from Danny Trevino, one of the buffet managers reporting to her. Danny states that one of the local produce suppliers he uses has offered Danny the use of season tickets to the local university football games. Danny likes football and would like to accept them.

1. Would you allow Danny to accept the tickets? Why or why not?

2. Would you allow your managers to accept a gift of any kind (including holiday gifts) from a vendor?

3. Draft a "gifts" policy that you would implement in your region. Would you be subject to the same policy?

KEY TERMS AND CONCEPTS

The following are terms and concepts discussed in the chapter that are important for you as a manager. To help you review, please define the terms below.

Popularity index	Fiscal year	Cost of food sold
Predicted number to be sold	Refusal hours	Beginning inventory
Standardized recipe	LIFO	Purchases
Recipe ready	FIFO	Goods available for sale
Working stock	Issue	Ending inventory
Safety stock	Purchase point	Cost of food consumed
Shelf life	As needed/just in time	Employee meals
Opportunity cost	Par level	Food cost %

MANAGING FOOD AND BEVERAGE PRICING*

Overview

This chapter shows you how to identify and use the menu formats you will most often encounter as a hospitality manager. Knowledge of these menu formats will help you reduce costs through effective utilization of food and beverage products as well as better utilization of your staff. In addition, the chapter examines and analyzes the factors that influence the prices you will charge for the menu items you will sell. Finally, the chapter explains the procedures used to assign individual menu item prices based on cost and collected sales data. By fully understanding the hospitality pricing process, you can help ensure that your menu items will generate the sales revenue needed to meet your profit goals.

Chapter Outline

Menu Formats

Factors Affecting Menu Pricing

Fun on the Web!

Assigning Menu Prices

Special Pricing Situations

Apply What You Have Learned

Key Terms and Concepts

---

* Authored by Jack E. Miller, Lea R. Dopson, and David K. Hayes.

Highlights

At the conclusion of this chapter, you will be able to:

- Choose and apply the best menu type to an operation.
- Identify the variables to be considered when establishing menu prices.
- Assign menu prices to menu items based on their cost, popularity, and ultimate profitability.

MENU FORMATS

If you have determined that your purchasing, receiving, storing, issuing, and production controls are well in line, you have an excellent chance of reaching your profit goals. It is possible to find, however, that even when these areas are properly controlled, food and beverage costs may still not be in line with your projections. When this is true, the problem may well lie in the fundamental areas of menu format, product pricing, or both. It makes good sense to analyze menu format first, as menu design decisions drive most pricing decisions.

Menus in foodservice establishments generally fall into one of the following three major categories:

1. Standard menu

2. Daily menu

3. Cycle menu

Any of these can be an asset to your effort to control food costs if it is used in the proper setting. The most commonly used menu is the standard menu.

Standard Menu

The standard menu is printed, recited by service staff, or otherwise communicated to the guest. The standard menu is fixed day after day. While you may periodically add or delete an item, the standard menu remains virtually constant. There are many operational advantages to a standard menu. First, the standard menu simplifies your ordering process. Since the menu remains constant each day, it is easy to know which products must be purchased to produce these specific menu items. Second, guests tend to have a good number of choices when selecting from a standard menu. This is true because virtually every item that can be produced by the kitchen is available for selection by each guest entering the operation.

A third advantage of the standard menu is that guest preference data are easily obtained since the total number of menu items that will be served stays constant and, thus, is generally smaller than in some alternative menu formats. As you learned in Chapter 19, menu item sales histories can be used to accurately compute percent selecting data and, thus, production schedules and purchasing requirements. In addition, standard menus often become marketing tools for your operation since guests soon become familiar with and return for their favorite menu items.

The standard menu is most typically found in the traditional restaurant or hotel segment of the hospitality industry. It tends to dominate those segments of the business where the guest selects the location of the dining experience, as contrasted with situations where a guest's choice is restricted. Examples of restricted-choice situations include a college dormitory cafeteria where students are required to dine in one location, a hospital where patients during their stay must choose their menu selections from that hospital only, or an elementary school cafeteria.

Despite its many advantages, the standard menu does have drawbacks from a control standpoint. First, standard menus are often not developed to utilize carryovers effectively. In fact, in many cases, items that are produced for a standard menu and remain unsold must be discarded, as the next day their quality will not be acceptable. An example would be a quick-service restaurant that produces too many hamburgers for lunch and does not sell all of them. Indeed, for some quick-service restaurants, a burger that is made but not sold within five minutes would be discarded. Contrast that cost control strategy with one that says that cooked burgers not sold within five minutes will be chopped and added to the house specialty chili, and it is easy to see how menu design and the items placed on the menu affect food cost control.

A second disadvantage of the standard menu is its lack of ability to respond quickly to market changes and product cost changes. A restaurant that does not list green beans on the menu cannot take advantage of the seasonal harvest of green beans, a time when they can be purchased extremely inexpensively. Conversely, if management has decided that its two house vegetables will be broccoli and corn, even considerable price increases in these two items will have to be absorbed by the operation since these two menu items are listed on the permanent menu. An extreme example of this kind of problem was found in a quick-service seafood restaurant chain that found itself paying almost three times what it had the previous year for a seafood item that constituted approximately 80% of its menu sales. A foreign government had restricted fishing for this product off its shores, and the price skyrocketed. This chain was nearly devastated by this turn of events. Needless to say, management quickly moved to add chicken and different seafood products to the menu in order to dilute the effect of this incredible price increase. Whenever possible, you should monitor food prices with an eye to making seasonal adjustments. The standard menu makes this quite difficult, although some restaurant groups respond to this problem by changing their standard menu on a regular basis. That is, they develop a standard menu for the summer, for example, and another for the winter. In this manner, they can take advantage of seasonal cost savings, add some variety to their menus, but still maintain the core menu items for which they are known.

Daily Menu

In some restaurants, you might elect to operate without a standard menu and, instead, implement a daily menu, that is, a menu that changes every day. This concept is especially popular in some upscale restaurants where the chef's daily creations are viewed with great anticipation, and some awe, by the eager guest. The daily menu offers some advantages over the standard menu, since management can respond very quickly to changes in ingredient or item prices. In fact, that is one of the daily menu's great advantages. In addition, carryovers are less of a problem because any product unsold from the previous day has at least the potential of being incorporated, often as a new dish, into today's menu.

Every silver lining has its cloud, however. For all its flexibility, the daily menu is recommended only for very special situations due to the tremendous control drawbacks associated with its implementation. First, item popularity data are difficult to obtain, since the items on any given day's menu may never have been served in that particular combination in that restaurant. Thus, the preparation of specific items in certain quantities is pure guesswork, and this is a dangerous way of determining production schedules. Second, it may be difficult for you to plan to have the necessary ingredients on hand to prepare the daily menu if the menu is not known ahead of time. How does one decide on Monday whether one should order tuna or sirloin steak for the menu on Thursday? Obviously, this situation requires that even the daily menu be planned far enough in advance to allow the purchasing agent to select and order the items necessary to produce the menu. Third, the daily menu may sometimes serve as a marketing tool, but can just as often serve as a disappointment to guests who had a wonderful menu item the last time they dined at this particular establishment and have now returned only to find that their favorite item is not being served. On the other hand, it is very unlikely that any guest will get bored with a routine at a daily menu restaurant, since the routine is, in fact, no routine at all.

Days	Cycle
1–7	A
8–14	B
15–21	C
22–28	D
29–35	A
36–42	B
43–49	C
50–56	D
57–63	A

FIGURE 20.1 Sample cycle menu rotation.

Both the standard and the daily menus have advantages and disadvantages. The cycle menu is an effort by management to enjoy the best aspects of both of these approaches and minimize their respective disadvantages.

Cycle Menu

A cycle menu is a menu in effect for a specific time period. The length of the cycle refers to the length of time the menu is in effect. Thus, we refer to a 7-day cycle menu, 21-day cycle menu, 30-day cycle menu, or one of any other length of time. Typically, the cycle menu is repeated on a regular basis. Thus, for example, a particular cycle menu could consist of four 7-day periods. If each of the four periods were labeled as A, B, C, and D, the cycle periods would rotate as illustrated in Figure 20.1.

Within each cycle, the menu items vary on a daily basis. For example, cycle menu A might consist of the following seven dinner items:

Day 1	Monday	Cheese Enchiladas
Day 2	Tuesday	Turkey and Dressing
Day 3	Wednesday	Corned Beef and Cabbage
Day 4	Thursday	Fried Chicken
Day 5	Friday	Chop Suey
Day 6	Saturday	Lasagna
Day 7	Sunday	Pot Roast

These menu items would be served again when cycle menu A repeated itself on days 29–35 as follows:

Day 29	Monday	Cheese Enchiladas
Day 30	Tuesday	Turkey and Dressing
Day 31	Wednesday	Corned Beef and Cabbage
Day 32	Thursday	Fried Chicken
Day 33	Friday	Chop Suey
Day 34	Saturday	Lasagna
Day 35	Sunday	Pot Roast

In the typical case, cycle menus B, C, and D would, of course, consist of different menu items. In this manner, no menu item would be repeated more frequently than desired by management.

Cycle menus make the most sense when your guests dine with you on a very regular basis, either through the choice of the individual, such as a college student or summer camper eating in a dining hall, or through the choice of an institution, such as a hospital or correctional facility feeding situation. In cases like this, menu variety is very important. The cycle menu provides a systematic method for incorporating that variety into the menu. At a glance, the foodservice manager can determine how often, for example, fried chicken will be served per week, month, or year, and how frequently bread dressing rather than saffron rice is served with baked chicken. In this respect, the cycle menu offers more to the guest than does the standard menu. With cycle menus, production personnel can be trained to produce a wider variety of foods than with the standard menu, thus improving their skills, but requiring fewer skills than might be needed with a daily menu concept.

Cycle menus have the advantage of being able to systematically incorporate today's carryovers into tomorrow's finished product. This is an important management advantage. Also, because of its cyclical nature, management should have a good idea of guest preferences and, thus, be able to schedule and control production to a greater degree than with the daily menu.

Purchasing, too, is simplified since the menu is known ahead of time, and menu ingredients that will be appear on all the different cycles can be ordered with plenty of lead time. Inventory levels are easier to maintain as well because, as is the case with the standard menu, product usage is well known.

To illustrate the differences and the impact of operating under the three different menu systems, consider the case of Larry, Moe, and Curly Jo, three foodservice operators who wish to serve roast turkey and dressing for their dinner entrée on a Saturday night in April. Larry operates a restaurant with a standard menu. If he is to print a standard menu that allows him to serve turkey in April, he may be required to have it available in January and June also. If he is to utilize any carryover parts of the turkey, he must incorporate a second turkey item, which also must be made available every day. Larry is not sure all the trouble and cost is worth it! In addition, consider the expense involved in a national chain restaurant with thousands of outlets. Reprints of menus in this situation are very costly. If Larry is the CEO (chief executive officer) of an organization such as this, any decision to change the standard menu is indeed a major undertaking.

Moe operates a restaurant with a daily menu. For him, roast turkey and dressing on a Saturday in April is quite easy. His problem, however, is that he has no idea how much to produce since he has never before served this item at this time of the year in his restaurant. Also, few of the guests he has served in the past year are likely to know about his decision to serve the menu item. What if no one orders it?

Curly Jo operates on a cycle menu. She can indeed put roast turkey and dressing on the cycle. If it sells well, she will keep it on the cycle. If it does not, it will be removed from the next cycle. Curly Jo makes a note to herself that she should record how well it sells and leave a space in the cycle for the utilization of any carryover product that might exist within the next few days. The advantages of the cycle menu, in this situation, are apparent.

Menu Specials

Regardless of the menu type used, you can generally incorporate minor menu changes on a regular basis. This is accomplished through the offering of daily or weekly menu specials, that is, menu items that will appear on the menu as you desire and be removed when they are either consumed or discontinued. These daily or weekly specials are an effort to provide variety, take advantage of low-cost raw ingredients, utilize carryover products, or test-market potential new menu items. The menu special is a powerful cost control tool. Properly utilized, it helps shape the future menu by testing guest acceptance of new menu items while, at the same time, providing opportunities for you to respond to the challenges of using carryover or new food and beverage products you have in inventory.

FACTORS AFFECTING MENU PRICING

A great deal of important information has been written in the area of menu pricing and strategy. For the serious foodservice operator, menu pricing is a topic that deserves its own significant research and study. Pricing is related to cost control by virtue of the basic formula from Chapter 17:

Revenue − Expense = Profit

When foodservice operators find that profits are too low, they frequently question whether prices (revenues) are too low. It is important to remember, however, that revenue and price are not synonymous terms. Revenue means the amount spent by *all* guests, while price refers to the amount charged to *one* guest. Thus, total revenue is generated by the following formula:

Price × Number Sold = Total Revenue

From this formula, it can be seen that there are two components of total revenue. While price is one component, the other is the number of items sold and, thus, guests served. It is a truism that as price increases, the number of items sold will generally decrease. For this reason, price increases must be evaluated based on their impact on total revenue and not price alone. Assume, for example, that you own a quick-service restaurant chain. You are considering raising the price of small drinks from $1.00 to $1.25. Figure 20.2 illustrates the possible effects of this price increase on total revenue in a single unit. Note especially that, in at least one alternative result, increasing price has the effect of actually decreasing total revenue. Experienced foodservice managers know that increasing prices without giving added value can result in higher prices but, frequently, lower revenue because of reduced guest counts. This is true because guests demand a good price/value relationship when making a purchase. The price/value relationship simply reflects guests' view of how much value they are receiving for the price they are paying.

Old Price	New Price	Number Served	Total Revenue	Revenue Result
$1.00		200	$200.00	
	$1.25	250	$312.50	Increase
	$1.25	200	$250.00	Increase
	$1.25	160	$200.00	No Change
	$1.25	150	$187.50	Decrease

FIGURE 20.2 Alternative results of price increases.

Perhaps no area of hospitality management is less understood than the area of pricing. This is not surprising when you consider the many factors that play a part in the pricing decision. For some foodservice operators, inefficiency in cost control is passed on to the guest in terms of higher prices. In fact, sound pricing decisions should be based on establishing a positive price/value relationship in the mind of the guest. Most foodservice operators face similar product costs when selecting their goods on the open market. Whether the product is oranges or beer, wholesale prices may vary only a little from supplier to supplier. In some cases, this variation is due to volume buying, while, in others, it is the result of the relationship established with the vendor. Regardless of their source, the fact remains that the variations are small relative to variations in menu pricing. This becomes easier to understand when you realize that selling price is a function of much more than product cost. It may be said that price is significantly affected by all of the following factors:

FACTORS INFLUENCING MENU PRICE

1. Local competition
2. Service levels
3. Guest type
4. Product quality
5. Portion size
6. Ambiance
7. Meal period
8. Location
9. Sales mix

Local Competition

This factor is often too closely monitored by the typical foodservice operator. It may seem to some that the average guest is vitally concerned with price and nothing more. In reality, small variations in price generally make little difference to the average guest. If a group of young professionals goes out for pizza and beer after work, the major determinant will not be whether the selling price for the beer is $3.00 in one establishment or $3.25 in another. Your competition's selling price is somewhat important when establishing price, but it is a well-known fact in foodservice that someone can always sell a lesser quality product for a lesser price. The price a competitor charges for his or her product can be useful information in helping you arrive at your own selling price. It should not, however, be the only determining factor in your pricing decision. Successful foodservice operators spend their time focusing on building guest value in their own operation and not in attempting to mimic the efforts of the competition.

Service Levels

Guests expect to pay more for the same product when service levels are higher. The can of soda sold from a vending machine is generally less expensive than one served by a human being. In a like manner, many pizza chains charge a lower price, for example, for a large pizza that is picked up by the guest than for that same pizza when it is delivered to the guest's door. This is as it should be. The hospitality industry is, in fact, a service industry. As the personal level of service increases, prices may also increase. This personal service may range from the delivery of products, as in the pizza example, to simply increasing the number of servers in a dining room and, thus, reducing the number of guests each must serve. This is not to imply that menu price increases based on service levels are reserved exclusively to pay for the labor required to increase those service levels. Guests are willing to pay more for increased service levels, but this higher price should provide for extra profit as well. In the hospitality industry, those companies that have been able to survive and thrive over the years have done so because of their uncompromising commitment to high levels of guest service. This trend will continue.

Guest Type

Some guests are simply less price sensitive than others. All guests, however, want value for their money. The question of what represents value varies by the type of clientele. An example of this can clearly be seen in the pricing decisions of convenience stores across the United States. In these facilities, food products such as sandwiches, fruit, drinks, cookies, and the like are sold at relatively high prices. The guests these stores cater to, however, value speed and convenience above all else. For this convenience and a wider range of products than would be found at most quick-service restaurants, they are willing to pay a premium price. In a like manner, guests at an expensive steakhouse restaurant are less likely to respond negatively to small variations in drink prices than are guests at a corner tavern. A thorough analysis of who their guests are and what they value most is critical to the success of foodservice operators.

Product Quality

In nearly every instance, the guest's quality perception of any specific product offered for sale in the foodservice business can range from very low to very high. This is not to say that the wholesomeness or safety of the product will vary. They should not. But the guest's perception of quality will be based on a variety of factors. As the product itself and those quality-influencing factors vary, so, too, does the guest's perception of quality. For example, when average foodservice guests think of a "hamburger," they actually think, not of one product, but of a range of products. A hamburger may be a rather small burger patty on a regular bun, wrapped in paper and served in a sack. If so, its price will be low and so perhaps may service levels and, thus, perceived quality. If, however, the guest's thoughts turn to an 8-ounce gourmet burger with avocado slices and alfalfa sprouts on a toasted whole-grain bun served in a white-tablecloth restaurant, the price will be much higher and so, probably, will service levels and perceived quality.

As an effective foodservice manager, you will choose from a variety of quality levels when developing product specifications and, consequently, planning menus and establishing prices. If you select the market's cheapest bourbon as your well brand, you will likely be able to charge less for drinks made from it than your competitor who selects a better brand. Your drink quality levels, however, may also be perceived by your guests as lower. To be successful, you should select the quality level that best represents your guests' anticipated desire as well as your goals, and then price your products accordingly.

Portion Size

Portion size plays a large role in determining menu pricing. It is a relatively misunderstood concept, yet it is probably the second most significant factor (next to sales mix) in overall pricing. The great chefs know that people "eat with their eyes first!" This relates to presenting food that is visually appealing. It also relates to portion size. A burger and fries that fill an 8-inch plate may well be lost on an 11-inch plate. Portion size, then, is a function of both food quantity and how it is presented. It is no secret why successful cafeteria chains use smaller than average dishes to plate their food. For their guests, the image of price / value comes across loud and clear.

In some dining situations, particularly in an "all you care to eat" operation, the previously mentioned principle again holds true. The proper dish size is just as critical as the proper size scoop or ladle when serving the food. Of course, in a traditional table service operation, management controls (or should control!) portion size. Simply put, the larger the portion size, the higher your costs. One very good way to determine whether portion sizes are too large is simply to watch the dishwashing area and see what comes back from the dining room as uneaten. In this regard, the dishroom operator becomes an important player in the cost control team.

Many of today's consumers prefer lighter food with more choices in fruits and vegetables. The portion sizes of these items can be increased at a fairly low increase in cost. At the same time, average beverage sizes are increasing, as are the size of side items such as French fries. Again, these tend to be lower cost items. This can be good news for the foodservice operator if prices can be increased to adequately cover the larger portion sizes.

Every menu item should be analyzed with an eye toward determining if the quantity being served is the "proper" quantity. You would, of course, like to serve this proper amount, but no more than that. The effect of portion size on menu price is significant, and it will be your job to establish and maintain strict control over proper portion size.

Ambiance

If people ate only because they were hungry, few restaurants would be open today. People eat out for a variety of reasons, some of which have little to do with food. Fun, companionship, time limitations, adventure, and variety are just a few reasons diners cite for eating out rather than eating at home. For the foodservice operator who provides an attractive ambiance, menu prices can be increased. In fact, the operator in such a situation is selling much more than food and, thus, justly deserves this increased price. In most cases, however, foodservice operations that count too heavily on ambiance alone to carry their business generally start well but are not ultimately successful. Excellent product quality with outstanding service goes much further over the long run than do clever restaurant designs. Ambiance may draw guests to a location the first time. When this is true, prices may be somewhat higher if the quality of products also supports the price structure.

Meal Period

In some cases, diners expect to pay more for an item served in the evening than for that same item served at a lunch period. Sometimes this is the result of a smaller "luncheon" portion size, but in other cases the portion size, as well as service levels, may be the same in the evening as earlier in the day. You must exercise caution in this area. Guests should clearly understand why a menu item's price changes with the time of day. If this cannot be answered to the guest's satisfaction, it may not be wise to implement a time-sensitive pricing structure.

Location

Location can be a major factor in determining price. One need look no further than America's many themed amusement parks or sports arenas to see evidence of this. Foodservice operators in these places are able to charge premium prices because they have, in effect, a monopoly on food sold to the visitors. The only all-night diner on the interstate highway exit is in much the same situation. Contrast that with an operator who is one of 10 seafood restaurants on restaurant row. It used to be said of restaurants that success was due to three things: location, location, and location! This may have been true before so many operations opened in the United States. There is, of course, no discounting the value of a prime restaurant location, and location alone can influence price. It does not, however, guarantee success. Location can be an asset or a liability. If it is an asset, menu prices may reflect that fact. If location is indeed a liability, menu prices may need to be lower to attract a sufficient clientele to ensure the operation's total revenue requirements.

Sales Mix

Of all the factors mentioned thus far, sales mix would most heavily influence the menu pricing decision, just as guest purchase decisions will influence total product costs. Recall that sales mix refers to the specific menu items selected by guests. Managers can respond to this situation by employing a concept called price blending. Price blending refers to the process of pricing products, with very different individual cost percentages, in groups with the intent of achieving a favorable overall cost situation. The ability to knowledgeably blend prices is a useful skill and one that is well worth mastering. As an example, assume that you are the operations vice president for a chain of upscale hamburger restaurants known as Texas Red's. Assume also that you hope to achieve an overall food cost of 40% in your units. For purposes of simplicity, assume that Figure 20.3 illustrates the three products you sell and their corresponding selling price if each is priced to achieve a 40% food cost.

In Chapter 19, you learned that the formula for computing food cost percentage is as follows:

$$\frac{\text{Cost of Food Sold}}{\text{Food Sales}} = \text{Food Cost \%}$$

This formula can be worded somewhat differently for a single menu item without changing its accuracy. Consider that:

$$\frac{\text{Cost of a Specific Food Item Sold}}{\text{Food Sales of That Item}} = \text{Food Cost \% of That Item}$$

It is important to understand that the food sales value in the preceding formula is a synonymous term to the selling price when evaluating the menu price of a single menu item. The principles of algebra allow you to rearrange the formula as follows:

$$\frac{\text{Cost of a Specific Food Item Sold}}{\text{Food Cost \% of That Item}} = \text{Food Sales (Selling Price) of That Item}$$

Thus, in Figure 20.3, the hamburger's selling price is established as

$$\frac{\$1.50}{0.40} = \$3.75$$

Notice that in Figure 20.3 all products are priced to sell at a price that would result in a 40% food cost. Certainly, under this system, sales mix, that is, the individual menu selections of guests, would not affect overall food cost %. The sales mix resulting from this pricing strategy could, however, have very damaging results on your profitability. The reason is very simple. If you use the price structure indicated previously, your drink prices are too low. Most guests expect to pay far in excess of 45 cents for a soft drink at a quick-service restaurant. You run the risk, in this example, of attracting many guests

Texas Red's Burgers			
Item	Item Cost	Desired Food Cost	Proposed Selling Price
Hamburger	$1.50	40%	$3.75
French Fries	0.32	40	0.80
Soft Drinks (12 oz.)	0.18	40	0.45
Total	2.00	40	5.00

FIGURE 20.3 Unblended price structure.

Texas Red's Burgers			
Item	Item Cost	Proposed Food Cost %	Proposed Selling Price
Hamburger	$1.50	60.2%	$2.49
French Fries	0.32	21.5	1.49
Soft Drinks (12 oz.)	0.18	16.5	1.09
Total	2.00	39.4	5.07

FIGURE 20.4 Blended price structure.

who are interested in buying only soft drinks at your restaurants. Your French fries may also be priced too low. Your burger itself, however, may be priced too high relative to your competitors. However, if you use the price-blending concept, and if you assume that each guest coming into your restaurants will buy a burger, French fries, and a soft drink, you can create a different menu price structure and still achieve your overall cost objective, as seen in Figure 20.4.

Note that, in this example, you would actually achieve a total food cost slightly lower than 40%. Your hamburger price is now less than $2.50 and in line with local competitors. Note also, however, that you have assumed each guest coming to Texas Red's will buy one of each item. In reality, of course, not all guests will select one of each item. Some guests will not elect fries, while others may stop in only for a soft drink. It is for this reason that guest selection data, discussed in Chapter 19, are so critical. These histories let you know exactly what your guests are buying when they visit your outlets. You can then apply percent selecting figures to your pricing strategy. To illustrate how this works, assume that you monitored a sample of 100 guests who came into one of your units and found the results presented in Figure 20.5.

As you can see from Figure 20.5, you can use the price-blending concept to achieve your overall cost objectives if you have a good understanding of how many people buy each menu item. In this example, you have achieved the 40% food cost you sought. It matters little if the burger has a 60.2% food cost if the burger is sold in conjunction with the sample number of soft drinks and fries. Obviously, there may be a danger if your guests begin to order nothing but hamburgers when they come to your establishment. Again, careful monitoring of guest preferences will allow you to make price adjustments,

Texas Red's Burgers

Total Sales: ___$449.25___   Guests Served: 100
Total Food Cost: $180.20   Food Cost %:   40.1%

Item	Number Sold	Item Cost	Total Food Cost	Selling Price	Total Sales	Food Cost %
Hamburger	92	$1.50	$138.00	$2.49	$229.08	60.2%
French Fries	79	0.32	25.28	1.49	117.71	21.5
Soft Drink (12 oz.)	94	0.18	16.92	1.09	102.46	16.5
Total			180.20		449.25	40.1

FIGURE 20.5 Sample sales mix data.

as needed, to keep your overall costs and prices in line. A word of caution regarding the manipulation of sales mix and price blending is, however, in order. Since price itself is one of the factors that impact percent selecting figures, a change in menu price may cause a change in item popularity. If, in an effort to reduce overall product cost percentage, you were to increase the price of soft drinks at Texas Red's, for example, you might find that a higher percentage of guests would elect *not* to purchase a soft drink. This could have the effect of actually increasing your overall product cost percentage since fewer guests would choose to buy the one item with an extremely low food cost percentage. The sales mix and the concept of price blending will have a major impact on your overall menu pricing philosophy and strategy.

FUN ON THE WEB!

Look up the following site to evaluate restaurants' pricing strategies based on reviewers' perceptions of service levels, guest type, product quality, ambiance, location, and much more!

www.zagat.com   First, choose a city. Then click on your specific restaurant criteria or just click on "Most Popular." Then click on any restaurant for a review that includes a features list, reviewers' ratings, prices, and much more. Look at several restaurant reviews. Evaluate whether you think the menu prices and the features/review are a "good fit," in your opinion. Spend about 30 minutes viewing this site.

ASSIGNING MENU PRICES

The methods used to assign menu prices are as varied as foodservice managers themselves. In general, however, menu prices are most often assigned on the basis of one of the following two concepts:

1. Product cost percentage

2. Product contribution margin

Product Cost Percentage

This method of pricing is based on the idea that product cost should be a predetermined percentage of selling price. As was illustrated earlier in this chapter, if you have a menu item that costs $1.50 (EP) to produce, and your desired cost percentage equals 40%, the following formula can be used to determine what the item's menu price should be:

$$\frac{\text{Cost of a Specific Food Item Sold}}{\text{Food Cost \% of That Item}} = \text{Food Sales (Selling Price) of That Item}$$

or

$$\frac{\$1.50}{0.40} = \$3.75$$

Thus, the recommended selling price, given a $1.50 product cost, is $3.75. If the item is sold for $3.75, then a 40% food cost should be achieved for that item. A check on this work can also be done using the food cost percentage formula:

$$\frac{\$1.50}{3.75} = 40\%$$

When management uses a predetermined food cost percentage to price menu items, it is stating its belief that product cost in relationship to selling price is of vital importance. Experienced foodservice managers know that a second method of arriving at appropriate selling prices based on predetermined food cost % goals can be employed. This method uses a cost factor or multiplier that can be assigned to each desired food cost percentage. This factor, when multiplied times the item's EP cost, will result in a selling price that yields the desired food cost percentage. Figure 20.6 details such a factor table.

In each case, the factor is arrived at by the following formula:

$$\frac{1.00}{\text{Desired Product Cost \%}} = \text{Pricing Factor}$$

Desired Product Cost %	Factor
20	5.000
23	4.348
25	4.000
28	3.571
30	3.333
33⅓	3.000
35	2.857
38	2.632
40	2.500
43	2.326
45	2.222

FIGURE 20.6 Pricing-factor table.

Thus, if one were attempting to price a product and achieve a product cost of 40%, the computation would be

$$\frac{1.00}{0.40} = 2.5$$

This pricing factor of 2.5, when multiplied by any product cost, will yield a selling price that is based on a 40% product cost. The formula is as follows:

Pricing Factor × Product Cost = Menu Price

To return to our example, you could use the previous version of the formula to establish your selling price if you hope to achieve a 40% product cost and if your item costs $1.50 to produce. The computation would be as follows:

2.5 × 1.50 = $3.75

As can be seen, these two methods of arriving at the proposed selling price yield the same results. One formula simply relies on division, while the other relies on multiplication. The decision about which formula to use is completely up to you. With either approach, the selling price will be determined with a goal of achieving a given product cost percentage for each item.

Product Contribution Margin

Some foodservice managers prefer an approach to menu pricing that is focused, not on product cost percentage, but rather on a menu item's contribution margin. Contribution margin is defined as the amount that remains after the product cost of the menu item is subtracted from the item's selling price. Contribution margin, then, is the amount that "contributes" to paying for your labor and other expenses and providing a profit. Thus, if an item sells for $3.75 and the product cost for this item is $1.50, the contribution margin would be computed as follows:

> **Selling Price − Product Cost = Contribution Margin**
>
> *or*
>
> **$3.75 − $1.50 = $2.25**

When this approach is used, the formula for determining selling price is

> **Product Cost + Contribution Margin Desired = Selling Price**

Establishing menu price is a matter of combining product cost with a predetermined contribution margin. Management's role here is to determine the desired contribution margin for each menu item. When using this approach, you would likely establish different contribution margins for various menu items or groups of items. For example, in a cafeteria where items are priced separately, entrées might be priced with a contribution margin of $2.50 each, desserts with a contribution margin of $1.25, and drinks, perhaps, with a contribution margin of $0.75. Those managers who rely on the contribution margin approach to pricing do so in the belief that the average contribution margin per item is a more important consideration in pricing decisions than food cost percentage.

Product Cost Percentage or Product Contribution Margin?

Proponents exist for both of these approaches to menu pricing. Indeed, there are additional methods that can be used for menu pricing, beyond the scope of an introduction to pricing theory. Some large foodservice organizations have established highly complex computer-driven formulas for determining appropriate menu prices. For the average foodservice operator, however, product cost percentage, contribution margin, or a combination of both will suffice when attempting to arrive at appropriate pricing decisions.

While the debate over the "best" pricing method is likely to continue for some time, you should remember to view pricing, not as an attempt to take advantage of the guest, but rather as an important process with an end goal of establishing a good price/value relationship in the mind of your guest.

Regardless of whether the pricing method used is based on food cost percentage or contribution margin, the selling price selected must provide for a predetermined operational profit. For this reason, it is important that the menu not be priced so low that no profit is possible or so high that you will not be able to sell a sufficient number of items to make a profit. In the final analysis, it is the market that will eventually determine what your sales will be on any given item.

Being sensitive to both required profit and your guests—their needs, wants, and desires—is very critical to a pricing philosophy. *Menu Pricing and Strategy* by Jack Miller and David Pavesic (New York: John Wiley & Sons, 1997), provides an excellent treatment of the menu development, marketing strategies, price support systems, and pricing strategies necessary to effectively design and assign price to a menu. It is highly recommended as an addition to your management library.

SPECIAL PRICING SITUATIONS

Some pricing decisions faced by foodservice managers call for a unique approach. In many cases, pricing is used as a way to influence guests' purchasing decisions or to respond to particularly difficult pricing situations. The following are examples of special pricing situations:

1. Coupons
2. Value pricing
3. Bundling
4. Salad bars and buffets
5. Bottled wine
6. Beverages at receptions and parties

Coupons

Coupons are a popular way to vary menu price. Essentially, there are two types of coupons in use in the hospitality industry. The first type generally allows the guest to get a free item when he or she buys another item. This has the effect of reducing by 50% the menu price of the couponed item. With the second type, some form of restriction is placed on the coupon's use. For example, the coupon may only be accepted at a certain time of day, or the reduction in price may be available only if the guest purchases a specific designated menu item. Whichever type is offered, coupons have the effect of reducing sales revenue from individual guests in the hope that the total number of guests will increase to the point that total sales revenue increases. Coupons are a popular marketing tool, but their use should be carefully evaluated in terms of effect on menu price, product cost percentage, and product contribution margin.

Value Pricing

Value pricing refers to the practice of reducing all or most prices on the menu in the belief that, as in couponing, total guest counts will increase to the point that total sales revenue also increases. A potential danger, of course, with value pricing is that if guest counts do not increase significantly, total sales revenue may, in fact, decline rather than increase. Many credit the Wendy's chain with establishing value pricing, but currently its use is widespread, as is evident by the large number of 99-cent menu items for sale in the major quick-service restaurant segment chains.

Bundling

Bundling refers to the practice of selecting specific menu items and pricing them as a group, in such a manner that the single menu price of the group is lower than if the items comprising the group were purchased individually. The most common example is the combination meals offered by many quick-service hamburger restaurants. In many cases, these bundled meals consist of a sandwich, French fries, and a drink. These bundled meals, often promoted as "value meals" or "combo" meals, encourage each individual guest to buy one of each menu item rather than only one or two of them. The bundled meal generally is priced so competitively that a strong value perception is established in the guest's mind.

When bundling, as in couponing or value pricing, lower menu prices are accepted by management in the belief that this pricing strategy will increase total sales revenue, and thus profit, by increasing the number of guests served.

Salad Bars and Buffets

The difficulty in establishing a set price for either a salad bar or a buffet is that total portion cost can vary greatly from one guest to the next. A person weighing 100 pounds will, most likely, consume fewer products from a buffet or an all-you-can-eat line than a 300-pound person will. The general rule, however, is that each of these guests will pay the same price to go through the salad bar or buffet line. Short of charging guests for the amount they actually consume (a technique that has been tried by some operators with little success), a method of determining a single selling price must be established. This price must be based on a known, overall cost for the average diner who selects the all-you-can-eat option.

This price may be different, of course, if your average clients weigh 300 pounds rather than 100 pounds. The point is that the selling price must be established and monitored so that either guest could be accommodated at a price you find acceptable. This can be accomplished rather easily if record keeping is accurate and timely. The secret to keeping the selling price low in a salad bar or buffet situation is to apply the ABC inventory approach. That is, A items, which are expensive, should comprise no more than 20% of the total product available. B items, which are moderate in price, should comprise about 30% of the item offerings. And C items, which are inexpensive, should comprise 50% of the offerings. Using this approach, a menu listing of items can be prepared to ensure that only items that stay within these predetermined ranges are offered for sale.

Regardless of the buffet items to be sold, their usage must be accurately recorded. Consider the situation of Mei, the manager of Lotus Gardens, a Chinese restaurant where patrons pay one price, but may return as often as they like to a buffet line. Mei finds that a form such as that presented in Figure 20.7 is helpful in recording both product usage and guests served. Note that Mei uses the ABC method to determine her menu items. She does so because total food costs on a buffet line or salad bar are a function of both *how much* is eaten and *what* is eaten. She also notes the amount of product she puts on the buffet to begin the dinner meal period (Beginning Amount), any additions during the meal period (Additions), and the amount of usable product left at the conclusion of the meal period (Ending Amount). From this information, Mei can compute her total product usage and, thus, her total product cost.

Based on the data in Figure 20.7, Mei knows that her total product cost for dinner on January 1 was $305.28. She can then use the following formula to determine her buffet product cost per guest:

$$\frac{\text{Total Buffet Product Cost}}{\text{Guests Served}} = \text{Buffet Product Cost per Guest}$$

$$or$$

$$\frac{\$305.28}{125} = \$2.44$$

Unit Name: **Lotus Gardens**　　　　　　Date: 1/1 (Dinner)

Item	Category	Beginning Amount	Additions	Ending Amount	Total Usage	Unit Cost	Total Cost
Sweet and Sour Pork	A	6 lb.	44 lb.	13 lb.	37 lb.	$4.40/lb.	$162.80
Bean Sprouts	B	3 lb.	17 lb.	2 lb.	18 lb.	1.60/lb.	28.80
Egg Rolls	B	40 each	85 each	17 each	108 each	0.56 each	60.48
Fried Rice	C	10 lb.	21.5 lb.	8.5 lb.	23 lb.	0.60/lb.	13.80
Steamed Rice	C	10 lb.	30 lb.	6.5 lb.	33.5 lb.	0.40/lb.	13.40
Wonton Soup	C	2 gal.	6 gal.	1.5 gal.	6.5 gal.	4.00/gal.	26.00
Total Product Cost							305.28

Total Product Cost:　　$305.28
Guests Served:　　　　125　　　　　　　　Cost per Guest:　$2.44

FIGURE 20.7 Salad bar or buffet product usage.

Thus, on her buffet, Mei had a portion cost per guest of $2.44. She can use this information to establish a menu price that she feels is appropriate. Assume, for example, that Mei uses the food cost percentage approach to establishing menu price and that she has determined a 25% food cost to be her goal. Using the pricing factor table in Figure 20.6, Mei would use the following formula to establish her per-person buffet price:

$2.44 (Per-Person Cost) × 4.00 (Pricing Factor) = $9.76

For marketing purposes, and to ensure her desired food cost percentage, Mei may well round her buffet selling price up, to say, $9.99 per person. The significant point to remember here is that the amount consumed by any *individual* guest is relatively unimportant. It is the consumption of the average, or typical, guest that is used to establish menu price.

It is to be expected that Mei's buffet product cost per guest will vary somewhat each day. This is not a cause for great concern. Minor variations in product cost per guest should be covered if selling price is properly established. By monitoring buffet costs on a regular basis, you can be assured that you can keep good control over both costs per guest and your most appropriate selling price.

Bottled Wine

Few areas of menu pricing create more controversy than that of pricing wines by the bottle. The reason for this may be the incredible variance in cost among different vintages, or years of production, as well as the quality of alternative wine offerings. If your foodservice operation will sell wine by the bottle, it is likely that you will have some wine products that appeal to value-oriented guests and other, higher priced wines that are preferred by guests seeking these superior wines. An additional element that affects wine pricing is the fact that many wines that are sold by the bottle in restaurants are also sold in retail grocery or liquor stores. Thus, guests have a good idea of what a similar bottle of wine would cost them if it were purchased in either of these locations. How you decide to price the bottled-wine offerings on your own menu will definitely affect your guest's perception of the price/value relationship offered by your operation.

Properly pricing wine by the bottle calls for skill and insight. Consider the case of Claudia, who owns and manages a fine-dining Armenian restaurant. Using the product cost percentage method of pricing, Claudia attempts to achieve an overall wine product cost in her restaurant of 25%. Thus, when pricing her wines and using the pricing factor table in Figure 20.6, Claudia multiplies the cost of each bottled wine she sells by four to arrive at her desired selling price. Following are the four wines she sells and the costs and prices associated with each type:

WINE	PRODUCT COST	SELLING PRICE	PRODUCT COST %
1	$ 4.00	$16.00	25%
2	6.00	24.00	25
3	15.00	60.00	25
4	20.00	80.00	25

Claudia decides that she would like to explore the contribution margin approach to wine pricing. She, therefore, computes the contribution margin (Selling Price —

Product Cost = Contribution Margin) for each wine she sells and finds the following results:

WINE	SELLING PRICE	PRODUCT COST	CONTRIBUTION MARGIN
1	$16.00	$ 4.00	$12.00
2	24.00	6.00	18.00
3	60.00	15.00	45.00
4	80.00	20.00	60.00

Her conclusion, after evaluating the contribution margin approach to pricing and what she believes to be her customers' perception of the price/value relationship she offers, is that she may be hurting sales of wines 3 and 4 by pricing these products too high, even though they are currently priced to achieve the same 25% product cost as wines 1 and 2. In the case of bottled wine, the contribution margin approach to price can often be used to your advantage. Guests are often quite price conscious when it comes to bottled wine. When operators seek to achieve profits guests feel are inappropriate, bottled-wine sales may decline. Following is an alternative pricing structure that Claudia has developed for use in her restaurant. She must, however, give this price structure a test run and monitor its effect on overall product sales and profitability if she is to determine whether this pricing strategy will be effective.

WINE	PRODUCT COST	SELLING PRICE	CONTRIBUTION MARGIN	PRODUCT COST %
1	$ 4.00	$19.00	$15.00	21.1%
2	6.00	22.00	16.00	27.3
3	15.00	33.00	18.00	45.5
4	20.00	39.00	19.00	51.3

Note that, while selling price has been increased in the case of wine 1, it has been reduced for wines 2, 3, and 4. Contribution margin still is higher for wine 4 than for wine 1. The difference is, however, not as dramatic as before. Product cost percentages have, of course, been altered due to the price changes Claudia is proposing. Note also that the price spread, defined as the range between the lowest and the highest priced menu item, has been drastically reduced. Where the price spread was previously $16.00 to $80.00, it is now $19.00 to $39.00. This reduction in price spread may assist Claudia in selling more of the higher priced wine because her guests may be more comfortable with the price/value relationship perceived under this new pricing approach. It is important to remember, however, that Claudia must monitor sales and determine if her new strategy is successful. In general, it may be stated that pricing bottled wine only by the product percentage method is a strategy that may result in overall decreased bottled-wine sales. In this specific pricing situation, the best approach to establishing selling price calls for you to evaluate both your product cost percentage and your contribution margin.

Beverages at Receptions and Parties

Pricing beverages for open-bar receptions and special events can be very difficult, but the reason for this is very simple. Each consumer group can be expected to behave somewhat differently when attending an open-bar or hosted-bar function. Clearly, we would not expect the guests at a formal wedding reception to consume as many drinks during a one-hour reception as a group of fun-loving individuals celebrating a sports victory.

Establishing a price per person in these two cases may well result in quite different numbers. One way to solve this problem is to charge each guest for what he or she actually consumes. In reality, however, many party hosts want their guests to consume beverage products without having to pay for each drink. When this is the case, you are required to charge the host either for the amount of beverage consumed or on a per-person, per-hour basis. When charging on a per-person, per-hour basis, you must have a good idea of how much the average attendee will consume during the length of the party or reception so that an appropriate price can be established.

For example, assume that you are the food and beverage director at the Carlton, a luxury hotel. Ms. Swan, a potential food and beverage guest, approaches you with the idea of providing a one-hour champagne reception for 100 guests prior to an important dinner that she is considering booking at your facility. The guest would like all of the attendees to drink as much champagne during the reception as they care to. Ms. Swan's specific question is: "How much will I be charged for the reception if 100 guests attend?" Clearly, an answer of "I don't know" or "It depends on how much they drink" is inappropriate. It is, of course, your business to know the answer to such questions, and you can know. If you are aware from past events and records you have kept on such events of what the average consumption *for a group of this type* has been previously, you can establish an appropriate price. To do so, records for this purpose must be maintained. Figure 20.8 is an example of one such device that can be used. Note that average consumption of any product type can be recorded. In this example, assume that you had recently recorded the data from the Gulley wedding, an event very similar to the one requested by Ms. Swan. In this case, a wedding reception, which also requested champagne, was sold to 97 guests. The product cost per guest for that event, based on your records in Figure 20.8, equaled $3.37.

Event: <u>Gulley Wedding</u>   Date: <u>1/1</u>

Unit Name: **The Carlton Hotel**

Beverage Type	Beginning Amount	Additions	Ending Amount	Total Usage	Unit Cost	Total Cost
Liquor A						
B						
C						
D						
E						
F						
G						
Beer A						
B						
C						
D						
E						
Wine A						
B						
C						
D						
Other: Champagne: A. Sparkling B. Sparkling Pink	8 bottles 8 bottles	24 24	9 11	23 21	6.00/btl. 9.00/btl.	$138.00 $189.00
Total Product Cost						327.00

Total Product Cost: <u>$327.00</u>

Guests Served: <u>97</u>   Cost per Guest: <u>$3.37</u>

Remarks: <u>Mild group; very orderly; no problems.</u>

FIGURE 20.8 Beverage consumption report.

Based on what you know about the drinking pattern of a similar group, you could use either the product cost percentage method or the contribution margin pricing method to establish your reception price. For purpose of illustration, assume that you used the product contribution margin approach to pricing alcoholic beverage receptions. Further, assume that the contribution margin desired per person served is $15.00. The computation of selling price using the contribution margin formula would be as follows:

> **Product Cost + Desired Contribution Margin = Selling Price**
>
> *In this example:*
>
> $\$\ \ 3.37$ (Per-Person Product Cost)
> $+\ \$15.00$ (Desired Contribution Margin)
> _____
> $=\ \$18.37$ Selling Price Per Person

Armed with this historical information, as well as that from other similar events, you can be well prepared to answer Ms. Swan's question: "How much will I be charged for the reception if 100 guests attend?"

Establishing product costs and then assigning reasonable menu prices based on these costs is a major component of your job as a foodservice manager. You must be able to perform this task well. Increasingly, however, the cost of labor, rather than the cost of products, has occupied a significant portion of the typical foodservice manager's cost control efforts. In fact, in some foodservice facilities, the cost of labor exceeds that of the food and beverage products sold. Because this area of cost control is so important, we will now turn our attention to the unique set of skills and knowledge you must acquire to adequately manage and control your labor costs.

APPLY WHAT YOU HAVE LEARNED

Dominic Carbonne owns Hungry Henry's pizza, a four-unit chain of take-out pizza shops in a city of 60,000 people (with an additional 25,000 college students attending the local State University). Recently, a new chain of pizza restaurants has opened in town. The products sold by this chain have lesser quality and use lesser quantity of ingredients (cheese, meat, and vegetable toppings, etc.), but are also priced 25% less than Hungry Henry's equivalent size. Dominic has seen his business decline somewhat since the new chain opened. This is especially true with the college students.

1.   How would you evaluate the new competitor's pricing strategy?

2.   What steps would you advise Dominic to take to counter this competitor?

3. Describe three specific strategies restaurants can use to communicate "quality," rather than "low price" to potential guests.

KEY TERMS AND CONCEPTS

The following are terms and concepts discussed in the chapter that are important for you as a manager. To help you review, please define the terms below.

Standard menu	Revenue versus price	Value pricing
Daily menu	Price/value relationship	Bundling
Cycle menu	Price blending	Vintage
Menu specials	Contribution margin	Price spread

CONTROLLING OTHER EXPENSES*

Overview

This chapter explains the management of foodservice costs that are neither food, beverage, nor labor. These costs can represent 10%, or even more, of an operation's sales revenue and must be controlled if your financial goals are to be achieved in the operation you manage. The chapter teaches how to identify the costs you can control, as well as those that are considered to be noncontrollable expenses. In addition, it details how to express other expenses in terms of both other expense per guest served and other expense as a percentage of sales revenue.

Chapter Outline

---

* Authored by Jack E. Miller, Lea R. Dopson, and David K. Hayes.

Highlights

At the conclusion of this chapter, you will be able to:

- Categorize other expenses in terms of being fixed, variable, or mixed.

- Differentiate controllable from noncontrollable other expenses.

- Compute other expense costs in terms of both cost per guest and percentage of sales.

MANAGING OTHER EXPENSES

Food, beverage, and payroll expenses represent the greatest cost areas you will encounter as a foodservice manager. There are, however, expenses you will encounter that are neither food, beverage, nor labor. These other expenses can account for significant financial expenditures on the part of your foodservice unit. Controlling these costs can be just as important to your success as controlling food, beverage, and payroll expenses. Remember that the profit margins in many restaurants are very small. Thus, the control of all costs is critically important. Even in those situations that are traditionally considered nonprofit, such as hospitals, nursing homes, and educational institutions, dollars that are wasted in the foodservice area are not available for use in other areas of the institution.

You must look for ways to control all of your expenses, but sometimes the environment in which you operate will act upon your facility to influence some of your costs in positive or negative ways. An excellent example of this is in the area of energy conservation and waste recycling. Energy costs are one of the other expenses we will examine in this chapter. In the past, serving water to each guest upon arrival in a restaurant was simply SOP (standard operating procedure) for many operations. The rising cost of energy has caused many foodservice operations to implement a policy of serving water on request rather than with each order, or of selling bottled water in lieu of "free" tap water. Guests have found these changes quite acceptable, and the savings in the expenses related to warewashing costs, equipment usage, energy, cleaning supplies, as well as labor, are significant. In a similar vein, many operators today are finding that recycling fats and oils, cans, jars, and paper can be good not only for the environment, but also for their bottom line. Recycling these items not only reduces your cost of routine garbage and refuse disposal, but, in some communities, the recycled materials themselves have a cash value.

Source reduction, that is, working with food manufacturers and wholesalers to reduce product packaging waste, is just one example of hospitality industry managers and suppliers working together to reduce costs. When product packaging and wrapping are held to a minimum, delivery and storage costs are reduced, thus reducing the price foodservice operators must pay to wholesalers. This impacts other expenses in that the quantity of trash generated by this packaging is reduced, which, in turn, reduces garbage pickup fees to the foodservice operator.

FUN ON THE WEB!

Look up the following sites for ideas on how to cut other expenses through energy conservation, source reduction, and recycling.

www.epa.gov/smallbiz   Check out this U.S. Environmental Protection Agency site, the Energy Star Small Business Web site. Click on "Restaurants" to find articles, products, and ideas on how to cut energy costs for restaurants.

www.cygnus-group.com   This is a great place to learn about source reduction, packaging, and waste reduction. Browse through this site for reports, articles, and links to other interesting sites.

www.dinegreen.com   This site is a good resource for recycling and other environmentally friendly practices. Learn about the Green Restaurant Association and find out what they do!

Other expenses can constitute almost anything in the foodservice business. If your restaurant is a floating ship, periodically scraping the barnacles off the boat is an "other expense." If an operator is serving food to oil field workers in Alaska, heating fuel for the dining rooms and kitchen is an other expense, probably a very large one! If a company has been selected to serve food at the Olympics in a foreign country, airfares for its employees may be a significant other expense.

Each foodservice operation will have its own unique list of required other expenses. It is not possible, therefore, to list all imaginable expenses that could be incurred by the foodservice operator. It is possible, nonetheless, to group them into categories that make them easier to manage and understand. Napkins, straws, paper cups, and plastic lids, for example, might all be listed under the heading *paper supplies*, while stir sticks, coasters, tiny plastic swords, small paper umbrellas, and the like used in a cocktail lounge might be grouped under the listing *bar supplies*. Groupings, if used, should make sense to the operator and should be specific enough to let the operator know what is in the category. While some operators prefer to make their own groups, the categories used in this text come from the *Uniform System of Accounts for Restaurants* (USAR) recommended for use by the National Restaurant Association. The categories may be identified either by name, such as paper supplies, or by number, such as category 7420. *The Uniform System of Accounts for Restaurants* lists categories both by name and by number. For the purpose of our discussion, titles alone will suffice.

The following list details many of the other expenses associated with these groupings:

1. Costs related to food and beverage operations

2. Costs related to labor

3. Costs related to facility maintenance

4. Occupancy costs

Again, it should be pointed out that you should recognize and monitor the other expense costs you incur in a way that is meaningful to you. Only then can you begin to truly manage, and thus control, these costs.

Costs Related to Food and Beverage Operations

Direct Operating Expenses

Uniforms
Laundry and dry cleaning
Linen rental
Linen
China and glassware
Silverware
Kitchen utensils
Auto and truck expenses
Cleaning supplies
Paper supplies
Guest supplies
Bar supplies
Menus and wine lists
Contract cleaning
Exterminating
Flowers and decorations

Parking lot expenses
Licenses and permits
Banquet expenses
Other operating expenses

Music and Entertainment

Musicians
Professional entertainers
Mechanical music
Contracted wire services
Piano rental and tuning
Films, records, tapes, and sheet music
Programs
Royalties to ASCAP, BMI
Booking agents fees
Meals served to musicians

Marketing

Selling and promotion
    Sales representative service
    Travel expense on solicitation
    Direct mail
    Telephone used for advertising and promotion
    Complimentary food and beverage (including gratis meals to
        customers)
    Postage
    Advertising
    Newspapers
    Magazines and trade journals
    Circulars, brochures, postal cards, and other mailing pieces
    Outdoor signs
    Radio and television
    Programs, directories, and guides
    Preparation of copy, photographs, etc.
Public relations and publicity
    Civic and community projects
    Donations

Souvenirs, favors, treasure chest items

Fees and commissions

Advertising or promotional agency fees

Research

Travel in connection with research

Outside research agency

Product testing

Utility Services

Electric current

Electric bulbs

Water

Removal of waste

Other fuel

Administrative and General Expenses

Office stationery, printing, and supplies

Data processing costs

Postage

Telegrams and telephone

Dues and subscriptions

Traveling expenses

Insurance—general

Commissions on credit card charges

Provision for doubtful accounts

Cash over (or short)

Professional fees

Protective and bank pickup services

Bank charges

Miscellaneous

Costs Related to Labor

Employee Benefits

FICA

Federal unemployment tax

State unemployment tax

Workmen's compensation

Group insurance

State health insurance tax

Welfare plan payments

Pension plan payments

Accident and health insurance premiums

Life insurance premiums

Employee meals

Employee instruction and education
   expenses

Employee Christmas and other parties

Employee sports activities

Medical expenses

Credit union

Awards and prizes

Transportation and housing

Costs Related to Facility Maintenance

Repairs and Maintenance

Furniture and fixtures

Kitchen equipment

Office equipment

Refrigeration

Air conditioning

Plumbing and heating

Electrical and mechanical

Floors and carpets

Buildings

Parking lot

Gardening and grounds maintenance

Building alterations

Painting, plastering, and decorating

Maintenance contracts

Autos and trucks

Other equipment and supplies

Occupancy Costs

Rent

    Rent—minimum or fixed

    Percentage rent

    Ground rental

    Equipment rental

    Real estate taxes

    Personal property taxes

    Other municipal taxes

    Franchise tax

    Capital stock tax

    Partnership or corporation license fees

    Insurance on building and contents

Interest

    Notes payable

    Long-term debt

    Other

Depreciation

    Buildings

    Amortization of leasehold

    Amortization of leasehold improvements

    Furniture, fixtures, and equipment

While there are many ways in which to consider other expenses, two views of these costs are particularly useful for the foodservice manager. They are:

1. Fixed, variable, or mixed

2. Controllable or noncontrollable

A short discussion of these two concepts will help you to understand other expenses.

(Source: *Uniform System of Accounts for Restaurants* published by the National Restaurant Association, http://www.restaurant.org.)

FIXED, VARIABLE, AND MIXED OTHER EXPENSES

As a foodservice manager, some of the costs you will incur will stay the same each month, while others may vary. For example, if you elect to lease a building to house a restaurant and cocktail lounge you want to operate, your lease payment may be such that you pay the same amount for each month of the lease. In other instances, the amount you pay for an expense will vary based on the success of your business. Expenses you incur for paper cocktail napkins used at the cocktail lounge in your restaurant, for example, will increase as the number of guests you serve increases and decrease as the number of guests you serve decreases. As an effective cost control manager, it is important to recognize the difference between costs that are fixed and those that vary with sales volume.

A fixed expense is one that remains constant despite increases or decreases in sales volume. A variable expense is one that generally increases as sales volume increases and decreases as sales volume decreases. A mixed expense is one that has properties of both a fixed and a variable expense. To illustrate all three of these expense types, consider Jo Ann's Hot Dogs Deluxe, a midsize, freestanding restaurant outside a shopping mall, where Jo Ann features upscale Chicago-style hot dogs.

Assume that Jo Ann's average sales volume is $136,000 per month. Assume also that rent for her building and parking spaces is fixed at $8,000 per month. Each month, Jo Ann computes her rent as a percentage of total sales, using the following standard cost percentage formula:

$$\frac{\text{Other Expenses}}{\text{Total Sales}} = \text{Other Expense Cost \%}$$

In this case, the other expense category she is interested in looking at is rent; therefore, the formula becomes:

$$\frac{\text{Rent Expenses}}{\text{Total Sales}} = \text{Rent Expense \%}$$

Jo Ann has computed her rent expense % for the last six months. The results are shown in Figure 21.1.

Note that Jo Ann's rent expense % ranges from a high of 6.67% (February) to a low of 4.88% (May), yet it is very clear that rent itself was a constant, or fixed, amount of $8,000 per month. Thus, rent, in this lease arrangement, is considered to be a fixed expense. It is important to note that, while the dollar amount of her rent expense is fixed, the rent % declines as volume increases. Thus, the rent payment, as a percentage

Month	Rent Expense	Sales	Rent %
January	$8,000	$121,000	6.61%
February	8,000	120,000	6.67
March	8,000	125,000	6.40
April	8,000	130,000	6.15
May	8,000	164,000	4.88
June	8,000	156,000	5.13
6-Month Average	8,000	136,000	5.88

FIGURE 21.1 Jo Ann's fixed rent.

of sales or cost per item sold, is not constant. This is true because, as sales volume increases, the number of guests contributing to rent expense also increases, so it takes a smaller dollar and percentage amount of each guest's sales revenue to generate the $8,000 fixed amount Jo Ann needs to pay her rent.

It makes little sense for Jo Ann to be concerned about the fact that her rent expense % varies by a great amount based on the time of the year. If Jo Ann is comfortable with the six-month average rent percentage (5.88%), she is in control of and managing her other expense category called rent. If Jo Ann feels that her rent expense % is too high, she has only two options. She must increase sales, and thereby reduce her rent expense %, or she must negotiate a lower monthly rental with her landlord. When rent is a fixed expense, as in this case, the expense, as expressed by the percentage of sales, may vary. The expense itself, however, is not affected by sales volume.

Some restaurant lease arrangements are based on the sales revenue an operator achieves in the leased facility. Assume, for example, that Jo Ann has a lease arrangement of this type, requiring Jo Ann to pay 5% of her monthly sales revenue as rent. If that were the case, Jo Ann's monthly lease payments would be completely variable, as displayed in Figure 21.2. Note that the dollar amount of Jo Ann's rent, in this case, varies a great deal. It ranges from a low of $6,000 (February) to a high of $8,200 (May). The percentage of her sales revenue that is devoted to rent, however, remains at 5%.

A third type of lease that is common in the hospitality industry illustrates the fact that some other expenses are mixed; that is, there is both a fixed and a variable

Month	Sales	Rent %	Rent Expense
January	$121,000	5.00%	$6,050
February	120,000	5.00	6,000
March	125,000	5.00	6,250
April	130,000	5.00	6,500
May	164,000	5.00	8,200
June	156,000	5.00	7,800
6-Month Average	136,000	5.00	6,800

FIGURE 21.2 Jo Ann's variable rent.

Month	Sales	Fixed Rent Expense	1% Variable Rent Expense	Total Rent Expense
January	$121,000	$5,000	$1,210	$6,210
February	120,000	5,000	1,200	6,200
March	125,000	5,000	1,250	6,250
April	130,000	5,000	1,300	6,300
May	164,000	5,000	1,640	6,640
June	156,000	5,000	1,560	6,560
6-Month Average	136,000	5,000	1,360	6,360

FIGURE 21.3 Jo Ann's mixed rent.

component to this type of expense. Figure 21.3 demonstrates such a lease type. In it, Jo Ann pays a flat lease amount of $5,000 per month plus 1% of total sales revenue.

In this arrangement, a major portion ($5,000) of Jo Ann's lease is fixed, while a smaller amount (1% of revenue) varies based on sales revenue. Mixed expenses of this type are common and include items such as energy costs, garbage pickup, some

Expense	As a Percentage of Sales	Total Dollars
Fixed Expense	Decreases	Remains the Same
Variable Expense	Remains the Same	Increases
Mixed Expense	Decreases	Increases

FIGURE 21.4 Fixed, variable, and mixed expense behaviors as sales volume increases.

franchise fees, and other expenses where the operator must pay a base amount and then additional amounts as usage or sales volume increases.

In summary, the total dollar amount of fixed expenses does not vary with sales volume, while the total dollar amount of variable expenses changes as volume changes. As a percentage of total sales, however, a fixed-expense % decreases as sales increase, and a variable-expense % does not change. A mixed expense has both fixed and variable components; therefore, as sales increase, the mixed-expense % decreases and total mixed expenses increase. Figure 21.4 shows how fixed, variable, and mixed expenses are affected as sales volume increases.

A convenient way to remember the distinction between fixed, variable, and mixed expenses is to consider a napkin holder and napkins on a cafeteria line. The napkin holder is a fixed expense. One holder is sufficient whether you serve 10 guests at lunch or 100 guests. The napkins themselves, however, are a variable expense. As you serve more guests (if each guest takes one napkin), you will incur a greater paper napkin expense. The cost of the napkin holder and napkins, if considered *together*, would be a mixed expense. For some very large restaurant chains, it makes sense to separate some mixed expenses into their fixed and variable components, while smaller operations may elect, as in the case of the napkin holder and napkins, to combine these expenses. The company you work for may make the choice of how you account for other expenses, or you may be free to consider other expense costs in a manner you feel is best for your own operation.

Effective managers know they should not categorize fixed, variable, or mixed costs in terms of being either "good" or "bad." Some expenses are, by their very nature, related to sales volume. Others are not. It is important to remember that the goal of management is not to reduce, but to increase variable expenses in direct relation to increases in sales volume. Expenses are required if you are to service your guests. In the example of the paper napkins, it is clear that management would prefer to use 100 napkins at lunch rather than 10. As long as the total cost of servicing guests is less than the amount spent, expanding the number of guests served will not only increase variable other expenses, but will increase profits as well. Thus, increasing variable costs is desirable if management increases them in a way that makes sense for both the operation and the satisfaction of the guest.

As we saw in the case of labor expense, the concept of fixed, variable, and mixed expense is quite useful. Variations in expense percentage that relate *only* to whether an expense is fixed, variable, or mixed should not be of undue concern to management. It is only when a fixed expense is too high or a variable expense is out of control that management should act. This is called the concept of management by exception. That is, if the expense is within an acceptable variation range, there is no need for management to intervene. You must take corrective action only when operational results are outside the range of acceptability. This approach keeps you from overreacting to minor variations in expense, while monitoring all important activities.

Examples of other fixed foodservice expenses include the areas of advertising (outdoor sign rentals), utilities (rest room lightbulbs), employee benefits (employee-of-the-month prize), repairs and maintenance (parking lot paving), and occupancy costs (interest due on long-term debt). Many foodservice operation other expenses, however, are related to sales volume, and, thus, management has some daily control over these items.

CONTROLLABLE AND NONCONTROLLABLE OTHER EXPENSES

While it is useful, in some cases, to consider other expenses in terms of their being fixed, variable, or mixed, it is also useful to consider some expenses in terms of their being controllable or noncontrollable. Consider, for a moment, the case of Steve, the operator of a neighborhood tavern/sandwich shop. Most of Steve's sales revenue comes from the sale of beer, sandwiches, and his special pizza.

Steve is, of course, free to decide on a weekly or monthly basis the amount he will spend on advertising. Advertising expense, then, is under Steve's direct control and, thus, would be considered a controllable expense. Some of his other expenses, however, are not under his control. Taxes on product sales are a familiar form of a noncontrollable expense. The state in which Steve operates charges a tax on all alcoholic beverage sales. As the state in which he operates increases the liquor tax, Steve is forced to pay more. In this situation, the alcoholic beverage tax would be considered a noncontrollable expense, that is, an expense beyond Steve's immediate control.

As an additional example, assume, for a moment, that you own a quick-service unit that sells takeout chicken. Your store is part of a nationwide chain of such stores.

Each month, your store is charged a $500 advertising and promotion fee by the regional headquarters' office. The $500 is used to purchase television advertising time for your company. This $500 charge, as long as you own the franchise, is a noncontrollable operating expense.

A noncontrollable expense, then, is one that the foodservice manager can neither increase nor decrease. A controllable expense is one in which decisions made by the foodservice manager can have the effect of either increasing or reducing the expense. Management has some control over controllable expenses, but has little or no control over noncontrollable expenses. Other examples of noncontrollable expenses include some insurance premiums, property taxes, interest on debt, and depreciation. In every one of these cases, the foodservice operator will find that even the best control systems will not affect the specific expense. Thus, as a manager, you should focus your attention on controllable rather than noncontrollable expenses.

MONITORING OTHER EXPENSES

When managing other expenses, two control and monitoring alternatives are available to you. They are:

1.   Other expense cost %
2.   Other expense cost per guest

Each alternative can be used effectively in specific management situations; thus, it is important for you to master both.

As you learned earlier in this chapter, the other expense cost % is computed as follows:

$$\frac{\text{Other Expenses}}{\text{Total Sales}} = \text{Other Expense Cost \%}$$

Thus, for example, in a situation where a restaurant you own incurs an advertising expense of $5,000 in a month, serves 10,000 guests, and achieves sales of $78,000 for that same month, you would compute your advertising expense percentage for that month as follows:

$$\frac{\$5,000}{\$78,000} = 6.4\%$$

Month	Total Sales	Linen Cost	Cost %
January	$ 68,000	$ 2,720	4.00%
February	70,000	2,758	3.94
March	72,000	2,772	3.85
April	71,500	2,753	3.85
May	74,000	2,812	3.80
Total	355,500	13,815	3.89

FIGURE 21.5 Chez Scot linen cost %.

The other expense cost per guest is computed as follows:

$$\frac{\text{Other Expenses}}{\text{Number of Guests Served}} = \text{Other Expense Cost per Guest}$$

In this example and using the preceding formula, you would compute your advertising expense cost per guest as follows:

$$\frac{\$5,000}{10,000} = \$0.50$$

As we have seen, the computation required to establish the other expense percentage requires the other expense category to be divided by total sales. In many cases, this approach yields useful management information. In some cases, however, this computation alone may not provide adequate information; therefore, using the concept of other expense cost per guest can be very useful. To illustrate, consider the following example: Scott operates Chez Scot, an exclusive, fine-dining establishment in a suburban area of a major city. One of Scott's major other expenses is linen. He uses both tablecloths and napkins. Scott's partner, Joshua, believes that linen costs are a variable operating expense and should be monitored through the use of a linen cost % figure. In fact, says Scott's partner, Figure 21.5 indicates that the linen cost % has been declining over the past five months; therefore, current control systems must be working. As is evident in Figure 21.5, the linen cost % has indeed been declining over the past five months.

Month	Linen Cost	Number of Guests Served	Cost per Guest
January	$ 2,720	2,566	$1.06
February	2,758	2,508	1.10
March	2,772	2,410	1.15
April	2,753	2,333	1.18
May	2,812	2,305	1.22
Total	13,815	12,122	1.14

FIGURE 21.6 Chez Scot linen cost per guest.

Scott, however, is convinced that there are control problems. He has monitored linen costs on a cost per guest basis. His information is presented in Figure 21.6, which validates Scott's fears. There is indeed a control problem in the linen area.

Figures 21.5 and 21.6 both show that a linen control problem does exist, yet Figure 21.6 shows it most clearly since it is plain that linen cost per guest has gone from $1.06 in January to its May high of $1.22.

Chez Scot is enjoying increased sales ($68,000 in January vs. $74,000 in May), but its guest count is declining (2,566 in January vs. 2,305 in May). The check average has obviously increased. This is a good sign, as it indicates that each guest is buying more food. The fact that fewer guests are being served should, however, result in a decrease in demand for linen and, thus, a decline in linen cost. In fact, on a per-person basis, linen costs are up. Scott is correct to be concerned about possible problems in the linen control area.

Other expense cost per guest may also be useful in a situation where the foodservice manager receives no sales figure. Consider a college dormitory feeding situation where paper products such as cups, napkins, straws, and lids are placed on the serving line to be used by the students eating their meals.

In this case, Juanita, the cafeteria manager, wonders whether students are taking more of these items than is normal. The problem is, of course, that she is not exactly sure what "normal" use is when it comes to supplying paper products to her students. Juanita belongs to a trade association that asks its members to supply annual cost figures to a central location where they are tabulated and sent back to the membership. Figure 21.7 shows the tabulations from five colleges in addition to those from Juanita's unit.

Institution	Cost of Paper Products	Number of Students	Paper Product Cost per Student
O. University	$140,592	8,080	$17.40
C. State University	109,200	6,500	16.80
P. University	122,276	7,940	15.40
University of T.	184,755	11,300	16.35
A. State University	61,560	3,600	17.10
5-University Average	123,676.60	7,484	16.53
Juanita's Institution	77,220	4,680	16.50

FIGURE 21.7 Average paper product cost.

Juanita has computed her paper products cost per student for the year and has found it to be higher than at P. University and the University of T., but lower than O. University, C. State University, and A. State University. Juanita's costs appear to be in line in the paper goods area. If, however, Juanita hopes to reduce paper products cost per student even further, she could, perhaps, call or arrange a visit to either P. University or University of T. to observe their operations or purchasing techniques.

The other expense cost per guest formula is of value when management believes it can be helpful or when lack of a sales figure makes the computation of other expense cost % impossible. Figure 21.8 presents a seven-column form that is useful in tracking both daily and cumulative cost per guest figures. It is maintained by inserting Other Expense Cost and Number of Guests Served in the first two sets of columns. The third set of columns, Cost per Guest, is obtained by using the other expense cost per guest formula.

REDUCING OTHER EXPENSES

Since other expenses can be broken down into four distinct areas, it is useful to consider these four areas when developing strategies for reducing overall other expense costs. It is important to remember that each foodservice manager faces his or her own unique set of other expenses. A restaurant on a beach in southern Florida may well experience the expense of hurricane insurance. A similar restaurant in Kansas

Weekday	Today	To Date	Today	To Date	Today	To Date
Monday	$ 145.50	$ 145.50	823	823	$0.18	$0.18
Tuesday	200.10	345.60	751	1,574	0.27	0.22
Wednesday	417.08	762.68	902	2,476	0.46	0.31
Thursday	0	762.68	489	2,965	0	0.26
Friday	237.51	1,000.19	499	3,464	0.48	0.29
Saturday	105.99	1,106.18	375	3,839	0.28	0.29
Sunday	0	1,106.18	250	4,089	0	0.27
Monday	157.10	1,263.28	841	4,930	0.19	0.26
Total	1,263.28		4,930		0.26	

FIGURE 21.8 Six-column cost of paper products.

would not. Each foodservice operation is unique. Those operators who are effective are constantly on the lookout for ways to reduce unnecessary additions to any other expense categories.

Reducing Costs Related to Food and Beverage Operations

In many respects, some of these other expenses should be treated like food and beverage expenses. For instance, in the case of cleaning supplies, linen, uniforms, and the like, products should be ordered, inventoried, and issued in the same manner used for food and beverage products. In general, fixed costs related to food and beverage operations can only be reduced when measuring them as a percentage of total sales. This is done, of course, by increasing the total sales figure. Reducing total variable cost expenses is generally not desirable, since, in fact, each additional sale will bring additional variable

Sales	Fixed Expense	Variable Expense (10%)	Total Other Expense	Other Expense Cost %
$ 1,000	$150	$ 100	$ 250	25.00%
3,000	150	300	450	15.00
9,000	150	900	1,050	11.67
10,000	150	1,000	1,150	11.50
15,000	150	1,500	1,650	11.00

FIGURE 21.9 Igloo's fixed and variable other expenses.

expense. In this case, while total variable expenses may increase, the positive impact of the additional sales on fixed costs will serve to reduce the overall other expense percentage.

To see how this is done, let's examine a shaved-ice kiosk called Igloo's located in the middle of a small parking lot. Figure 21.9 demonstrates the impact of volume increases on both total other expense and other expense cost %. In this example, some of the other expenses related to food and beverage operations are fixed and others are variable. The variable portion of other expense, in this example, equals 10% of gross sales. Fixed expenses equal $150.

While variable expense increases from $100 to $1,500, total other expense percentage drops from 25% of sales to 11% of sales. Thus, to reduce the percentage of costs related to food and beverage operations, increases in sales are quite helpful! If all other expenses related to food and beverage operations were 100% variable, however, this strategy would not have the effect of reducing other expense cost %, since the dollar amount of other expenses would increase proportionately to volume increases, and total other expense cost % would be unchanged.

Reducing Costs Related to Labor

Some labor-related expenses can be considered partially fixed and some partially variable. To help reduce other expense costs related to labor, it is necessary for you to eliminate wasteful labor-related expense. Examples include the cost of advertising, hiring, and training new employees because of excessive employee turnover. It also means implementing cost-reducing hiring practices, such as preemployment drug screening that may result in lower health insurance premiums for employees, if these benefits are provided. Proper employment practices also impact the worker's compensation and unemployment tax rates you may pay. Remember that, in many states, the rate you pay for these two insurance programs is determined, in part, by your history of work-related injuries and employment separations. Careful hiring and providing excellent training to reduce the costs associated with worker's compensation claims and unemployment compensation will save you money and reduce your other expense related to labor.

Conversely, those operators who attempt to reduce other expenses related to labor too much, by not providing adequate health care, retirement savings programs such as 401(k)s, or sick leave benefits, will find that the best employees prefer to work elsewhere. This would leave the operator with a less productive workforce than would otherwise be possible. In many ways, employees will be your most valuable assets. Effective management is not magic. If your employees feel they are treated fairly, they will be motivated to do their best. If they do not, they will not. Reducing employee benefits while attempting to retain a well-qualified workforce is simply management at its worst!

Reducing Costs Related to Facility Maintenance

Any employee knows that keeping his or her tools clean and in good working order will make them last longer and perform better. The same is true for foodservice facilities. A properly designed and implemented preventative maintenance program can go a long way toward reducing equipment failure and, thus, decreasing equipment and facility-related costs. Proper care of mechanical equipment not only prolongs its life, but also actually reduces operational costs. As prices for water, gas, and kwh (kilowatt hours—the measure of electrical usage) needed to operate facilities continues to rise, you must implement a facility repair and maintenance program that seeks to discover and treat minor equipment and facility problems before they become major problems.

One way to help ensure that costs are as low as possible is to use a competitive-bid process before awarding contracts for services you require. For example, if you hire a carpet cleaning company to clean your dining room carpets monthly, it is a good idea to annually seek competitive bids from new carpet cleaners. This can help to reduce your costs by ensuring that the carpet cleaner you select has given you a price that is competitive with other service providers. In the area of maintenance contracts, for areas such as the kitchen or for mechanical equipment, elevators, or grounds, it is

Item Inspected	Inspection Date	Inspected By	Action Recommended
A. Refrigerator #6	1/1	D.H.	Replace gasket
B. Fryer	1/7	D.H.	Inspected, no maintenance needed
C. Ice Machine	1/9	D.H.	Drain, de-lime
D.			
E.			

FIGURE 21.10 Equipment inspection report.

recommended that these contracts be bid at least once per year. This is especially true if the dollar value of the contract is large.

Air-conditioning, plumbing, heating, and refrigerated units should be inspected at least yearly, and kitchen equipment, such as dishwashers, slicers, and mixers, should be inspected at least monthly for purposes of preventative maintenance. A form such as the one in Figure 21.10 is useful in this process.

Some foodservice managers operate facilities that are large enough to employ their own facility maintenance staff. If this is the case, make sure these employees have copies of the operating and maintenance manuals of all equipment. These documents can prove invaluable in the reduction of equipment and facility-related operating costs.

Reducing Occupancy Costs

Occupancy costs refer to those expenses incurred by the foodservice unit that are related to the occupancy of and payment for the physical facility it occupies.

For the foodservice manager who is not the owner, the majority of occupancy costs will be noncontrollable. Rent, taxes, and interest on debt are real costs but are beyond the immediate control of the manager. However, if you own the facility you manage, occupancy costs are a primary determinant of both profit on sales and return on dollars invested. When occupancy costs are too high because of unfavorable rent or lease arrangements or due to excessive debt load, the foodservice operation's owner may face extreme difficulty in generating profit. Food, beverage, and labor costs can only be managed to a point; beyond that, efforts to reduce costs will result in decreased guest satisfaction. If occupancy costs are unrealistically high, no amount of effective cost control can help "save" the operation's profitability.

Total other expenses in an operation can range from 5% to 20% or more of the gross sales. These expenses, while considered as minor expenses, can be extremely important to overall operational profitability. This is especially true in a situation where the number of guests you serve is fixed, or nearly so, and the prices you are allowed to charge for your products is fixed also. In a case such as this, your ability to control other expenses is vital to your success.

APPLY WHAT YOU HAVE LEARNED

Kathy Waldo owns her own catering business. She provides her full-time employees with good health insurance benefits. Part-time employees do not receive the benefit. This year, Kathy's health insurer advises Kathy that insurance rates for her employees will increase 25% next year.

Kathy had planned on giving both full- and part-time employees a wage increase on January 1, but finds that the increased cost of the health care premiums for her full-time employees will take all of the funds she had budgeted for the wage increases.

1.   If you were Kathy, would you give your part-time employees a wage increase? If so, how?

2.   What steps can Kathy take next year to help control her health insurance coverage costs?

3.   What specific types of employees value health insurance coverage more than hourly pay rate?

KEY TERMS AND CONCEPTS

The following are terms and concepts discussed in the chapter that are important for you as a manager. To help you review, please define the terms below.

Other expenses	Variable expense	Controllable expense
SOP	Mixed expense	401(k)
Source reduction	Management by exception	kwh
Fixed expense	Noncontrollable expense	Occupancy costs

ANALYZING RESULTS USING THE INCOME STATEMENT*

Overview

This chapter explains what you will do to analyze the cost effectiveness of your operation. It teaches you how to read and use the income statement, a financial document also known as the profit and loss (P&L) statement. Included are techniques to analyze your sales volume as well as expense levels, including food, beverage, labor, and other expenses. Finally, the chapter shows you how to review the income statement to analyze your overall profitability.

Chapter Outline

---

* Authored by Jack E. Miller, Lea R. Dopson, and David K. Hayes.

Highlights

At the conclusion of this chapter, you will be able to:

- Prepare an income (profit and loss) statement.
- Analyze sales and expenses using the P&L statement.
- Evaluate a facility's profitability using the P&L statement.

INTRODUCTION TO FINANCIAL ANALYSIS

Far too many foodservice managers find that they collect information, fill out forms, and enter and receive numbers from their computers with little regard for what they should do with all these data. Some have said that managers often make poor decisions because they lack information, but when it comes to the financial analysis of an operation, the opposite is usually true. Foodservice managers, more often than not, find themselves awash in numbers! It will be your job to sift through this information and select for analysis those numbers that can shed light on exactly what is happening in your operation. This information, in an appropriate form, is necessary not only to effectively operate your business, but also to serve many interest groups that are directly or indirectly involved with the financial operation of your facility. Local, state, and federal financial records relating to taxes and employee wages will have to be submitted to the government on a regular basis. In addition, records showing the financial health of an operation may have to be submitted to new suppliers in order to establish credit worthiness. Also, if a foodservice operation has been established with both operating partners and investors, those owners and investors will certainly require accurate and timely updates that focus on the financial health of the business. Owners, stockholders, and investment bankers may all have an interest in the day-to-day efficiency and effectiveness of management. For each of these groups, and, of course, for the foodservice organization's own management, an accurate examination of operational efficiency is critical.

As a professional foodservice manager, you will be very interested in examining your cost of doing business. Documenting and analyzing sales, expenses, and profits is sometimes called cost accounting, but a more appropriate term for the process is managerial accounting, a term that reflects the importance managers place on this process.

In this text, we use the term managerial accounting when referring to documenting, analyzing, and managing sales, expenses, and profit data.

It is important for you to be keenly aware of the difference between bookkeeping, the process of simply recording and summarizing financial data, and the actual analysis of that data. As an example, electronic cash registers or computer-based point of sales (POS) systems can be programmed to provide data about food and beverage sales per server. Management can track, per shift, the relative sales effort of each service employee. If this is done with the goal of either increasing the training of the less productive server or rewarding the most productive one, the cash register or POS system has, in fact, added information that is valuable and has assisted in the unit's operation. If, on the other hand, this information is dutifully recorded on a daily basis, filed away or sent to a regional office, and is then left to collect dust, the cash register or POS system has actually harmed the operation by taking management's time away from the more important task of running the business. It has converted the manager's role from that of cost analyst to one of a book (record) keeper only. This type of situation must be avoided at all cost if you are to maximize your effectiveness by being in the production area or dining room during service periods and not by staying in the office catching up on your paperwork. Bookkeeping is essentially the summarizing and recording of data. Managerial accounting involves the summarizing, recording, and, most important, the analysis of those data.

You do not have to be a certified management accountant (CMA) or a certified public accountant (CPA) to analyze data related to foodservice revenue and expense. While this is not meant to discount the value of an accounting professional who assists the foodservice manager, it is important to establish that it is the professional foodservice manager, not an outside expert, who is most qualified to assess the effectiveness of the foodservice team in providing the service levels desired by management and in controlling production-related costs. The analysis of operating data, a traditional role of the accountant, must also be part of your role as manager. The process is not complex and, in fact, is one of the most fun and creative aspects of a foodservice manager's job.

A good foodservice manager is, in fact, a manager first and not an accountant. It is important, however, for you to be able to read and understand financial information and be able to converse intelligently and confidently with the many parties outside the operation who will read and use the information generated by accountants. This information will be crucial to the operation's overall health and success and can provide the needed data that will assist in sharpening the quality of your management decisions. It is also important to remember that, by Federal law, it is management that is called upon to verify the accuracy of the work of the accountant, and not the reverse.

UNIFORM SYSTEM OF ACCOUNTS

Financial statements related to the operation of a foodservice facility are of interest to management, stockholders, owners, creditors, governmental agencies, and, often, the general public. To ensure that this financial information is presented in a way that is both useful and consistent, uniform systems of accounts have been established for many areas of the hospitality industry. The National Restaurant Association, for example, has developed the *Uniform System of Accounts for Restaurants* (USAR). Uniform systems of accounts also exist for hotels, clubs, nursing homes, schools, and hospitals. Each system seeks to provide a consistent and clear manner in which to record sales, expenses, and overall financial condition for a specific type of organization. Sales categories, expense classifications, and methods of computing relevant ratios are included in the uniform systems of accounts. These uniform systems are typically available from the national trade associations involved with each hospitality segment.

It is important to note that the uniform systems of accounts attempt to provide operator guidelines rather than mandated methodology. Small foodservice operations, for example, may use the *Uniform System of Accounts for Restaurants* in a slightly different way than will large operations. In all cases, however, operators who use the uniform system of accounts "speak the same language," and it is truly useful that they do so. If each operator prepared financial statements in any manner he or she elected, it is unlikely that the many external audiences who must use them could properly interpret these statements. Thus, an effective manager will secure a copy of the appropriate uniform system of accounts for his or her operation and become familiar with its basic formats and principles.

FUN ON THE WEB!

Look up the following sites to review and obtain copies of the uniform system of accounts for restaurants, lodging facilities, and clubs.

www.restaurant.org   Click on "Store," then click on "Business and Finance."
   Next, click on "Uniform System of Accounts for Restaurants" to see a synopsis
   of the book and place an order.
www.ei-ahla.org   Click on "Books." Scroll down to "Financial Management,"
   and click on "Uniform System of Accounts for the Lodging
   Industry 9th Edition" to see a synopsis of the book and place an order.
www.clubnet.com   Click on "Education," then click on "Books and Publications
   (Bookmart)." Click on "Browse List of Books" at the top of the page. Then,

scroll down to find the "Uniform System of Financial Reporting for Clubs," and click on it to see a synopsis of the book and place an order.

INCOME STATEMENT (USAR)

The income statement, often referred to as the profit and loss (P&L) statement, is the key management tool for cost control. Essentially, the P&L statement seeks to show revenue and expense in a level of detail determined by management after review of the appropriate uniform system of accounts.

The word profit can mean many things; therefore, the profit and loss statement can be somewhat confusing if you are not familiar with it. For example, some operators consider profit to be what they earned before they pay taxes, while other managers reserve the term for their after-tax figure. A tight definition of what is meant by profit must be established for each P&L statement if it is to be helpful. In all cases, however, a purpose of the profit and loss statement is to identify net income, or the profit generated after all appropriate expenses of the business have been paid.

Figure 22.1 details two years' P&L statements for Joshua's Inc., a foodservice complex that includes a cocktail lounge, two dining areas, and banquet space. Joshua's fiscal year, that is, his accounting year, begins on October 1 and concludes on September 30 of the next year. This fiscal year coincides with the beginning of his busy season and, thus, gives him a logical starting point. As can be seen in Figure 22.1, last year Joshua's generated $2,306,110 in total sales and achieved a net income of $101,772. This year, when the corporation generated total sales revenue of $2,541,206, Joshua achieved a net income of $114,923. The question Joshua must ask himself, of course, is, "How good is this performance?"

It is important to note that each operation's P&L statement will look slightly different. All of them, however, typically take a similar approach to reporting revenue and expense. Note that, while the detail is much greater, the layout of Joshua's P&L is similar in structure to the abbreviated P&L presented as Figure 17.3. Both statements list revenue first, then expense, and finally the difference between the revenue and expense figures. If this number is positive, it represents a profit. If expenses exceed revenue, a loss, represented by a negative number or a number in brackets, is shown. Operating at a loss is, for some unknown reason, often referred to as operating "in the red" or "shedding red ink." Regardless of the color of the ink, operating at a loss can cause an operator to shed a few tears!

To ensure that *your* operation does not produce a loss, you need to know some important components of the *Uniform System of Accounts for Restaurants*. The USAR can better be understood by dividing it into three sections: gross profit, operating

Joshua's Inc.
Last Year versus This Year

	Last Year	%	This Year	%
SALES:				
Food	$1,891,011	82.0%	$2,058,376	81.0%
Beverage	415,099	18.0	482,830	19.0
Total Sales	**2,306,110**	**100.0**	**2,541,206**	**100.0**
COST OF SALES:				
Food	712,587	37.7	767,443	37.3
Beverage	94,550	22.8	96,566	20.0
Total Cost of Sales	**807,137**	**35.0**	**864,009**	**34.0**
GROSS PROFIT:				
Food	1,178,424	62.3	1,290,933	62.7
Beverage	320,549	77.2	386,264	80.0
Total Gross Profit	**1,498,973**	**65.0**	**1,677,197**	**66.0**
OPERATING EXPENSES:				
Salaries and Wages	641,099	27.8	714,079	28.1
Employee Benefits	99,163	4.3	111,813	4.4
Direct Operating Expenses	122,224	5.3	132,143	5.2
Music and Entertainment	2,306	0.1	7,624	0.3
Marketing	43,816	1.9	63,530	2.5

FIGURE 22.1 Joshua's income statement (P&L).

expenses, and nonoperating expenses. Referring to Figure 22.1, the gross profit section consists of Sales through Total Gross Profit, the operating expenses section covers Operating Expenses through Operating Income, and the nonoperating expenses section includes Interest through Net Income. These three sections are arranged on the income

	Last Year	%	This Year	%
Joshua's Inc.				
Last Year Versus This Year				
Utility Services	73,796	3.2	88,942	3.5
Repairs and Maintenance	34,592	1.5	35,577	1.4
Administrative and General	66,877	2.9	71,154	2.8
Occupancy	120,000	5.2	120,000	4.7
Depreciation	41,510	1.8	55,907	2.2
Total Operating Expenses	1,245,383	54.0	1,400,769	55.1
Operating Income	253,590	11.0	276,428	10.9
Interest	86,750	3.8	84,889	3.3
Income Before Income Taxes	166,840	7.2	191,539	7.5
Income Taxes	65,068	2.8	76,616	3.0
Net Income	101,772	4.4	114,923	4.5

Prepared By: M. Chaplin, CPA
Modified By: L. Dopson, Ed.D.

FIGURE 22.1 (Continued).

statement *from most controllable to least controllable* by the foodservice manager. The gross profit section consists of food and beverage sales and costs that can and should be controlled by the manager on a daily basis. The majority of this book is devoted to controlling these items. The operating expenses section is also under the control of the manager but more so on a weekly or monthly basis (with the exception of wages, which you can control daily). Consider the Repairs and Maintenance category. Although repairs will be needed when equipment breaks down, maintenance is typically scheduled on a monthly basis. The manager can control, to some extent, how employees use the equipment, but he or she cannot control or predict the breakdown of equipment when it occurs. The third section of the USAR is the nonoperating expenses section. It is this section that is least controllable by the foodservice manager. Interest paid to creditors for short-term or long-term debt is due regardless of the ability of the manager to control operations.

Furthermore, taxes are controlled by the government; to paraphrase Benjamin Franklin, the only sure things in life are death and taxes. So, the foodservice manager has little control over the amount of money "Uncle Sam" gets every year. Knowing the three sections of the income statement allows you to focus on those things over which you have the most control as a foodservice manager. This book helps you to focus on these controllable areas so that you can better manage your time and make the most out of your efforts to control costs.

Note also that each revenue and expense category in Figure 22.1 is represented in terms of both its whole dollar amount and its percentage of total sales. All ratios are calculated as a percentage of total sales except the following:

Food costs are divided by food sales.
Beverage costs are divided by beverage sales.
Food gross profit is divided by food sales.
Beverage gross profit is divided by beverage sales.

Food and beverage items use their respective food and beverage sales as the denominator so that these items can be evaluated separately from total sales. Since food costs and beverage costs are the most controllable items on the income statement, Joshua needs to separate these sales and costs out of the aggregate and evaluate these items more carefully. Notice also that Joshua's accountant presents this year's P&L statement along with last year's, as this can help Joshua make comparisons and analyze trends in his business.

Another facet of the uniform system of accounts that you should know is the supporting schedule. The income statement as shown in Figure 22.1 is an aggregate statement. This means that all details associated with the sales, costs, and profits of the foodservice establishment are *summarized* on the P&L statement. Although this summary gives the manager a one-shot look at the performance of the operation, the details are not included directly on the statement. These details can be found in supporting schedules. Each line item on the income statement should be accompanied by a schedule that outlines all of the information that you, as a manager, need to know to operate your business successfully. For example, Direct Operating Expenses could have an accompanying schedule that details costs incurred in uniforms, laundry and linen, china and glassware, silverware, etc. These expenses can also be broken down by percentage of total direct operating expenses. In addition, the schedule should have a column in which notes can be taken regarding the costs. Figure 22.2 is an example of a schedule that would accompany Joshua's Income Statement shown in Figure 22.1. Note that the Total Direct Operating Expenses, $132,143, in the schedule, is taken

Type of Expense	Expense	% of Direct Operating Expenses	Notes
Uniforms	$13,408	10.15%	
Laundry and Linen	40,964	31.00	
China and Glassware	22,475	17.01	Expense is higher than budgeted because china shelf collapsed on March 22.
Silverware	3,854	2.92	
Kitchen Utensils	9,150	6.92	
Kitchen Fuel	2,542	1.92	
Cleaning Supplies	10,571	8.00	
Paper Supplies	2,675	2.02	
Bar Expenses	5,413	4.10	
Menus and Wine Lists	6,670	5.05	Expense is lower than budgeted because the new wine supplier agreed to print the wine lists free of charge.
Exterminating	1,803	1.36	
Flowers and Decorations	9,014	6.82	
Licenses	3,604	2.73	
Total Direct Operating Expenses	132,143	100.00	

FIGURE 22.2 Direct operating expenses schedule.

directly from this year's Direct Operating Expenses on the income statement. The type of information and the level of detail that are included on the schedules are left up to you, based on what is appropriate for your operation. It is in the schedules that you collect the information you need to break down sales or costs and determine problem areas and potential opportunities for improving each item on the income statement.

The P&L statement is one of several documents that can help evaluate profitability. The P&L statement alone, however, can yield important information that is critical to the development of your future management plans and budgets. The analysis of P&L statements is a fun and very creative process if basic procedures are well understood. In general, managers who seek to uncover all that their P&L will tell them undertake the following areas of analysis:

1.   Sales/volume
2.   Food expense
3.   Beverage expense
4.   Labor expense
5.   Other expense
6.   Profits

Using the data from Figure 22.1, each of these areas will be reviewed in turn.

ANALYSIS OF SALES/VOLUME

As discussed earlier in this text, foodservice operators can measure sales in terms of either dollars or number of guests served. In both cases, an increase in sales volume is usually to be desired. A sales increase or decrease must, however, be analyzed carefully if you are to truly understand the revenue direction of your business. Consider the sales portion of Joshua's P&L statement, as detailed in Figure 22.3. Based on the data from Figure 22.3, Joshua can compute his overall sales increase or decrease using the following steps:

Sales	Last Year	% of Sales	This Year	% of Sales
Food Sales	$1,891,011	82.0%	$2,058,376	81.0%
Beverage Sales	415,099	18.0	482,830	19.0
Total Sales	2,306,110	100.0	$2,541,206	100.0

FIGURE 22.3 Joshua's P&L sales comparison.

Sales	Last Year	This Year	Variance	Variance %
Food Sales	$1,891,011	$2,058,376	$167,365	+ 8.9%
Beverage Sales	415,099	482,830	67,731	+16.3
Total Sales	2,306,110	2,541,206	235,096	+10.2

FIGURE 22.4 Joshua's P&L sales variance.

Step 1. Determine sales for this accounting period.

Step 2. Calculate the following: this period's sales minus last period's sales.

Step 3. Divide the difference in Step 2 above by last period's sales to determine percentage variance.

For Joshua, the percentage variance is as indicated in Figure 22.4.

To illustrate the steps outlined using total sales as an example, we find:

Step 1. $2,541,206

Step 2. $2,541,206 − $2,306,110 = $235,096

Step 3. $235,096/$2,306,110 = 10.2%

It appears that Joshua has achieved an overall increase in sales of 10.2%.

There are several ways Joshua could have experienced total sales increases in the current year. These are:

1. Serve the same number of guests at a higher check average.

2. Serve more guests at the same check average.

3. Serve more guests at a higher check average.

4. Serve *fewer* guests at a *much* higher check average.

To determine which of these alternatives is indeed the case, Joshua must use a sales adjustment technique.

Assume, for a moment, that Joshua raised prices for food and beverage by 5% at the beginning of this fiscal year. If this was the case, and he wishes to determine fairly his sales increase, he must adjust for that 5% menu price increase. The procedure he would use to adjust sales variance for known menu price increases is as follows:

Sales	Last Year	Adjusted Sales (Last Year × 1.05)	This Year	Variance	Variance %
Food Sales	$1,891,011	$1,985,561.60	$2,058,376	$ 72,814.40	+ 3.67%
Beverage Sales	415,099	435,853.95	482,830	46,976.05	+10.78
Total Sales	2,306,110	2,421,415.50	2,541,206	119,790.50	+ 4.95

FIGURE 22.5 Joshua's P&L sales comparison with 5% menu price increase.

Step 1. Increase prior-period (last year) sales by amount of the price increase.

Step 2. Subtract the result in Step 1 from this period's sales.

Step 3. Divide the difference in Step 2 by the value of Step 1.

Thus, in our example, Joshua would follow the steps outlined previously to determine his real sales increase. In the case of total sales, the procedure would be as follows:

Step 1. $2,306,110 × 1.05 5 $2,421,415.50

Step 2. $2,541,206 − $2,421,415.50 = $119,790.50

Step 3. $119,790.50/$2,421,415.50 = 4.95%

Figure 22.5 details the results that are achieved if this 5% adjustment process is completed for all sales areas. Joshua's total sales figure would be up by 4.95% if he adjusted it for a 5% menu price increase.

There is still more, however, that the P&L can tell Joshua about his sales. If he has kept accurate guest count records, he can compute his sales per guest figure (see Chapter 18). With this information, he can determine whether his sales are up because he is serving more guests, or because he is serving the same number of guests but each one is spending more per visit, or because some of both has occurred. In fact, if each guest is spending quite a bit more per visit, Joshua may even have experienced a decrease in total guest count yet an increase in total sales. If this were the case, he would want to know about it, since it is unrealistic to assume that revenue will continue to increase over the long run if the number of guests visiting his establishment is declining.

Other Factors Influencing Sales Analysis

In some foodservice establishments, other factors must be taken into consideration before sales revenue can be accurately analyzed. Consider the situation you would face

	Last Year	This Year	Variance	Variance %
Total Sales (October)	$17,710.00	$17,506.00	$-204	-1.2%
Number of Operating Days	22 days	21 days	1 day	
Average Daily Sales	$ 805.00	$ 833.62	$28.62	+3.6

FIGURE 22.6 Hot Dog! sales data.

if you owned a restaurant across the street from a professional basketball stadium. If you were to compare sales from this May to sales generated last May, the number of home games in May for this professional team would have to be determined before you could make valid conclusions about guest count increases or decreases. Also, if a foodservice facility is open only Monday through Friday, the number of operating days in two given accounting periods may be different for the facility. When this is the case, percentage increases or decreases in sales volume must be based on average daily sales, rather than the total sales figure. To illustrate this, consider a hot dog stand that operates in the city center Monday through Friday only. In October of this year, the stand was open for 21 operating days. Last year, however, because of the number of weekend days in October, the stand operated for 22 days. Figure 22.6 details the comparison of sales for the stand, assuming no increase in menu selling price this year compared with last year.

While, at first glance, it appears that October sales this year are 1.2% lower than last year, in reality, average daily sales are up 3.6%! Are sales for October up or down? Clearly, the answer must be qualified in terms of monthly or daily sales. For this reason, effective foodservice managers must be careful to consider all of the relevant facts before making determinations about sales direction.

Every critical factor must be considered when evaluating sales revenue, including: the number of operating meal periods or days; changes in menu prices, guest counts, and check averages; and special events. Only after carefully reviewing all details can you truly know whether your sales are going up or down.

ANALYSIS OF FOOD EXPENSE

In addition to sales analysis, the P&L statement, whether weekly, monthly, or annual, can provide information about other areas of operational interest. For the effective foodservice manager, the analysis of food expense is a matter of major concern.

	Last Year	% of Food Sales	This Year	% of Food Sales
Food Sales	$1,891,011	100.0%	$2,058,376	100.0%
Cost of Food Sold				
Meats and Seafood	$ 297,488	15.7	$ 343,063	16.7%
Fruits and Vegetables	94,550	5.0	127,060	6.2
Dairy	55,347	2.9	40,660	2.0
Baked Goods	16,142	0.9	22,870	1.1
Other	249,060	13.2	233,790	11.4
Total Cost of Food Sold	712,587	37.7	767,443	37.3

FIGURE 22.7 Joshua's P&L food expense schedule.

Figure 22.7 details the food expense portion of Joshua's P&L as outlined in the food expense schedule.

It is important to remember that the numerator of the food cost % equation is cost of food sold, while the denominator is total food sales, rather than total food and beverage sales. With total cost of food sold this year of $767,443 and total food sales of $2,058,376, the total food cost % is 37.3% ($767,443 / $2,058,376 = 37.3%). A food cost percentage can be computed in a similar manner for each food subcategory. For instance, the cost percentage for the category Meats and Seafood for this year would be computed as follows:

$$\frac{\text{Meats and Seafood Cost}}{\text{Total Food Sales}} = \text{Meats and Seafood Cost \%}$$

or

$$\frac{\$343,063}{\$2,058,376} = 16.7\%$$

At first glance, it appears that Joshua has done well for the year and that his total cost of goods sold expense has declined 0.4%, from 37.7% overall last year to 37.3%

Category	Last Year %	This Year %	Variance
Meats and Seafood	15.7%	16.7%	+1.0%
Fruits and Vegetables	5.0	6.2	+1.2
Dairy	2.9	2.0	−0.9
Baked Goods	0.9	1.1	+0.2
Other	13.2	11.4	−1.8
Total Cost of Food Sold	**37.7**	**37.3**	**−0.4**

FIGURE 22.8 Joshua's P&L variation in food expense by category.

this year. This is true. Closer inspection, however, indicates that, while the categories Dairy and Other showed declines, Meats and Seafood, Fruits and Vegetables, and Baked Goods showed increases.

Figure 22.8 shows the actual differences in food cost percentage for each of Joshua's food categories. While it is true that Joshua's overall food cost percentage is down by 0.4%, the variation among categories is quite marked. It is clearly to his benefit to subcategorize food products so that he can watch for fluctuations within and among groups, rather than merely monitor his overall increase or decrease in food costs. Without such a breakdown of categories, he will not know exactly where to look if costs get too high.

It would also be helpful for Joshua to determine how appropriate the inventory levels are for each of his product subgroups so that he can adjust the inventory sizes accordingly. To do this, Joshua must be able to compute his food inventory turnover.

Food Inventory Turnover

Inventory turnover refers to the number of times the total value of inventory has been purchased and replaced in an accounting period. Each time the cycle is completed once, we are said to have "turned" the inventory. For example, if you normally keep $100 worth of oranges on hand at any given time and your monthly usage of oranges is $500, you would have replaced your orange inventory five times in the month. The formula used to compute inventory turnover is as follows:

$$\frac{\text{Cost of Food Consumed}}{\text{Average Inventory Value}} = \text{Food Inventory Turnover}$$

Note that it is cost of food consumed, rather than cost of food sold, that is used as the numerator in this ratio. This is because all food inventory should be tracked so that you can better determine what is sold, wasted, spoiled, pilfered, or provided to employees as employee meals.

Stated another way, inventory turnover is a measure of how many times the inventory value is purchased and sold to guests. In the foodservice industry, we are, of course, interested in high inventory turnover as it relates to increased sales. It simply makes sense that if a 5% profit is made on the sale of an inventory item, we would like to sell (turn) that item as many times per year as possible. If the item were sold from inventory only once a year, one 5% profit would result. If the item turned 10 times, a 5% profit on each of the 10 sales would result. However, you have to be sure that a high inventory turnover is caused by increased sales and not by increased food waste, food spoilage, or employee theft.

To compute his inventory turnover for each of his food categories, Joshua must first establish his average inventory value for each category. The average inventory value is computed by adding the beginning inventory for this year to the ending inventory for this year and dividing by 2 using the following formula:

$$\frac{\text{Beginning Inventory Value} + \text{Ending Inventory Value}}{2}$$
$$= \text{Average Inventory Value}$$

From his inventory records, Joshua creates the data recorded in Figure 22.9.

To illustrate the computation of average inventory value, note that Joshua's Meats and Seafood beginning inventory for this year was $16,520, while his ending inventory for that category was $14,574. His average inventory value for that category is $15,547 [($16,520 + $14,574)/2 = $15,547]. All other categories and the total average inventory value are computed in the same manner.

Now that Joshua has determined the average inventory values for his food categories, he can compute the inventory turnovers for each of these. Cost of food consumed is identical to cost of food sold when no reduction is made in cost as a result of employee meals. That is the case at Joshua's facility because he charges employees full price for menu items that he sells to them as employee meals. Therefore, employee meals are included in regular food cost because a normal food sales price is charged for

Inventory Category	This Year Beginning Inventory	This Year Ending Inventory	Average Inventory Value
Meats and Seafood	$16,520	$14,574	$15,547
Fruits and Vegetables	1,314	846	1,080
Dairy	594	310	452
Baked Goods	123	109	116
Other	8,106	9,196	8,651
Total	26,657	25,035	25,846

FIGURE 22.9 Joshua's P&L average inventory values.

Inventory Category	Cost of Food Consumed	Average Inventory	Inventory Turnover
Meats and Seafood	$343,063	$15,547	22.1
Fruits and Vegetables	127,060	1,080	117.6
Dairy	40,660	452	90.0
Baked Goods	22,870	116	197.2
Other	233,790	8,651	27.0
Total	767,443	25,846	29.7

FIGURE 22.10 Joshua's P&L food inventory turnover.

these meals. Figure 22.10 shows the result of his computation using the food inventory turnover formula for this year.

To illustrate, Joshua's Meats and Seafood inventory turnover is 22.1 ($343,063/ $15,547 = 22.1). That is, Joshua purchased, sold, and replaced his meat and seafood inventory, on average, 22 times this year, which was nearly twice per month. Note that all other food categories and the total inventory turnover are computed in the same manner. Note also that in categories such as fruits and vegetables, dairy, and baked goods, the turnovers are very high, reflecting the perishability of these items. Joshua's overall inventory turnover is 29.7. If Joshua's *target* inventory turnover for the year was 26 times, then he should investigate why his actual inventory turnover is higher. It could be because of his increase in sales, which is a good sign of his restaurant's performance. Or it could be due to wastage, pilferage, and spoilage. He should use inventory turnover, in this case, to help him determine how he can more effectively control his costs in the future.

ANALYSIS OF BEVERAGE EXPENSE

Joshua's P&L (Figure 22.1) indicates beverage sales for this year of $482,830. With total sales of $2,541,206, beverages represent 19% of Joshua's total sales ($482,830/$2,541,206 = 19%). Also from Figure 22.1, beverage costs for this year are $96,566; thus, Joshua's beverage cost percentage, which is computed as cost of beverages divided by *beverage* sales, is 20% ($96,566/$482,830 = 20%). To completely analyze this expense category, Joshua would compute his beverage cost percentage, compare that to his planned expense, and compute a beverage inventory turnover rate using the same formulas as he did for his food products.

If an operation carries a large number of rare and expensive wines, it will find that its beverage inventory turnover rate is relatively low. Conversely, those beverage operations that sell their products primarily by the glass are likely to experience inventory turnover rates that are quite high. The important concept here is to compute the turnover rates at least once per year (or more often if needed) to gauge whether inventory sizes should be increased or decreased. High beverage inventory turnovers accompanied by frequent product outages may indicate inventory levels that are too low, while low turnover rates and many slow-moving inventory items may indicate the need to reduce beverage inventory levels.

Figure 22.1 shows that last year's beverage cost percentage was 22.8% ($94,550/$415,099 = 22.8%). A first glance would indicate that beverage costs have been reduced, not in total dollars spent since sales were higher this year than last year, but in percentage terms. In other words, a beverage cost % of 22.8% last year versus 20% this year indicates a 2.8% overall reduction.

Assume, for a moment, however, that Joshua raised drink prices by 10% this year over last year. Assume also that Joshua pays, on average, 5% more for beverages this year compared with last year. Is his beverage operation more efficient this year than last year, less efficient, or the same? To determine the answer to this important question in the beverage or any other expense category, Joshua must make adjustments to both his sales and his cost figures. Similar to the method for adjusting sales, the method for adjusting expense categories for known cost increases is as follows:

Step 1. Increase prior-period expense by amount of cost increase.

Step 2. Determine appropriate sales data, remembering to adjust prior-period sales, if applicable.

Step 3. Divide costs determined in Step 1 by sales determined in Step 2.

Thus, in our example, Joshua's beverage expense last year, adjusted for this year's costs, would be $94,550 \times 1.05 = $99,277.50$.

His beverage sales from last year, adjusted for this year's menu prices, would be $415,099 \times 1.10 = $456,608.90$.

His last year's beverage cost percentage, adjusted for increases in this year's costs and selling prices, would be computed as follows:

$$\frac{\$99,277.50}{\$456,608.90} = 21.7\%$$

In this case, Joshua's real cost of beverage sold has, in fact, declined this year, although not by as much as he had originally thought. That is, a 21.7% adjusted cost for last year versus a 20% cost for this year equals a reduction of 1.7%, not 2.8% as originally determined.

All food and beverage expense categories must be adjusted in terms of both costs and selling price if effective comparisons are to be made over time. When older foodservice managers remember back to the time when they purchased hamburger for $0.59 per pound, it is important to recall that the 1/4-pound hamburger they made from it may have sold for $0.59 also! The 25% resulting product cost percentage is no different from today's operator paying $3.00 a pound for ground beef and selling the resulting 1/4-pound burger for $3.00. Notice that it is not possible to compare efficiency in food and beverage usage from one time period to the next unless you are making that comparison in equal terms. As product costs increase or decrease and as menu prices change, so, too, will food and beverage expense percentages change. As an effective manager, you must determine if variations in product percentage costs are caused by real changes in your operation or by differences in the price you pay for your products as well as your selling price for those products.

ANALYSIS OF LABOR EXPENSE

From Figure 22.1, it is interesting to note that, while the total dollars Joshua spent on labor increased greatly from last year to this year, his labor cost percentage increased only slightly. This was true in the Salaries and Wages category as well as the Employee Benefits category. Recall that, whenever your labor costs are not 100% variable costs, increasing sales volume will help you decrease your labor cost percentage, although the total dollars you spend on labor will increase. The reason for this is simple. When total dollar sales volume increases, fixed labor cost percentages will decline. In other words, the dollars paid for fixed labor will consume a smaller percentage of your total revenue. Thus, as long as any portion of total labor cost is fixed (manager's salaries for example), increasing volume will have the effect of reducing labor cost percentage. Variable labor costs, of course, will increase along with sales volume increases, but the percentage of revenue they consume should stay constant.

When you combine a declining percentage (fixed labor cost) with a constant percentage (variable labor cost), you should achieve a reduced overall percentage, although your total labor dollars expended can be higher. Serving additional guests may cost additional labor dollars. That, in itself, is not a bad thing. In most foodservice situations, you *want* to serve more guests. If labor expenses are controlled properly, you will find that an increase in the number of guests and sales will result in an appropriate increase in the labor costs required to service those guests. You must be careful, however, to always ensure that increased costs are appropriate to increases in sales volume.

Remember, too, that declining costs of labor are not always a sign that all is well in a foodservice unit. Declining costs of labor may be the result of significant reductions in the number of guests served. If, for example, a foodservice facility produces poor-quality products and gives poor service, guest counts can be expected to decline, as would the cost of labor required to service those guests who do remain. Labor dollars expended by management would decline, but it would be an indication of improper, rather than effective, management. An effective foodservice manager seeks to achieve declines in operational expense because of operational efficiencies, not reduced sales.

Figure 22.11 details the labor cost portion of Joshua's P&L. Note that all the labor-related percentages he computes are based on his total sales, that is, the combination of his food and beverage sales. This is different from computing expense percentages

Labor Cost	Last Year	% of Total Sales	This Year	% of Total Sales
Salaries and Wages	$641,099	27.8%	$714,079	28.1%
Employee Benefits	99,163	4.3	111,813	4.4
Total Labor Cost	740,262	32.1	825,892	32.5

FIGURE 22.11 Joshua's P&L labor cost.

such as food and beverage because food cost percentage is determined using food sales as the denominator and beverage cost percentage computations use beverage sales as a denominator. Thus, Joshua's salaries and wages expense % for this year is computed as follows:

$$\frac{\text{Salaries and Wages Expense}}{\text{Total Sales}} = \text{Salaries and Wages Expense \%}$$

$$or$$

$$\frac{\$714,079}{\$2,541,206} = 28.1\%$$

A brief examination of the labor portion of Joshua's P&L would indicate an increase in both dollars spent for labor and labor cost percentage. Just as adjustments must be made for changes in food and beverage expenses before valid expense comparisons can be made, so too must adjustments be made for changes, if any, in the price an operator pays for labor. In Joshua's case, assume that all employees were given a COLA (cost of living adjustment), or raise, of 5% at the beginning of this year. This, coupled with an assumed 10% menu price increase, would indeed have the effect of changing overall the labor cost % even if his labor productivity did not change.

From Figure 22.11, Joshua can see that his actual labor cost % increased from 32.1% last year to 32.5% this year, an increase of 0.4%. To adjust for the changes in the cost of labor and his selling prices, if these indeed occurred, Joshua uses the techniques previously detailed in this chapter. Thus, based on the assumption of a 5% increase in the cost of labor and a 10% increase in selling price, he adjusts both sales and cost of labor, using the same steps as those employed for adjusting food or beverage cost percentage, and computes a new labor cost for last year as follows:

Step 1. Determine sales adjustment:

$$\$2,306,110 \times 1.10 = \$2,536,721$$

Step 2. Determine total labor cost adjustment:

$$\$740,262 \times 1.05 = \$777,275.10$$

Step 3. Compute adjusted labor cost percentage:

$$\frac{\$777,275.10}{\$2,536,721} = 30.6\%$$

As can be seen, *last year* Joshua's P&L would have indicated a 30.6% labor cost percentage if he had operated under this year's increased costs and selling prices. This is certainly an area that Joshua would want to investigate. The reason is simple. If he were exactly as efficient this year as he was last year, and if he assumed a 10% menu price increase and a 5% labor cost increase, Joshua's cost of labor for this year should have been computed as follows:

> **This Year's Sales × Last Year's Adjusted Labor Cost %**
>
> **= This Year's Projected Labor Cost**
>
> *or*
>
> **$2,541,206 × 0.306 = $777,609.04**

Put in another way, if Joshua were as efficient with his labor this year as he was last year, he would have expected to spend 30.6% of sales for labor this year, given his 5% payroll increase and 10% menu price increase. In actuality, Joshua's labor cost was $48,282.96 higher ($825,892 actual this year − $777,609.04 projected = $48,282.96). A variation this large should obviously be of concern to Joshua and should be examined closely.

Increases in payroll taxes, benefit programs, and employee turnover can all affect labor cost percentage. Although, for our example, we assumed that employee benefits (including payroll taxes, insurance, etc.) increased at the same 5% rate as did salaries and wages, these are, of course, different expenses and may increase at rates higher or lower than salary and wage payments to employees. Indeed, one of the fastest increasing labor-related costs for foodservice managers today is cost of medical insurance benefit programs. These programs are needed to attract the best employees to the hospitality industry, but they can be expensive.

Controlling and evaluating labor cost is an important part of your job as a hospitality manager. In fact, many managers feel it is more important to control labor costs than product costs because, for many of them, labor and labor-related costs comprise a larger portion of their operating budgets than do food and beverage products.

ANALYSIS OF OTHER EXPENSE

The analysis of other expenses should be performed each time the P&L is produced. Figure 22.12 details other expenses from Joshua's P&L statement.

Joshua's other expenses consist of both operating expenses (excluding salaries and wages and employee benefits) and nonoperating items, and he must review these carefully. In Joshua's operation, these costs have increased from last year's levels. Note

Other Expenses	Last Year	% of Total Sales	This Year	% of Total Sales
Operating Expenses (excluding salaries and wages and employee benefits)				
Direct Operating Expense	$122,224	5.3%	$132,143	5.2%
Music and Entertainment	2,306	0.1	7,624	0.3
Marketing	43,816	1.9	63,530	2.5
Utility Services	73,796	3.2	88,942	3.5
Repairs and Maintenance	34,592	1.5	35,577	1.4
Administrative and General	66,877	2.9	71,154	2.8
Occupancy	120,000	5.2	120,000	4.7
Depreciation	41,510	1.8	55,907	2.2
Nonoperating Expenses				
Interest	$86,750	3.8%	$ 84,889	3.3%
Income Taxes	65,068	2.8	76,616	3.0

FIGURE 22.12 Joshua's P&L other expenses.

that his Repairs and Maintenance category is also higher this year than it was last year. This is one area in which he both expects and approves a cost increase. It is logical to assume that kitchen repairs will increase as a kitchen ages. In that sense, a kitchen is much like a car. Even with a good preventative maintenance program, Joshua does not expect an annual decline in kitchen repair expense. In fact, he would be somewhat surprised and concerned should this category be smaller this year than in the previous year because his sales were higher and probably caused more wear and tear on his kitchen equipment. In the same way, his contributions to his state and national association political action funds, charged to Administrative and General expense, were up significantly. This is due to Joshua's belief that he, as part of the hospitality industry, needs to make his voice heard to his local and national political leaders. Joshua is a strong believer in taking a leadership role in his association on the local, state, and national levels. Indeed, one of his goals is to someday serve on the board of his national association! He knows that membership in that organization gives him a voice straight to the nation's law- and policymakers.

An analysis of other expenses proves difficult for Joshua since he is not sure how he compares with others in his area or with operations of a similar nature. For comparison purposes, he is, however, able to use industry trade publications to get national averages on other expense categories. One helpful source Joshua can use is an annual publication, *The Restaurant Industry Operations Report*, published by the National Restaurant Association and prepared by Deloitte & Touche (it can be ordered through www.restaurant.org). For operations that are a part of a corporate chain, unit managers can receive comparison data from district and regional managers, who can chart performance against those of other units in the city, region, state, and nation.

ANALYSIS OF PROFITS

As can be seen in Figure 22.1, profits for Joshua's, Inc., refer to the net income figure at the bottom of the income statement. Joshua's net income for this year was $114,923, or 4.5% of total sales, and his total sales for this year were $2,541,206. His profit percentage using the profit margin formula is as follows:

$$\frac{\text{Net Income}}{\text{Total Sales}} = \text{Profit Margin}$$

$$or$$

$$\frac{\$114,923}{\$2,541,206} = 4.5\%$$

Profit margin is also known as return on sales (ROS). For the foodservice manager, perhaps no figure is more important than the ROS. This percentage is the most telling indicator of a manager's overall effectiveness at generating revenues and controlling costs in line with forecasted results. While it is not possible to state what a "good" return on sales figure should be for all restaurants, industry averages, depending on the specific segment, range from 1% to over 20%. Some operators prefer to use operating income (see Figure 22.1) as the numerator for profit margin instead of net income. This is because interest and income taxes are considered nonoperating expenses and, thus, not truly reflective of a manager's ability to generate a profit.

Joshua's results this year represent an improvement over last year's figure of $101,772, or 4.4% of total sales. Thus, he has shown improvement both in the dollar size of his net income and in the size of the net income as related to total sales. He realizes, however, that increased sales, rather than great improvements in operational efficiency, could have caused this progress, because his sales volume this year was greater than his sales volume last year. To analyze his profitability appropriately, he must determine how much of this increase was due to increased menu prices as opposed to increased guest count or check average. Joshua's improvement in net income for the year can be measured by the following formula:

$$\frac{\text{Net Income This Period} - \text{Net Income Last Period}}{\text{Net Income Last Period}}$$

$$= \text{Profit Variance \%}$$

or

$$\frac{\$114,923 - \$101,772}{\$101,772} = +12.9\%$$

How much of this improvement is due to improved operational methods versus increased sales will depend, of course, on how much Joshua actually did increase his sales relative to increases in his costs. Monitoring selling price, guest count, sales per guest, operating days, special events, and actual operating costs is necessary for accurate profit comparisons. Without knowledge of each of these areas, the effective analysis of profits becomes a risky proposition.

APPLY WHAT YOU HAVE LEARNED

Terri Settles is a Registered Dietitian (R.D.). She supervises five hospitals for Maramark Dining Services, the company the hospitals have selected to operate their foodservices. Her company produces a monthly and annual income statement for each hospital.

1. Discuss five ways in which income statements can help Terri do her job better.

2. What would a hospital do with "profits" or surpluses made in the foodservice area?

3. What effect will "profit" or "loss" have on the ability of Terri's company to continue to manage the foodservices for these hospitals?

KEY TERMS AND CONCEPTS

The following are terms and concepts discussed in the chapter that are important for you as a manager. To help you review, please define the terms below.

Cost accounting	Gross profit section (of the USAR)	Aggregate statement
Managerial accounting		Supporting schedules
Bookkeeping	Operating expenses section (of the USAR)	Inventory turnover
Uniform system of accounts		COLA
	Nonoperating expenses section (of the USAR)	Return on sales (ROS)

PLANNING FOR PROFIT *

Overview

This chapter shows you how to analyze your menu so you can identify which individual menu items make the most contribution to your profits. In addition, it teaches you how to determine the sales dollars and volume you must achieve to break even and to generate a profit in your operation. Finally, the chapter shows you how to establish an operating budget and presents techniques you can use to monitor your effectiveness in staying within that budget.

Chapter Outline

Financial Analysis and Profit Planning

Menu Analysis

Cost/Volume/Profit Analysis

The Budget

Developing the Budget

Monitoring the Budget

Fun on the Web!

Apply What You Have Learned

Key Terms and Concepts

---

* Authored by Jack E. Miller, Lea R. Dopson, and David K. Hayes.

Highlights

At the conclusion of this chapter, you will be able to:

- Analyze a menu for profitability.
- Prepare a cost/volume/profit analysis.
- Establish a budget and monitor performance to the budget.

FINANCIAL ANALYSIS AND PROFIT PLANNING

In addition to analyzing the P&L statement, you should also undertake a thorough study of three areas that will assist you in planning for profit. These three areas of analysis are:

1. Menu analysis
2. Cost/volume/profit (CVP) analysis
3. Budgeting

Whereas menu analysis concerns itself with the profitability of the menu items you sell, CVP analysis deals with the sales dollars and volume required by your foodservice unit to avoid an operating loss and to make a profit. The process of budgeting allows you to plan your next year's operating results by projecting sales, expenses, and profits to develop a budgeted P&L statement.

Many foodservice operators practice the activity of "hoping for profit" instead of "planning for profit." Although hoping is an admirable pursuit when playing the lottery, it does little good when managing a foodservice operation. Therefore, planning is the key to ensuring that owners achieve the profit goals that will keep them in business.

MENU ANALYSIS

A large number of methods have been proposed as being the best way to analyze the profitability of a menu and its pricing structure. The one you choose to use, however, should simply seek to answer the question: "How does the sale of this menu item contribute to the overall success of my operation?" It is unfortunate, in many ways, that the discussion of menu analysis typically leads one to elaborate mathematical formulas and computations. This is, of course, just one component of the analysis of a menu. It is not, however, nor should it ever be, the only component.

Consider the case of Danny, who operates a successful family restaurant, called "The Mark Twain," in rural Tennessee. The restaurant has been in his family for three generations. One item on the menu is mustard greens with scrambled eggs. It does not sell often, but both mustard greens and eggs are ingredients in other, more popular items. Why does Danny keep the item in a prominent spot on the menu? The answer is simple and has little to do with finance. The menu item was Danny's grandfather's favorite. As a thank-you to his grandfather, who started the business and inspired Danny to become service and guest oriented, the menu item survives every menu reprint.

Menu analysis, then, is about more than just numbers. It involves marketing, sociology, psychology, and emotions. Remember that guests respond not to weighty financial analyses, but rather to menu copy, the description of the menu items, the placement of items on the menu, their price, and their current popularity. While the financial analysis of a menu is indeed done "by the numbers," you must realize that those numbers are just one part, albeit an important part, of the total menu analysis picture.

For the serious foodservice manager, the analysis of a menu deserves special study. Many components of the menu, such as pricing, layout, design, and copy, play an important role in the overall success of a foodservice operation. The foodservice manager who does not seek to understand how a menu truly works is akin to the manager who does not seek to understand the essential components of making a good cup of coffee!

If you investigate the menu analysis methods that have been widely used, you will find that each seeks to perform the analysis using one or more of the following operational variables with which you are familiar:

- Food cost percentage
- Popularity
- Contribution margin
- Selling price
- Variable expenses
- Fixed expenses

Three of the most popular systems of menu analysis, shown in Figure 23.1, are discussed here because they represent the three major philosophical approaches to menu analysis. The matrix analysis referenced in Figure 23.1 provides a method for comparisons between menu items. A matrix allows menu items to be placed into categories based on whether they are above or below menu item averages such as food cost %, popularity, and contribution margin.

Method	Variables Considered	Analysis Method	Goal
1. Food cost %	a. Food cost % b. Popularity	Matrix	Minimize overall food cost %
2. Contribution margin	a. Contribution margin b. Popularity	Matrix	Maximize contribution margin
3. Goal value analysis	a. Contribution margin % b. Popularity c. Selling price d. Variable cost % e. Food cost %	Algebraic equation	Achieve predetermined profit % goals

FIGURE 23.1 Three methods of menu analysis.

Each approach to menu analysis has its proponents and detractors, but an understanding of each will help you as you attempt to develop your own philosophy of menu analysis.

Food Cost Percentage

Menu analysis that focuses on food cost percentage is the oldest and most traditional method used. When analyzing a menu using the food cost percentage method, you are seeking menu items that have the effect of *minimizing your overall food cost percentage*. The rationale for this is that a lowered food cost percentage leaves more of the sales dollar to be spent for other operational expenses. A criticism of the food cost percentage approach is that items that have a higher food cost percentage may be removed from the menu in favor of items that have a lower food cost percentage but that, when purchased by guests, also contribute fewer dollars to overall profit.

To illustrate the use of the food cost percentage menu analysis method, consider the case of Maureen, who operates a steak and seafood restaurant near the beach in a busy resort town. Maureen sells seven items in the entrée section of her menu. The items and information related to their cost, selling price, and popularity are presented in Figure 23.2.

To determine her average selling price, Maureen divides total sales by the total number of items sold. In this case, the computation is $11,583/700 = $16.55. To

Menu Item	Number Sold	Selling Price	Total Sales	Item Cost	Total Cost	Item Contribution Margin	Total Contribution Margin	Food Cost %
Strip Steak	73	$17.95	$ 1,310.35	$ 8.08	$ 589.84	$ 9.87	720.51	45%
Coconut Shrimp	121	16.95	2,050.95	5.09	615.89	11.86	1,435.06	30
Grilled Tuna	105	17.95	1,884.75	7.18	753.90	10.77	1,130.85	40
Chicken Breast	140	13.95	1,953.00	3.07	429.80	10.88	1,523.20	22
Lobster Stir-Fry	51	21.95	1,119.45	11.19	570.69	10.76	548.76	51
Scallops/Pasta	85	14.95	1,270.75	3.59	305.15	11.36	965.60	24
Beef Medallions	125	15.95	1,993.75	5.90	737.50	10.05	1,256.25	37
Total	700		11,583.00		4,002.77		7,580.23	
Weighted Average	100	16.55	1,654.71	5.72	571.82	10.83	1,082.89	35

FIGURE 23.2 Maureen's menu analysis worksheet.

determine her total food cost percentage, she divides total cost by total sales. The computation is $4,002.77/$11,583 = 35%. The columns titled Item Contribution Margin and Total Contribution Margin are not used in the food cost percentage approach to menu analysis and, thus, are not discussed till later in this chapter.

To analyze her menu using the food cost percentage method, Maureen must segregate her items based on the following two variables:

1. Food cost percentage
2. Popularity (number sold)

Since Maureen's overall food cost is 35%, she determines that any individual menu item with a food cost percentage above 35% will be considered *high* in food cost percentage, while any menu item with a food cost below 35% will be considered low. In a similar vein, with a total of 700 entrées served in this accounting period and seven possible menu choices, each menu item would sell 700/7, or 100 times, if all were equally popular. Given that fact, Maureen determines that any item sold more than 100 times during this week's accounting period would be considered high in popularity, while any item selling less than 100 times would be considered low in popularity. Having made these determinations, Maureen can produce a matrix labeled as follows:

		POPULARITY	
		LOW	HIGH
FOOD COST %	High	Square 1	Square 2
		High food cost %, low popularity	High food cost %, high popularity
	Low	Square 3	Square 4
		Low food cost %, low popularity	Low food cost %, high popularity

Based on the number sold and food cost percentage data in Figure 23.2, Maureen can classify her menu items in the following manner:

SQUARE	CHARACTERISTICS	MENU ITEM
1	High food cost %, low popularity	Strip steak, lobster stir-fry
2	High food cost %, high popularity	Grilled tuna, beef medallions
3	Low food cost %, low popularity	Scallops/pasta
4	Low food cost %, high popularity	Coconut shrimp, chicken breast

Note that each menu item inhabits one, and only one, square. Using the food cost percentage method of menu analysis, Maureen would like as many menu items as possible to fall within square 4. These items have the characteristics of being low in food cost percentage but high in guest acceptance. Thus, both coconut shrimp and chicken breast have below-average food cost percentages and above-average popularity. When developing a menu that seeks to minimize food cost percentage, items in the fourth square are highly desirable. These, of course, are kept on the menu. They should be well promoted and have high menu visibility. Promote them to your best guests and take care not to develop and attempt to sell a menu item that is similar enough in nature that it could detract from the sales of these items.

The characteristics of the menu items that fall into each of the four matrix squares are unique and, thus, should be managed differently. Because of this, each individual square requires a special marketing strategy, depending on their square location. These strategies can be summarized as shown in Figure 23.3.

It can be quite effective to use the food cost percentage method of menu evaluation. It is fast, logical, and time tested. Remember that if you achieve too high a food cost percentage, you run the risk that not enough percentage points will remain to generate a profit on your sales. There are other factors to consider, however. You should be cautioned against promoting low-cost items with low selling prices at the expense of items that have a higher food cost percentage but that also contribute greater gross profits. For example, most foodservice operators would say it is better to achieve a 20% food cost than a 40% food cost. Consider, however, a chicken dish that sells for $5.00 and costs you just $1.00 to make. This item yields a 20% food cost ($1.00/$5.00 = 20%),

Square	Characteristics	Problem	Marketing Strategy
1	High food cost %, low popularity	Marginal due to both high product cost and lack of sales	a. Remove from the menu. b. Consider current food trends to determine if the item itself is unpopular or if its method of preparation is. c. Survey guests to determine current wants regarding this item. d. If this is a high-contribution-margin item, consider reducing price and/or portion size.
2	High food cost %, high popularity	Marginal due to high product cost	a. Increase price. b. Reduce prominence on the menu. c. Reduce portion size. d. "Bundle" the sale of this item with one that has a lower cost and, thus, provides better overall food cost %.
3	Low food cost %, low popularity	Marginal due to lack of sales	a. Relocate on the menu for greater visibility. b. Take off the regular menu and run as specials. c. Reduce menu price. d. Eliminate other unpopular menu items in order to increase demand for this one.
4	Low food cost %, high popularity	None	a. Promote well. b. Increase visibility on the menu.

FIGURE 23.3 Analysis of food cost matrix results.

and there are $4.00 ($5.00 − $1.00 = $4.00) remaining to pay for the labor and other expenses of serving this guest. Compare that to the same guest buying steak for $10.00 that cost you $4.00 to make. Your food cost percentage would be 40% ($4.00/$10.00 = 40%). In this case, however, there would be $6.00 ($10.00 − $4.00 = $6.00) remaining to pay for the labor and other expenses of serving this guest. For this reason, some operators prefer to analyze their menus using the contribution margin matrix.

Contribution Margin

When analyzing a menu using the contribution margin approach, the operator seeks to produce a menu that maximizes the overall contribution margin. Recall from Chapter 20 that contribution margin per menu item is defined as the amount that remains after the product cost of the menu item is subtracted from the item's selling price. Contribution margin is the amount that you will have available to pay for your labor and other expenses and to keep for your profit. Thus, from Figure 23.2, if an item on Maureen's menu, such as strip steak, sells for $17.95 and the product cost for the item is $8.08, the contribution margin per menu item would be computed as follows:

> **Selling Price − Product Cost = Contribution Margin per Menu Item**
> *or*
> **$17.95 − $8.08 = $9.87**

When contribution margin is the driving factor in analyzing a menu, the two variables used for the analysis are contribution margin and item popularity. To illustrate the use of the contribution margin approach to menu analysis, the data in Figure 23.2 are again used. In this case, Maureen must again separate her items based on high or low popularity, which gives the same results as those obtained when using the food cost percentage method; thus, any item that sells 700/7, or 100 times, or more is considered to be a high-popularity item, while any menu choice selling less than 100 times would be considered low in popularity. To employ the contribution margin approach to menu analysis, Maureen computes her average item contribution margin. When computing average contribution margin for the entire menu, two steps are required. First, to determine the total contribution margin for the menu, the following formula is used:

> **Total Sales − Total Product Costs = Total Contribution Margin**
> *or*
> **$11,583.00 − $4,002.77 = $7,580.23**

Because 700 total menu items were sold, you can determine the average contribution margin per *item* using the following formula:

$$\frac{\text{Total Contribution Margin}}{\text{Number of Items Sold}} = \text{Average Contribution Margin per Item}$$

or

$$\frac{\$7,580.23}{700} = \$10.83$$

To develop the contribution margin matrix, you proceed along much the same lines as with the food cost percentage matrix. In this case, average item popularity is 100 and average item contribution margin is $10.83. The matrix is developed as follows:

		POPULARITY	
		LOW	HIGH
	High	*Square 1*	*Square 2*
CONTRIBUTION		High contribution margin,	High contribution margin,
MARGIN		low popularity	high popularity
	Low	*Square 3*	*Square 4*
		Low contribution margin,	Low contribution margin,
		low popularity	high popularity

Maureen now classifies her menu items according to the contribution margin matrix in the following manner:

SQUARE	CHARACTERISTICS	MENU ITEM
1	High contribution margin, low popularity	Scallops/pasta
2	High contribution margin, high popularity	Coconut shrimp, chicken breast
3	Low contribution margin, low popularity	Strip steak, lobster stir-fry
4	Low contribution margin, high popularity	Grilled tuna, beef medallions

Again, each menu item finds itself in one, and only one, matrix square. Using the contribution margin method of menu analysis, Maureen would like as many of her menu items as possible to fall within square 2, that is, high contribution margin and high popularity. From this analysis, Maureen knows that both coconut shrimp and chicken breast yield a higher than average contribution margin. In addition, these items sell very well. Just as Maureen would seek to give high menu visibility to items with low food cost percentage and high popularity when using the food cost percentage method of menu analysis, she would seek to give that same visibility to items with high contribution margin and high popularity when using the contribution margin approach.

Square	Characteristics	Problem	Marketing Strategy
1	High contribution margin, low popularity	Marginal due to lack of sales	a. Relocate on menu for greater visibility. b. Consider reducing selling price.
2	High contribution margin, high popularity	None	a. Promote well. b. Increase prominence on the menu.
3	Low contribution margin, low popularity	Marginal due to both low contribution margin and lack of sales	a. Remove from menu. b. Consider offering as a special occasionally, but at a higher menu price.
4	Low contribution margin, high popularity	Marginal due to low contribution margin	a. Increase price. b. Reduce prominence on the menu. c. Consider reducing portion size.

FIGURE 23.4 Analysis of contribution margin matrix results.

Each of the menu items that fall in the other squares requires a special marketing strategy, depending on its square location. These strategies can be summarized as shown in Figure 23.4.

A frequent, and legitimate criticism of the contribution margin approach to menu analysis is that it tends to favor high-priced menu items over low-priced ones, since higher priced menu items, in general, tend to have the highest contribution margins. Over the long term, this can result in sales techniques and menu placement decisions that tend to put in the guest's mind a higher check average than the operation may warrant or desire.

The selection of either food cost percentage or contribution margin as a menu analysis technique is really an attempt by the foodservice operator to answer the following questions:

- Are my menu items priced correctly?
- Are the individual menu items selling well enough to warrant keeping them on the menu?
- Is the overall profit margin on my menu items satisfactory?

Because of the limitations of matrix analysis, neither the matrix food cost nor the matrix contribution margin approach is tremendously effective in analyzing menus. This is the case because the axes on the matrix are determined by the mean (average) of food cost percentage, contribution margin, or sales level (popularity). When this is done, some items will *always* fall into the less desirable categories. This is so because, in matrix analysis, high food cost percentage, for instance, really means food cost percentage *above* that operation's average. Obviously, then, some items must fall below the average regardless of their contribution to operational profitability. Eliminating the poorest items only shifts other items into undesirable categories. To illustrate this drawback to matrix analysis, consider the following example. Assume that Homer, one of Maureen's competitors, sells only four items, as follows:

HOMER'S #1 MENU	
ITEM	NUMBER SOLD
Beef	70
Chicken	60
Pork	15
Seafood	55
Total	200
Average sold	50 (200/4)

Homer may elect to remove the pork item, since its sales range is below the average of 50 items sold. If Homer adds turkey to the menu and removes the pork, he could get the following results:

HOMER'S #2 MENU	
ITEM	NUMBER SOLD
Beef	65
Chicken	55
Turkey	50
Seafood	30
Total	200
Average sold	50 (200/4)

As can be seen, the turkey item drew sales away from the beef, chicken, and seafood dishes and did not increase the total number of menu items sold. In this case, it is now the seafood item that falls below the menu average. Should it be removed because its sales are below average? Clearly, this might not be wise. Removing the seafood item might serve only to draw sales *from* the remaining items to the seafood replacement item. Obviously, the same type of result can occur when you use a matrix to analyze food cost percentage or contribution margin. As someone once stated, half of us are always below average in anything. Thus, the matrix approach *forces* some items to be below average. How, then, can an operator answer questions related to price, sales volume, and overall profit margin? One answer is to avoid the overly simplistic and ineffective matrix analysis and employ all, or even part, of a more effective method of menu analysis called goal value analysis.

Goal Value Analysis

Goal value analysis was introduced by Dr. David Hayes and Dr. Lynn Huffman in an article titled "Menu Analysis: A Better Way," (Hayes & Huffman, 1985) published by the respected hospitality journal *The Cornell Quarterly*. Ten years later, at the height of what was known as the "value pricing" (i.e., extremely low pricing strategies used to drive significant increases in guest counts) debate, goal value analysis proved its effectiveness in a second article, "Value Pricing: How Low Can You Go?" (Hayes & Huffman, 1995), which was also published in *The Cornell Quarterly*.

Essentially, goal value analysis uses the power of an algebraic formula to replace less sophisticated menu average techniques. Before the widespread introduction of computerized spreadsheet programs, some managers found the computations required to use goal value analysis challenging. Today, however, such computations are easily made. The advantages of goal value analysis are many, including ease of use, accuracy, and the ability to simultaneously consider more variables than is possible with two-dimensional matrix analysis. Mastering the power of goal value analysis can truly help you design menus that are effective, popular, and, most important, profitable.

Goal value analysis evaluates each menu item's food cost percentage, contribution margin, and popularity and, unlike the two previous analysis methods introduced, includes the analysis of the menu item's nonfood variable costs as well as its selling price. Returning to the data in Figure 23.2, we see that Maureen has an overall food cost % of 35%. In addition, she served 700 guests at an entrée check average of $16.55. If we knew about Maureen's overall fixed and variable costs, we would know more about the profitability of each of Maureen's menu items. One difficulty, of course, resides in the assignment of nonfood variable costs to individual menu items. The issue is complex. It is very likely true, for example, that different items on Maureen's menu require differing amounts of labor to prepare. For instance, the strip steak on her menu is purchased precut and vacuum-sealed. Its preparation simply requires opening the steak package, seasoning the steak, and placing it on a broiler. The lobster stir-fry, on the other hand, is a complex dish that requires cooking and shelling the lobster, cleaning and trimming the vegetables, then preparing the item when ordered by quickly cooking the lobster, vegetables, and a sauce in a wok. Thus, the *variable* labor cost of preparing the two dishes is very different. It is assumed that Maureen responds to these differing costs by charging more for a more labor-intensive dish and less for one that is less labor intensive. Other dishes require essentially the same amount of labor to prepare; thus, their variable labor costs figure less significantly in the establishment of price. Because that is true, for analysis purposes, most operators find it convenient to assign variable costs to individual menu items based on menu price. For example, if labor and other variable costs are 30% of total sales, all menu items may be assigned that same variable cost percentage of their selling price.

For the purpose of her goal value analysis, Maureen determines her total variable costs. These are all the costs that vary with her sales volume, excluding the cost of the food itself. She computes those variable costs from her P&L statement and finds that they account for 30% of her total sales. Using this information, Maureen assigns a variable cost of 30% of selling price to each menu item.

Having compiled the information in Figure 23.2, Maureen can use the algebraic goal value formula to create a specific goal value for her entire menu, then use the same formula to compute the goal value of each individual menu item. Menu items that achieve goal values higher than that of the overall menu goal value will contribute greater than average profit percentages. As the goal value for an item increases, so, too, does its profitability percentage. The overall menu goal value can be used as a "target" in this way, assuming that Maureen's average food cost %, average number of items sold per menu item, average selling price (check average), and average variable cost % all meet the overall profitability goals of her restaurant. The goal value formula is as follows:

$$A \times B \times C \times D = \text{Goal Value}$$

where

$A = 1.00 - \text{Food Cost \%}$

$B = \text{Item Popularity}$

$C = \text{Selling Price}$

$D = 1.00 - (\text{Variable Cost \%} + \text{Food Cost \%})$

Note that A in the preceding formula is really the contribution margin percentage of a menu item and that D is the amount available to fund fixed costs and provide for a profit after all variable costs are covered.

Item	Food Cost % (in decimal form)	Number Sold	Selling Price	Variable Cost % (in decimal form)
Strip Steak	0.45	73	$17.95	0.30
Coconut Shrimp	0.30	121	16.95	0.30
Grilled Tuna	0.40	105	17.95	0.30
Chicken Breast	0.22	140	13.95	0.30
Lobster Stir-Fry	0.51	51	21.95	0.30
Scallops/Pasta	0.24	85	14.95	0.30
Beef Medallions	0.37	125	15.95	0.30

FIGURE 23.5   Maureen's goal value analysis data.

Maureen uses this formula to compute the goal value of her *total menu* and finds that:

$$A \quad \times B \ \times C \quad \times D \qquad\qquad\quad = \text{Goal Value}$$
$$(1.00 - 0.35) \times 100 \times \$16.55 \times [1.00 - (0.30 + 0.35)] = \text{Goal Value}$$

or

$$0.65 \qquad \times 100 \times \$16.55 \times 0.35 \qquad\qquad = 376.5$$

According to this formula, any menu item whose goal value equals or exceeds 376.5 will achieve profitability that equals or exceeds that of Maureen's overall menu. The computed goal value carries no unit designation; that is, it is neither a percentage nor a dollar figure because it is really a numerical target or score. Figure 23.5 details the goal value data Maureen needs to complete a goal value analysis on each of her seven menu items.

Figure 23.6 details the results of Maureen's goal value analysis. Note that she has calculated the goal values of her menu items and ranked them in order of highest to lowest goal value. She has also inserted her overall menu goal value in the appropriate rank order.

Note that the grilled tuna falls slightly below the profitability of the entire menu, while the strip steak and lobster stir-fry fall substantially below the overall goal value

Rank	Menu Item	A	B	C	D	Goal Value
1	Chicken Breast	$(1 - 0.22)$	140	$13.95	$1 - (0.30 + 0.22)$	731.2
2	Coconut Shrimp	$(1 - 0.30)$	121	16.95	$1 - (0.30 + 0.30)$	574.3
3	Scallops/Pasta	$(1 - 0.24)$	85	14.95	$1 - (0.30 + 0.24)$	444.3
4	Beef Medallions	$(1 - 0.37)$	125	15.95	$1 - (0.30 + 0.37)$	414.5
	Overall Menu (Goal Value)	**$(1 - 0.35)$**	**100**	**$16.55**	**$1 - (0.30 + 0.35)$**	**376.5**
5	Grilled Tuna	$(1 - 0.40)$	105	17.95	$1 - (0.30 + 0.40)$	339.3
6	Strip Steak	$(1 - 0.45)$	73	17.95	$1 - (0.30 + 0.45)$	180.2
7	Lobster Stir-Fry	$(1 - 0.51)$	51	21.95	$1 - (0.30 + 0.51)$	104.2

FIGURE 23.6 Goal value analysis results.

score. Should these two items be replaced? The answer, most likely, is no *if* Maureen is satisfied with her current target food cost percentage, profit margin, check average, and guest count. Every menu will have items that are more or less profitable than others. In fact, some operators develop items called loss leaders. A loss leader is a menu item that is priced very low, sometimes even below total costs, for the purpose of drawing large numbers of guests to the operation. If, for example, Maureen has the only operation in town that serves outstanding lobster stir-fry, that item may, in fact, contribute to the overall success of the operation by drawing people who will buy it, while their fellow diners may order items that are more profitable.

The accuracy of goal value analysis is well documented. Used properly, it is a convenient way for management to make shorthand decisions regarding required profitability, sales volume, and pricing. Because all of the values needed for the goal value formula are readily available, management need not concern itself with puzzling through endless decisions about item replacement.

Items that do not achieve the targeted goal value tend to be deficient in one or more of the key areas of food cost percentage, popularity, selling price, or variable cost percentage. In theory, all menu items have the potential of reaching the goal value, although management may determine that some menu items can indeed best serve the operation as loss leader–type items. For example, examine the goal value analysis results for the item, strip steak:

A		× B	× C	× D		= Goal Value
Strip Steak (1.00 − 0.45)		× 73	× $17.95	× [1 − (0.30 + 0.45)]		= Goal Value
0.55		× 73	× $17.95	× 0.25		= 180.2

This item did not meet the goal value target. Why? There can be several answers. One is that the item's 45% food cost is too high. This can be addressed by reducing portion size or changing the item's recipe, since both of these actions have the effect of reducing the food cost % and, thus, increasing the A value. A second approach to improving the goal value score of the strip steak is to work on improving the B value, that is, the number of times the item is sold. This may be done through merchandising or, since it is one of the more expensive items on the menu, incentives to waitstaff for upselling this item. Variable C, menu price, while certainly in line with the rest of the menu, can also be adjusted upward; however, you must remember that adjustments upward in C may well result in declines in the number of items sold (B value)! Increases in the menu price will also have the effect of *decreasing* the food cost % of the menu item. This is because item food cost % = food cost of the item/menu price. Obviously, the changes you undertake as a result of menu analysis are varied and can be complex. As you gain experience in knowing the tastes and behavior of your guests, however, your skill in menu-related decision making will quickly improve.

Sophisticated users of the goal value analysis system can, as suggested by Lendal Kotschevar, Ph.D., in his book, *Management By Menu*, (Kotschevar, 2001) modify the formula to increase its accuracy and usefulness even more. In the area of variable costs, a menu item might be assigned a low, medium, or high variable cost. If overall variable costs equal 30%, for example, management may choose to assign a variable cost of 25% to those items with very low labor costs attached to them, 30% to others with average labor costs, and 35% to others with even higher costs. This adjustment affects only the D variable of the goal value formula and can be accommodated quite easily.

Goal value analysis will also allow you to make better decisions more quickly. This is especially true if you know a bit of algebra and realize that anytime you determine a desired goal value *and* when any three of the four variables contained in the formula are known, you can solve for the fourth unknown variable by using goal value as the numerator and placing the known variables in the denominator. Figure 23.7 shows you how to solve for each unknown variable in the goal value formula.

To illustrate how the information in Figure 23.7 can be used, let's return to the information in Figure 23.6 and assume that, in Maureen's case, she feels the 12-ounce

Known Variables	Unknown Variables	Method to Find Unknown
A, B, C, D	**Goal Value (GV)**	$A \times B \times C \times D$
B, C, D, GV	A	$\dfrac{GV}{B \times C \times D}$
A, C, D, GV	B	$\dfrac{GV}{A \times C \times D}$
A, B, D, GV	C	$\dfrac{GV}{A \times B \times D}$
A, B, C, GV	D	$\dfrac{GV}{A \times B \times C}$

FIGURE 23.7 Solving for goal value unknowns.

strip steak she is offering may be too large for her typical guest and that is why its popularity (B value) is low. Thus, Maureen elects to take three actions:

1. She reduces the portion size of the item from 12 ounces to 9 ounces, resulting in a reduction in her food cost from $8.08 to $6.10.

2. Because she knows her guests will likely be hesitant to pay the same price for a smaller steak, she also *reduces* the selling price of this item by $1.00 to $16.95. She feels that this will keep the strip steak from losing any popularity resulting from the reduction in portion size. Her new food cost percentage for this item is 36% ($6.10 / $16.95 = 36%).

3. Since the labor required to prepare this menu item is so low, she assigns a below-average 25% variable cost to its D value.

Maureen now knows three of the goal value variables for this item and can solve for the fourth. Maureen knows her A value $(1.00 - 0.36)$, her C value ($16.95), and her D value $[1.00 - (0.25 + 0.36)]$. The question she would ask is this, "Given this newly structured menu item, how many must be sold to make the item achieve the targeted goal value?" The answer requires solving the goal value equation for B, the number sold. From Figure 23.7, recall that, if B is the unknown variable, it can be computed by using the following formula:

$$\frac{\text{Goal Value}}{A \times C \times D} = B$$

In this case:

$$\frac{376.5}{(1.00 - 0.36) \times \$16.95 \times [1.00 - (0.25 + 0.36)]} = B$$

Thus:

$$89 = B$$

According to the formula, 89 servings of strip steak would have to be sold to achieve Maureen's target goal value. Again, goal value analysis is a very useful estimation tool for management. You can use it to establish a desired food cost percentage, target popularity figure, selling price, or variable cost percentage.

Goal value analysis is also powerful because it is not, as is matrix analysis, dependent on past performance to establish profitability but can be used by management to establish *future* menu targets. To explain, assume, for a moment, that Maureen wishes to achieve a greater profit margin and a $17.00 entrée average selling price for next year. She plans to achieve this through a reduction in her overall food cost to 33% and her other variable costs to 29%. Her overall menu goal value formula for next year, assuming no reduction or increase in guest count, would be as follows:

$$A \qquad \times B \ \times C \qquad \times D \qquad\qquad = \text{Goal Value}$$
$$(1.00 - 0.33) \times 100 \times \$17.00 \times [1.00 - (0.29 + 0.33)] = \text{Goal Value}$$

or

$$0.67 \qquad \times 100 \times \$17.00 \times 0.38 \qquad\qquad = 432.8$$

Thus, each item on next year's menu should be evaluated with the new goal value in mind. It is important to remember, however, that Maureen's actual profitability will be heavily influenced by sales mix. Thus, all pricing, portion size, and menu placement decisions become critical. Note that Maureen can examine each of her menu items and determine whether she wishes to change any of the item's characteristics in order to meet her goals. It is at this point that she must remember that she is a foodservice operator and not merely an accountant. A purely quantitative approach to menu analysis is neither practical nor desirable. Menu analysis and pricing decisions are always a matter of experience, skill, and educated predicting because it is difficult to know in advance how changing any one menu item may affect the sales mix of the remaining items.

COST/VOLUME/PROFIT ANALYSIS

Each foodservice operator knows that some accounting periods are more profitable than others. Often, this is because sales volume is higher or costs are lower during certain periods. The ski resort that experiences tremendous sales during the ski season but has a greatly reduced volume or may even close during the summer season is a good example. Profitability, then, can be viewed as existing on a graph similar to Figure 23.8.

The x axis represents the number of covers sold in a foodservice operation. The y axis represents the costs/revenues in dollars. The Total Revenues line starts at 0 because if no covers are sold, no dollars are generated. The Total Costs line starts farther up the y axis because fixed costs are incurred even if no covers are sold. The point at which the two lines cross is called the break-even point. At the break-even point, operational expenses are exactly equal to sales revenue. Stated in another way, when sales volume in your operation equals the sum of your total variable and fixed costs, your break-even point

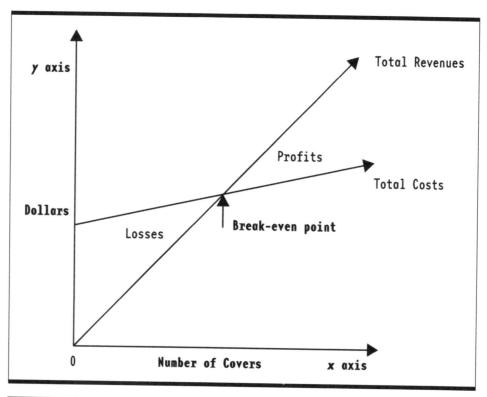

FIGURE 23.8 Cost/volume/profit graph.

has been reached. Below the break-even point, costs are higher than revenues, so losses occur. Above the break-even point, revenues exceed costs, so profits are made. Most operators would like to know their break-even point on a daily, weekly, or monthly basis. In effect, by determining the break-even point, the operator is answering the question: "How much sales volume must I generate before I begin to make a profit?"

Beyond the break-even point, you will want to answer another question: "How much sales dollars and volume must I generate to make *my desired* profit level?" To answer this question, you must conduct a cost/volume/profit (CVP) analysis. A CVP analysis helps predict the sales dollars and volume required to achieve desired *profit* (or break-even) based on your known costs.

The answer to these questions may be found either by constructing a CVP graph or by arithmetical calculation. While there are advantages to both methods, the arithmetical calculation is typically the most accurate. CVP calculations can be done either on the dollar sales volume required to break even or achieve the desired profit, or on the basis of the number of guests (covers) required.

Consider the case of Jennifer, who operates an Asian restaurant in a suburban northwestern city. Based on her income statement and sales records of last month, Jennifer has converted her P&L statement to a contribution margin income statement, as shown in Figure 23.9. A contribution margin income statement simply shows P&L items in terms of sales, variable costs, contribution margin, fixed costs, and profit.

As discussed in Chapter 21, foodservice expenses can generally be classified as either fixed or variable. Of course, some expenses have both a fixed and a variable component and, thus, are, in reality, mixed. For the purpose of engaging in a CVP analysis, however, it is necessary for the operator to assign costs to either a fixed or a variable category, as Jennifer has done. In addition, the contribution margin for her overall operation is defined as the dollar amount that *contributes* to covering fixed costs and providing for a profit. Contribution margin is calculated for Jennifer's as follows:

> **Total Sales — Variable Costs = Contribution Margin**
> *or*
> **$125,000 — $50,000 = $75,000**

Jennifer can also view her contribution margin income statement in terms of per-unit (guest) and percentage sales, variable costs, and contribution margin, as shown in Figure 23.10.

Jennifer's

Total Sales	$125,000	Sales per Guest	$12.50
Variable Costs	50,000	Guests Served	10,000
Contribution Margin	75,000		
Fixed Costs	60,000		
Before-Tax Profit	15,000		
Taxes (40%)	6,000		
After-Tax Profit	9,000		

FIGURE 23.9 Contribution margin income statement.

Jennifer's

Total Sales	$125,000
Variable Costs	50,000
Contribution Margin	75,000
Fixed Costs	60,000
Before-Tax Profit	15,000
Taxes (40%)	6,000
After-Tax Profit	9,000

Per Unit (Guest)		Percentage
SP	$12.50	100%
VC	5.00	40
CM	7.50	60
Guests Served		10,000

FIGURE 23.10 Contribution margin income statement with per-unit and percentage calculations.

Notice the dark box in Figure 23.10 that includes per unit (guest) and percentage calculations for sales per guest (selling price, SP), variable costs (VC), and contribution margin (CM). Note also that fixed costs are not calculated as per unit or as a percentage of sales. This is because fixed costs do not vary with sales volume increases.

To calculate these numbers, the following steps apply:

Step 1. Divide total sales, variable costs, and contribution margin by the number of guests served (units) to get per-unit (guest) values:

SP/Unit = $125,000/10,000 Units = $12.50

VC/Unit = $50,000/10,000 Units = $5.00

CM/Unit 5 = $75,000/10,000 Units = $7.50

SP/Unit − VC/Unit = CM/Unit

$12.50 − $5.00 = $7.50

Step 2. Divide VC/Unit by SP/Unit, and CM/Unit by SP/Unit to get percentage values:

SP% = 100%

VC% = $5.00/$12.50 = 40%

CM% = $7.50/$12.50 = 60%

SP% − VC% = CM%

100% − 40% = 60%

Once Jennifer's P&L statement has been converted to a contribution margin income statement and per-unit values and percentages have been calculated, she can proceed to determine her operational break-even point and sales required to achieve her desired profit. She wants to do this based both on dollar sales and on the number of guests (units) required to do so.

To determine the dollar sales required to break even, Jennifer uses the following formula:

$$\frac{\text{Fixed Costs}}{\text{Contribution Margin \%}} = \text{Break-Even Point in Sales}$$

or

$$\frac{\$60,000}{0.60} = \$100,000$$

Thus, Jennifer must generate $100,000 in sales per month *before* she begins to make a profit. At a sales volume of less than $100,000, she would be operating at a loss. In terms of the number of guests that must be served in order to break even, Jennifer uses the following formula:

$$\frac{\text{Fixed Costs}}{\text{Contribution Margin per Unit (Guest)}} = \text{Break-Even Point in Guests Served}$$

or

$$\frac{\$60,000}{\$7.50} = 8,000 \text{ Guests (Covers)}$$

Now, assume that Jennifer has decided that next month she will plan for $12,000 in after-tax profits. To determine sales dollars and covers to achieve her after-tax profit goal, Jennifer uses the following formula:

$$\frac{\text{Fixed Costs} + \text{Before-Tax Profit}}{\text{Contribution Margin \%}} = \text{Sales Dollars to Achieve Desired After-Tax Profit}$$

Jennifer knows that her after-tax-profit goal is $12,000, but the preceding formula calls for *before-tax profit*. To convert her after-tax profit to before-tax profit, Jennifer must compute the following:

$$\frac{\text{After-Tax Profit}}{1 - \text{Tax Rate}} = \text{Before-Tax Profit}$$

or

$$\frac{\$12,000}{1 - 0.40} = \$20,000$$

Now that Jennifer knows her before-tax profit of $20,000, she can calculate her sales dollars to achieve her desired after-tax profit as follows:

$$\frac{\text{Fixed Costs} + \text{Before-Tax Profit}}{\text{Contribution Margin \%}} = \text{Sales Dollars to Achieve Desired After-Tax Profit}$$

or

$$\frac{\$60,000 + \$20,000}{0.60} = \$133,333.33$$

Thus, Jennifer must generate $133,333.33 in sales per month to achieve her desired after-tax profit of $12,000. In terms of calculating the number of guests that must be served in order to make her profit, Jennifer uses the following formula:

$$\frac{\text{Fixed Costs} + \text{Before-Tax Profit}}{\text{Contribution Margin per Unit (Guest)}}$$
= Guests to Be Served to Achieve Desired After-Tax Profit

or

$$\frac{\$60,000 + \$20,000}{\$7.50} = 10,666.67 \text{ Guests (Covers), Round up to 10,667 Guests}$$

When calculating sales and covers to achieve break-even and desired after-tax profits, you can easily remember which formulas to use if you know the following:

1.   Contribution margin % is used to calculate sales *dollars*.
2.   Contribution margin per *unit* is used to calculate sales volume in *units* (guests).

Once you fully understand the CVP analysis concepts, you can predict any sales level for break-even or after-tax profits based on your selling price, fixed costs, variable costs, and contribution margin. You can also make changes in your selling prices and costs to improve your ability to break even and achieve desired profit levels. This is where menu pricing and cost controls concepts covered in this text come into play. As you make changes in your control areas, you will be able to manage your operation efficiently so that losses can be prevented and planned profits can be achieved.

Minimum Sales Point

Every foodservice operator should know his or her break-even point. The concept of minimum sales point is related to this area. Minimum sales point (MSP) is the dollar sales volume required to justify staying open for a given period of time. The information necessary to compute MSP is as follows:

1.   Food cost %
2.   Minimum payroll cost for the time period
3.   Variable cost %

Fixed costs are eliminated from the calculation because, even if the volume of sales equals zero, fixed costs still exist and must be paid. Consider the situation of Richard, who is trying to determine whether he should close his steakhouse at 10:00 P.M. or 11:00 P.M. Richard wishes to compute the sales volume necessary to justify staying open the additional hour. He can make this calculation because he knows that his food cost equals 40%, his minimum labor cost to stay open for the extra hour equals $150, and his variable costs (taken from his P&L statement) equal 30%. In calculating MSP, his Food Cost % 1 Variable Cost % is called his minimum operating cost. Richard applies the MSP formula as follows:

$$\frac{\text{Minimum Labor Cost}}{1 - \text{Minimum Operating Cost}} = \text{MSP}$$

or

$$\frac{\text{Minimum Labor Cost}}{1 - (\text{Food Cost\%} + \text{Variable Cost\%})} = \text{MSP}$$

In this case, the computation would be as follows:

$$\frac{\$150}{1 - (0.40 + 0.30)} = \text{MSP}$$

$$\frac{\$150}{1 - 0.70} = \text{MSP}$$

or

$$\frac{\$150}{0.30}$$

thus

$$\$500 = \text{MSP}$$

If Richard can achieve a sales volume of $500 in the 10:00 P.M. to 11:00 P.M. time period, he should stay open. If this level of sales is not feasible, he should consider closing the operation at 10:00 P.M. Richard can use MSP to determine the hours his operation is most profitable. Of course, some operators may not have the authority to close the operation, even when remaining open is not particularly profitable. Corporate policy, contractual hours, promotion of a new unit, competition, and other factors must all be taken into account before the decision is made to modify operational hours.

THE BUDGET

In most managerial settings, you will be responsible for preparing and maintaining a budget for your foodservice operation. This budget, or financial plan, will detail the operational direction of your unit and your expected financial results. The techniques used in managerial accounting will show you how close your actual performance was when compared to your budget, while providing you with the information you need to make changes to your operational procedures or budget. This will ensure that your operation achieves the goals of your financial plan. It is important to note that the budget should not be a static document. It should be modified and fine-tuned as managerial accounting presents data about sales and costs that affect the direction of the overall operation.

For example, if you own a dance club featuring Latin music, and you find that a major competitor in your city has closed its doors, you may quite logically determine that you want to revise upward your estimate of the number of guests who will come to your club. This would, of course, affect your projected sales revenue, your costs, and your profitability. Not to do so might allow you to meet and exceed your original sales goals but would ignore a significant event that very likely will affect your financial plan for the club.

In a similar manner, if you are the manager of a delicatessen specializing in salads, sliced meats, and related items, and you find through your purchase orders that the price you pay for corned beef has tripled since last month, you must adjust your budget or you will find that you have no chance of staying within your food cost guidelines. Again, the point is that the foodservice budget should be closely monitored through the use of managerial accounting, which includes the thoughtful analysis of the data this type of accounting provides.

Just as the P&L tells you about your past performance, the budget is developed to help you achieve your future goals. In effect, the budget tells you what must be done if predetermined profit and cost objectives are to be met. In this respect, you are attempting to modify the profit formula, as presented in Chapter 17. With a well-thought-out and attainable budget, your profit formula would read as follows:

Budgeted Revenue − Budgeted Expense = Budgeted Profit

To prepare the budget and stay within it assures you predetermined profit levels. Without such a plan, you must guess about how much to spend and how much sales you should anticipate. The effective foodservice operator builds his or her budget, monitors it closely, modifies it when necessary, and achieves the desired results. Yet, many operators do not develop a budget. Some say that the process is too time consuming. Others feel that a budget, especially one shared with the entire organization, is too revealing. Budgeting can also cause conflicts. This is true, for example, when dollars budgeted for new equipment must be used for either a new kitchen stove or a new beer-tapping system. Obviously, the kitchen manager and the beverage manager may hold different points of view on where these funds can best be spent!

Despite the fact that some operators avoid budgets, they are extremely important. The rationale for having and using a budget can be summarized as follows:

1. It is the best means of analyzing alternative courses of action and allows management to examine these alternatives prior to adopting a particular one.

2. It forces management to examine the facts regarding what is necessary to achieve desired profit levels.

3. It provides a standard for comparison essential for good controls.

4. It allows management to anticipate and prepare for future business conditions.

5. It helps management to periodically carry out a self-evaluation of the organization and its progress toward its financial objectives.

6. It provides a communication channel whereby the organization's objectives are passed along to its various departments.

7. It encourages department managers who have participated in the preparation of the budget to establish their own operating objectives and evaluation techniques and tools.

8. It provides management with reasonable estimates of future expense levels and serves as an instrument for setting proper prices.

9. It communicates to owners and investors the realistic financial performance expectations of management.

Budgeting is best done by the entire management team, for it is only through participation in the process that the whole organization will feel compelled to support the budget's implementation. Foodservice budgets can be considered as one of three main types:

1. Long-range budget
2. Annual budget
3. Achievement budget

Long-Range Budget

The long-range budget is typically prepared for a period of three to five years. While its detail is not great, it does provide a long-term view about where the operation should be going. It is also particularly useful in those cases where additional operational units may increase sales volume and accompanying expense. Assume, for example, that you are preparing a budget for a corporation you own. Your corporation has entered into an agreement with an international franchise company to open 45 cinnamon bun kiosks in malls across the United States and Canada. You will open a new store approximately every month for the next four years. To properly plan for your revenue and expense in the coming four-year period, a long-range budget for your company will be much needed.

Annual Budget

The annual, or yearly, budget is the type most operators think of when the word budget is used. As it states, the annual budget is for a one-year period or, in some cases, one season. This would be true, for example, in the case of a religious summer camp that is open and serving meals only while school is out of session and campers are attending, or a ski resort that opens in late fall but closes when the snow melts.

It is important to remember that an annual budget need not follow a calendar year. In fact, the best time period for an annual budget is the one that makes sense for your own operation. A college foodservice director, for example, would want a budget that covers the time period of a school year, that is, from the fall of one year through the spring of the next. For a restaurant whose owners have a fiscal year different from a calendar year, the annual budget may coincide with either the fiscal year or the calendar, as the owners prefer.

It is also important to remember that an annual budget need not consist of 12, one-month periods. While many operators prefer one-month budgets, some prefer budgets consisting of 13, 28-day periods, while others use quarterly (three-month) or even weekly budgets to plan for revenues and costs throughout the budget year.

Achievement Budget

The achievement budget is always of a shorter range, perhaps a month or a week. It provides current operating information and, thus, assists in making current operational decisions. A weekly achievement budget might, for example, be used to predict the number of gallons of milk needed for this time period or the number of servers to be scheduled on Tuesday night.

DEVELOPING THE BUDGET

Some managers think it is very difficult to establish a budget, and, thus, they simply do not take the time to do so. Creating a budget is not that complex. You can learn to do it and do it well. To establish any type of budget, you need to have the following information available:

1. Prior-period operating results
2. Assumptions of next-period operations
3. Goals
4. Monitoring policies

To examine how prior-period operating results, assumptions of next-period operations, and goals drive the budgeting process, we will consider the case of Levi, who is preparing the annual foodservice budget for his 100-bed extended-care facility.

Prior-Period Operating Results

Levi's facility serves patient meals to an average occupancy of 80%, and he serves approximately 300 additional meals per day to staff and visitors. His department is allotted a flat dollar amount by the facility's administration for each meal he serves. His operating results for last year are detailed in Figure 23.11. Patient and additional

---

Patient Meals Served: 29,200   Revenue per Meal: $3.46

Additional Meals Served: 109,528

Total Meals Served: 138,728

	Amount	Percentage
Total Department Revenue	$480,000	100%
Cost of Food	192,000	40%
Cost of Labor	153,600	32%
Other Expenses	86,400	18%
Total Expenses	432,000	90%
Profit	48,000	10%

FIGURE 23.11 Levi's last-year operating results.

meals served were determined by actual count. Revenue and expense figures were taken from Levi's income (P&L) statements at the year's end. It is important to note that Levi must have this information if he is to do any meaningful profit planning. Foodservice unit managers who do not have access to their operating results are at a tremendous managerial disadvantage. Levi has his operational summaries and the data that produced them. Because he knows how he has operated in the past, he is now ready to proceed to the assumptions section of the planning process.

Assumptions of Next-Period Operations

If Levi is to prepare a budget with enough strength to serve as a guide and enough flexibility to adapt to a changing environment, he must factor in the assumptions he and others feel will affect the operation. While each management team will arrive at its own conclusions given the circumstances of the operation, in this example, Levi makes the following assumptions regarding next year:

1.   Food costs will increase by 3%.

2.   Labor costs will increase by 5%.

3.   Other expenses will rise by 10% due to a significant increase in utility costs.

4.   Revenue received for all meals served will be increased by no more than 1%.

5.   Patient occupancy of 80% of facility capacity will remain unchanged.

Levi would be able to establish these assumptions through discussions with his suppliers and union leaders, his own records, and, most important, his sense of the operation itself. In the commercial sector, when arriving at assumptions, operators must also consider new or diminished competition, changes in traffic patterns, and national food trends. At the highest level of foodservice management, assumptions regarding the acquisition of new units or the introduction of new products will certainly affect the budget process. As an operator, Levi predicts items 1, 2, and 3 by himself, while his supervisor has given him input about items 4 and 5. Given these assumptions, Levi can establish operating goals for next year.

Establishing Operating Goals

Given the assumptions he has made, Levi can now determine actual operating goals for the coming year. He will establish them for each of the following areas:

1.   Meals served

2.   Revenue

3. Food costs

4. Labor costs

5. Other expenses

6. Profit

Meals Served Given the assumption of no increase in patient occupancy, and in light of his results from last year, Levi budgets to prepare and serve 29,200 patient meals. He feels, however, that he can increase his visitor and staff meals somewhat by being more customer service driven and by offering a wider selection of items on the facility's cycle menu. He decides, therefore, to raise his goal for additional meals from the 109,528 served last year to 115,000 for the coming year. Thus, his budgeted total meals to be served will equal 144,200 meals (29,200 + 115,000 = 144,200).

Revenue Levi knows that his total revenue is to increase by only 1%. His revenue per meal will thus be $3.46 × 1.01 = $3.49. With 144,200 meals to be served, Levi will receive $503,258 (144,200 × $3.49 = $503,258) if he meets his meals-served budget.

Food Costs Since Levi is planning to serve more meals, he expects to spend more on food. In addition, he assumes that this food will cost, on average, 3% more than last year. To determine a food budget, Levi computes the estimated food cost for 144,200 meals as follows:

1. Last Year's Food Cost per Meal = Last Year's Cost of Food/Total Meals Served = $192,000/138,728 = $1.38.

2. Last Year's Food Cost per Meal + 3% Estimated Increase in Food Cost = $1.38 × 1.03 = $1.42 per Meal.

3. $1.42 × 144,200 Meals to be Served this Year = $204,764 Estimated Cost of Food this Year.

Labor Costs Since Levi is planning to serve more meals, he expects to spend more on labor cost. In addition, he assumes that this labor will cost, on average, 5% more than last year. To determine a labor budget, Levi computes the estimated labor cost for 144,200 meals to be served as follows:

1. Last Year's Labor Cost per Meal = Last Year's Cost of Labor/Total Meals Served = $153,600/138,728 = $1.11 per Meal.

2. Last Year's Labor Cost per Meal + 5% Estimated Increase in Labor Cost = $1.11 × 1.05 = $1.17 per Meal.

3. $1.17 × 144,200 Meals to Be Served this Year = $168,714 Estimated Cost of Labor this Year.

Other Expenses Since Levi assumes a 10% increase in other expenses, they are budgeted as last year's amount plus an increase of 10%. Thus, $86,400 × 1.10 = $95,040.

Based on his assumptions about next year, Figure 23.12 details Levi's budget summary for the coming 12 months.

Profit Note that the increased costs Levi will be forced to bear, when coupled with his minimal revenue increase, will cause his profit to fall from $48,000 for last year to a projected $34,740 for the coming year. If this is not acceptable, Levi must either increase his revenue beyond his assumption or look to his operation to reduce costs.

Levi has now developed concrete guidelines for his operation. Since his supervisor approves his budget as submitted, Levi is now ready to implement and monitor this new budget.

MONITORING THE BUDGET

An operational plan has little value if management does not use it. In general, the budget should be monitored in each of the following three areas:

1. Revenue
2. Expense
3. Profit

| Patient Meals Budgeted: | 29,200 | Budgeted Revenue per Meal: $3.49 |

Patient Meals Budgeted: 29,200   Budgeted Revenue per Meal: $3.49

Additional Meals Budgeted: 115,000

Total Meals Budgeted: 144,200

	Amount	**Percentage**
Total Budgeted Revenue	$503,258	100.0%
Budgeted Expenses		
Cost of Food	204,764	40.7%
Cost of Labor	168,714	33.5%
Other Expense	95,040	18.9%
Total Expense	468,518	93.1%
Profit	34,740	6.9%

FIGURE 23.12 Levi's budget for next year.

Revenue Analysis

If revenue should fall below projected levels, the impact on profit can be substantial. Simply put, if revenues fall far short of projections, it will be difficult for you to meet your profit goals. If revenue consistently exceeds your projections, the overall budget must be modified or the expenses associated with these increased sales will soon exceed budgeted amounts. Effective managers compare their actual revenue to that which they have projected on a regular basis.

It is clear that increases in operational revenue should dictate proportional increases in variable expense budgets, although fixed expenses, of course, need not be adjusted for these increases. For those foodservice operations with more than one meal period, monitoring budgeted sales volume may mean monitoring each meal period. Consider the case of Rosa, the night (p.m.) manager of a college cafeteria. She feels that she is busier than ever, but her boss, Lois, maintains that there can be no

Meal Period	Budget	Actual	% of Budget
A.M.	$480,500	$166,698	35%
P.M.	350,250	248,677	71
Total	830,750	415,375	50

FIGURE 23.13 College cafeteria revenue budget summary.

increase in Rosa's labor budget since the overall cafeteria sales volume is exactly in line with budgeted projections. Figure 23.13 shows the complete story of the sales volume situation at the college cafeteria after the first six months of the fiscal year. Note that the year is half (or 50%) completed at the time of this analysis.

Based on the sales volume she generates, Rosa *should* have an increase in her labor budget for the P.M. meal period. The amount of business she is generating in the evenings is substantially higher than budgeted. Note that she is one-half way through her budget year, but has already generated 71% of the annual revenue forecasted by the budget. This, however, does not mean that the labor budget for the entire cafeteria should be increased. In fact, the labor budget for the A.M. shift should likely be reduced, as those dollars are more appropriately needed in the evening meal period.

Some foodservice operators relate revenue to the number of seats they have available in their operation. As a result, they sometimes budget based on sales per seat, the total revenue generated by a facility divided by the number of seats in the dining area(s). Since the size of a foodservice facility affects both total investment and operating costs, this can be a useful number. The formula for the computation of sales per seat is as follows:

$$\frac{\text{Total Sales}}{\text{Available Seats}} = \text{Sales per Seat}$$

To illustrate this, assume that, if Rosa's cafeteria has 120 seats, her P.M. sales per seat thus far this year would be as follows:

$$\frac{\$248,677}{120} = \$2,072.31$$

The A.M. sales per seat, given the same number of seats, would be computed as follows:

$$\frac{\$166,698}{120} = \$1,389.15$$

As can be seen, Rosa's sales per seat are much higher than that of her A.M. counterpart. Of course, part of that may be due to the fact that evening menu items in the cafeteria may sell for more, on average, than do breakfast items.

When sales volume is lower than originally projected, management must seek ways to increase revenue or reduce costs. As stated earlier, one of management's main tasks is to generate guests, while the employee's main task is to service these guests to the best of his or her ability. There are a variety of methods used for increasing sales volume, including the use of coupons, increased advertising, price discounting, and specials. For the serious foodservice manager, a thorough study of the modern techniques of foodservice marketing is mandatory if you are to be ready to meet all the challenges you may face.

Expense Analysis

Effective foodservice managers are careful to monitor operational expense because costs that are too high or too low may be cause for concern. Just as it is not possible to estimate future sales volume perfectly, it is also not possible to estimate future expense perfectly, since some expenses will vary as sales volume increases or decreases. To know that an operation spent $800 for fruits and vegetables in a given week becomes meaningful only if we know what the sales volume for that week was. Similarly, knowing that $500 was spent for labor during a given lunch period can be analyzed only in terms of the amount of sales achieved in that same period. To help them make an expense assessment quickly, some operators elect to utilize the yardstick method of calculating expense standards so determinations can be made as to whether variations in expenses are due to changes in sales volume or other reasons such as waste or theft.

To illustrate the yardstick method, consider the case of Marion, who operates a college cafeteria during nine months of the year in a small southeastern city. Marion has developed both revenue and expense budgets. His problem, however, is that variations in revenue cause variations in expense. This is true in terms of food products, labor, and other expenses. As a truly effective manager, he wishes to know whether changes in his actual expenses are due to inefficiencies in his operation or to normal sales variation.

To begin his analysis, Marion establishes a purchase standard for food products using a seven-step model.

Developing Yardstick Standards for Food

Step 1. Divide total inventory into management-designated subgroups, for example, meats, produce, dairy, and groceries.

Step 2. Establish dollar value of subgroup purchases for prior accounting period.

Step 3. Establish sales volume for the prior accounting period.

Step 4. Determine percentage of purchasing dollar spent for each food category.

Step 5. Determine percentage of revenue dollar spent for each food category.

Step 6. Develop weekly sales volume and associated expense projection. Compute % cost to sales for each food grouping and sales estimate.

Step 7. Compare weekly revenue and expense to projection. Correct if necessary.

To develop his yardstick standards for food, Marion collects data from last year as shown in Figure 23.14.

Assuming that Marion has created a revenue estimate of $52,000 per month for this year and that he was satisfied with both last year's food cost percentage and profits, he can now follow the steps outlined previously to establish his yardstick standards for food. Marion estimates a weekly sales volume of $52,000/4, or $13,000, for this year.

		Last School Year (9 Months)
Total Sales: $450,000		Average Sales per Month: $50,000
Purchases		
Meats	$ 66,600	
Fish/Poultry	36,500	
Produce	26,500	
Dairy	20,000	
Groceries	18,300	
Total	167,900	

FIGURE 23.14 Marion's college cafeteria food data.

Marion's Yardstick Standards for Food

Step 1. Meats
         Fish/Poultry
         Produce
         Diary
         Groceries

Step 2. Meats $66,600
         Fish/Poultry   36,500
         Produce   26,500
         Diary   20,000
         Groceries   18,300
         Total   167,900

Step 3. $450,000 total revenue in prior period (9 months)

Step 4. Meats $66,600/167,900 =   39.7%
         Fish/Poultry   36,500/167,900 =   21.7
         Produce   26,500/167,900 =   15.8
         Diary   20,000/167,900 =   11.9
         Groceries   18,300/167,900 =   10.9
         Total   167,900   100.0

Step 5. Meats $66,600/450,000 =   14.8%
         Fish/Poultry   36,500/450,000 =   8.1
         Produce   26,500/450,000 =   5.9
         Diary   20,000/450,000 =   4.4
         Groceries   18,300/450,000 =   4.1
         Total   167,900   37.3%

Step 6.

CATEGORY	% COST TO TOTAL COST	% COST TO TOTAL SALES	WEEKLY SALES ESTIMATE				
			$11,000	$12,000	$13,000	$14,000	$15,000
Meat	39.7%	14.8%	$1,628	$1,776	$1,924	$2,072	$2,220
Fish/Poultry	21.7	8.1	891	972	1,053	1,134	1,215
Produce	15.8	5.9	649	708	767	826	885
Diary	11.9	4.4	484	528	572	616	660
Groceries	10.9	4.1	451	492	533	574	615
Total	100.0	37.3	4,103	4,476	4,849	5,222	5,595

Note that, to compute the data for Step 6, you must multiply % cost to total sales by the weekly sales estimate. When using the sales estimate of $13,000 for a week, for example, the meat budget would be computed as follows:

$$0.148 \times \$13,000 = \$1,924$$

Fish/poultry would be computed as follows:

$$0.081 \times \$13,000 = \$1,053$$

Step 7. Analysis

Marion can now compare his budgeted expense with actual perfor mance over several volume levels. In a week in which sales volume equals $14,000, for example, Marion would expect that total meat used for that period, according to his yardstick measure, would equal approximately $2,072 ($14,000 × 14.8% = $2,072). If his usage exceeds $2,072 for the period, he would know exactly where to direct his attention. Using the yardstick system, Marion can easily monitor any expense over any number of differing volume levels. The yardstick method of purchase estimation is especially helpful for those operations that experience great variation in sales volume. A hotel that has a slow season and a busy season, for instance, will find that the use of this method is quite helpful in estimating the money needed for inventory acquisition.

Developing Yardstick Standards for Labor Just as Marion used the yardstick method to estimate food expense at varying sales volume levels, he can also used it to estimate labor cost expenditures at those various levels. To develop a labor yardstick, he follows these steps:

Step 1. Divide total labor cost into management-designated subgroups, for example, cooks, warewashers, and bartenders.

Step 2. Establish dollar value spent for each subgroup during the prior accounting period.

Step 3. Establish sales volume for the prior accounting period.

Step 4. Determine percentage of labor dollar spent for each subgroup.

Step 5. Determine percentage of revenue dollar spent for each labor category.

Step 6. Develop weekly sales volume and associated expense projection. Compute % cost to sales for each labor category and sales estimate.

Step 7. Compare weekly revenue and expense to projection. Correct if necessary.

Marion collected labor-related data from last year's operation as shown in Figure 23.15.

```
                                              Last School Year (9 Months)

Total Sales: $450,000                   Average Sales per Month: $50,000

Labor Costs
   Management            $ 40,000
   Food Production         65,000
   Service                 12,000
   Sanitation              18,000
   Total                  135,000
```

Marion's college cafeteria labor data.

It is important to note that Marion can develop a labor yardstick based on guests served, labor hours worked, or, as is his preference, labor cost percentage. To develop the labor standard based on labor cost percentage, Marion follows the seven-step process outlined below.

Marion's Yardstick Standards for Labor

Step 1. Management
         Food Production
         Service
         Sanitation

Step 2. Management $40,000
         Food Production   65,000
         Service   12,000
         Sanitation   18,000
         Total   135,000

Step 3. $450,000 total revenue in prior period (9 months)

Step 4. Management $40,000/135,000 =   29.6%
         Food Production   65,000/135,000 =   48.2
         Service   12,000/135,000 =   8.9
         Sanitation   18,000/135,000 =   13.3
         Total   135,000   100.0%

Step 5. Management $40,000/450,000 =   8.9%
         Food Production   65,000/450,000 =   14.4
         Service   12,000/450,000 =   2.7
         Sanitation   18,000/450,000 =   4.0
         Total   135,000   30.0%

Step 6.

CATEGORY	% COST TO TOTAL COST	% COST TO TOTAL SALES	WEEKLY SALES ESTIMATE				
			$11,000	$12,000	$13,000	$14,000	$15,000
Management	29.6%	8.9%	$979	$1,068	$1,157	$1,246	$1,335
Food Production	48.2	14.4	1,584	1,728	1,872	2,016	2,160
Service	8.9	2.7	297	324	351	378	450
Sanitation	13.3	4.0	440	480	520	560	600
Total	100.0	30.0	3,300	3,600	3,900	4,200	4,500

Note that, to compute the data for Step 6, Marion must multiply % cost to total sales by his weekly sales estimate. Using the sales estimate of $13,000, for example, the management portion of the budget would be computed as follows:

$$0.089 \times \$13,000 = \$1,157$$

The food production cost expense estimate, based on the same weekly sales, would be computed as follows:

$$0.144 \times \$13,000 = \$1,872$$

Step 7. Analysis

It is now easy for Marion to identify exactly where his labor variations, if any, are to be found.

The yardstick method may, of course, be used for any operational expense, be it food, labor, or one of the many other expenses you will incur. In all cases, however, you must monitor your actual expenditures as they relate to budgeted expenditures, while keeping changes in sales volume in mind.

Profit Analysis

As business conditions change, changes in the budget are to be expected. This is because budgets are based on a specific set of assumptions, and, as these assumptions change, so, too, does the budget that follows from the assumptions. Assume, for example, that you budgeted $1,000 in January for snow removal from the parking lot attached to the restaurant you own in New York State. If unusually severe weather causes you to spend $2,000 for snow removal in January instead, the assumption (normal levels of snowfall) was incorrect and the budget will be incorrect as well.

Budgeted profit must be realized if the operation is to provide adequate returns for owner and investor risk. Consider the case of James, the operator of a foodservice establishment with excellent sales but below-budgeted profits. James budgeted a 5% profit on $2,000,000 of sales; thus, $100,000 profit ($2,000,000 × 0.05 = $100,000) was anticipated. In reality, at year's end, James achieved only $50,000 profit, or 2.5% of sales ($50,000/$2,000,000 = 2.5%). If the operation's owners feel that $50,000 is an adequate return for their risk, James' services may be retained. If they do not, he may lose his position, *even though* the operation is profitable. Remember that your goal is not merely to generate a profit, but rather to generate budgeted profit. You will be rewarded when you meet this goal. A primary goal of management is to generate the profits necessary for the successful continuation of the business. Budgeting for these profits is a fundamental step in the process. Analyzing the success of achieving budget forecasts is of tremendous managerial concern. If profit goals are to be met, safeguarding your operational revenue is critical. It is to that task that we now turn our attention. The proper collection and accounting for guest payment of services is one of the final steps in a successful food and beverage cost control system.

FUN ON THE WEB!

Look up the following sites to review and obtain copies of books related to accounting and cost control.

www.amazon.com
www.borders.com
www.barnesandnoble.com

Search these sites using key words such as:

Food and beverage management
Food and beverage cost control
Menu pricing
Menu design
Hospitality accounting
Restaurant management

APPLY WHAT YOU HAVE LEARNED

Ananda Fields is the CEO of a company that operates a very large number of quick-service restaurants. Recently, competitors have been increasing sales at their restaurants at a faster rate than at Ananda's. Joseph Smith, Vice President of Operations, is encouraging Ananda to introduce a new line of higher priced, higher quality, and higher contribution margin items to increase sales and improve profits. Sonya Miller, her V.P. for Marketing, is recommending Ananda introduce a "value" line of products that would be priced very low, but significantly increase traffic to the stores.

1.   Do you think more customers would be attracted using Joseph's recommendation or Sonya's?

2.   What factors would cause Ananda to choose one V.P.'s menu recommendation over the other's?

3.   What impact will Ananda's menu decision have on the image projected by her stores? What can she do to influence this image?

KEY TERMS AND CONCEPTS

The following are terms and concepts discussed in the chapter that are important for you as a manager. To help you review, please define the terms below.

Matrix analysis	Contribution margin for overall operation	Budget
Contribution margin per menu item	Minimum sales point (MSP)	Long-range budget
Cost/volume/profit (CVP) analysis	Minimum operating cost	Annual budget
Contribution margin income statement	Goal value analysis	Achievement budget
	Loss leaders	Sales per seat
	Break-even point	Yardstick method

FREQUENTLY USED FORMULAS
FOR MANAGING OPERATIONS*

MANAGING REVENUE AND EXPENSE

Revenue − Expense = Profit

Revenue − Desired Profit = Ideal Expense

$$\frac{Part}{Whole} = Percent$$

$$\frac{Expense}{Revenue} = Expense\,\%$$

$$\frac{Profit}{Revenue} = Profit\,\%$$

$$\frac{Desired\ Profit}{Revenue} = Desired\ Profit\,\%$$

Revenue − (Food and Beverage Cost + Labor Cost + Other Expense) = Profit

$$\frac{Food\ and\ Beverage\ Cost}{Revenue} = Food\ and\ Beverage\ Cost\,\%$$

$$\frac{Labour\ Cost}{Revenue} = Labor\ Cost\,\%$$

$$\frac{Other\ Expense}{Revenue} = Other\ Expense\,\%$$

$$\frac{Total\ Expense}{Revenue} = Total\ Expense\,\%$$

---

* Authored by Jack E. Miller, Lea R. Dopson, and David K. Hayes.

$$\frac{Profit}{Revenue} = Profit\,\%$$

$$\frac{Actual}{Budget} = \%\ of\ Budget$$

DETERMINING SALES FORECASTS

$$\frac{Total\ Sales}{Number\ of\ Guests\ Served} = Average\ Sales\ per\ Guest$$

Sales This Year − Sales Last Year = Variance

$$\frac{Sales\ This\ Year - Sales\ Last\ Year}{Sales\ Last\ Year} = Percentage\ Variance$$

$$\frac{Variance}{Sales\ Last\ Year} = Percentage\ Variance$$

(Sales This Year/Sales Last Year) − 1 = Percentage Variance

Sales Last Year + (Sales Last Year × Increase Estimate) = Revenue Forecast

Sales Last Year × (1 + % Increase Estimate) = Revenue Forecast

Guest Count Last Year + (Guest Count Last Year × % Increase Estimate)
 = Guest Count Forecast

Guests Last Year + (1 + % Increase Estimate) = Guest Count Forecast

Last Year's Average Sales per Guest + Estimated Increase in Sales per Guest
 = Sales per Guest Forecast

$$\frac{Revenue\ Forecast}{Guest\ Count\ Forecast} = Average\ Sales\ per\ Guest\ Forecast$$

MANAGING THE COST OF FOOD

$$\frac{Total\ Number\ of\ a\ Specific\ Menu\ Item\ Sold}{Total\ Number\ of\ All\ Menu\ Items\ Sold} = Popularity\ Index$$

Number of Guests Expected × Item Popularity Index
 = Predicted Number of That Item to be Sold

$$\frac{\text{Yield Desired}}{\text{Current Yield}} = \text{Conversion Factor}$$

$$\frac{\text{Ingredient Weight}}{\text{Total Recipe Weight}} = \% \text{ of Total}$$

Desired Servings × Oz. per Portion = Oz. Required

% of Total × Total Amount Required = New Recipe Amount

Par Value − On-Hand + Special Order = Order Amount

Unit Price × Number of Units = Extended Price

Item Amount × Item Value = Item Inventory Value

  Beginning Inventory
+ Purchases
  Goods Available for Sale
− Ending Inventor
  Cost of Food Consumed
− Employee Meals
  Cost of Food Sold

  Beginning Inventory
+ Purchases
  Goods Available for Sale
− Ending Inventory
− Value of Transfers Out
+ Value of Transfers In
  Cost of Food Consumed
− Employee Meals
  Cost of Food Sold

$$\frac{\text{Cost of Food Sold}}{\text{Food Sales}} = \text{Food Cost \%}$$

$$\frac{\text{Purchases Today}}{\text{Sales Today}} = \text{Cost \% Today (Six-Column Food Cost \% Estimate)}$$

$$\frac{\text{Purchases to Date}}{\text{Sales to Date}} = \text{Cost \% to Date (Six-Column Food Cost \% Estimate)}$$

MANAGING THE COST OF BEVERAGES

$$\frac{\text{Cost of Beverages Sold}}{\text{Beverages Sales}} = \text{Beverage Cost \%}$$

Beginning Inventory
$+$ Purchases
Goods Available for Sale
$-$ Ending Inventory
$-$ Transfers from Bar
$+$ Transfers to Bar
Cost of Beverages Sold

$$\frac{\text{Item Dollar Sales}}{\text{Total Beverage Sales}} = \text{Item \% of Total Beverage Sales}$$

MANAGING THE FOOD AND BEVERAGE PRODUCTION PROCESS

Prior-Day Carryover $+$ Today's Production

$= $ Today's Sales Forecast $\pm$ Margin of Error

$$\frac{\text{Issues Today}}{\text{Sales Today}} = \text{Beverage Cost Estimate Today}$$

$$\frac{\text{Issues to Date}}{\text{Sales to Date}} = \text{Beverage Cost Estimate to Date}$$

$$\frac{\text{Issues to Date} - \text{Inventory Adjustment}}{\text{Sales to Date}} = \text{Cost of Beverages Sold}$$

$$\frac{\text{Cost in Product Category}}{\text{Total Cost in All Categories}} = \text{Proportion of Total Product Cost}$$

$$\frac{\text{Product Loss}}{\text{AP Weight}} = \text{Weight \%}$$

$$1.00 - \text{Waste} = \text{Yield \%}$$

$$\frac{\text{EP Required}}{\text{Yield \%}} = \text{AP Required}$$

$$\text{EP Required} = \text{AP Required} \times \text{Yield \%}$$

$$\frac{\text{EP Weight}}{\text{AP Weight}} = \text{Product Yield \%}$$

$$\frac{\text{AP Price per Pound}}{\text{Product Yield \%}} = \text{EP Cost (per Pound)}$$

$$\frac{\text{Actual Product Cost}}{\text{Attainable Product Cost}} = \text{Operational Efficiency Ratio}$$

$$\frac{\text{Cost as per Standardized Recipes}}{\text{Total Sales}} = \text{Attainable Food Cost \%}$$

MANAGING FOOD AND BEVERAGE PRICING

$$\text{Revenue} - \text{Expense} = \text{Profit}$$

$$\text{Price} \times \text{Number Sold} = \text{Total Revenues}$$

$$\frac{\text{Cost of Food Sold}}{\text{Food Sales}} = \text{Food Cost \%}$$

$$\frac{\text{Cost of a Specific Food Item Sold}}{\text{Food Sales of That Item}} = \text{Food Cost \% of That Item}$$

$$\frac{\text{Cost of a Specific Food Item Sold}}{\text{Food Cost \% of That Item}} = \text{Food Sales (Selling Price) of That Item}$$

$$\frac{1.00}{\text{Desired Product Cost \%}} = \text{Pricing Factor}$$

$$\text{Pricing Factor} \times \text{Product Cost} = \text{Menu Price}$$

$$\text{Selling Price} - \text{Product Cost} = \text{Contribution Margin}$$

$$\text{Product Cost} + \text{Contribution Margin Desired} = \text{Selling Price}$$

$$\frac{\text{Total Buffet Product Cost}}{\text{Guest Served}} = \text{Buffet Product Cost per Guest}$$

MANAGING THE COST OF LABOR

$$\frac{\text{Output}}{\text{Input}} = \text{Productivity Ratio}$$

$$\frac{\text{\# of Employees Separated}}{\text{\# of Employees in Workforce}} = \text{Employee Turnover Rate}$$

$$\frac{\text{\# of Employees Involuntarily Separated}}{\text{\# of Employees in Workforce}} = \text{Involuntary Employee Turnover Rate}$$

$$\frac{\text{\# of Employees Voluntarily Separated}}{\text{\# of Employees in Workforce}} = \text{Voluntary Employee Turnover Rate}$$

$$\frac{\text{Cost of Labor}}{\text{Total Sales}} = \text{Labor Cost \%}$$

$$\frac{\text{Total Sales}}{\text{Labor Hours Used}} = \text{Sales per Labor Hour}$$

$$\frac{\text{Cost of Labor}}{\text{Guests Served}} = \text{Labor Dollars per Guest Served}$$

$$\frac{\text{Guests Served}}{\text{Cost of Labor}} = \text{Guests Served per Labor Dollar}$$

$$\frac{\text{Guests Served}}{\text{Labor Hours Used}} = \text{Guests Served per Labor Hour}$$

$$\frac{\text{Number of Estimated Guests Served}}{\text{Guests Served per Labor Dollar}} = \text{Estimated Cost of Labor}$$

$$\text{Forecasted Total Sales} \times \text{Labor Cost \% Standard} = \text{Cost of Labor Budget}$$

$$\frac{\text{Forecasted Number of Guests Served}}{\text{Guests Served per Labor Hour Standard}} = \text{Labor Hours Budget}$$

$$\frac{\text{Actual Amount}}{\text{Budgeted Amount}} = \text{\% of Budget}$$

CONTROLLING OTHER EXPENSES

$$\frac{\text{Other Expense}}{\text{Total Sales}} = \text{Other Expense Cost \%}$$

$$\frac{\text{Other Expense}}{\text{Number of Guests Served}} = \text{Other Expense Cost per Guest}$$

ANALYZING RESULTS USING THE INCOME STATEMENT

$$\frac{\text{Food Category Cost}}{\text{Total Food Sales}} = \text{Food Category Cost Percentage}$$

$$\frac{\text{Cost of Food Consumed}}{\text{Average Inventory Value}} = \text{Food Inventory Turnover}$$

$$\frac{\text{Beginning Inventory Value} + \text{Ending Inventory Value}}{2} = \text{Average Inventory Value}$$

$$\frac{\text{Cost of Beverages Consumed}}{\text{Average Beverage Inventory Value}} = \text{Beverage Inventory Turnover}$$

This Year's Sales × Last Year's Adjusted Labor Cost Percentage

= This Year's Projected Labor Cost

$$\frac{\text{Net Income}}{\text{Total Sales}} = \text{Profit Margin (Return on Sales)}$$

$$\frac{\text{Net Income This Period} - \text{Net Income Last Period}}{\text{Net Income Last Period}} = \text{Profit Variance \%}$$

PLANNING FOR PROFIT

Selling Price − Product Cost = Contribution Margin per Menu Item

Total Sales − Total Product Costs = Total Contribution Margin

$$\frac{\text{Total Contribution Margin}}{\text{Number of Items Sold}} = \text{Average Contribution Margin per Item}$$

A × B × C × D = Goal Value

where

A = 1.00 − Food Cost %

B = Item Popularity

C = Selling Price

D = 1.00 − (Variable Cost % + Food Cost %)

Total Sales − Variable Costs = Contribution Margin

Selling Price − Variable Cost/Unit = Contribution Margin/Unit

SP/Unit − VC/Unit = CM/Unit

SP% − VC% = CM%

100% − VC% = CM%

$$\frac{\text{Fixed Costs}}{\text{Contribution Margin \%}} = \text{Break-Even Point in Sales}$$

$$\frac{\text{Fixed Costs}}{\text{Contribution Margin per Unit (Guest)}} = \text{Break-Even Point in Guests Served}$$

$$\frac{\text{After-Tax Profit}}{(1 - \text{Tax Rate})} = \text{Before-Tax Profit}$$

$$\frac{\text{Fixed Costs} + \text{Before-Tax Profit}}{\text{Contribution Margin \%}}$$

$$= \text{Sales Dollars to Achieve Desired After-Tax Profit}$$

$$\frac{\text{Fixed Costs} + \text{Before-Tax Profit}}{\text{Contribution Margin per Unit (Guest)}}$$

$$= \text{Guests to be Served to Achieve Desired After-Tax Profit}$$

$$\frac{\text{Minimum Labor Cost}}{1 - \text{Minimum Operating Cost}} = \text{MSP}$$

or

$$\frac{\text{Minimum Labor Cost}}{1 - (\text{Food Cost \%} + \text{Variable Cost \%})} = \text{MSP}$$

$$\text{Budgeted Revenue} - \text{Budgeted Expense} = \text{Budgeted Profit}$$

$$\frac{\text{Total Sales}}{\text{Available Seats}} = \text{Sales per Seat}$$

MAINTAINING AND IMPROVING THE REVENUE CONTROL SYSTEM

Product Issues = Guest Charges = Sales Receipts = Sales Deposits

Documented Product Requests = Product Issues

Product Issues = Guest Charges

Guest Charges = Sales Receipts

Sales Receipts = Sales Deposits

INDEX